The Fuzzy Systems Handbook

A Practitioner's Guide to Building, Using, and Maintaining Fuzzy Systems

Second Edition

Earl Cox
Chappaqua, New York

AP PROFESSIONAL
AP Professional is a Division of Academic Press
San Diego London Boston
New York Sydney Tokyo Toronto

This book is printed on acid-free paper. ∞

ACADEMIC PRESS
525 B Street, Suite 1900, San Diego, CA 92101-4495, USA
1300 Boylston Street, Chestnut Hill, MA 02167, USA
http://www.apnet.com

ACADEMIC PRESS
24–28 Oval Road, London NW1 7DX
http://www.hbuk.co.uk/ap/

Library of Congress Cataloging-in-Publication Data
Cox, Earl.
The fuzzy systems handbook / Earl Cox. — 2nd ed.
p. cm.
Includes bibliographical references and index.
ISBN 0-12-194455-7 (acid-free paper). — ISBN 0-12-194456-5 (acid-free paper : CD-Rom)
1. System design. 2. Adaptive control systems. 3. Fuzzy systems.
I. Title.
QA76.9.S88C72 1998
003'.7—dc21 98-23214
CIP

Printed in the United States of America

98 99 00 01 02 CP 9 8 7 6 5 4 3 2 1

This second edition is, like the first, dedicated to my wonderful wife, Marilyn.
Without her encouragement, insight, and patience (or, perhaps, impatience)
this book would not have been possible.

***

Contents

Figures

Code Listings

Foreword

My 1965 paper on fuzzy sets was motivated in large measure by the conviction that traditional methods of systems analysis are unsuited for dealing with systems in which relations between variables do not lend themselves to representation in terms of differential or difference equations. Such systems are the norm in biology, sociology, economics, and, more generally, in fields in which the systems are humanistic rather than mechanistic in nature.

Traditional methods of analysis are oriented toward the use of numerical techniques. By contrast, much of human reasoning involves the use of variables whose values are fuzzy sets. This observation is the basis for the concept of a linguistic variable—a variable whose values are words rather than numbers. The concept of a linguistic variable has played and is continuing to play a key role in the applications of fuzzy set theory and fuzzy logic.

The use of linguistic variables represents a significant paradigm shift in systems analysis. More specifically, in the linguistic approach, the focus of attention in the representation of dependencies shifts from difference and differential equations to fuzzy if–then rules in the form *if X is A then Y is B*, where *X* and *Y* are linguistic variables and *A* and *B* are their linguistic values. For example, *if pressure is high then volume is low.* Such rules serve to characterize imprecise dependencies and constitute the point of departure for the construction of what might be called the calculus of fuzzy rules, or CFR for short. In this perspective, *The Fuzzy Systems Handbook, Second Edition,* may be viewed as an up-to-date, informative, and easy-to-follow introduction to the methodology of fuzzy if–then rules and its applications. A key aspect of this methodology is that its role model is the human mind.

Although I am not an impartial observer, I have no doubt that in the years ahead CFR will become an essential tool in the representation and analysis of dependencies that (a) are intrinsically imprecise or (b) can be treated imprecisely to exploit the tolerance for imprecision. At this juncture, many of the applications of fuzzy set theory and fuzzy logic in the realm of knowledge systems fall into the first category. By contrast, most of the applications in the realm of control fall into the second category. This is especially true of the applications of CFR in the design of consumer products such as washing machines, cameras, air conditioners, and vacuum cleaners. In such products, the methodology of fuzzy if–then rules serves to increase the machine intelligence quotient (MIQ) and lower the cost of products by making it possible to program their behavior in the language of fuzzy rules—and thus lessen the need for precision in data gathering and data manipulation. In this context, the use of linguistic variables and fuzzy rules may be viewed as a form of data compression.

The numerous applications of fuzzy set theory and fuzzy logic described in *The Fuzzy Systems Handbook* underscore the unreality of traditional methods in dealing with decision-making in an environment of uncertainty and imprecision. A prominent flaw of such methods relates to the assumption that probabilities can be assumed to be expressible as real numbers. In fact, most real-world probabilities are fuzzy rather than crisp. As a case in point, consider the following problem: I want to insure my car against theft. In this connection, I have to know the probability that my car might be stolen. What is it? The answer is: There is no way for estimating the probability in question as a real number. Fuzzy set theory and fuzzy logic it make possible to estimate the probability in question as a fuzzy number. The fuzzy probability, then, could be used to compute the fuzzy expected loss and thus ultimately lead to a crisp decision.

Many of the real-world problems arising in the analysis and design of decision, control, and knowledge systems are far from simple. Fuzzy set theory and fuzzy logic do not replace the existing methods for dealing with such problems. Rather, they provide additional tools that significantly enlarge the domain of problems that can be solved. Earl Cox's *The Fuzzy Systems Handbook* is an important contribution in this regard. Oriented toward applications, it is certain to have a wide-ranging impact by showing how fuzzy set theory and fuzzy logic can be employed to exploit the tolerance for imprecision, and how they make it possible to address problems that lie beyond the reach of traditional methods.

Lotfi A. Zadeh

Acknowledgments

The book represents the composite knowledge of many friends and associates over the years that I have been designing and using fuzzy logic systems. Foremost among them is Lotfi Zadeh, the founder of fuzzy set theory. Lotfi's encouragement, criticism, and critical view of fuzzy systems as well as his review of early handbook manuscripts, were all invaluable. A decidedly long lasting and pervasive influence on my view of the potential of fuzzy logic and its application to information systems belongs to my friend and fuzzy logic pioneer, Peter Llewelyn Jones. Bart Kosko, with whom I have had a lively debate concerning the merits of linguistic variables versus a more rigorous mathematical view of fuzzy systems, contributed much to the initial explanation of fuzzy systems as universal approximators (all errors of interpretation, however, are mine). I owe a special thanks to Steve Marsh, Jim Huffman, and Bob Seaton at Motorola's Center for Emerging Computer Technologies in Austin, Texas—their vision and work in fuzzy logic as well as their (often revealing) feedback on my views about fuzzy system modeling helped shape many of my ideas about the interpretation and direction of fuzzy systems. Others who provided invaluable insights, ideas, and criticisms include Tom Parish at the TPIS Group in Austin, Arnold Siegel at GEC Marconi, and Tomohiro Takagi, deputy director of LIFE in Yokohama, Japan. Some of the ideas on how to handle output fuzzy set saturation in dense, complex fuzzy systems, as well as ways to encode a fuzzy explanatory facility, belong to my friend Roger Stein, a senior analyst at Moody's Investors Service in New York.

This book grew out of a series of articles on fuzzy logic that I wrote for *AI Expert* magazine in the latter part of 1992 and 1993, from my monthly *AI Apprentice* columns in the same magazine, and from a series of articles on conventional and adaptive fuzzy systems that appeared in *IEEE Spectrum* magazine. Alan Zeichick, then editor of *AI Expert*, and Mike Reisenman, senior editor for *IEEE Spectrum*, both contributed editorial suggestions that improved the readability of these articles and helped shape the way in which this book is written.

Preface

Understanding comes from doing.

Mr. Robert Miller
My twelfth-grade physics teacher
Glen Burnie Senior High, Class of 1964

Life is just one damn thing after another.

Elbert Hubbard (1859–1915)
A Thousand and One Epigrams, p. 137

While several books are available today that address the mathematical and philosophical foundations of fuzzy logic, unfortunately, none provides the practicing knowledge engineer, system analyst, and project manager with specific, practical information about fuzzy system modeling. Those few books that include applications and case studies concentrate almost exclusively on engineering problems such as pendulum balancing, truck backer–uppers, cement kilns, antilock braking systems, image pattern recognition, and digital signal processing. Yet applying fuzzy logic to engineering problems represents only a fraction of its real potential. As a method of encoding and using human knowledge in a form very similar to the way experts think about complex problems, fuzzy systems provide the facilities necessary to break through the computational bottlenecks associated with traditional decision support and expert systems. Additionally, fuzzy systems provide a rich and robust method of building systems that include multiple conflicting, cooperating, and collaborating experts. (This is a capability that generally eludes not only symbolic expert system users but also analysts who have turned to such related technologies as neural networks and genetic algorithms.)

Although fuzzy logic has been described and examined for nearly thirty years, it has only recently appeared in the popular and technical press. As Figure 1 illustrates, the commercial interest in fuzzy logic was quite low for a long time after it was first introduced by Lotfi Zadeh in 1965. Only in the 1986–87 time frame, when the Sendai subway system's fuzzy logic–based Automatic Train Operator (ATO) made front-page news was there a significant jump in both the public's awareness of fuzzy logic and the emergence of companies specializing in fuzzy system products and tools.

From the late 1980s until the middle 1990s, several fuzzy logic tool companies appeared, including Togai Infralogic in Irvine, California, and Aptronix in San Jose, California. Large chip-manufacturing and control companies also entered the market. Motorola, in Austin, Texas, formed a fuzzy logic development group, produced fuzzy logic training courses, and held regular conferences and seminars on the technology. The IEEE (the International Institute for Electronic and Electrical Engineers) held the first of its ongoing conferences in fuzzy

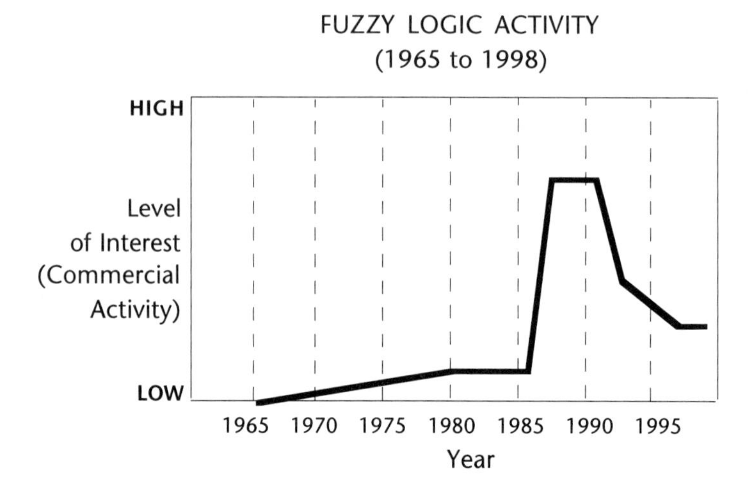

Figure 1 Relative Interest in Fuzzy Logic (Over Thirty Years)

control systems, and *Computer Design* magazine began a series of conferences aimed at fuzzy system tools and commercial applications. About this time Simon and Schuster released *Fuzzy Logic*, an excellent book by Dan McNeil and Paul Freiberger, and Disney Hyperion published Bart Kosko's seminal book, *Fuzzy Thinking*. For those of us in the fuzzy logic community, it appeared as though the technology was finally beginning to take off. We were mistaken.

Today the situation is only marginally better than it was in the dark ages before 1986. While many more technical fuzzy logic books from a wide range of academics have appeared, the momentum of the market place has slowly subsided. Togai Infralogic and Aptronix went out of business (although Aptronix has reappeared recently as a strong player in fuzzy Java software). Motorola has moved from fuzzy logic into more conventional control systems. *Computer Design* magazine has abandoned its annual "Fuzzy Tools" conference, and few articles on the technology now appear in its pages. These days, fuzzy logic is increasingly clumped under an umbrella classification called "computational intelligence," a hodge-podge of related artificial intelligence and computer science topics ranging from object-oriented design to such diverse technologies and disciplines as neural networks, genetic algorithms, evolutionary programming, chaos theory, and artificial life. While its neural networks and genetic programming cousins have continued a steady climb in commercial activity (see Figure 2), fuzzy logic has fallen on hard times.

Fuzzy logic continues to play an important role in many aspects of control engineering, especially in the Far East. But the infusion of fuzzy logic into decision support and expert systems, as well as into a broad spectrum of information management applications has never materialized. The application of fuzzy logic in the areas of information technology, decision support, database analysis, data mining, and knowledge discovery has been largely ignored by both the commercial vendors of decision support products and the knowledge engineers who use or build them. Fuzzy logic has not found its way into the information modeling field

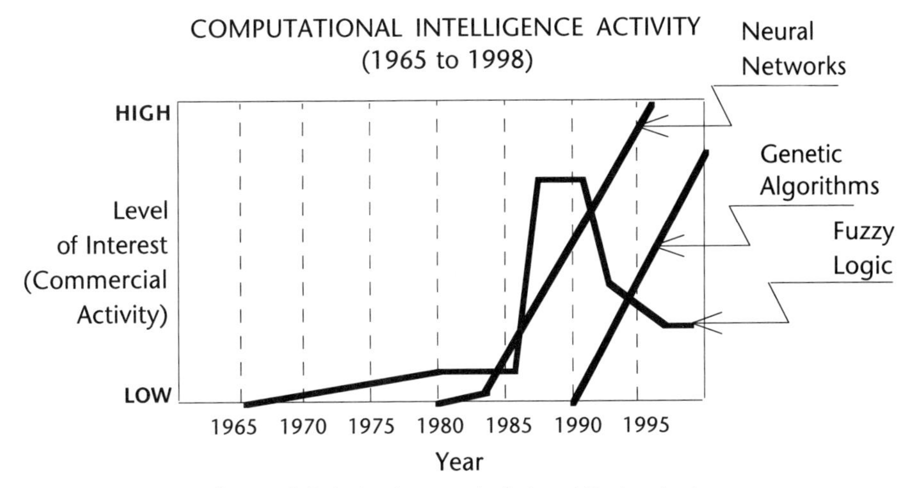

Figure 2 Relative Interest in Related Technologies

due to a number of factors that are rapidly changing—unfamiliarity with the concept, a predilection for the use of confidence factors and Bayesian probabilities among most knowledge engineers (stemming from the early successes of expert systems such as MYCIN, PROSPECTOR, and XCON), and a suspicion that there is something fundamentally wrong with a reasoning system that announces its own imprecision. Fuzzy logic is the essential oxymoron.

Fuzzy Decision Systems: The Early Days

Fuzzy logic is a technology that has patiently bided its time. Today, in the world of highly complex international business systems, webs of communications networks, high-density information overloads, and the recognition that many seemingly simple problems belie a deep nonlinearity, fuzzy logic is proving itself as a powerful tool in knowledge modeling. Fuzzy logic will soon usher in the second wave of intelligent systems.

I have good reason to believe this prediction.

A little more than sixteen years ago, while marketing a project management and enterprise modeling system in the United Kingdom, I was introduced to the idea of fuzzy logic by my friend Peter Llewelyn Jones. Peter is the author of Reveal, undoubtedly the first commercial fuzzy expert system and, with Ian Graham, the author of *Expert Systems: Knowledge, Uncertainty, and Decision*,[1] one of the very first books on fuzzy information systems. Developed in the late 1970s, and available in the early 1980s, Reveal ran under Tymshare's VM/CMS interactive operating system (those were the days when a large microcomputer, as they were called in the early 80s, had 16K of memory, a 60×80 character video screen, and a five-inch, low-density floppy disk). Figure 3 is a very-high level schematic of the system's organization.

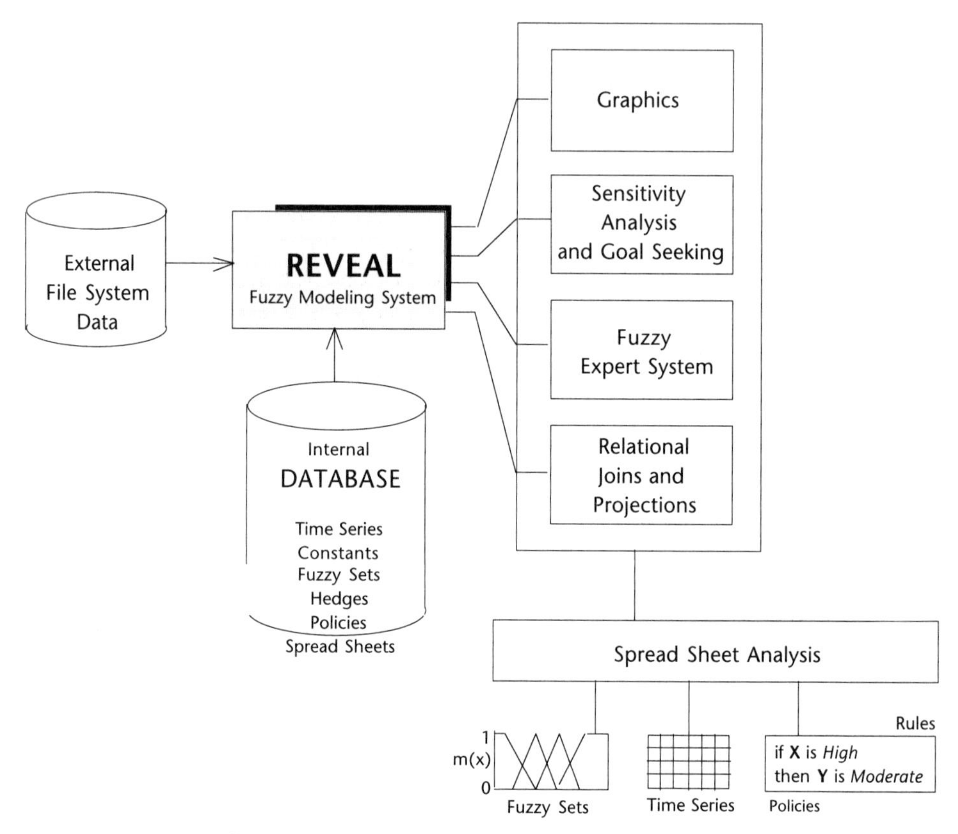

Figure 3 The REVEAL Fuzzy Modeling Environment

Reveal was a powerful time-series modeling system with its own spread sheet reporting, analysis language, and integrated graphics, as well as goal seeking and sensitivity analysis features. The heart of Reveal was its approximate reasoning capabilities, combining fuzzy sets, hedges, and fuzzy if–then–else rules into nonprocedural collections called policies. As a system whose natural unit of data was a series of data elements, Reveal's policies executed rules that automatically looped over multiple series of data, handled fuzzy lead and lag relationships, and supported multiple methods of correlation and defuzzification. This was a product ideally suited to complex, nonlinear, knowledge-intensive problems, and, indeed, Reveal was applied to a wide spectrum of client problems in project management, financial analysis, market positioning, new product pricing, and computer system configuration.

But I knew absolutely nothing about fuzzy logic until 1981. Then, sitting one evening in a crowded pub just off Fleet Street, tucked neatly into the outskirts of Covent Gardens, about a block from our office on the Waterloo Bridge, Peter explained in clear and convincing terms just why fuzzy logic, in the more general form of approximate reasoning, was an important emerging technology. From my background in project management, I immediately understood the real importance of fuzzy logic. Like Paul on the road to Damascus, a brilliant light went off in my mind, and I left the pub an eager devotee to the cult of fuzzy logic. Following in the footsteps of all naive revolutionaries, I expected the world to welcome my insights and revealed truths with open arms. However, in spite of its evident potential and the success of many projects, Reveal was eventually shelved by the new owners of the time-sharing service, and fuzzy logic remained, for a long, long time, the arcane study of Lotfi Zadeh and his slowly, but steadily, increasing circle of believers (usually graduate students who remained well within the sheltering walls of Evans Hall, high on a hill at the University of California at Berkeley).

In the years since I was introduced to and began using fuzzy logic, I have seen first-hand the power and breadth that fuzzy decision and expert systems bring to a wide spectrum of unusually difficult problems. I have architected, designed, and programmed three fuzzy expert production systems. These tools have been successfully applied to large, real-world applications in such areas as transportation, managed health care, financial services, insurance risk assessment, database information mining, company stability analysis, multi-resource and multi-project management, fraud detection in managed health care, acquisition suitability studies, new product marketing, and sales analysis. Generally, the final models were less complex, smaller, and easier to build, implement, maintain, and extend than similar systems built using conventional symbolic expert systems.

Nature of this Book

You can use this book as a guide in evaluating possible fuzzy system projects, designing model policies, developing the underlying programming code, integrating model components, and evaluating overall model performance. The book assumes no previous familiarity with fuzzy logic, first-order predicate calculus, or Bayesian probabilities, but it does presuppose a moderately deep familiarity with business system analysis, high school algebra, and C++ programming. In general, the concentration of concepts, examples, and application code listings is geared toward real-world implementation issues. The case studies are also drawn from actual fuzzy models in such areas as new product pricing, risk assessment, and inventory management. While a few of the examples have been slightly simplified for illustrative reasons,[2] I have attempted to retain the flavor of robustness, error tolerance, and thoroughness that are required in actual commercial applications.

This book also examines the architecture, function, and operation of rule-based fuzzy systems in a business environment. In general, we will be concerned with understanding the basics of fuzzy logic and fuzzy systems, selecting the appropriate problems and domain representation, implementing the design of fuzzy sets, learning the use of operators, and interpreting model output. We will pay less attention to the mathematical principles underlying the nature of fuzzy logic, thus making the book a minimally mathematical tour of how to design, build, deliver, and maintain fuzzy systems. A comprehensive bibliography provides an annotated listing of books that can provide a complete explanation of the mathematics underlying fuzzy logic.

I have conceptually divided this book into two major parts. The first part, Chapters 1 through 6, deals with the fundamental issues of imprecision, fuzzy sets, various fuzzy operators, and fuzzy propositions (also known as rules). This first part also discusses how a final result is resolved from a fuzzy model. The second part, Chapters 7 through 9, addresses the idea of fuzzy models and fuzzy systems by exploring the nature of fuzzy models and how they are represented, the architecture of operation of several fuzzy system case studies, and the outline of a general methodology for developing a fuzzy system. The book concludes with a description of the fuzzy modeling software in Chapter 10, a glossary of terms, and a bibliography.

On a final note, this is not a book on programming nor is it a book on how to create and use a particular fuzzy model. In order to effectively compile and integrate the C/C++ procedures included with the book, you should be an experienced programmer. Additionally, to compile and use the Windows/95 DLL procedures, you must understand how dynamic link libraries manage their memory, how byte alignment is used, and how name mangling in C++ is suppressed. You must also know how to specify the location of all necessary *include* files (such as *.hpp* files), how to create a *.def* (definition) file, how to specify a prototype definition, and how to call and use your compiler environment to link the compiled procedures.

The programs included with this book are provided as examples only, and they may or may not be suitable for any particular business, scientific, or engineering purpose.

Adapting and Using the C++ Code Library

C++ is a general-purpose programming language with intrinsic facilities for bundling code packages into objects. It is a vehicle for object-oriented programming (OOP) and a language for actualizing a system described through an object-oriented design (OOD) methodology. In and of itself, however, C++ does not enforce an object-oriented approach to programming. With these concepts in mind, you should understand that *the library code in this book is written in C++ but is not organized into objects*. This is a controversial yet purposeful design based on three distinct, but related, architectural goals.

First, the code is organized as 32-bit dynamic link libraries (DLLs) with a seamless call-level interface for Microsoft Visual Basic 5.0 and Microsoft Visual C++ (4.0 and above), as well as for most other rapid application development environments supporting access to 32-bit DLLs (such as Powersoft's PowerBuilder and Borland's Delphi). Maintaining all service calls at the subroutine level provides the widest possible accessibility across any object environment. This permits, for instance, the services to be used in a variety of C++ object models, in the Visual Basic COM representation, and in Delphi's object model. In fact, this book contains a complete description of the C/C++ DLL entry points in a Visual Basic 5.0 module file. You can execute these procedures from within Visual Basic with a simple CALL statement.

Second, the included source code provides a wealth of specialized fuzzy logic programs that can be used to create (derive) object representations. As a programmer and developer, I have often found other object code libraries difficult to understand, integrate, and maintain. I have thus decided to present the basic services in disarticulated procedures with the realization that application designers and programmers can assemble them into their own objects. In this way, you can decide how to represent fuzzy sets, how to implement hedges, what to make public, and what to hide through encapsulation. Such a simple fuzzy class might look like this:

```
class FuzzySet
  {
   protected:
     char      FDBid[IDENLEN+1],
               FDBdesc[IDENLEN+1];
     int       FDBgentype;
     mbool     FDBempty;
     int       FDBorder;
     double    FDBdomain[2],
               FDBparms[4];
     float     FDBalfacut,
               FDBvector[VECMAX];
   public:
     mbool     PE  FzyAboveAlfa();
     float     PE  FzyAND();
     void      PE  FzyApplyAlfa();
     FDB*      PE  FzyHedge();
     :
     :
  };
```

This brief example omits some basic C++ object controls such as constructors and destructors as well as deep and shallow object copying procedures. From this basic *FuzzySet* object, an application system can create, through inheritance, such elements as predicate fuzzy sets (*pFuzzySet*), consequent fuzzy sets (*cFuzzySet*), and solution fuzzy sets (*sFuzzySet*).

Third, the code in this book should be accessible to both C and C++ programmers. While C++ has emerged as the standard language of choice for the design and development of complex systems, the exclusive use of object-oriented constructs would have unnecessarily restricted the book's audience. A large audience of readers using C or recently moving into C++ should find the modeling code easy to understand, modify, and integrate with their own code. C++ code is, by nature, difficult to understand. The use of operator overloading, the implicit availability of class variables in procedures, inheritance, and other C++ language characteristics make for code that is sometimes impenetrable for even experienced C programmers. Thus, both C and C++ programmers can understand, adapt and use this code. Experienced C++ programmers should also be able to cannibalize this code and create their own object class libraries. In nearly all cases, when given a choice between self-documentation and "efficiency" or "compactness," I chose to write simple, straightforward code.

The Microsoft Windows/95 Dynamic Link Library C++ code included in this book has been compiled and tested under Microsoft's Visual C++ 4.0 32-bit compiler. The 16-bit version of the code, intended for DOS applications (not Windows 3.1) has been compiled and tested under the Microsoft Visual C++ 1.52 16-bit compiler. Since this fuzzy logic code is extracted from a larger library of modeling software used to build production decision and expert systems, the code in the book also contains some functions not directly associated with fuzzy logic (such as the error message handler, line entokeners, and string managers).

If you find a problem with any part of the code please contact the author (see "Contacting the Author" later in this preface). If at all possible, I will repair the problem quickly and ship you an updated procedure (and, at the same time, make the fix available to all other users through the Metus Systems FTP site).

The Graphical Representation of Fuzzy Sets

The code libraries included with this book contain several programs that draw fuzzy sets. In keeping with the general principle of maintaining platform independence, these programs write character-based graphs to a specified file. An individual fuzzy set, when plotted, will appear as the following:

```
FuzzySet:   near 2*MfgCosts
Description:
 1.00                         ..
 0.90                      ...  ....
 0.80                    ..         .
 0.70                   .            ..
 0.60                 ..               .
 0.50                .                  .
 0.40               .                    .
 0.30              .                      .
 0.20             .                        .
 0.10           ..                          ..
 0.00...........                              ........................
     0---|---|---|---|---|---|---|---|---|---|---|---|---|---|---|---0
   16.00   18.50   21.00   23.50   26.00   28.50   31.00   33.50   36.00
   Domained:        16.00 to      36.00
```

When multiple fuzzy sets are printed on the same graph, each is represented by its own character symbol. A legend following the graph indicates which fuzzy set is represented by which symbol. The following shows a PI (bell-shaped) fuzzy set and the new fuzzy set created by applying the hedge *somewhat*:

```
Hedge 'somewhat' applied to Fuzzy Set "PI.Curve"

      1.00                                       *
      0.90                                     *** ***
      0.80                                    *.     .*
      0.70
      0.60                                   *.       .*
      0.50                                  *           *
      0.40                                   .         .
      0.30                                  *           *
      0.20                                 * .         . *
      0.10                                  .           .
      0.00************************.              .************************
           0---|---|---|---|---|---|---|---|---|---|---|---|---|---|---|---0
         16.00   20.00   24.00   28.00   32.00   36.00   40.00   44.00   48.00
     .    FuzzySet:    PI.Curve
        Description:
     *   FuzzySet:    PI.Curve
        Description: somewhat PI.Curve
        Domained:          16.00 to       48.00
```

To extract a table of domain values with their corresponding fuzzy membership values for use in graphical tools such as Mathematica, MatLab, or for use in Microsoft's Windows environment, use the *FzyExtractSetdata* program.

Contacting the Author

I welcome comments on this book, as well as error notifications, complaints, or recommendations. I especially want to hear about any problems you have with the code (other than difficulties associated with running your compiler), and I would also be interested in hearing about how you are using this code. I can be reached via the Internet as well as America Online at the following addresses:

Internet `ecox@metus.com`
AOL `earlcox@aol.com`

The Metus Systems Group maintains a web page on fuzzy systems and computational intelligence. Metus also manages an anonymous FTP site for technical abstracts, shared software, and technical publications. The web page will also contain downloadable updates and fixes to the source code contained in this book. You can access the Metus homepage at:

```
http://www.metus.com
```

The web page provides a subscription service for businesses and consultants involved in building operational fuzzy, neural, and genetic systems in information technology. The page also provides access to a quarterly online newsletter covering a wide spectrum of topics in the application of fuzzy systems to business and government planning.

Icons and Topic Symbols

Throughout the book you will encounter a series of "sidebars" covering a wide variety of special topics. Each topic is set off from the main text and is introduced by a specific icon.

Code Discussion and Cross-Reference. The CD icon indicates that this section deals with code models, programs, and software in general. This icon usually means that a concept discussion is being tied back to the actual supporting code in one or more of the libraries.

System Internals. The plumbing icon indicates that this section discusses system internals. System internals refer to the data structures, control blocks, inter-program linkages, process flow, and other details associated with the way fuzzy modeling systems work.

Key Topics. A key icon indicates a detailed discussion on some important aspect of the system. The nature of key topics covers a broad spectrum: from hints and techniques that may not be obvious to detailed and off-line explorations of concepts and philosophies.

System Construction. The hammer icon represents discussions centered around the construction of models, applications, and high-level software systems. This topic is not specifically concerned with function code or data structures.

Practical Hints and Techniques. The wingnut ("nuts and bolts") icon indicates a discussion about the practical considerations of the current topic. This includes programming techniques, system design considerations, methodology approaches, etc.

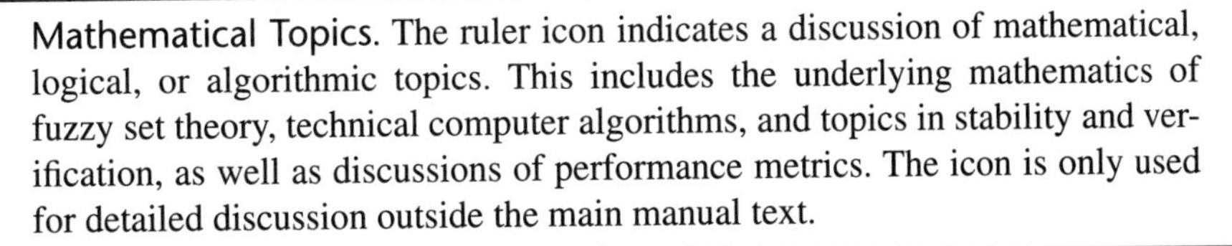

Mathematical Topics. The ruler icon indicates a discussion of mathematical, logical, or algorithmic topics. This includes the underlying mathematics of fuzzy set theory, technical computer algorithms, and topics in stability and verification, as well as discussions of performance metrics. The icon is only used for detailed discussion outside the main manual text.

Danger! Warning! The lightning bolt icon indicates a topic that could be potentially dangerous to model integrity, validation, or data. This icon is also used to highlight harmful side effects that are not obvious. You should read every danger topic carefully.

Reminders and Warnings. The reminder icon indicates a discussion of important, but generally not obvious, points that the user must consider when exploiting some portion of the fuzzy modeling technology.

Symbol	Use
~	set NOT (also complement or inversion)
$\cap$	set AND (also intersection operator)
$\cup$	set OR (also union operator)
$\aleph$	higher dimensional fuzzy space
[x,x,x]	fuzzy membership value
$\in$	member of a set (general membership)
poss(x)	the possibility of event x
prob(x)	the probability of event x
{x}	crisp or Boolean membership function
•	dyadic operator
$\xi(x)$	expected value of a fuzzy region
μ	fuzzy membership function
$\propto$	proportionality

$\mu[x]$	membership or truth function in fuzzy set
$\Re$	element from domain of fuzzy set
$\otimes$	Cartesian product or space
$\varnothing$	empty or null set
$\supset$	implication
$\wedge$	logical AND
$\vee$	logical OR
Σ	summation

Notes

1. 1988. Published by Chapman and Hall, London.

2. I wanted to say "for pedagogical purposes" but resisted the temptation to slip into unnecessary jargon. That's another characteristic of this book: I have avoided unnecessary and pointedly academic jargon and have chosen straightforward English instead. A complete polysyllabic glossary is included for academic readers who are able to actually read the handbook and thus have been bewildered and disoriented.

The Fuzzy Systems Handbook

Second Edition

One

Introduction

Men, gud sig forbarme, sligt noget gor man da ikke!
But good God, people don't do such things!

Henrik Ibsen (1828–1906)
Hedda Gabler (1890), Act 4.

φιλοκαλουμεν γαρ μετ' ευτελειας
και φιλοσοφουμεν ανευ μαλακιας
The love of what is beautiful does not lead to extravagance;
The love of the things of the mind does not make us soft.

Pericles (c.495-429 BC)
Funeral Oration, Athens 430BC,
in Thucydides, *History of the Peloponnesian War*, bk 2, ch. 40, sect. 1

In this introductory chapter we discuss the rationale behind fuzzy system modeling, introduce the ideas of a fuzzy set and a linguistic variable, differentiate between fuzzy logic and the more general area of approximate reasoning, briefly trace the history of fuzzy set theory, and outline the benefits of using fuzzy logic in information system modeling. We also review the reasons why manufacturers often reject fuzzy system solutions and we show why many of these reasons are based on misinformation or extrapolations from extreme cases.

The following discussions of fuzzy system benefits and why fuzzy solutions are often rejected assume some knowledge of fuzzy sets and fuzzy logic. If you want a tune-up on these two topics, you may want to return to this material after reading Chapter Two, Fuzziness and Certainty, and Chapter Three, Fuzzy Sets.

Fuzzy System Models

Nearly all the attention that fuzzy logic receives has been in the area of process and control engineering, brought on by the boom of fuzzy logic products developed in Japan. This focus is understandable since fuzzy logic provides a valuable technology in the design of embedded machine intelligence. With fuzzy logic, manufacturers can significantly reduce development time, model highly complex nonlinear systems, deploy advanced systems using control engineers rather than control scientists, and implement controls using less expensive chips and sensors. Yet control systems exploit only a small fraction of fuzzy logic's representational power.

You can find a larger and much more powerful application of fuzzy logic in the expert and decision support systems. These next generation systems fuse fuzzy logic, production rules, and a wide spectrum of inferencing techniques. Coupling fuzzy logic with expert system technology provides a mechanism for producing fuzzy models that address important classes of problems in information decision support. Fuzzy expert systems model the world in terms of the semantics associated with the underlying variables, thus providing a much closer relationship between real world phenomena and computer models. This handbook's goal is two-fold: first, to explain how fuzzy models work, and second, to provide a tool kit you can use to build these models.

Logic, Complexity, and Comprehension

To understand how fuzzy systems provide superior information modeling, we need to go back to its origins. As conceived by Lotfi Zadeh, its inventor, fuzzy logic provides a method of reducing as well as explaining system complexity. Zadeh, originally an engineer and systems scientist, was concerned with the rapid decline in information afforded by traditional mathematical models as the complexity of the target system increased. Much of this complexity, he realized, came from the way in which the variables of the system were represented and manipulated. Since these variables could only represent the state of a phenomenon as either existing or not existing, the mathematics necessary to evaluate operations at various "boundary" states became increasing complex. At some point this complexity overwhelmed the information content of the model itself, leaving only a morass of equations and little knowledge about the underlying process.

Zadeh stated this idea succinctly in his "principle of incompatibility":

> [A]s the complexity of a system increases, our ability to make precise and yet significant statements about its behavior diminishes until a threshold is reached beyond which precision and significance (or relevance) become almost mutually exclusive characteristics.
>
> —Lotfi Zadeh
> *Outline of a New Approach to the Analysis of Complex Systems and Decision Processes*

In this view of modeling complex systems, the underlying mechanics are represented linguistically rather than mathematically. Zadeh makes a case that humans reason not in terms of discrete symbols and numbers but in terms of fuzzy sets. These fuzzy terms define general categories, but not rigid, fixed collections. The transition from one category—concept, idea, or problem state—to the next is gradual with some states having greater or lesser membership in the one set and then another. From this idea of elastic sets, and drawing on the works of Max Black and Jan Lukasiewicz, Zadeh proposed the concept of a fuzzy set.

The Idea of Fuzzy Sets

Although we will be investigating the nature of fuzzy sets in Chapter Two, Fuzziness and Certaingy, on page 45, for the moment, a brief introduction is sufficient. Fuzzy sets are functions that map a value, which might be a member of a set, to a number between zero and one, indicating its actual degree of membership. A degree of zero means that the value is not in the set, and a degree of one means that the value is completely representative of the set. This mapping produces a curve across the members of the set. As a simple example, consider the idea of a *Long* project. Figure 1.1 illustrates how this concept might be represented.

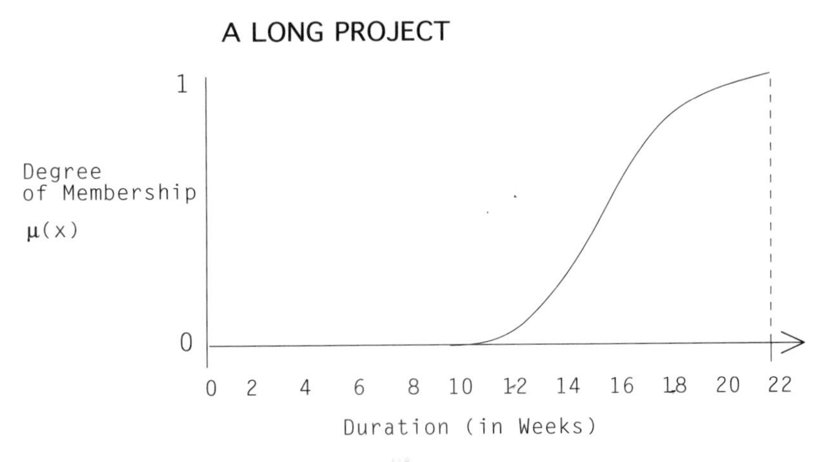

Figure 1.1 The Idea of a *Long* Project

The members of this set (shown along the horizontal or X-axis) are duration periods, in weeks, for a project. The fuzzy set indicates (along the vertical or Y-axis) to what degree a project of a specified duration is a member of the set of *Long* projects. As the number of weeks increases, our belief that the project is *Long* increases. A project ten weeks in duration would not be considered *Long*, a project fifteen to seventeen weeks in duration would have a moderate membership in the set of *Long* projects, and a project of more than twenty weeks in duration is most certainly a *Long* project. Thus, a fuzzy set describes the semantics of a particular model's parameters, and the definition of what is a *Long* project differs with each model.

Fuzzy sets are highly model dependent. They have meaning only within a specific context. Thus, the actual definition of what is a Long project depends on the context in which it used. For some models, a one- or two-day project might be Long, and for others, such as projects performed by government contractors, the idea of Long only begins to make sense at some distant point in the future (such as after a change in the current administration).

In most models, the parameters (or variables) are decomposed into a number of different, overlapping fuzzy sets. This overlap reflects the very nature of fuzzy information: that values (elements) can belong, to various degrees, to many fuzzy sets at the same time. Real world models often have five to seven underlying fuzzy sets for important model parameters. The combination of these fuzzy sets is called the "term set." Figure 1.2 shows a simple term set for the *Project.duration* variable.

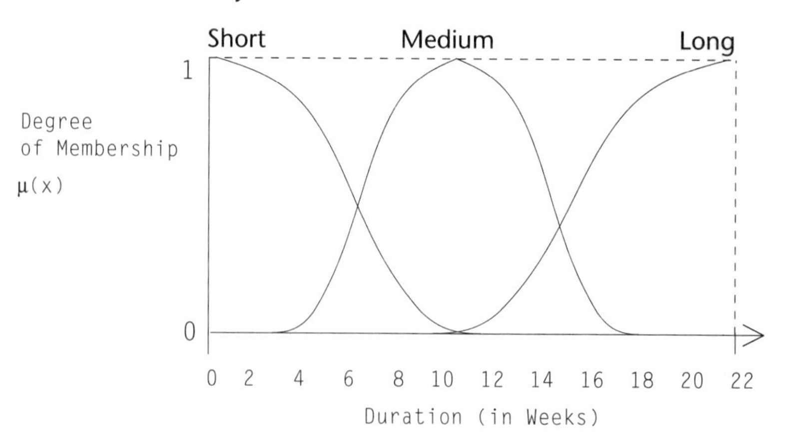

Figure 1.2 The Fuzzy Sets Describing Project Duration

Linguistic Variables

The center of the fuzzy modeling technique is the idea of a *linguistic variable*. At its root, a linguistic variable is the name of a fuzzy set. In the previous example, the fuzzy set

Long is a simple linguistic variable and can be used in a rule-based system to make decisions based on the length of a particular project:

```
if project.duration is LONG
   then the completion.risk is INCREASED;
```

But a linguistic variable also carries with it the concept of fuzzy set qualifiers (often called *hedges*). These qualifiers change the shape of fuzzy sets in predictable ways and function in the same fashion as adverbs and adjectives in the English language. Figure 1.3 illustrates the structure of a linguistic variable.

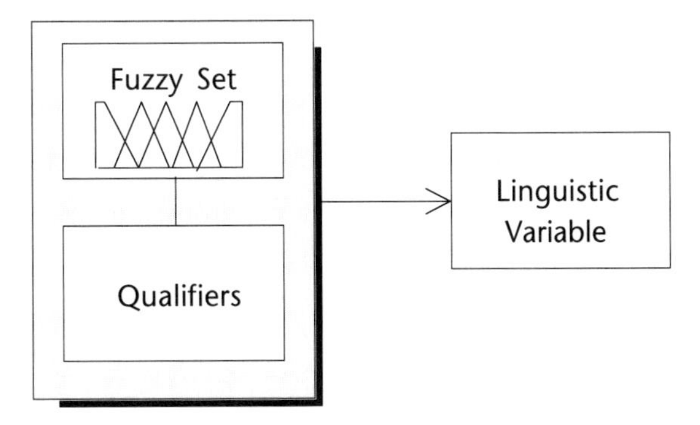

Figure 1.3 Structure of a Linguistic Variable

Linguistic variables allows the knowledge engineer to write expressive statements about related concepts. The following are linguistic variables using the fuzzy set LONG—*very* LONG, *somewhat* LONG, *slightly* LONG, and *positively not very* LONG. We interpret these expressions using the same rules of precedence as English; thus, *not very Long* and *very not Long* are two distinct statements. Linguistic variables permit the fuzzy modeling language to directly express the shades of semantic meanings used by experts, as is illustrated in the following rule:

```
if project.duration is positively not very long
   then the completion.risk is somewhat reduced;
```

Fuzzy numbers are also linguistic variables. You can create a fuzzy number by approximating a numeric value, either a variable or a numeric constant. Linguistic variables such as *Near MfgCosts* (near the current manufacturing costs) or *About MiddleAge* take the form of triangular, bell-shaped, or trapezoidal fuzzy sets. Figure 1.4 illustrates the linguistic variable *MiddleAged* as a fuzzy number placed on the domain of the variable *Client.Age*.

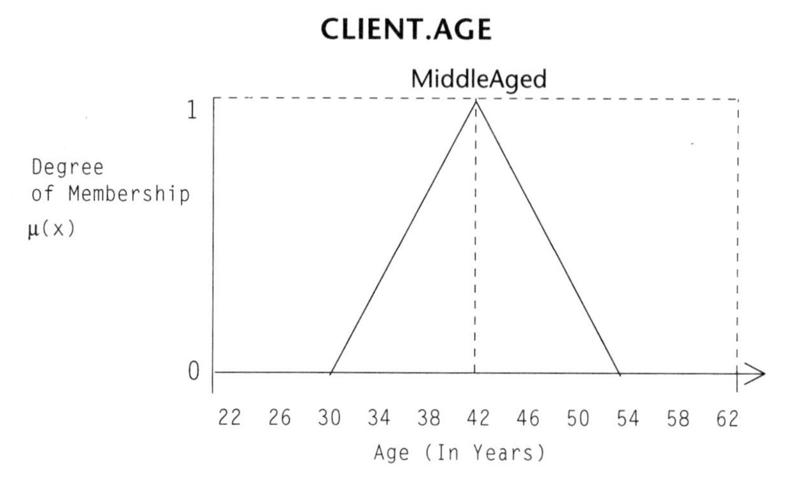

Figure 1.4 The Linguistic Variable *MiddleAged*

Other types of qualifiers that can be applied to fuzzy sets include modifiers for usuality, frequencies, and counts (so we can say *most Long projects are usually Late*). This has important representational implications for time-series-based fuzzy models. Hedges and other qualifiers are discussed in Fuzzy Set Hedges on page 217.

A linguistic variable encapsulates the properties of approximate or imprecise concepts in a systematic and computationally useful way. It reduces the apparent complexity of describing a system by matching a semantic tag to the underlying concept. Yet a linguistic variable always represents a fuzzy space, which is another way of saying that when we evaluate a linguistic variable, we come up with a fuzzy set.

Approximate Reasoning

Fuzzy logic, in the sense of fuzzy set theory, provides the underpinnings for designing and writing fuzzy models. Fuzzy set theory, and the broader area of fuzzy logic itself, is not a specific method for any particular application any more than Boolean logic or probability is a logic for a particular kind of application. The theory of fuzzy sets supports a more general theory of fuzzy logic or the fuzzy statement calculus. This, in turn, supports the logical constructs used to create and manipulate fuzzy systems. This has the more general name of *approximate reasoning*. Figure 1.5 shows how these platforms are related.

In this handbook, fuzzy logic and approximate reasoning are often used inter-changeably to indicate the process of expressing imprecise or approximate concepts and relationships. In nearly all important cases, this equivalence holds; however, you should be aware

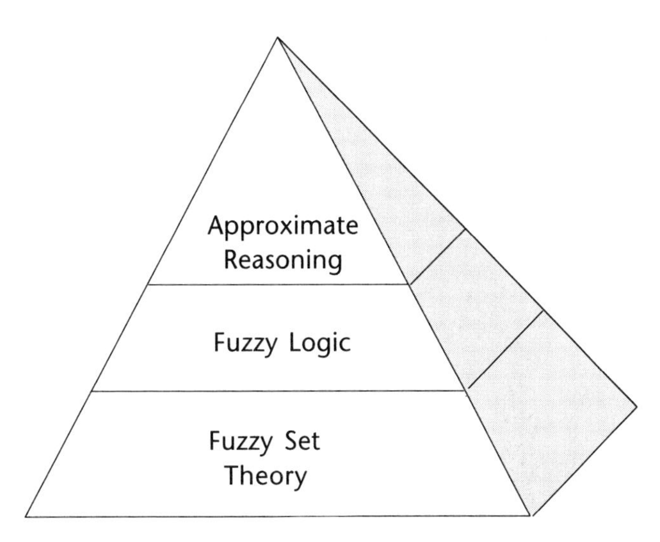

Figure 1.5 Levels of Logic Supporting Approximate Reasoning

that fuzzy logic is a more formal representation of fuzzy set theory than of approximate reasoning. As a method of encoding knowledge through both conditional and unconditional fuzzy rules, approximate reasoning draws not only on the underlying mathematics of fuzzy logic, but incorporates a set of heuristics that lies outside fuzzy set theory but appears to work consistently and predictably. The use of hedge qualifiers, already mentioned, is an example of a concept that belongs to the approximate reasoning platform but not at the fuzzy logic or fuzzy set theory level.

Benefits of Fuzzy System Modeling

Fuzzy models have important implications for the corporate bottom-line: reduced mean-time-between-failure (MTBF), improved mean-time-to-repair (MTTR), the easier and more stable extensibility of existing systems, closer agreement with real-world processes, and improved understandability. Systems scientists, academics, and the technical press often fail to appreciate that 80% to 90% of the work done in most organizations involves the maintenance and support of existing applications. This has a significant impact on the way management views, approves, and evaluates advanced technologies and supports future applications in organizations that are regularly using advanced technology. These organizations require skilled staff such as maintenance analysts and programmers to maintain and extend the current applications; help-desk and on-site technicians to answer questions, resolve problems, and provide feedback to the support staff; and a training staff to conduct regular classes in the use of the applications. Since fuzzy systems are less externally complex, they can be understood easier, problems can be isolated and fixed sooner (thus the reduced MTTR), and with fewer rules the MTBF is generally improved.

The following sections cover some of the benefits derived from using fuzzy models in decision support and expert systems. We will return to these benefits throughout the handbook as we discuss the way in which fuzzy logic works and the mechanism for creating and implementing fuzzy models.

THE ABILITY TO MODEL HIGHLY COMPLEX BUSINESS PROBLEMS

Many business problems are deceptively complex, which is a principal reason why the current breed of expert systems has failed to grow rapidly in the business community. Essentially, all the easy problems were solved quickly. Conventional systems then ran up against highly nonlinear and computationally complex problems that they simply could not address without consuming prohibitive amounts of computer power—if they could address them at all. This computational and complexity boundary has been responsible for the failure of many expert systems. Fuzzy systems, on the other hand, are universal approximators and are well suited to modeling highly complex, often nonlinear problem spaces. Thus, they are able to approximate the behavior of systems displaying a variety of poorly understood and/or nonlinear properties. Fuzzy rule-based systems usually execute faster than conventional rule-based systems and require fewer rules. With the ability to explain their reasoning, they provide an ideal way of addressing difficult problems.

IMPROVED COGNITIVE MODELING OF EXPERT SYSTEMS

For many knowledge engineers, a significant benefit of fuzzy system modeling is the ability to encode knowledge directly in a form very close to the way experts themselves think about the underlying problems and the details of the decision process. A continual inability of conventional expert and decision support systems to represent problems close to their actual real-world behaviors forces experts to crisply dichotomize rules at (usually) artificial boundaries. This process not only leads to an unnecessary multiplication of rules, but also robs an expert of his/her ability to effectively articulate a solution to a complex problem. (And, of course, that's where the payoff lies—today's expert systems have already proven themselves at representing simple problems.)

In the rush to build and use expert systems, we often overlook a single important question: *What makes an expert an expert?* Specialized knowledge, of course. Generally, however, it is the ability to make diagnostic or prescriptive recommendations in imprecise terms. In other words, experts can solve general classes of problems. As an example, consider an expert Credit Analyst's rule for granting approval:

```
if the peratio is high
      and the company is highly profitable
      and sales are increasing
            then credit approval is fairly safe;
```

But if we now insist that the expert define the boundaries of *High* for the profit to earnings ratio, describe the exact meaning and limits for *Highly Profitable*, and specify the minimum rate for *Increasing* to be true, we are forcing the expert to deconstruct his or her expertise into fragments of knowledge. This fragmentation is a major contributing factor to the very poor performance of many systems.

With their ability to directly model such imprecise information, fuzzy systems reduce overall cognitive dissonance[2] in the modeling process. The way an expert thinks about a problem's solution is much closer to the way it is represented in the machine. A fuzzy system captures expertise close to the expert's own cognitive model of the problem, making the knowledge acquisition process easier, more reliable, and less prone to unrecognized errors and ambiguities. The model validation process is enhanced since the rules are understood by the expert in a terminology and representation that closely parallels his or her own thought processes.

THE ABILITY TO MODEL SYSTEMS INVOLVING MULTIPLE EXPERTS

In the expert system literature (books, articles, and technical papers) there is almost always an unstated assumption that a single expert exists or that all the experts in a field are continuously in at least close agreement. In the real world of decision modeling, however, this is seldom the case. Experts in many disciplines disagree significantly. This disagreement can polarize experts in such areas an econometric modeling, battlefield and strategic mobility models, and models concerned with government-level social programs. Even everyday business decisions such as

- ✔ How to price a new product.
- ✔ When and to what degree to grant credit to an organization.
- ✔ How to allocate resources to an enterprise's project pool.
- ✔ What constitutes safety and suitability in a portfolio mixture.
- ✔ When is some activity unusual or anomalous.

have no simple solution and involve conflicting views from a variety of experts. Such questions form the core of most real-world decision models, so that any model that is not a toy and attempts to represent complex business systems will encounter experts in the same domain who significantly disagree about the meaning of or restrictions on an important model variable. Fuzzy systems are well suited to representing multiple cooperating, collaborating, and even conflicting experts. The new product pricing model (developed for a British retailer and discussed later in this handbook) is one example. The following are the first four rules from the model (shown exactly as they appeared in the rule base):

```
our price must be low
our price must be high
our price must be around 2*mfgcosts
if the competition.price is not very high
   then our price should be near the competition.price
```

Note that the first three rules do not have an **if** statement since they are unconditional fuzzy propositions and represent conflicting opinions about the positioning of the product price. In fact, these were the actual positions taken by the marketing, financial, and production control managers.

Reduced Model Complexity

Fuzzy models, as we will see, require fewer rules than conventional systems, and these rules are closer to the way knowledge is expressed in natural language. This has two important side benefits for model maintenance engineers. First, a model can generally be modified with fewer induced errors. Second, the relative simplicity of a fuzzy model means that logic or structural problems can be located and fixed in a minimum amount of time. These factors greatly improve overall MTTR. This same ease of maintenance and understandability also means that the model can be validated with greater precision and for a wider variety of input cases, thus significantly both increasing confidence in the model and reducing MTBF.

Improved Handling of Uncertainty and Possibilities

The representation of uncertainty in expert and decision support systems is an area of continual debate. The methods employed by most conventional systems are Bayesian probabilities and some form of confidence or certainty factors. Although fuzzy logic represents uncertainty, and imprecision as an intrinsic part of the model, both of these alternative approaches rely on the assignment of uncertainty values outside the model itself. In this aspect, fuzzy logic provides a better, more consistent, and more mathematically sound method of handling uncertainties.

While certainty factors[3] proved their usefulness in early expert system applications such as MYCIN, they are an essentially ad-hoc approach to belief management, subject to unpredictable interpretations, and are assigned external to the model rather than as part of the knowledge representation itself. Let me give a brief example.The following is a single-rule expert system to predict the weight of an individual given his or her height:

```
if height is Tall then weight is Heavy;
```

where *Tall* and *Heavy* are fuzzy sets. A given height will produce an estimated weight with its degree of membership in the fuzzy set (its compatibility with our belief in the implication function between the two concepts). No such capability exists with certainty factors. While fuzzy logic will predict and generate an answer, certainty factors are applied to an answer you already know. Using certainty factors, a weight estimating model would contain many, many rules, which would look something like this:

```
if height is > 4.5 and height < 5.5
   then weight is 180, CF=.82;
```

This tells us little about the intrinsic relationship between height and weight. The certainty is simply tacked on to the answer. The combination of certainty-based propositions using intersections, unions, and complements also tends to be arbitrary.

Common Objections to Fuzzy Logic

Before moving on to the actual nature of fuzzy logic and its justification as a modelling tool, I want to address many of the objections to the use of fuzzy logic. Some of these are clearly misguided, but most result from simple ignorance. There is, in fact, a mythology of folk tales and apocryphal stories surrounding fuzzy logic. In this mythology the "One True God of Science and Fact" is bedeviled by the "Angel of Vagueness and Anti-Reason." Few of these mythologies have any basis in truth.

I am constantly amazed at the amount of negative and often hostile information about fuzzy logic earnestly offered by scientists, knowledge engineers, business analysts, and technical managers, all of whom have almost no knowledge of the actual workings and applications of fuzzy logic. Much of this, I suspect, is due to one fact: the appearance of the word *fuzzy* coupled with the word *logic*. Not since Galileo Galilei sent a telegram to the medieval church with the message "The Earth Goes Around the Sun. Stop. Ptolemy is Wrong. Stop." has a single concept raised the hackles of so many scientists and academics. Today's crowd of academics and scientists, unfortunately, are taking the same approach to fuzzy logic that the church took to the heliocentric theory of the solar system: issue condemnations, deny its validity, reject its adherents, and laugh at the entire apparatus. The church cardinals could have peered through Galileo's telescope to see the moons of Jupiter, the myriad stars in the Milky Way, and the dark spots that moved around the sun. Its scientists could have used reason and deductive logic to ascertain the truth about the physical world. But they chose dogma and the comfort of tradition. In the same way, the opponents of fuzzy logic could examine the principles of fuzzy set theory and fuzzy logic to decide whether or not the field offers a valid and important description of phenomena that is not otherwise addressed by mathematical models, probability, and a host of ad-hoc uncertainty mechanisms. But they have not chosen that approach.

A lack of symmetry in knowledge makes a discussion of fuzzy logic with mathematicians, statisticians, and knowledge engineers very difficult. Although experts in fuzzy logic are generally competent to discuss and use calculus, ordinary and partial differential equations, statistics, and probability, as well as information decision theory, their opposition usually has dismissed fuzzy logic (almost "out of hand") and has not bothered to learn its principles, its theories, or its uses across a wide spectrum of applications. A mathematician in the investments department of a major New York bank (who proudly boasted that he knew nothing about fuzzy logic and really didn't want to know anything about it) actually told me,

> "If you can't explain fuzzy logic to me in one or two sentences, then it's probably hogwash, and I'm not interested."

This same mathematician, I suspect, would be hard-pressed in explaining the principles of portfolio optimization, or even calculus, statistics, solid geometry, or differential equations in one or two sentences. I doubt that even the principles of simple arithmetic could be explained in one or two sentences. Reasoned debate about fuzzy logic is thus clouded in prejudice, assumption, and misinformation. This lack of concern for the underlying reasoning mechanism and descriptive space of fuzzy logic in the face of its evident success in a broad range of applications, especially in Japan and Europe, is puzzling. Perhaps this says more about the American technological mind, the exercise of the scientific method in academia and business, and the complacency of American entrepreneurial spirit, than it does about either the merits of fuzzy logic or the adaptive energies of the Japanese and European scientists.

In the words of a leading expert in the application and use of fuzzy logic[4]:

> There Is Nothing Fuzzy About Fuzzy Logic.

Fuzzy systems are based on well-defined and provable propositions of fuzzy set theory. Fuzzy rule-based systems employ both fuzzy *modus ponens* and fuzzy *modus tollens* to make inferences based on the degree of truth in the rule antecedents. The fuzziness of a fuzzy system involves its tolerance for dealing in a rigorous and predictable way with imprecision in the underlying model parameters. This imprecision has nothing to do with missing data, cloudiness in the knowledge base, the probability that an event will occur, or sloppiness in the model design. Rather, the imprecision rests in the natural and real-world imprecision associated with nearly all natural phenomena. As an example, a fuzzy proposition such as

```
if costs[t] are High
   and delta(income,4) is Rapidly Declining
       then margins[t+2] are Reduced
```

measures the degree to which a particular value for *costs* is considered *High* and the degree to which a particular change in incomes over the past four time periods is considered *Rapidly Declining*. The truth of this antecedent is then used as evidence that margins in two time periods in the future will be *Reduced*. Of course, since the rules of a fuzzy model are run in parallel and their consequent actions combined through a special kind of aggregation, this rule only indicates that "some evidence exists" that the margins are reduced. In any case, the underlying fuzzy sets define the semantics of the model as well as the precise relationship between data points and set memberships. These fuzzy sets translate data points into combined collections of fuzzy sets. The only actual imprecision or fuzziness is associated with the very high level of structure represented by the fuzzy sets themselves. And, as we can see, these sets are always concrete and deterministic.

What Can Fuzzy Logic Do?

Another issue concerning the application of fuzzy logic is often raised. This has the form

Can fuzzy logic do X?

where "X" is some knowledge management task. Example questions include: Can fuzzy logic predict the stock market? Can fuzzy logic find the best stocks for my portfolio? Can fuzzy logic do risk assessment? These are, of course, the wrong questions. This questions are equivalent to asking,

Can Boolean logic do X?
Can Algebra do X?

The questions are wrong because, as a logical system of reasoning, fuzzy logic by itself can't actually *do* anything, just like Boolean logic can't actually do anything other than define the allowable effect of its operators. Fuzzy logic is a method of encoding and using imprecise information. It also defines what effect various operators have on one or more fuzzy sets. Fuzzy logic is only useful and powerful when combined with analytical methodologies, machine reasoning techniques, and the decision support apparata associated with conventional expert systems. Thus, the strength of the modeling tool that incorporates fuzzy logic determines what you can and can't do in terms of problem solving. Users of conventional expert and decision support systems are already familiar with this restriction. What you can model with a conventional system depends on its features and capabilities. The question, then, is: Does the fuzzy system have the features and capabilities I need to model a particular problem? This book should help you answer this question, because to find an answer, you must, of course, have a rather deep understanding of both your problem and the nature of fuzzy reasoning.

Reasons to Reject Fuzzy System Solutions

Business analysts and executives share a common practice with their government counterparts: they consistently reject solutions based on fuzzy logic in favor of those using mathematical and conventional expert systems. Lately, they have even rejected fuzzy logic in favor of solutions involving such computational intelligence approaches as neural networks, genetic algorithms, and the chaos theory. The fact that a manager would prefer to use, without understanding its underlying principles, chaotic models over fuzzy systems lends support to the notion that a large proportion of this rejection is due directly to the pejorative connotations carried by the term "fuzzy" in fuzzy logic. Lotfi Zadeh, the inventor of fuzzy logic, has this to say about this hostility, "Much of the opposition to fuzzy logic is based on misconceptions. What is generally the case is that the degree of opposition to fuzzy logic bears an inverse relation to familiarity with it."

Managers also reject fuzzy systems based on hear-say, reports in popular publications, and fragmentary knowledge about the real workings of fuzzy logic. The following sections include some frequent reasons given for selecting another modeling scheme over fuzzy logic. A brief discussion of its merits, if any, follows the reason, but a complete discussion of the invalidity of each excuse is deferred to the general text of book. This is also a rather personal list of the reasons used to reject fuzzy models. Consequently, the discussion following each excuse often contains an example of both how the reason was offered and how I interpreted the context and the knowledge of the individual.

The Precise Organization

"We're a very precise organization; there's just no room for fuzzy logic." I recently heard a variation of this argument from the head of research at a large chemical company in Michigan. After conducting presentations in fuzzy logic and fuzzy modeling (with many actual case examples) over a three-day trip, I called to check on the status of a target project in product risk assessment and market share analysis. I was told that the company's technical staff had been consulted (which is always a bad sign on any call to a prospective client). The technical staff's department head indicated that fuzzy logic was basically a tool for building signal switches (see the next section, "Fuzzy Logic Is a Control Engineering Tool on page 14). In any case, the company took pride in its analytical capabilities, demanding high precision from their corporate models. As a precise organization, I was told, they would have very little room for fuzzy logic. Since I've had eighteen years of experience in building fuzzy information models and reviewed twenty-three case studies in the presentations, I was only slightly amazed to hear her ask, "Have you ever built a fuzzy model that was not used in control engineering?" I decided that doing business with this company was probably not worth the trouble, and, offering my apologies for disturbing her, slowly hung up the phone.

Confusing fuzzy logic with fuzzy or rough results is, all in all, a fairly frequent response. In one form or another, the term *fuzzy logic* invites the belief that the modeling process generates imprecise answers. The prospective client supposes that a product pricing model produces an approximate price, something that is a rough estimate of what the price would be if we had used a mathematical model or a conventional expert system. In fact, one of my clients, an electronics company in New Jersey with many large government contracts, made a proposal to the Air Force on a ground-to-air communication system involving a fuzzy tactical deployment manager and a fuzzy frequency hopper. The Air Force rejected the proposal completely, saying that fuzzy logic was "the poor man's expert system" and indicating that fuzzy logic did not meet their standards for rigor and precision. When asked exactly what features of fuzzy models they objected to, the Air Force contract officers could only reply that the entire concept of fuzzy logic was unacceptable.

Fuzzy Logic Is a Control Engineering Tool

"We need to make information-based decisions. Fuzzy logic is a control engineering tool." Two events of recent history have cast, in the minds of many, the idea that fuzzy logic is a tool used by control engineers. The first event occurred in 1986 and 1987 with the full scale operation of the fuzzy-logic-controlled Sendai subway system in Japan. The

Automatic Train Operator (ATO) using a fuzzy system developed by Hitachi performed better than any human operator. The subway train, in fact, had a better on-time schedule history, used less energy, and ran smoother than the same train operated by a human. In less than a year, hundreds of fuzzy-logic-controlled products became available on the Japanese market. Today, Japanese companies produce thousands of fuzzy products. A brief list of some common fuzzy consumer products include:

Mitsubishi fuzzy logic transmission
Canon autofocusing mechanism
Minolta subject tracking system (Maxxum 7xi)
Panasonic Electronic Image Stabilizer (a fuzzy gyroscope)

Articles in trade journals and mainstream magazines such as *Business Week* stressed the innovations of fuzzy logic used by the Japanese in a wide range of products. Companies such as Omron produced fuzzy chips, which are microchips that perform thousands of fuzzy inferences per second. This lead to the second event that popularized fuzzy logic as a tool for control engineers. In the United States, software companies, formed by engineers who learned fuzzy logic as a control discipline, began offering tools to create fuzzy systems for families of microprocessors. The first was Togai Infralogic in Irvine, California, and the second was Aptronix in San Jose, California. Aptronix was partially funded and supported by Motorola's semiconductor group in Austin, Texas, and generated code for their HC11 and HC40 microprocessors.

However, this control orientation is an illusion. As originally envisioned by Lotfi Zadeh, fuzzy logic provides a methodology for analyzing complex information systems. Prior to the Sendai subway debut, most of the research in fuzzy logic focused on information decision making. In fact, the only organization of fuzzy researchers in the United States at that time was the North American Fuzzy Information Processing Society (NAFIPS). You need only examine their annual conference literature to see that the preponderance of papers concerned information and decision theory. For example, here is a selection of papers from the seventh annual meeting of NAFIPS (held in 1988):

- ✔ B. Bouchon, *Stability of Linguistic Modifiers with Regard to Generalized Modus Ponens*
- ✔ B. Buckles, F. Petry, J. Pillai, *Uncertainty Models of Network Databases*
- ✔ T.C. Chang, *Project Resource Allocation with Fuzzy Logic*
- ✔ F. Choobineh, W. Dong, F. Wong, X. Liu, *Project Evaluation and Review in an Uncertain Environment (Fuzzy PERT)*
- ✔ R. Li, E.S. Lee, *Analysis of Fuzzy Queues by Imbedded Markov Chains*
- ✔ K. Oh, A. Kandel, *New Reasoning Method for Fuzzy Expert Systems*
- ✔ E. Rudensteiner, L. Bic, *Extending Fuzzy Relational Query Languages by Aggregates*

- ✔ B. Schott, T. Whalen, *Using Guidance of Backchaining in a Fuzzy Expert Decision Support System*
- ✔ D. Schwartz, *A Computationally Simple Interpretation of Linguistic Decision Rules*
- ✔ T. Walen, B. Schott, *Prejudice, Denial, and Modus Ponens in Forward Inference*
- ✔ L.A. Zadeh, *Qualitative Systems Analysis Based on Fuzzy Logic*

Of the sixty-nine papers included in the annual proceeding, only five (about 7% of the total number) concerned fuzzy control. The remainder focused on fuzzy knowledge, fuzzy decision making, and expert systems that used fuzzy logic. But, by late 1988, news of the Sendai subway system and the proliferation of Japanese products was already changing the flavor of fuzzy logic research in the United States. The First Industrial Conference on Fuzzy Systems, held in Austin, Texas, in June 1991 and sponsored by MCC, was dominated by control applications. I was the only speaker to address the issues of fuzzy information decision making (and to a highly unmotivated audience). Over the next several years, the IEEE conferences on fuzzy and neural systems have concentrated almost exclusively on control applications. Only in the past few years, with significant urging on my part, has the annual fuzzy systems conference sponsored by *Computer Design* magazine included a tract on fuzzy information modelng.

In fact, fuzzy logic as an information decision tool can be traced back to the late 1970s. The first commercially available fuzzy modeling tool, called Reveal, was built as an expert and decision support system. Developed by Peter Llywellyn Jones in 1978 at Decision Products, and marketed in the United Kingdom by Tymshare and Unilever in the early 1980s, Reveal was a financial and general planning tool based on the row and column spread sheet concept (long before personal computer spread sheets where available). Available across a fully interactive time-sharing system, Reveal had its own time-series relational database with joins and projections; a complete financial modeling language with goal seeking and sensitivity analysis, a fuzzy modelling capability that included such advanced concepts as hedges, fuzzy relational joins, and fuzzy comparisons. Although Reveal was removed from the market when McDonnell Douglas bought Tymshare in the middle 1980s, it was a highly successful product used in a large number of real-world applications.

More important than the historical perspective on fuzzy information modeling is the current use of approximate reasoning in solving mission-critical business problems. I have been involved in the development of fuzzy modeling tools and actual business applications for over eighteen years. These real-world fuzzy models include many varied projects: intelligent project management and project risk assessment; leveraged buy-out (LBO) and company acquisitions advisor; financial statement analyzer; fleet container management, planning, and forecasting; integrated MRP and production scheduling system; portfolio asset allocations; investments strategy, safety, and suitability model; software complexity analysis; fuzzy database retrieval system, a managed health care provider fraud and abuse detection system; product pricing and positioning models for the pharmaceutical, chemical, and retailing industries; criminal (suspect) identification system; just-in-time inventory control model; and capital budgeting and cost allocation systems.

Complex Time-Series Modeling

"We have complex, chaotic time series problems with leads and lags. I understand that fuzzy logic can only model simple series, like seasonal temperature and humidity fluctuations." Several clients have been skeptical of fuzzy logic's ability to handle complex time-series models. In business, many problem solutions require the analysis of chaotic time series with feedback cycles as well as multiple lead and lag relationships. Much of this skepticism stems from one of the few examples of fuzzy time series analysis in literature. This reference appeared in two places. The first appeared in USC-SIPI Report #169 (January 1991), *Generating Fuzzy Rules from Numerical Data, with Applications,* by Li-Xin Wang and Jerry M. Mendel. Part 5, entitled "Application to Time-Series Prediction," modeled the monthly mean temperature of St. Louis, Missouri, from January 1945 through December 1974. The second appeared in the book *Adaptive Fuzzy Systems and Control: Design and Stability Analysis*, by Li-Xin Wang (1994). Time series analysis techniques are discussed in section 4.5, "Modeling the Mackey-Glass Chaotic Time Series by FBF Expansion," and section 5.4, "Application to Time-Series Prediction." This last section duplicates the time series temperature prediction of the report from the University of Southern California Signal and Image Processing Institute. The idea persists, however, that fuzzy logic can solve simple sinusoidal (seasonal) time series but is not suited for more complex systems. At a 1995 conference on fuzzy applications in San Francisco, the director of a medical research facility indicated that he had rejected fuzzy logic because he had a very difficult time-series problem involving cardiac pulse monitoring under electrical stimulation. This situation posed a feed-back problem with delays, and he understood that fuzzy logic could not handle this kind of problem. His complete knowledge about the time-series modeling capabilities of fuzzy logic came from only the simple temperature forecasting example. He was not even aware of the chaotic time-series examples in both of these sources. He was also following the fallacy of viewing fuzzy logic as a specific technique or a specific implementation rather than as a logical methodology.

The Power of Conventional Expert Systems

"I can do the same kind of model using a conventional expert system." This is a fairly common response to fuzzy expert systems. During a presentation in 1989 to the director of expert systems at a very large insurance company in New York City, I was confronted with a variation on this in the form of a question, "If fuzzy logic is so powerful, why doesn't everyone else have it?" Here, *everyone else* meant all the other expert system vendors (at that time, AI Corporation, Inference, and Aion Corporation). The implication, of course, was that the other vendors had no need for fuzzy logic because their systems could solve problems just as easily using conventional (Boolean) expert system technologies. At the heart of this reason is a basic redundant statement: conventional expert systems are successful at solving problems that can be solved by conventional expert systems. But they are also very brittle, generally fail to scale-up well, and perform very poorly in real-world problems involving highly nonlinear relationships, multiple conflicting and collaborating experts, or systems that must accumulate evidence for and against a set of solution states. These are exactly the kinds of systems that expert systems based on fuzzy logic handle well. You will find a rather complete discussion of this issue later in this book.

The Precision of Mathematical Models

"I can do the same kind of thing using a precise, mathematical model." Rejecting fuzzy systems in favor of a precise mathematical model is a different form of both the previous and the first objections: that fuzzy systems are insufficiently precise to handle the exacting and quantitatively demanding needs of the client's organization. Once more, the idea that Boolean logic's intrinsic precision offers a better model of observed phenomena than fuzzy logic is generally based on a reaction to both the word "fuzzy" and the unspoken belief that the model generates incomplete, vague, rough, or approximate answers. Lotfi Zadeh addresses this issue in the following way:

> As a logic of approximate reasoning, fuzzy logic is perceived by some to be in conflict with the deeply entrenched Cartesian tradition of respect for what is rigorous, quantitative, and precise. In reality, parts of fuzzy logic—and especially its foundations—are quite precise. But what is true is that the quest for precision is not the dominating force in fuzzy logic—as it is in the traditional logical systems. Thus, in many of its applications, the success of fuzzy logic is measured pragmatically by the performance of the system in which it is employed rather than by the degree to which it meets a high standard of rigor and precision.

More important to the issue of precision, however, is the realization that fuzzy systems are universal approximators. This means that we can approximate a function to any degree of accuracy by changing the rules and the level of granularity in the fuzzy descriptors. Thus, a fuzzy system can represent, as precisely as we want, the same system dynamics as a conventional mathematical model.

Fuzzy Model Stability

"There is no proof for stability in a fuzzy model." In some respects, this statement provides a valid reason to be careful in employing a fuzzy model. Although, as we will see, it is no more significant than the care exercised in employing any other kind of complex model. Indeed, there is currently no mathematically rigorous proof that a fuzzy model is stable. This means that we cannot prove, *a priori*, that a fuzzy model will perform consistently within the behavior we observe for the validation set. Some control engineers interpret this situation as a fatal weakness in fuzzy models. Because stability methods exist for conventional process control approaches (PID and state-space, as an example), the lack of a stability proof for a fuzzy model means that we must rely on empirical evidence to test the pathological behavior of these systems. In the process engineering field, engineers test the long-term operational stability of their control models through a mathematical process that measures the convergence of behavior. They assume that the mathematical stability shown for a control device will be the same stability exhibited by the actual device in operation. This is seldom the case, but a test of stability does show that the system does not contain any regions of chaotic behavior. This realization is comforting, emotionally and legally.

Business systems are often several orders of magnitude more complex than engineering control systems. A proof of mathematical stability has never been an issue with business decision support and expert systems, just like we can't expect a mathematical proof of stability and correctness in a software program. Business and general information decision systems are validated through exhaustive and rigorous testing. Of course, this approach is no different than one you would take when building any kind of complex business system. Nor is it any different than the approach taken for business models using neural networks, genetic algorithms, and chaos theory. Thus, the issue of stability is a concern both in fuzzy system research and in control engineering, but it is not a deciding issue against the use of fuzzy logic in decision support and expert systems.

Fuzzy Model Execution Speed

"I understand fuzzy models take longer to run than conventional models." In fact, a fuzzy rule–based system usually *does* take longer, on a rule-by-rule basis, to execute than conventional expert system rules. Fuzzy systems need time to "fuzzify" the data points (mapping data values into the vocabulary fuzzy sets), to create intermediate fuzzy regions through hedging operations, to correlate the consequent fuzzy set, and to aggregate the correlated fuzzy set into the under-generation fuzzy region. However, to compensate, fuzzy rule bases are generally an order of magnitude smaller than the corresponding rule bases in conventional systems. For example, a vehicle risk underwriting system developed for a major insurance company in Hartford, Connecticut, contained 181 rules in a conventional expert system (ADS from Aion Corporation). The equivalent fuzzy model contained seventeen rules and produced the same answer. This rule reduction results from set theoretic representation of the model in fuzzy system.

The Vehicle Risk Underwriting Model (A Brief Explanation)

The basic vehicle risk underwriting model consisted of a single dependent variable (the risk rating) and twelve independent variable including age, length of time that policy has been with company, miles driven per year, miles driven to work, number of chargeable accidents in the past two years, any DWI/DUI arrests, and number of moving violations in the past year. The conventional expert system approach calibrated the underwriting risk based on a combination of these factors. A somewhat typical rule might be:

```
if age is between(16,21)
   and milesperyear > 10000 and milestowork > 40
   and (accidents > 2 or violation > 2)
   and DWIcount > 0
       then Risk = Risk + MAXRISKAMOUNT;
```

On the other hand, the fuzzy model's approach viewed the problem from the semantics of risk rather than from the numerical boundaries of risk components. As an example, Figure 1.6 shows one fuzzy set defining the risk associated with the age of the driver.

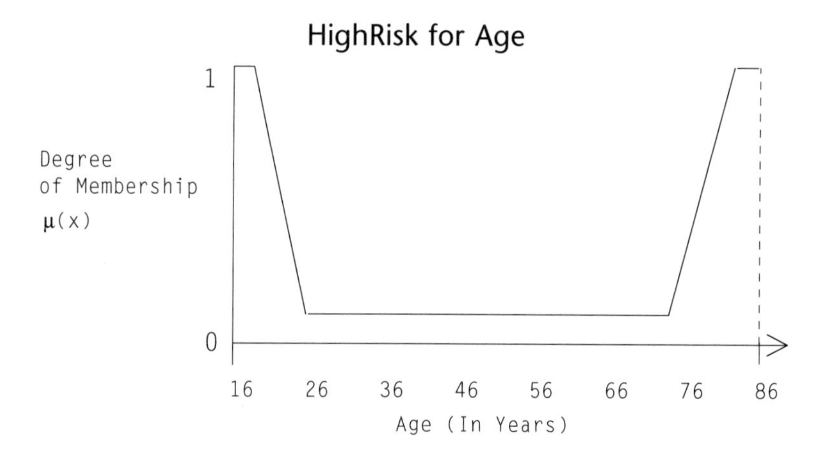

Figure 1.6 High Risk for Age Fuzzy Set

Instead of organizing the problem into a set of rules that check each possible combination of variables, the fuzzy model aggregates the risk based on the product of the fuzzy spaces associated with the variables. A few of the fuzzy underwriting rules look like this:

```
if age is a HIGH.RISKforAge then Risk is Increased;
if milesperyear is LARGE.RISKforDISTANCE
   then RISK is INCREASED
```

The consequent fuzzy region, *Increased*, is one of several fuzzy sets that increase, decrease, or make no change to the current risk assessment. Figure 1.7 shows the fuzzy term set associated with the risk solution variable.

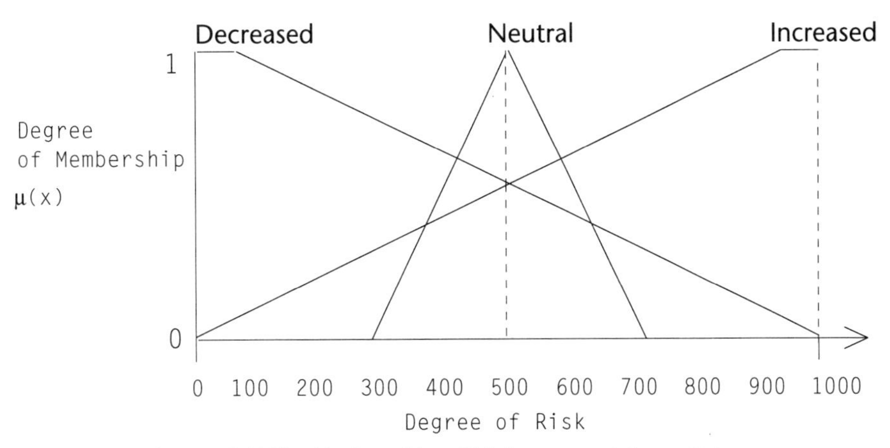

Figure 1.7 The Underwriting Risk Assessment Fuzzy Sets

Using an additive form of the fuzzy model, the underwriting system combines the factors contributing to insurance risk into a single output fuzzy region. Then this region is defuzzified to find the degree of risk associated with the current prospect (or client for renewal purposes). Figure 1.8 illustrates how the combined solution fuzzy region appears after the model executes.

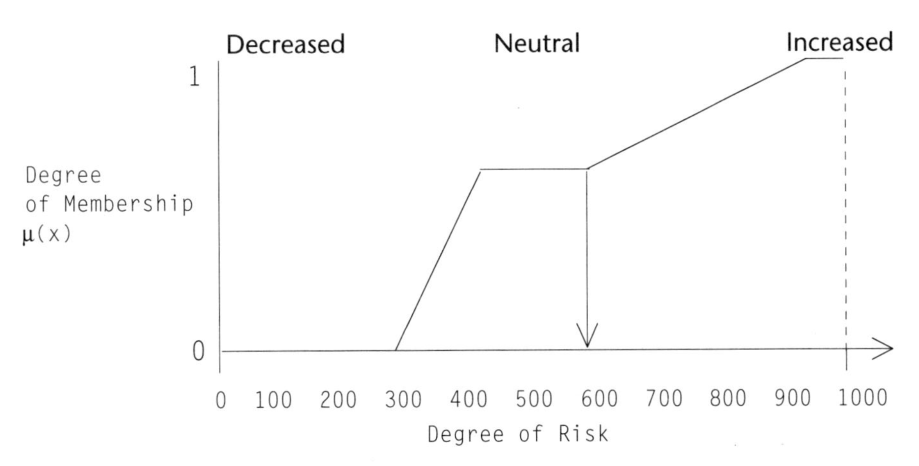

Figure 1.8 The Combined Risk Assessment Value

In this example, the preponderance of evidence is neutral, with a movement toward an increased risk. Therefore, defuzzification yields a slightly increased risk in underwriting this particular candidate.

Fuzzy Set Discovery and Correctness

"I hear that developing the fuzzy sets is too hard. It's too subjective and error prone." A common misconception about fuzzy systems arises from what is viewed as subjectiveness and arbitrariness of fuzzy sets. Knowledge engineers who have never developed a fuzzy system often imagine that finding the correct fuzzy set shapes (and the correct amount of overlap among fuzzy sets defining the same concept) is an arduous and error-prone task. However, this is not the case. In fact, knowledge engineers experienced in building and tuning fuzzy models find that:

- ✔ Fuzzy set discovery is a fairly straight-forward process and actually aids in the knowledge acquisition process.
- ✔ Fuzzy systems are remarkably tolerant of estimations in the fuzzy set morphology and, to some extent, the degrees that fuzzy sets overlap.
- ✔ Fuzzy set shapes can be refined quickly, bringing the model prototype into alignment with reality in an easy, step-by-step fashion.

You can find both knowledge mining methodologies and neural network approaches to decomposing a fuzzy system's variables into a set of underlying fuzzy sets. However, these two avenues are more appropriate to control applications where the semantics of the fuzzy sets are generally of little or no concern. In business models, the meaning and relation between fuzzy sets is quite important and stems from the very nature of the information richness in the system itself. But this does not mean that eliciting the fuzzy sets is a difficult task. Since the sets reflect the semantics of the model, they will "fall out," so to speak, from the natural progression of defining the model's process characteristics.

As an example, in a project risk assessment model, we extract from the project managers and the corporation's senior management exactly what is meant by "short," "medium," and "long" duration projects. Figure 1.9 shows how these three concepts are tied together to represent the semantics for project duration.

The shapes in figure 1.9 were elicited through a process of knowledge discovery, generally between the system builder and the expert or team of experts. A discussion between the knowledge engineer (Susan) and the project manager (Russell) might go something like this:

Susan: Tell me, Russ, how do you measure a long project?

Russell: That's pretty simple. Most of our projects, as we can see by the graph (Figure 1.9), fall between six and eighteen weeks, roughly from one and a half months to four and a half months. When a

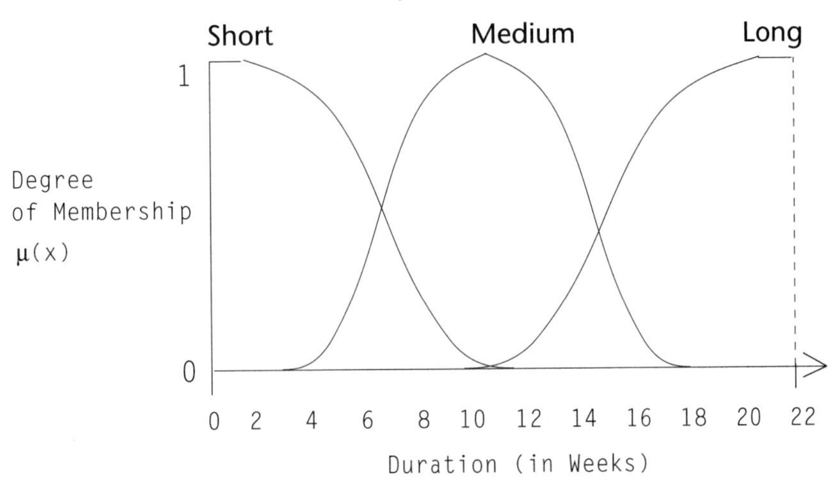

Figure 1.9 Project Duration Semantics

project goes over about fourteen weeks, we have to be very careful about the allocation of resources and capital, and we must decide whether we really want to manage a project that goes on that long.

Susan: So once a project goes over three and a half months, you immediately consider it "long."

Russell: Well, I wouldn't say *immediately*. But, it becomes long awfully fast after that.

Susan: So a sixteen-week project is really not long.

Russell: Well, it's long all right, but not long enough to trigger most of those administrative controls that might consider it fairly long. But some other manager might consider it only a little long.

Susan: Then a twenty-week project is not long?

Russell: (Pondering this a moment.) No, I'd say that a five-month project is getting to be pretty long for us.

Susan: By "getting to be" you mean that it's not actually what you would call long but that it is starting . . .

Russell: Exactly. We'd begin to pay special attention to this project. If a project's duration is much more than twelve weeks, we'd want to move it to a special list of critical projects. Management needs to monitor the rate of capital and human resource consumption, and also any event slippage or requests for specification changes, no matter how small, from the end user.

The fuzzy sets define how the model user perceives the three categories of project duration: short, medium, and long. Fuzzy sets do not exist in a context-free space; they have meaning only when associated with specific states in the model itself. This association means that their structure is determined by considering what the membership grade means at each point along the axis. Consider the concept of a *Long* project. How difficult is the process of deriving a representation of *Long*? In Figure 1.10, we can see four other fuzzy sets that could possibly represent the idea of *Long* (in addition to the sigmoid set of Figure 1.9).

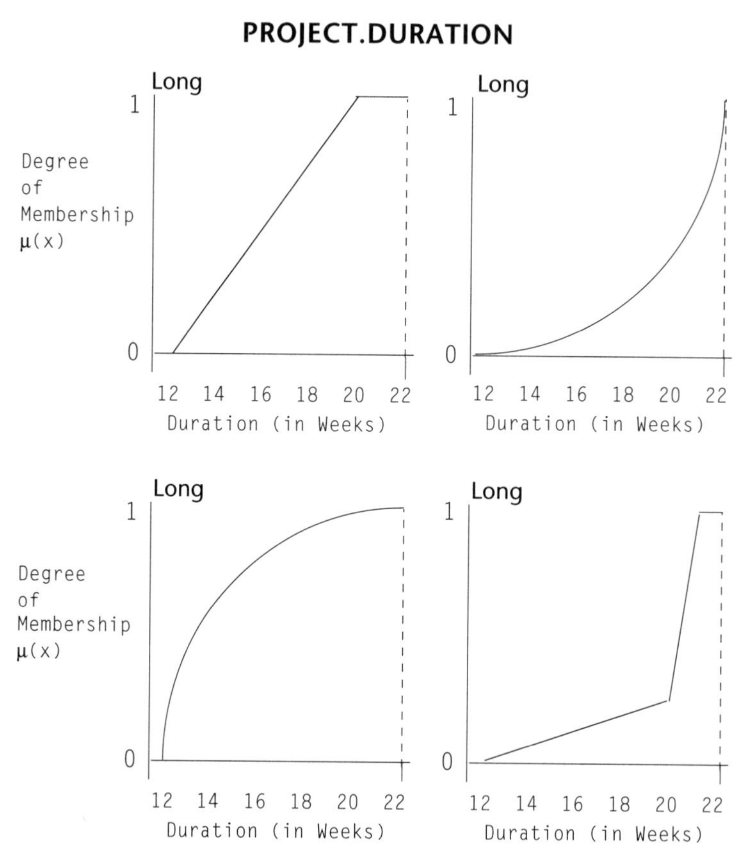

Figure 1.10 Different Semantics for LONG Projects

Each of these fuzzy sets in Figures 1.9 and 1.10 might be an acceptable representation of *Long*. The actual way we encode the semantics of *Long*, however, depends on the way the expert (in this case, the project, department, or senior management) measures the degree to which a project is considered long as the duration increases. In Figure 1.9, the idea of *Long* gradually increases as the project duration increases. Its S-curve representation means that the compatibility with *Long* has a high inflection point, where 50% of the projects above the point are significantly long, and 50% of the projects below the point are moderately or weakly long. In Figure 1.10, the first fuzzy set (at the top left) indicates a linear relationship between duration and the idea of a long project. The greater the duration, the stronger the membership in the *Long* fuzzy set. The next two fuzzy sets (top right and bottom left) describe highly nonlinear relationships, where the idea of long takes on a high degree of membership in *Long* almost immediately, as soon as the duration moves beyond eighteen weeks. And the final representation (on the bottom right), illustrates an irregular fuzzy set. In this set, the concept of long remains quite low but still increases appreciably until we reach a threshold of approximately twenty weeks, when the concept makes a sharply rapid rise into the *Long* fuzzy set. Each of these sets (and many other set shapes are also possible) represents the ways in which the idea of a project with a particular duration can be mapped into the fuzzy set of *Long* projects.

A slightly more difficult task, but straightforward nonetheless, in fuzzy knowledge engineering involves the estimation of fuzzy numbers. In the real world of information modeling, nearly all quantities are fuzzy; that is, they have some degree of uncertainty, ambiguity, noise, or expectancy.

Producing Fuzzy Numbers

There are two ways to produce fuzzy numbers in the context of a model: either through a static fuzzy set definition or dynamically through an approximation hedge. The static definitions depend on the triangular, trapezoidal, or bell-shaped fuzzy set definitions (for example, Gaussian or Gaussian-like functions such as pi, beta, and Weibul). As illustrated in Figure 1.11, these fuzzy sets usually become part of a variable's term set.

In Figure 1.11 the fuzzy sets *Mature*, *MiddleAged*, and *Senior* are fuzzy numbers. *Young* and *Old* are fuzzy quantities but are not numbers, since they have theoretically (if not actually) infinite lower or upper bounds.

Dynamic fuzzy numbers are produced by applying approximation hedges to numeric quantities. Approximation hedges include *around*, *near*, *about*, and *close to*. Examples of rules using dynamic fuzzy numbers (taken from a new product pricing model) include:

```
our price must be above near 2*MfgCosts;
if the competition.price is not very high
   then our price should be near the competition.price
```

Fuzzy numbers provide models with measures of certainty in the underlying control variables. They also provide methods to expand or restrict the operation of models, based on the degree to which we tolerate the diffusion of numerical concepts.

Fuzzy numbers have several properties: their bandwidth or horizontal domain, called expectancy; their degree of spatial ambiguity, generally a measure of kurtosis; and their

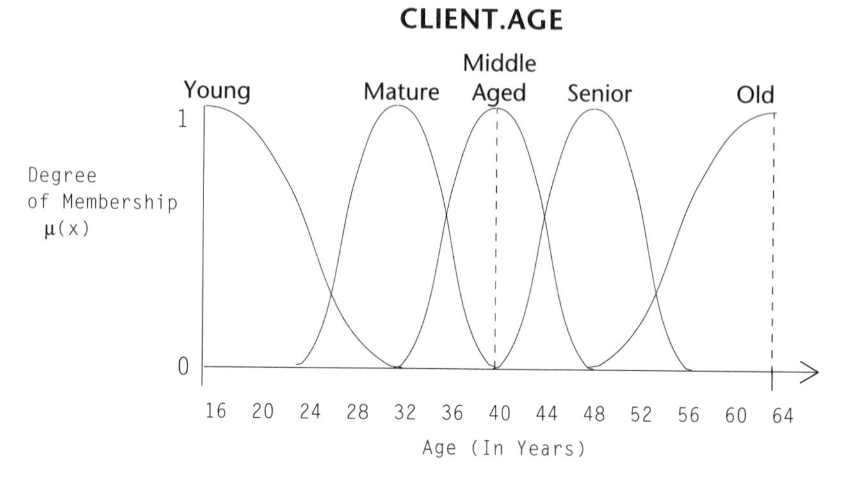

Figure 1.11 Fuzzy Numbers in a Variable's Term Set

orientation preference or skew. A *crisp number* is simply a fuzzy number with an absolute degree of certainty in its quantitative properties. As an example, Figure 1.12 shows two fuzzy numbers representing twice a particular manufacturing cost (arithmetically derived by solving the equation *2*MfgCosts*).

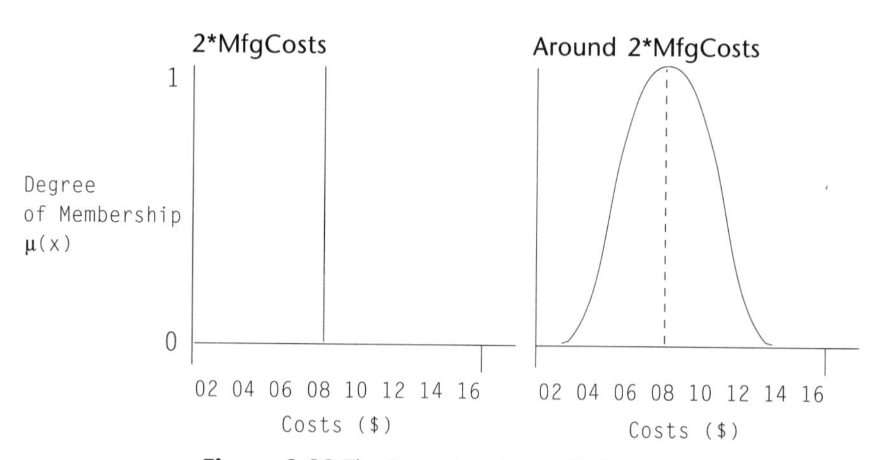

Figure 1.12 The Fuzzy Number: *2*MfgCosts*

Fuzzy numbers and fuzzy quantifiers (non-fuzzy numbers) provide the basis for bringing information models into agreement with the real-world. The use of fuzzy numbers greatly simplifies the task of writing rules to encode behaviors that include the intrinsic uncertainties associated with everyday processes. You can find a more complete discussion of fuzzy numbers and approximation fuzzy sets in Fuzzy Numbers on page 100.

Tuning and Validating Fuzzy Systems

"I've heard that fuzzy models are just too difficult to tune. You can never get them to work properly." This statement also holds a small kernel of truth: fuzzy systems involve a larger number of control points than conventional expert and decision support systems. And, because the rules in a fuzzy expert system are executed in parallel, the organization of the system involves the refinement of parallel processing units (called *policies* in the Metus/IMS fuzzy modeling system). Figure 1.13 schematically represents some of the various tuning options available in a fuzzy model at the knowledge base or rule level.

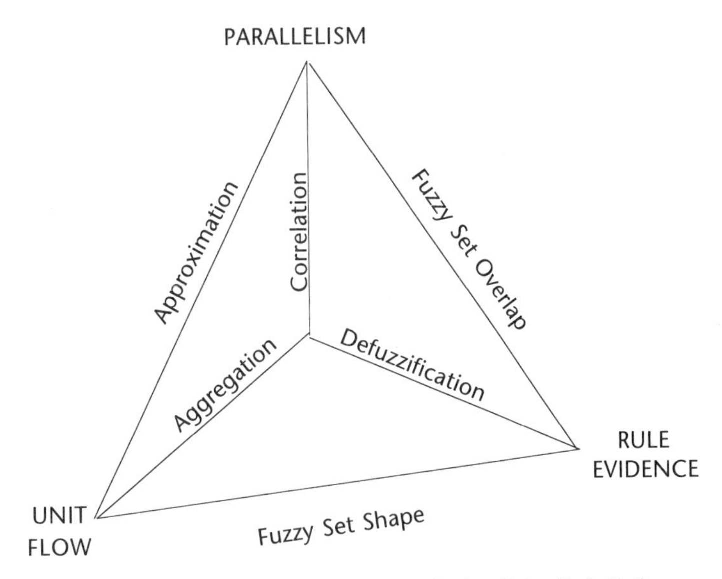

Figure 1.13 Tuning a Fuzzy Knowledge Base (Rule Set)

The overall complexity of fuzzy information models increases because fuzzy rules are executed in parallel and are organized in process units (such as policies) in order to accumulate evidence for or against a particular unit's objective (solution variable). At the same time, as illustrated in Figure 1.14, we can also tune the entire model itself through fine-grain adjustment of such features as the fine-grain decomposition of model parameters or variables into fuzzy sets; the inferencing technique used to actually extract knowledge from the rule set; and our choice of union, intersection, and complement operators.

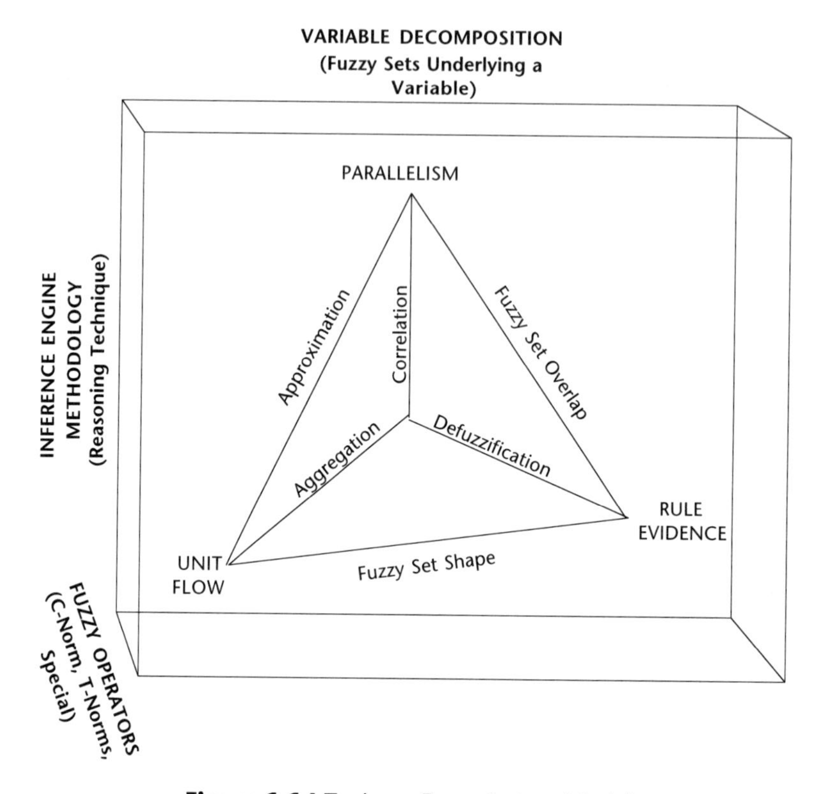

Figure 1.14 Tuning a Fuzzy System Model

The many degrees of freedom available to the model builder for tuning a fuzzy system not only provide a rich assortment of tuning techniques, but also extend the calibration capabilities of knowledge engineering to a level of detail not available in conventional expert systems. On the other hand, this same array of macro- and micro-tuning facilities also makes building a fuzzy model a difficult task for the inexperienced engineer. Much of the stable, long-term operational controls associated with fuzzy modeling are derived from the underlying richness of approximate reasoning. Combining and selecting the proper values for these tuning parameters often involves a certain amount of model building experience. But all model building capabilities, except for trivial problems, are ultimately derived from experience. You should not reject the benefits of fuzzy logic just because using it often implies a relatively significant learning curve. Instead, you should consider the learning curve an integral part of the model construction experience.

Fuzzy model validation presents some additional difficulties over conventional expert and decision support systems. Generally, validation, like stability, in fuzzy models is determined the same way as in conventional systems: through test case analysis, pathological data stressing, and reasonable solution checks with experts. Fuzzy models, however, present some intrinsic difficulties in applying logical construction analysis to fuzzy rules. These difficulties arise from fuzzy logic's failure to obey boolean logic's "Law of the Excluded Middle." Fuzzy rules can therefore appear contradictory. For example, the following rules are part of an actual product pricing model:

```
our price must be High;
our price must be Low;
our price must be around 2*MfgCosts;
```

Fuzzy rules can also seem completely contradictory and illogical when cast in conventionally crisp environments. As an example:

```
if EarningsPerShare are Weak
   and EarningsPerShare are Strong
      then CompanyStrength is Robust;

if ProjectDuration is Short
   and ProjectDuration is not Short
      then ProjectRisk is Moderate;
```

A fuzzy *And* combines the two antecedent propositions in a way that is not possible in Boolean systems. Because fuzzy numbers are spread out over numerical space, we expect to find some residual values in regions that overlap the two predicates; thus, the idea of expectancy occurs once more. We can see this idea with a little more clarity through the following simple fuzzy rule:

```
If A is 5 and A is 7 then B is 12;
```

This rule is not executable in conventional (or crisp) expert systems since Boolean logic tells us that the value for variable *A* can not be simultaneously 5 and 7. In fuzzy logic, however, we can establish a truth for the predicate by finding the degree to which the current value for *A* is inside the fuzzy numbers *5* and *7*. Figure 1.15 shows the representation of the fuzzy numbers *About 5* and *About 7* as triangular membership functions.

Because fuzzy numbers are not discrete points in space, a quantity can coexist in multiple numbers at the same time. Figure 1.16 shows how variable *A*'s value of 6.5 has a nonzero membership in *About 5* and *About 7*. We can see that 6.5 is moderately a member of *About 5,* but is strongly a member of *About 7, (as we might expect).*

When we evaluate this rule using the fuzzy *And,* thereby taking the minimum of the two values, a membership of m[.64] is found. This rule will execute with a significant degree of truth, rather than the zero degree of truth in a crisp system. As a result of this innate ambiguity in numbers, as well as the ability to postulate completely reasonable and logically sound rules that reflect logical contradictions, some of the tools and techniques used to find reasoning errors in conventional systems are not applicable in fuzzy systems.

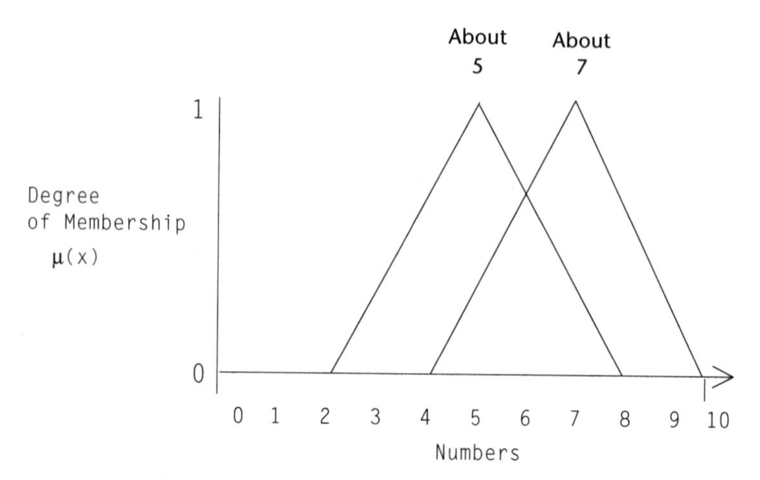

Figure 1.15 The Fuzzy Numbers 5 and 7

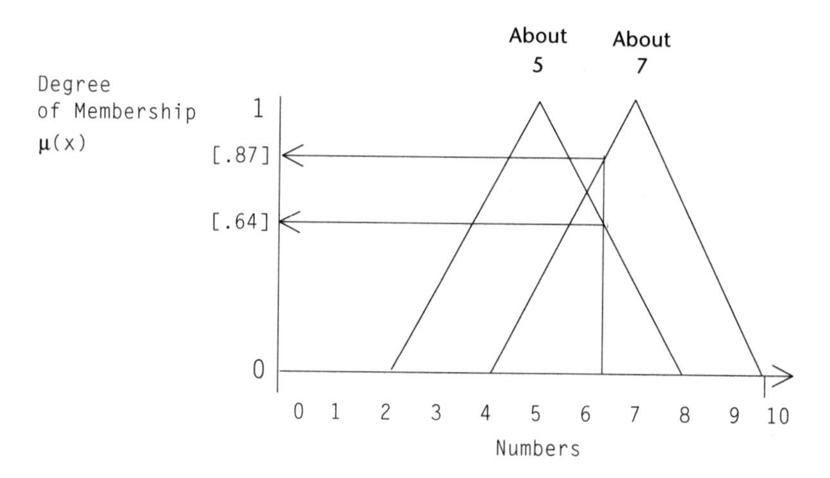

Figure 1.16 Locating a Value in Multiple Fuzzy Numbers

The Somewhat Ad-Hoc Nature of Defuzzification

"Solutions in a fuzzy system are too ad-hoc. No mathematical basis exists for defuzzification." We now arrive at a fundamental problem in fuzzy modeling: the method of deriving a value from a fuzzy system. This is the problem of *defuzzification*. The result of executing a set of fuzzy rules is a fuzzy set whose shape represents the aggregation of each rule's consequent set (modified by the degree of truth in the rule's premise). As a general case, we need a single value for each solution (outcome) variable in the model. The problem then is: how can we find a single number that best represents the information encoded in the output fuzzy set? Over the years, quite a number of approaches to defuzzification have been developed. These approaches attempt to find the value of the solution by applying various operations to the solution fuzzy set. You should note, however, that the technique of defuzzification is not supported by axiomatic principles derived from the foundations of fuzzy set theory itself.

Nearly all models use two principal methods of defuzzification: the center of gravity (or *centroid*) and the maximum of the output membership function. As Figure 1.17 illustrates, these techniques produce entirely different results.

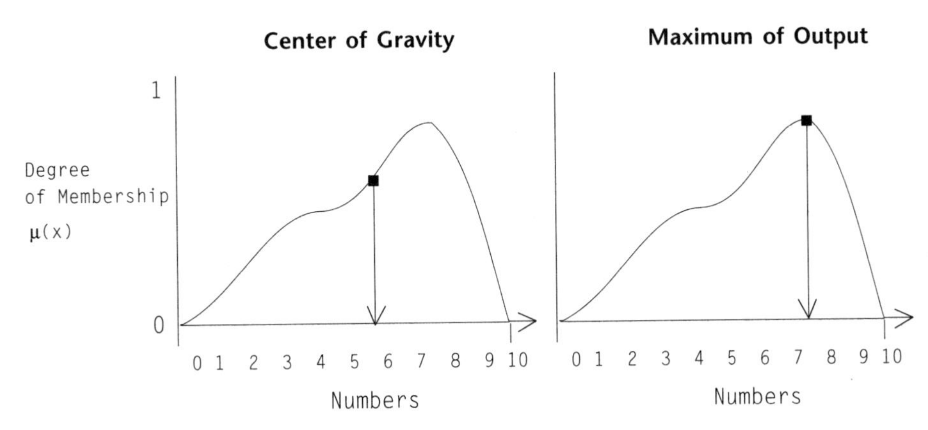

Figure 1.17 Two Common Methods of Defuzzification

Both of these techniques produce reasonable results when applied to specific kinds of fuzzy models. The center of gravity technique is the most widely used, since it combines evidence from all the rules and yields an answer that is weighted by the degree of truth in the aggregate. It has its mathematical foundations in the evidence mass functions of Bayesian probability. The center of gravity is, essentially, the weighted average of the output membership function. The result R is calculated using the formula in Equation 1.2.

$$R = \frac{\sum_{i=0}^{N} d_i \times \mu_i}{\sum_{i=0}^{N} \mu_i}$$

Equation 1.2 Defuzzification of Fuzzy Output

In equation 1.1, d_i is the underlying domain value, and μ_i is the value for that domain point in the membership function. On the other hand, the maximum is generally computed as the domain value that has the maximum membership value (with some interpolation arithmetic used for values that fall on a plateau or when the output fuzzy set is multimodal).

Although defuzzification is not derived from fundamental principles drawn from fuzzy set theory, it is based on convincing, time-proven heuristics that produce accurate solutions. The combination of centroid defuzzification with the additive aggregation technique forms the basis of Dr. Bart Kosko's Standard Additive Model (SAM). Dr. Kosko has shown that a fuzzy system employing the standard additive model can approximate any continuous function. And work by Dr. Fred Watkins indicates that the same approach might be used to model certain noncontinuous functions as well. This unifying methodological approach to fuzzy systems provides a mathematical foundation for the defuzzification process.

We should not lose sight of the relationship between the calculus of information and the heuristics of developing a result. In a large conventional expert or decision support system, the results are derived from the interplay of many rules and procedures. The method of extracting a solution value in such systems is often less understood and more ad-hoc than the compact mathematics of defuzzification.

The Problem of Combinatorial Explosion

"I guess fuzzy logic might be fine for some simple problems, but we have very large, difficult problems. Fuzzy models simply cannot handle problems of our size." The dual problems of combinatorial rule explosion and scalability are often a concern for business analysts and technical managers unfamiliar with the techniques used in building information-based fuzzy models. This misunderstanding is primarily due to much published work, by academic "experts" in fuzzy logic, claiming that fuzzy systems become unmanageable as the number of variables increases due to a combinatorial explosion in the rule set. In real-world fuzzy models, however, rule explosion is seldom a problem. Furthermore, recent work by practicing fuzzy scientists, in particular Dr. Bill Combs at Boeing in Seattle, Washington, has shown that fuzzy models can be design so that the rules increase linearly with dimensionality, rather than exponentially. (A complete description of Bill Combs' discovery, "The Combs Method for Rapid Inference" appears in Appendix A. Working examples using Microsoft Excel spread sheets are also located on the accompanying CD-ROM.)

In order to understand the problem solved by the Combs' method and see how dimensionality affects conventional fuzzy rule–based systems, let's examine the problem in a little detail. For each independent variable, the rule size grows geometrically according to the rate indicated in Equation 1.3

$$r_v^k = \prod_{i=1}^{N} f_i$$

Equation 1.3 Rule Dimensionality Growth

As an example, consider a simple system of two independent (*X* and *Y*) variables and one dependent (*Z*) or outcome variable. If the independent variables are decomposed into five and seven fuzzy sets (respectively), then there are thirty-five possible rules to describe the behavior of the system. The relationship between fuzzy sets and rules is normally shown in a Fuzzy Associative Memory (FAM). A fuzzy associative memory encodes the fuzzy rules. Each dimension of the matrix represents the fuzzy sets assigned to an independent variable. Thus, a two-dimensional FAM can show three variables: two independent and one dependent. Figure 1.18 shows the rule matrix for the two input and one output fuzzy systems.

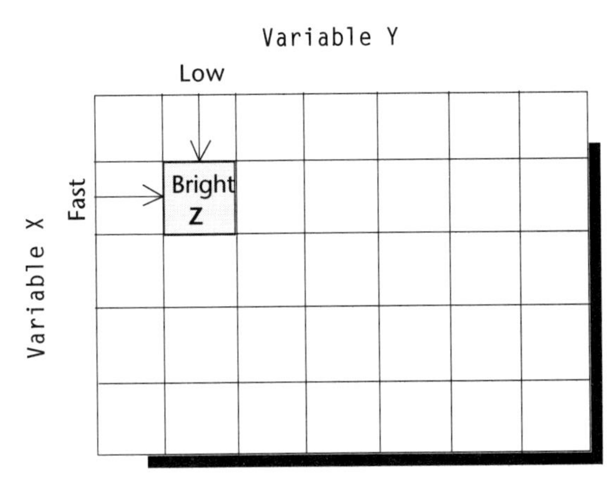

Figure 1.18 The 2x1 Fuzzy Associative Memory [FAM]

In our example, the horizontal axis represents each fuzzy set associated with the variable *Y* and the vertical represents the fuzzy sets associated with the variable *X*. At their intersection, is the solution fuzzy set associated with variable *Z*. We can read the matrix rules in this fashion:

if *X* is *Fast* and *Y* is *Low* then *Z* is *Bright*;

The problem with combinatorial rule explosion becomes evident when we add one more independent variable (*W*). Each two-dimensional FAM is now nested inside the cell of a higher order dimensional array. Figure 1.19 illustrates this arrangement.

So, with just three fuzzy sets in the new variable, the model jumps to 105 rules (see Equation 1.3). This is in accordance with the multiplicative increase in rules defined earlier in Equation 1.2.

$$r_v^k = \prod_{i=1}^{N} f_i = 5_X \times 7_Y \times 3_W = 105$$

Equation 1.4 Rules as a Function of Fuzzy Sets

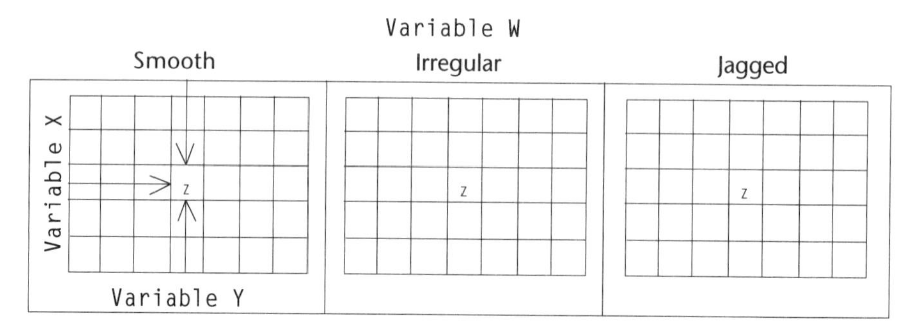

Figure 1.19 A 3x1 Fuzzy Associative Memory

If the new variable had a somewhat more realistic term set of five rules, the matrix would grow to 175 rules. Once more, these are the predicate terms in the rule. You read a rule in the three dimensional FAM matrix as:

if *X* is *Fast* and *Y* is *Low* and *W* is *Smooth* then *Z* is *Bright*;

Thus, as the number of variables in the model increases, the number of rules necessary to describe the complete behavior of the model appears to grow geometrically. A model with nine variables, decomposed into seven fuzzy sets each, would require 7^9 (or 40,353,607) rules to completely describe all possible behaviors in the model. The combinatorial rule growth assumes three factors:, that you are defining a behavior for each possible fuzzy state in the model, that the model variables are decomposed into a significant number of fuzzy sets, and that the model is driven by a fuzzy associative memory matrix and not by knowledge-engineered rules. In actual practice, a fuzzy business model has a sparse rule set driven by fuzzy sets that are shared by variables and that often describe large qualitative spaces in the application domain. As an example, in a new product pricing model, the fuzzy description underlying the concept of *Price* simply consists of two highly overlapping fuzzy sets. Figure 1.20 shows this decomposition into the *High* and *Low* sets.

Business models also tend to rely on fuzzy quantities or fuzzy numbers to a much greater degree than control models do. These fuzzy sets are dynamically created, usually through a hedge operation, and are then deleted from the active system. This process significantly reduces the number of rules needed to capture a particular behavior.

Unlike control engineering applications, which often have real-time or time-critical performance components, business applications can take advantage of very high speed workstations and personal computers to execute a large number of rules in a time frame that might be prohibitive in a process control environment. A portfolio asset allocation system might execute over fifteen hundred fuzzy rules in four or five seconds in order to determine the proper mix of equity concentrations and their associated strategic risks. This time frame is extraordinarily long for a real-time processor, but is insignificant for an ordinary business system.

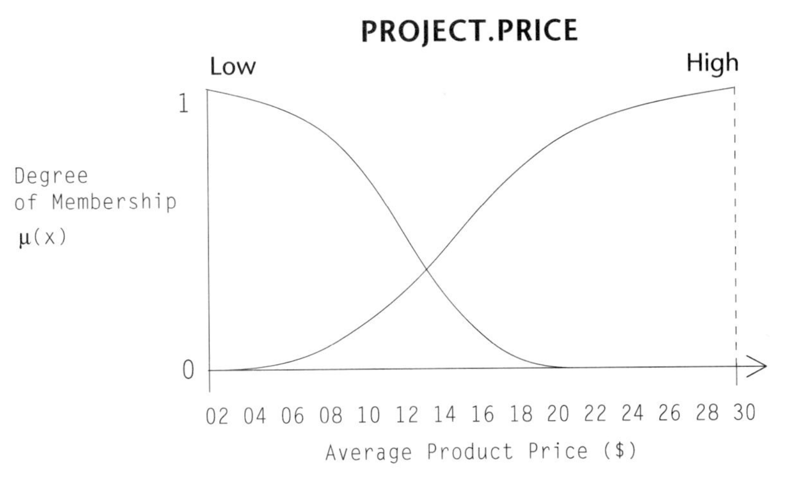

Figure 1.20 Product.Price Fuzzy Term Set

Finally, business applications generally take advantage of structural decomposition on a system level to break a problem into a set of independent units. These units, called *policies*, reduce the computational complexity of the overall system. Figure 1.21 shows a typical unit flow in a large fuzzy system.

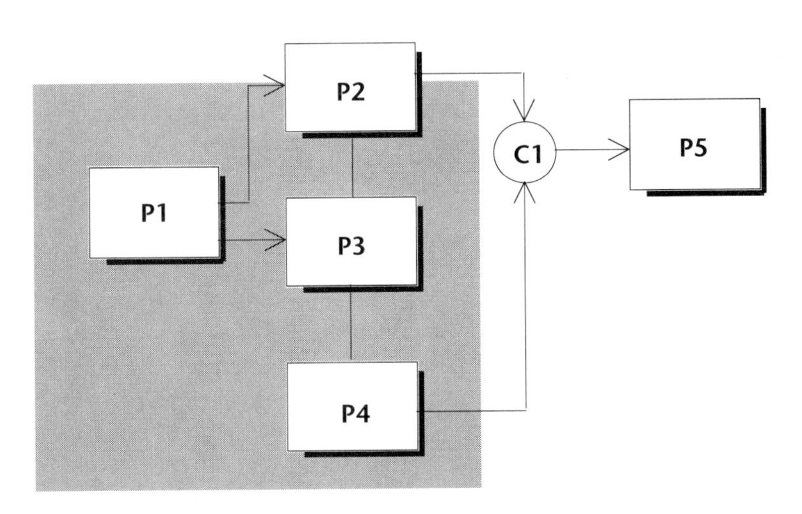

Figure 1.21 Policy (Unit) flow in a Fuzzy System

Policies—model subroutines, if you wish—solve small parts of the overall problem, passing on either defuzzified results or outcome fuzzy sets to the next stage in the overall problem. Synchronization connectors (*C1*, in Figure 1.1) bring together the results of many policies for a final analysis. Using this design strategy, many large fuzzy business systems in finance, econometrics, managed health care, transportation, and manufacturing contain dozens of variables whose behavior is well-defined by a relatively moderate number of rules.

Some Actual Fuzzy System Models

After discussing the benefits of fuzzy logic and the reasons why fuzzy systems are often unfairly rejected as solutions to complex business problems, I'll now conclude this chapter with a brief survey of actual fuzzy systems I have designed and delivered over the past seventeen years. This is not a comprehensive list, nor are these examples intended as complete case studies (which are available from *www.metus.com/casestudies* on the World Wide Web.)

Company Acquisition and Credit Analysis

After the "October Crash," several investment and general brokerage firms began examining different ways to isolate candidate companies for investment or acquisition. The fuzzy analysis model we built operates in two stages: a fuzzy database manager and a fuzzy policy analyzer. The database component allows analysts, using the same linguistic vocabulary, to interrogate large DB2 databases. The data query facility uses a proprietary FuzzySQL system to support queries such as:

```
select companyid
   from ieh001.findata
   where companysize is below large
       and sales are increasing
       and revenues are significant
       and peratio is acceptable;
```

The database query calculates the truth of the query predicate, creates a new column containing this truth, and sorts the results in descending order by this truth. Once a set of candidate companies has been isolated, this information is fed into a fuzzy financial analyzer that combines information from a few sources: the company's P&L; operating statements; and source and application of funds statements with public data from Standard and Poor's, Dunn and Bradstreet, and Valueline.

The model in Figure 1.22 ranks companies in terms of risk, stability, viability, and actual worth. This application has been described in *"Applications of Fuzzy System Models"* in the October 1992 issue of *AI Expert*. It has also appeared in the proceedings of the IEEE 1991 *AI on Wall Street* conference.

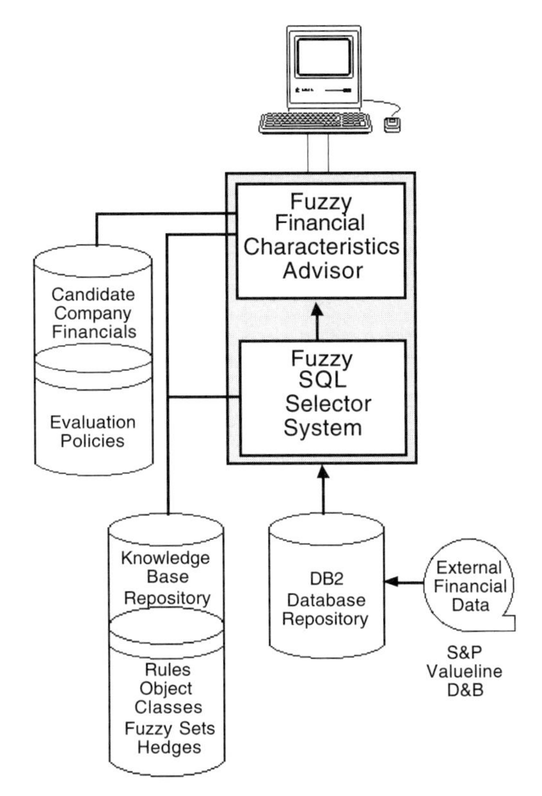

Figure 1.22 The Company Acquisition and Stability Model

Credit Authorization

A major credit card issuing company employs a combination of fuzzy logic and neural networks to authorize nonroutine purchases on its credit cards. The combination (hybrid) system also detects irregular purchasing patterns and can flag possibly fraudulent use of cards. An important feature of the credit card authorization network is the fuzzy time series analysis that tracks the spending curve in terms of fuzzy changes and fuzzy amounts.

Criminal Identification System

Developed for a European police force, this criminal identification system allows victims of crimes to interrogate a database of convicted and known criminals in fuzzy terms (the mugger, as an example, was *Fairly old*, *Rather tall*, *Somewhat heavy*, etc.) The fuzzy model also incorporated perspective shifts so that the interrogator could say: *Old from a young per-*

spective, *Tall from a short perspective*, and so forth. (A description of this application appeared in a special issue on fuzzy logic in *Computer Design* magazine in March 1992.)

Mainframe DASD Planning

A large New Jersey insurance company uses a fuzzy model to track and simulate the use of direct access storage device (DASD) space on the six mainframes in its central data center. This facility reads JES sub-system allocation records and evaluates how disk space is used by a mix of applications. From this information, it makes recommendations about how disk space should be allocated to jobs, it determines where bottlenecks are occurring, and it projects the amount of DASD that will be necessary to handle a new mix of jobs or a workload growth of a specific percent.

Expense Auditing

Two metropolitan New York banks use a fuzzy model to analyze and evaluate trip expense reports. The model uses intelligence from the bank's own auditing department to establish a fuzzy metric based on several factors: the type, length, and purpose of the trip; the type of transportation segments used (air travel, rental car, etc.); the hotel accommodations; and the distance to the site. The model also considers such factors as the title of the attendee. (Executive vice presidents *always* spend more than systems programmers—it's an underlying principle of the known universe!)

Financial Statement Advisor

Developed for a credit reporting company, this fuzzy model reads a corporation's financial statements for the past nine quarters, calculates all the necessary ratios and statistics, and then ranks the company according to its credit worthiness and its position relative to its peers. This model runs on an IBM mainframe and is interconnected to FOCUS, a fourth-generation data management system.

Container Management System

This transhemispheric container management system was developed for a consortium of fourteen shipping lines in London, and runs on an international information network. This system not only solves a supply-and-demand model for 280,000[plus] containers of fifteen different types across ninety-three ports in Europe, the Far East, and the United States, but it also provides utilization forecasting, "graveyard" balancing, and ship loading details. The container management system handles requests from users of containers throughout the world. The system develops a least-cost and a least-risk recommendation. The model makes extensive use of fuzzy logic in representing nearly all the operating characteristics of the model: (high or low tariffs, distances to the next port, yard size, probable number of containers on site, probable arrival time and transit time of the vessel, probable number of containers in for repair, costs of leasing, costs of maintaining inventory, etc. The model accesses eighteen spindles (racks with multiple disks) of IMS/DL1 databases files in real-time. While a large consulting firm in France estimated that the solution to this problem using dynamic programming would take 172 hours on their Cray II, the fuzzy model produces both solutions in less than eight minutes.

Intelligent Project Management

During the middle to late 1980s the SRMS project management and enterprise modeling system incorporated advanced approximate reasoning (fuzzy logic) techniques to plan the logical structure of projects, calculate cash flows, handle the allocation of resources, and support project management and tracking against the underlying relational database repository. Through a random, Monte Carlo modeling facility, fuzzy time and duration quantifiers could be used to describe a critical path network. The probable (possible) extent of the project was calculated by sampling the high-level Cartesian space created by the fuzzy network topology. The availability, quantities, and available durations of resources were specified with fuzzy quantifiers. Networks were adjusted based on limited approximate resource commitments across simultaneously occurring project events. At a higher level of abstraction, project managers and staff officers could interrogate the project repository using a FuzzySQL system, such as:

```
select projects
where duration is LONG and budget is EXPENSIVE

select projects
where taskcount is HIGH
and (budgetspent/budgetallocated) is LARGE
and (currentdue-originaldue) is VERY LATE;
```

This system allowed managers to retrieve and view projects that were close to the semantic meaning of such concepts as *Late*, *Expensive*, *Critical*, *High*, etc. The SRMS system also maintained a complete tracking history so that managers could implement such fuzzy concepts as *Rapidly changing*, *Increasing costs*, *Wide variance*, and so forth. (This application has been described in *"Applications of Fuzzy System Models"* in the October 1992 issue of *AI Expert*.)

Integrated MRP and Production Scheduler

A major manufacturer uses a fuzzy logic–based, materials-requirement, planning system to schedule the ordering, stocking, and staging of parts in its assembly plant. This ordering system is coupled with an IDMS database as part of an integrated manufacturing system. The underlying fuzzy policies include the concepts of service levels, opening on-hand stock, items produced per capita of the workforce per hour, and the cost of materials (labor, direct expenses, spoilage, and indirect costs). The fuzzy model contains fuzzy logic policies for such functions as customer demands, actual sales, the strategic sales plan, a sales review and feedback controller, production planning, material requirements, overtime and shift requirements, and labor levels, as well as finished stock. As an example, the customer demand policy has rules like this:

```
if price is reasonable and quality is good
   and previous service is high then
       {customer demand is about the sales plan,
       demand variance is low}
```

This model replaces exclusively mathematical inventory and production models with a more robust and flexible fuzzy logic approach. The model responds better to fluctuating sales, orders, and profits than conventional linear programming systems. (The initial prototype for this model appeared in *Knowledge Engineering and Decision Support: Theory and Applications*, edited by M. Small, International Computer Limited, 1984.)

Organizational Dynamics

Developed for the management of a large transportation company, this organizational dynamics fuzzy model simulates the effect of increased or decreased management control, bureaucracy, and regulations on worker productivity. The basis for the model was an attempt by management to understand how to best negotiate new contracts and control the collective bargaining process.

Managed Health Care—Provider Fraud Detection

Originally developed for a major Connecticut insurance company, the health provider fraud detection system is now independently marketed by IBM's insurance industry sector. The original fraud detection system, as developed by the author, uses a set of fuzzy behavior patterns to measure the performance of health care providers against the behavior of their peers in a particular demographic or geographic region. Providers are categorized by size, specialty, and geographic region (for example, we might examine all small-clinic osteopaths in the Chicago metropolitan area).

The fraud detection system is a rather deep time-series fuzzy model, which examines the behavior of service providers across several dimensions of total aggregated costs, the vector of behavior patterns, and some fundamental statistical variances. The system tracks the change in behavior over time in order to spot anomalous trends as early as possible (it is better to correct a small problem than to prosecute a large problem.) Figure 1.23 shows a brief schematic of the system.

The fraud program is actually a nonparametric, anomaly-detecting system that uses the law of large numbers and certain principles of entropy associated with information theory to self-adjust in each demographic area. (This application has been described briefly in *"Applications of Fuzzy System Models"* in the October 1992 issue of *AI Expert*.)

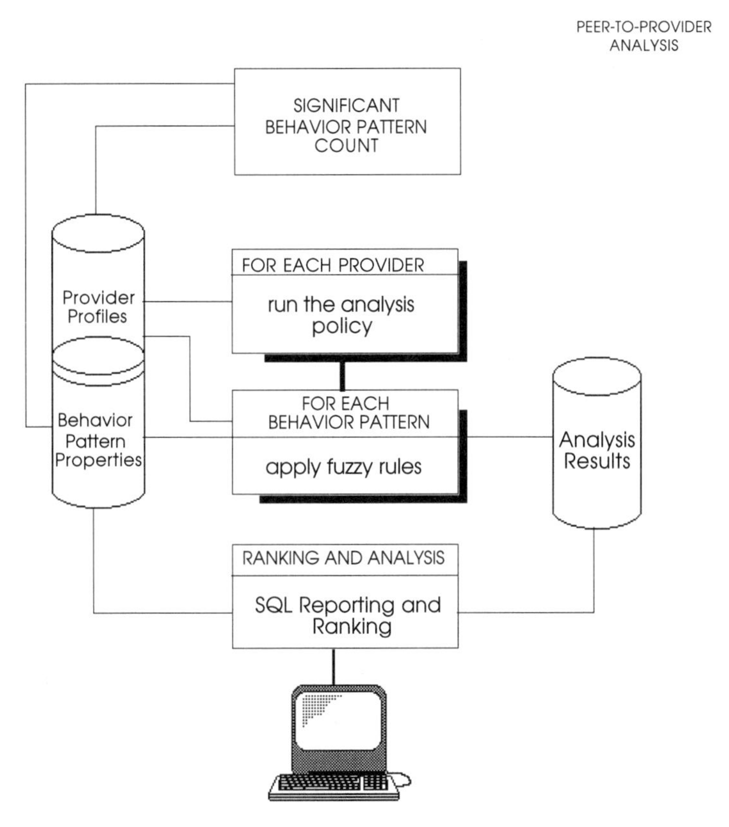

Figure 1.23 The Organization of the Provider Fraud Detection System

Loan Evaluation Advisor

A large bank in the Washington D.C. area employs a fuzzy model to evaluate commercial loan applications. The analysis compares the application profile with the company's financial behavior over the past twenty-four months. A fuzzy approach allows the loan officer to deal with fluctuations in account balances, over-drafts, delinquent payments, the number of pre-existing loans, collateral appraisals, and so forth, in approximate terms rather than in sharp cut-off points. This approach provides a much clearer view of the company's assets, its financial management, and its potential for future stability (the measure of whether a significant loan should be made).

Portfolio Safety and Suitability Model

Written for a large Wall Street investment firm, the safety and suitability model attempts to detect problems in portfolio management and composition before they are brought to arbitration or handled through a conventional law suit. As Figure 1.24 illustrates schematically, this is a large, multifaceted fuzzy system integrating several factors: the client's financial capabilities, the client's stated investment objectives, situational constraints with the intrinsic risks inherent in the portfolio's instrument mix and the financial consultant's management technique and behavior.

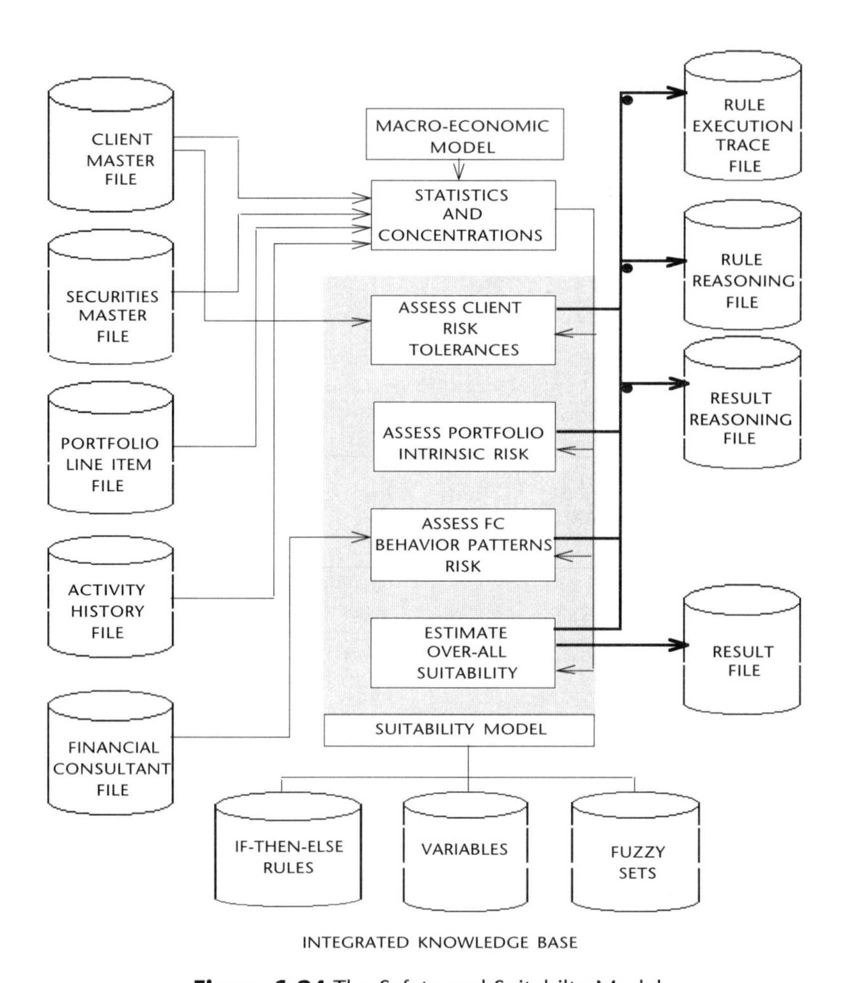

Figure 1.24 The Safety and Suitabilty Model

A fuzzy rule base measures the degree to which a client's objectives and financial resources are in concert with both the risk imposed by the portfolio mix and the financial manager's investment strategy. The model not only evaluates the safety of the portfolio but also balances this factor against its suitability for a particular client's investment strategy.

For example, a client with a strategy of high-risk investment would find a very safe portfolio to be very unsuitable.

Product Pricing Model

Developed for a major British retailing firm, this very successful product pricing model establishes a price for new products, re-prices products based on changes in the market or in demographics, and predicts changes in the pricing of competitive products within the same or adjacent market regions. The model uses advanced fuzzy logic techniques, such as the Delphi Method, to develop a consensual recommendation from multiple co-operating, collaborating, and conflicting experts in the fields of finance, marketing, sales, manufacturing, transportation, and administration. (This application has been described and dissected in *"Problem Solving with Fuzzy Logic"* in the March 1992 issue of *AI Expert*. The basic pricing model is also discussed in some detail throughout this handbook.)

Risk Underwriting

Several large insurance corporations are using fuzzy models to evaluate the risks associated with automobile, commercial property, and health underwriting. Fuzzy logic provides a broader, more sensitive approach to risk assessment in these cases.

Systems Complexity Analysis

A software system complexity analysis incorporates a fuzzy logic policy that considers function point, code density, and the total operational interface to decide on a software project complexity rating.

Notes

3. Zadeh, L.A., "Outline of a New Approach to the Analysis of Complex Systems and Decision Processes," *IEEE Trans. Systems, Man, and Cybernetics*, SMC-3 (1973), pp. 28–44.

4. Cognitive dissonance is the lack of harmony or agreement between an idea or problem state and the way it is expressed in the software modeling system. The higher the dissonance between the way a problem is expressed and the way it is coded, the more difficulty an expert will have in verifying that the machine's knowledge parallels his/her own knowledge.

5. We will discuss the issues associated with probabilities and uncertainty representations in Chapter 2.

6. The author.

Two

Fuzziness and Certainty

Contra negantem principia non est disputandum
You cannot argue with someone who denies the first principles

Auctoritates Aristotelis
(A compilation of medieval propositions)

Mehr Licht!
More Light!

Johann Wolfgang von Goethe (1749–1832)
Dying words.

This chapter examines the general nature of imprecise information. We look at the ideas of imprecision, inexactness, ambiguity, uncertainty, and vagueness as they pertain to our descriptions of physical and conceptual phenomena. Although a more detailed exploration of fuzziness is handled in the next chapter, we introduce the idea of fuzziness as an intrinsic property of continuous problem states. Finally, the conflicting concepts of probability, possibility, and fuzziness are discussed in relation to their ability to represent a variety of uncertainties.

The Different Faces of Imprecision

A great deal of confusion exists, even in the fuzzy logic community, about the exact meaning of fuzziness and what it attempts to represent. Much of this uncertainty is associated with the ease with which we use related concepts as intentional substitutes for intrinsic imprecision (the fundamental property of fuzziness). Table 2.1 illustrates how we apply some of the same descriptions to the different kinds of imprecision.

Measurement (Knowledge)	Probability	Description
Uncertain Inaccurate Inexact Imprecise Confidence	Uncertain Probable	Fuzzy Vague Undecidable Imprecise Possible

Table 2.1 Ways of Describing Types of Imprecision

Table 2.1 clearly shows that we often use the same term to describe imprecision and uncertainty in only slightly related areas of measurement. Imprecision in measurement is associated with a lack of knowledge. Imprecision as a form of probability is associated with an uncertainty about the future occurrence of an event. Imprecision in description, the type of imprecision addressed by fuzzy logic, is connected with the intrinsic or built-in imprecision that belongs to the event itself. In this section, we will look at some of these closely allied terms. In fact, some of these terms can be used interchangeably with fuzziness under some special circumstances. We will discuss the ideas of probability and the more general concepts of belief systems later in this chapter.

INEXACTNESS

The idea of inexactness corresponds to a precision in our ability to measure. In conducting scientific experiments—measuring the radius of a proton, the heat output from cold fusion, or the thermodynamic equilibrium of a distant star—many researchers measure the same data points and compare their results. Every physical measurement, however, is subject to a degree of uncertainty since instruments are affected by both internal and external influences. Heat, magnetic fields, dampness, friction, vibration, and fluctuations in electrical currents, to name a few, all contribute to uncertainty in our measuring devices. A magnetometer, for example, can be affected by high-voltage power lines, emissions from a laptop computer, or variations in the earth's own magnetic field.

Even when the device itself has a high precision, the human side of the measuring equation can often be at fault. Through a phenomenon known as *parallax*, we sometimes fail to read a physical device correctly. Parallax is the apparent change in the position of an object caused by an actual change in the position of the observer. We can see this phenomenon clearly by examining the path of an electrically charged particle through a three-dimensional detector. Figure 2.1 shows the path when viewed from slightly above the detector.

The particle path begins at the upper right-hand corner and spirals in a helical path to the exact center of the bottom plate (at coordinates [0,0,0]). Now, by simply shifting our viewpoint slightly below and to the right, we introduce a new degree of uncertainty about

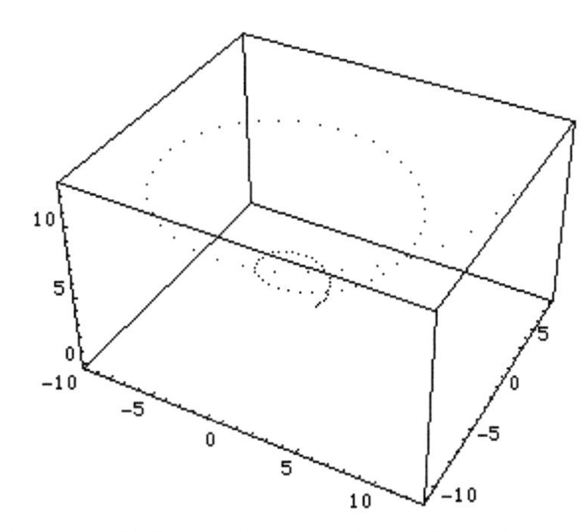

Figure 2.1 The Particle Trace Path Viewed from Above

the apparent particle path (see also the discussion of visual ambiguity on page 50). Although the final particle point has an absolute position of (0,0,0) in the three-dimensional space, Figure 2.1 illustrates how this change in our perspective makes a change in the apparent position of the trace. Does the particle path terminate at (0,0,0) or (0,0,-5)?

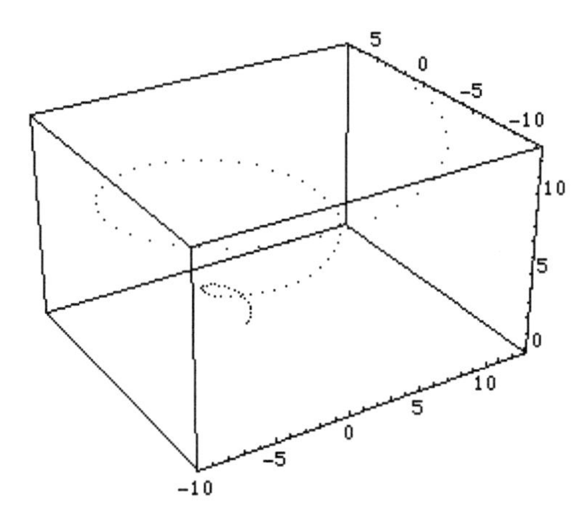

Figure 2.2 The Particle Trace Path Viewed with Some Ambiguity

We are confronted with compromises in precision and measurements every day. What we see is largely a degree of position and instrumentation. Figure 2.1 shows the same particle trail viewed from below. Notice how the path seems to change its position in space.

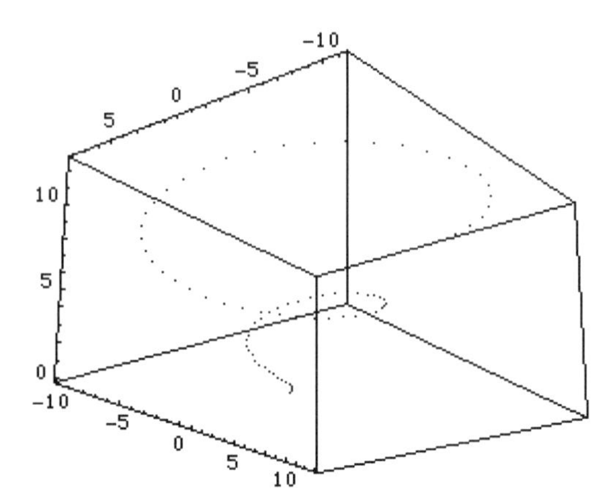

Figure 2.3 The Particle Trace Path Viewed from Below

These influences and human errors cannot be avoided. We compensate for such induced inexactness by making many measurements of the same quantity. Thus, we arrive at a statistically plausible estimate of the quantity's true value.

PRECISION AND ACCURACY

In the physical world, precision and accuracy are related concepts. Precision is the degree of exactness with which we can measure a quantity. This exactness is related both to the fineness of our instruments as well as to the statistical correctness of our measurements. The precision of any device is limited to the fundamental granularity of its scale—that is, the finest division of the device restricts our measurement. A meter ruler is divided into millimeter increments. The millimeter is thus the limit of our accuracy with such a ruler; anything below a millimeter is an estimate.[1] For those old enough to remember using a slide rule,[2] the idea of estimating the distance between two thin lines was a common experience.

Accuracy and Imprecision

Accuracy is the degree to which a measurement corresponds with the standard value of a quantity. Accuracy is the difference between experimental measurements and the standard value for a quantity. Thus, if we measure the density of ammonia in grams per cubic centimeter as 0.7701, but know that the actual density is $0.7708\ \text{g/cm}^{-3}$, then the accuracy of our measurement is $.0007$. This error in accuracy may have nothing to do with the precision of the instrument. A highly precise instrument can yield inaccurate results due to environmental factors or observer error, or simply through a lack of proper calibration. Thus, uncertainties in a measurement can affect accuracy, but precision is not affected since it is based on the granularity of the device.

Measurement Imprecision and Intrinsic Imprecision

How is this kind of imprecision related to fuzzy logic? Fuzzy logic addresses the issues associated with intrinsic imprecision rather than those directly concerned with the failure of our measuring devices or the inaccuracy of measurements. Intrinsic imprecision is associated with a description of the properties of a phenomenon, and not with our measurement of the properties using some external device. We will see later that these issues occasionally collide, and we can often gain a better understanding of physical systems by relaxing the rigid precision of our system metrics. In general, however, the type of imprecision and inexactness that our fuzzy models use is independent of the measuring systems associated with their control and solution variables. This kind of imprecision is not diminished by increasing the precision of these metrics.

AMBIGUITY

A close semantic relationship exists between the idea of ambiguity and that of fuzziness; in fact, some fuzzy states can be highly ambiguous. (For an example, see Chapter 4, *Fuzzy Set Operators*, Counterintuitives and the Law of Noncontradiction on page 186). Ambiguity connotes the property of possessing several distinct but plausible and reasonable interpretations. These interpretations can have different belief states—we can believe in one interpretation more strongly or less strongly than others. Ambiguity in meaning is a common occurrence in everyday language. As an example, you're driving down a busy street with your spouse and ask,

> "Do I turn left here?"
>
> He or she answers "Right."

That could mean turn right or, idiomatically, "yes, you are correct, turn left." Depending on your experience, mood, and relationship with your spouse, you can favor one interpretation over another and take appropriate action. That is, you have a higher level of belief in one state over the other (assuming you don't just decide to keep driving!)

Semantic Ambiguity

Like the preceding example, semantic ambiguity, that of meaning, is the most pervasive both in everyday language and in the knowledge elicitation process. Generally, though, semantic ambiguity is distinct from fuzziness when it deals with the interpretation of disjoint and discrete concepts. The statement

> *The food is hot*

is ambiguous, but not fuzzy, since *hot* could mean the food's temperature, the amount of spice (if the food was a bowl of chili), or its trendiness (such as cajun escargot, currently the rage in New York's TriBeCa district). There are several distinct and possible interpretations and, depending on the context, some might be more plausible than others. Once we have established a domain of discourse, however—such as temperature—ambiguity approaches fuzziness.

Ambiguity becomes *equivalent*, for most uses, with fuzziness when we attempt to define the degree to which the food is hot. How cool can the food become and still qualify as hot? Is there a precise point at which the food is not hot and then becomes hot when this threshold is crossed? This is a fuzzy metric. At some point, as the cooling process or the heating process continues, we would say that the food is, to some degree, both cool and warm or warm and hot—that is, an ambiguity arises concerning the proper description of the food's temperature.

Visual Ambiguity

A form of perceptual ambiguity similar to perceptual undecidability (discussed in the next section) also exists. A familiar form of visual ambiguity, illustrated in Figure 2.4, occurs in the Möbius strip.[3] The Möbius strip is formed by taking a rectangular strip of paper, giving it a half-twist, and connecting the edges.

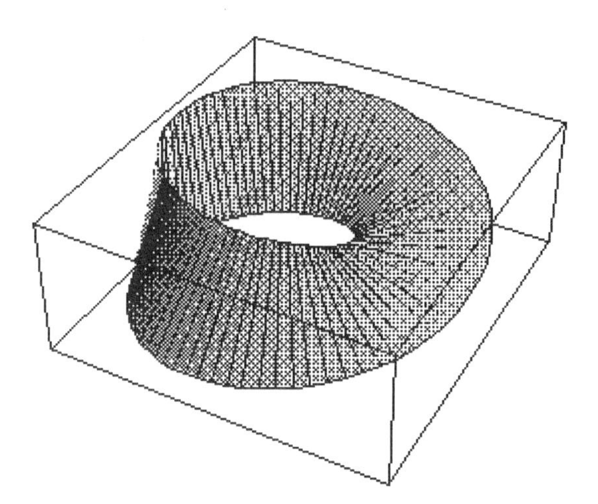

Figure 2.4 The Möbius Strip

When people first look at a Möbius strip, they usually have an unsettling experience. The form looks like a bent, curved belt. If you trace the surface with your finger, however, you discover that the strip has only *one* side. No matter how we examine the Möbius strip, we are faced with an object that is essentially two-dimensional in our three-dimensional space.

In the past, knowledge engineers and analysts seldom faced real-world problems of visual ambiguity. Recently, however, the ergonomics of designing effective and usable graphical user interfaces involves decisions about the organization and design of icons, dialogue boxes, and similar active screen components. In this environment, the use of color (for color-blind users), the use of icons (for the sight-impaired), and the use of dialogue boxes can introduce ambiguous visual stimuli.

Structural Ambiguity

The relationship between components of a system can reflect a high level of ambiguity. This ambiguity can be attached to the description of the system (in which case it becomes an ambiguity of meaning) or it can be actually associated with the construction of the system itself. With the increase in the deployment of object-oriented systems, this kind of ambiguity is often associated with the inheritance of property mechanisms. This is especially pernicious and frequent in systems that permit multiple inheritance; for example, a child or subclass occurrence can have more than one parent. The object-oriented system schematic shown in Figure 2.5 illustrates the source of possible structural ambiguity. The shaded subclass on the right inherits properties from two parent classes. When the parent classes contain attributes with the same name, then an analyst cannot decide from the structure itself which property is inherited.

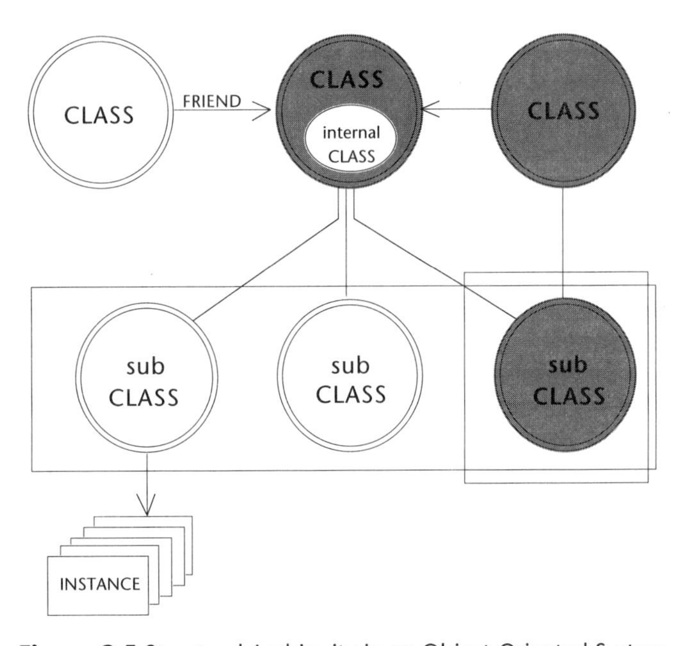

Figure 2.5 Structural Ambiguity in an Object-Oriented System

Semantic, visual, and structural ambiguities represent common and persistent problems in the knowledge acquisition and system design processes. In fuzzy system modeling, we are (on the whole) concerned with identifying semantically ambiguous concepts and converting them to fuzzy sets so they can be used in the model. This conversion process, as we discussed earlier, retains the apparently ambiguous language of the expert while preserving the computational power of the computer model.

Undecidability

A deeper form of ambiguity, called *undecidability,* is associated with our ability to discriminate between different states of an event. Undecidable phenomena are more common than you may initially think, as they represent a large class of problems called "paradoxes." In an undecidable state, we are unable to make a reasoned choice between the ambiguous representations that exist in the model. The undecidability generally stems from the properties of the model itself, not from any lack of knowledge on our part. A familiar and interesting visual paradox is the Nekker cube, illustrated in Figure 2.6.

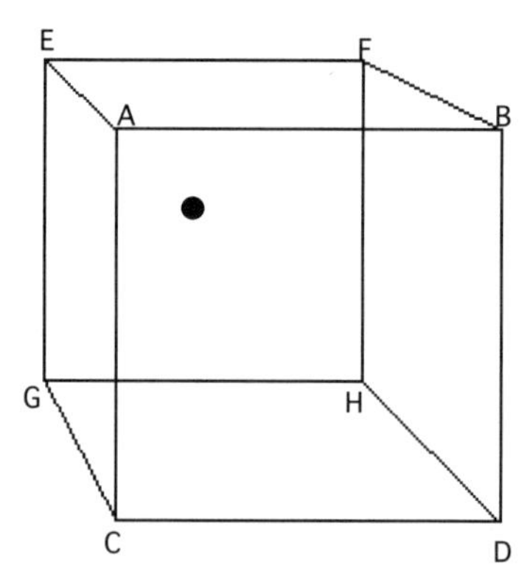

Figure 2.6 The Visual Ambiguity of a Nekker Cube

Our minds continuously switch between two undecidable representations: a cube facing to the upper left with the face EFGH in front, and a cube facing to the lower right with the face ABCD in front. In the absence of any clues, we are free to interpret the cube in either orientation—but not both at the same time. This kind of undecidability is psychological and not physical. A related and equally profound type of undecidability is associated with patterns that can have several interpretations. The Kaniza square, illustrated in Figure 2.7, is an example of perceptual ambiguity.

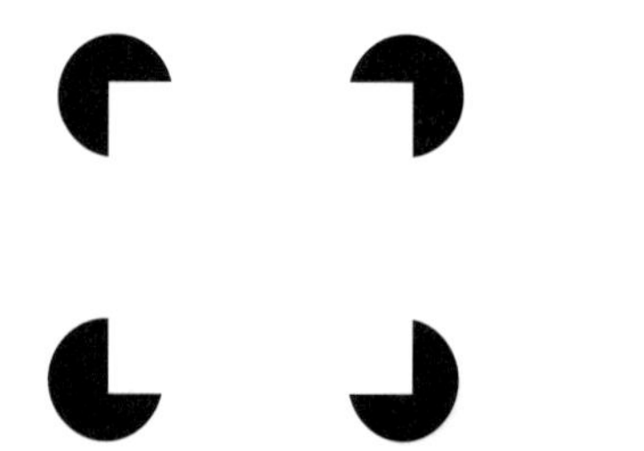

Figure 2.7 The Kaniza Square Phenomenon

This image, a bright square bounded by the black circles at each corner, does not actually exist. Even knowing that the square is an illusion, we cannot look at this figure without seeing the square.

But there are other, real-world kinds of undecidable ambiguity that are closer to the concept of fuzziness. Take the often-used metaphor, shown in Figure 2.8, for evaluating a bad situation:

Is the glass half full or half empty?

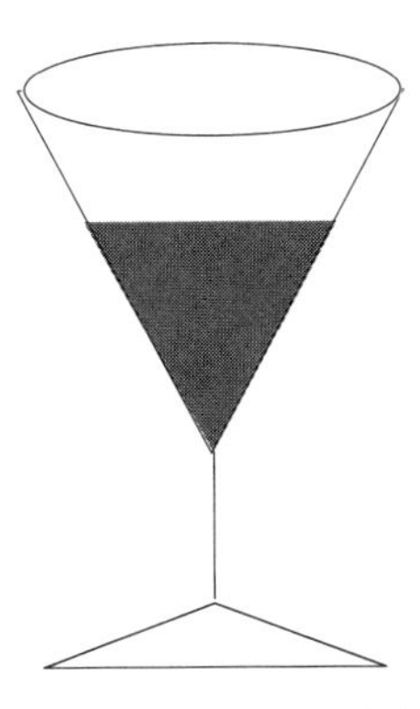

Figure 2.8 Descriptive Ambiguity of the Half-full Glass

A glass that is half filled (see Figure 2.8) is both half full and half empty, so that both states exist at the same time. This is a natural consequence of the state of the system. No measurement can resolve the question (assuming we stay within the accuracy of the system calibration metric that initially established the half-full/half-empty dichotomy). This is exactly the kind of ambiguity that fuzzy logic handles well. In fact, as illustrated in Figure 2.9, the half-full or half-empty dichotomy can be recast in terms of the appropriate fuzzy sets *Full* and *Empty.*

When a twelve-ounce glass contains six ounces, we have the same degree of membership in the fuzzy sets *Empty* and *Full*. This is a measure of the intrinsic ambiguity inherent in the phenomenon itself, independent of any external measurement or observer considerations (allowing for parallax and errors in measurement, of course.) As Figure 2.10 shows, the change in ambiguity is, itself, a fuzzy concept. Thus, as an example, around the point of maximum ambiguity (the cross-over point for *Empty* and *Full*) there is a region first of high, then a region of moderate, and then a region of low ambiguity.

This kind of ambiguity is far from being unique to half-empty glasses; we encounter it in a wide range of everyday phenomena. Gradual and often ambiguous changes from high to low, fast to slow, and long to short are found throughout econometric, public planning, manufacturing, retailing, and other business decision models. Because real-world processes share ambiguous states at many critical values, fuzzy systems provide the ideal way to represent and evaluate these processes. We will have much more to say about this topic later.

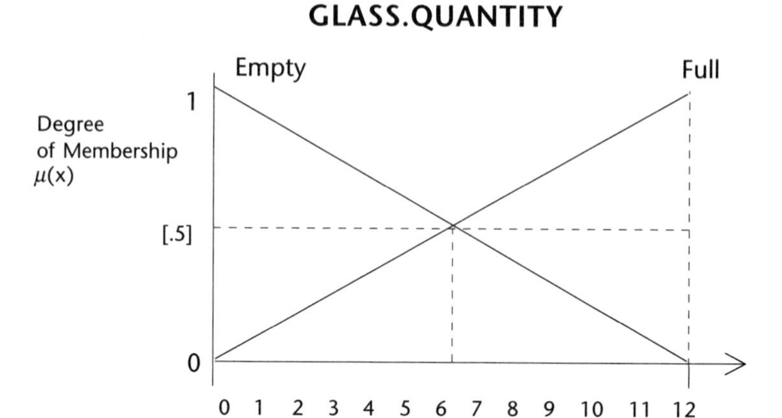

Figure 2.9 Fuzzy Sets for the Half-Full and Half-Empty Glass Problem

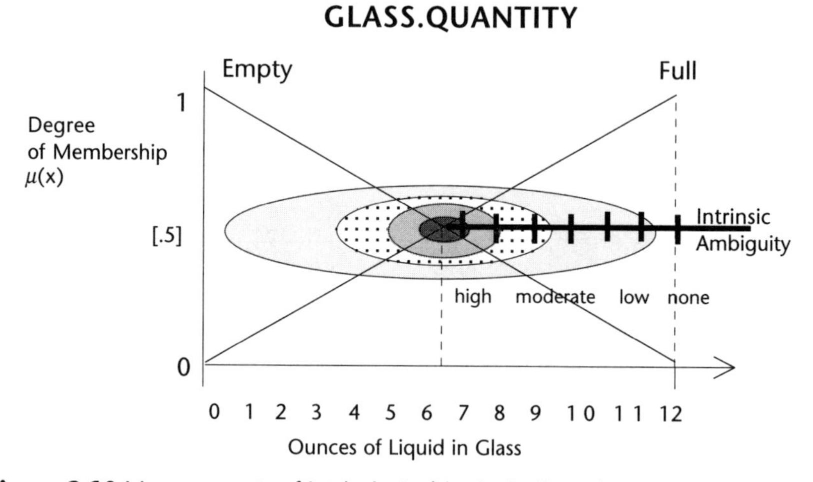

Figure 2.10 Measurements of Intrinsic Ambiguity in Complementary Fuzzy Regions

Vagueness

In trying to understand the ideas underlying fuzziness and ambiguity, we must address the notion of vagueness. The dictionary defines *vague* as

> *lacking precision or sharpness of definition; indistinct; blurred; lacking in character and purpose, or addicted to haziness in thought [L., vagus, wandering,—vagari, to wander]*

and this sounds a lot like the definition of fuzziness—lacking in precision or sharpness of definition. Zadeh, as well as many other theorists, views fuzziness as a possibilistic (i.e., nonprobabilistic) species of vagueness. It is this idea of measurement, the application of a metric to the vagueness, that accounts for the utility of fuzziness rather than pure vagueness. In fact, an unmeasured fuzzy or vague region is simply empty (we say that it is *indistinct*). As an example, the statement,

Quarterly profits are very low

can be considered indistinct if the term "low" is not mapped to an underlying metric. That is, what property of profits is applied to the imprecise region *low* to determine the degree to which profits are low? If, as an example, we use discounted retained earnings to measure profitability, then *low* reflects a correspondence between the concept *profitability* and the retained (after-tax, after-adjustment) earnings—profits can then be quantified. Once quantified, the term assumes a degree of computational utility promoting it from simply indistinct to fuzzy.

In literature, philosophers and philologists have wrestled with the ideas of ambiguity and vagueness in terms that closely parallel the way that Zadeh initially defined fuzziness. In a study of vagueness, M. Black makes several distinctions between vagueness and ambiguity, but settles on a definition of vagueness that is similar to fuzziness. W. Alston has developed a metric that uses a degree of vagueness with properties that map closely with Zadeh's semantic interpretation of fuzziness. In these cases, however, the researchers were not concerned with exploring system complexity, and made no attempt to combine their studies with an algebra of vague sets. Linking vagueness with a computational capability moves fuzzy logic and fuzzy set theory from a philosophical concept to a working system for describing complex systems.

Fuzzy Logic and Interval Arithmetic

In addition to its confusion with probability (See "Fuzzy Logic and Probability" on page 57. later in this chapter), fuzzy logic is often confused with some form of interval arithmetic or categorical calculus. This mistake commonly results from a surface knowledge of fuzzy logic and a poor understanding of the concept of a fuzzy set. It is the fuzzy set, after all, that underlies fuzzy set theory and is often the most visible feature of both the underlying logic and the systems that employ it. To many observers, membership in a fuzzy set is simply a way of assigning an ordinal rank or categorical classification. They ask, with some justification, what is the difference between the fuzzy systems approach to describing a variable such as *Speed* and the interval arithmetic approach? Both give us semantic labels for a group of domain values. But, to answer that question, we need to look at the knowledge representation schemes employed by each approach. Consider the following two representations of *Speed*. Figure 2.11 uses interval segmentation.

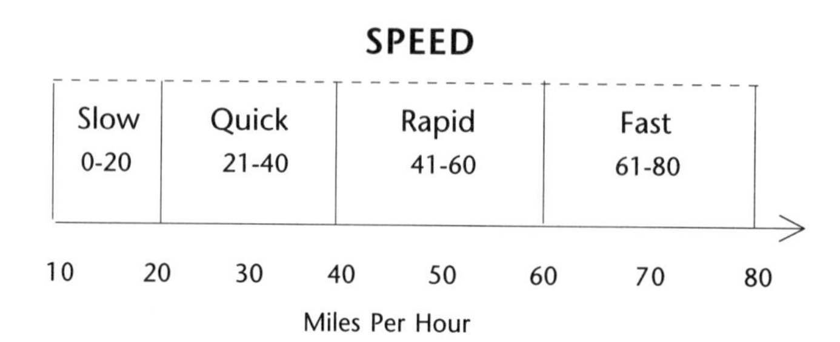

Figure 2.11 Interval Categories for the Parameter Speed

With the interval approach, knowledge engineers can treat the segments as semantic descriptors, similar to the way they use fuzzy sets. However, the intervals lack the generality and the representational power of true fuzzy sets. We can see this clearly in Figure 2.12, which shows a fuzzy term set decomposition for *Speed*.

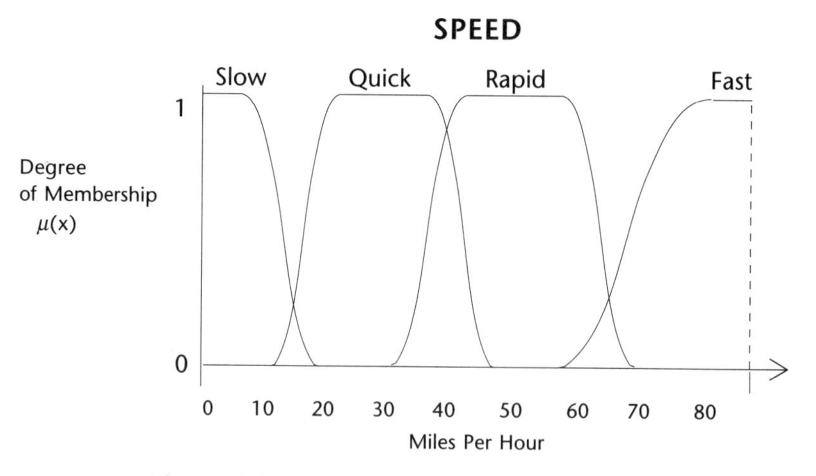

Figure 2.12 Fuzzy Sets for the Parameter Speed

In the interval category approach, we still have crisp boundaries at the extremes of each interval. Furthermore, within a category or interval, we lack an intrinsic measure of the degree to which a value is actually a member of that interval. And we lack a method of spreading values out into neighboring interval space, which would allow us to measure the degree to which an individual value is also a member of neighboring intervals. It is, of course, the degree of membership coupled with an overlap in the semantic definitions of *Slow, Quick, Rapid,* and *Fast* that actually defines the real nature of these concepts. Naturally, if we add degrees of membership to our intervals and relax their rigid boundaries, we will gain this kind of representation. But at that point, we have, for all intents and purposes, turned our intervals into fuzzy sets.

Fuzzy Logic and Probability

This next section on probability and fuzzy logic is not intended either as a complete discussion of probability or as a complete analysis of the two concepts. Rather, it is intended as a high-level perspective, both on the fundamental differences between the two approaches and on the kind of uncertainty they each measure. With these differences in mind, the interested reader can consult the wealth of books on statistics and probability to make further connections between the two techniques.

An amazing amount of literature has been generated discussing the relationship between fuzzy logic and probability. Incredibly, the preponderance of this material is written by academics with little fundamental knowledge of fuzzy set theory. They continually assert that fuzzy logic is simply another form of probability, that probability is the only method of describing uncertainty, and that anything that can be done with fuzzy logic can be done better with probability. Other academics charge that fuzzy logic is not a logic at all, but rather some form of pernicious heuristic that clouds the analytic mind and destroys the objective representation of scientific data. The debate between proponents and opponents of fuzzy logic seems to veer between hysteria, unsupported hype, and intellectual blindness. In this next section, we examine the real difference between fuzziness and probability.

What Is Probability?

Some philosophers of mathematics argue that we have never understood the meaning of probability. Such a meaning eludes us for two reasons: the quantum information theory (entropy) and the nature of uncertainty associated with probability. In either case, of course, probability represents an attempt to explain how events occur in a random space. The underlying "first principle" of probability is randomness. This randomness presupposes our ability to measure and order the random space. There are, in general, two schools of thought about the structure and application of probability. We can call the first the *frequentist approach* and the second the *subjective approach.*

Frequentist Probabilities

The frequentist approach to probability relies completely on a description of population behaviors; ultimately the frequency with which an event X can be expected to occur in a large population. Let's look at the population of randomly selected adult men at the intersection of 42nd Street and Lexington Avenue in New York City. Our blindfolded census taker grabs a passer-by every few seconds, and if the individual is an adult male, the individual's height is recorded. After a while we discover that male adults have a distribution of heights, most of them around 5'8". The frequency histogram created by the measurement takes the form of a bell curve. Figure 2.13 shows the properties of this bell curve.

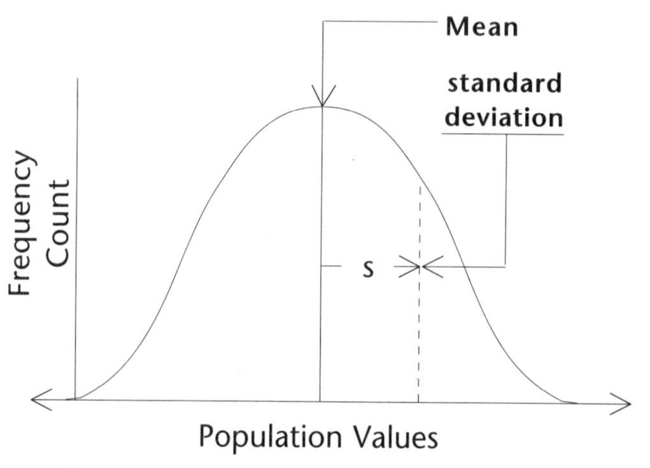

Figure 2.13 Properties of a Statistical Distribution

The bell-shaped curve is called a *Gaussian* distribution. Its properties are measured by its mean (the average value) and the standard deviation (a measure of the compactness of the population around the mean). If we know the mean and standard deviation of a Gaussian population, we can predict the probability of an event in the underlying population. Thus, if we select the next passer-by, we find there is a very high probability that he will be somewhere near 5'8" in height, and a moderate probability that he will be 5'6" in height, and so forth. In this frequentist approach to probability, a knowledge of the population is essential. We declare a prediction of the future on a knowledge of the past (that is, the history of the population's behavior).

Subjective Probabilities

Unlike the frequentist approach, subjective probability combines a purely frequency-based approach with subjective or intuitive estimates of an event's likelihood. At the heart of the subjective approach is Bayes Theorem, named after Thomas Bayes (1702-61) whose principal work, *Essay towards solving a problem in the doctrine of chances* (1763), showed a workable method for combining frequency distributions with newly acquired evidence. Using Bayes Theorem (see Bayes Theorem and Fuzzy Probability on page 63), we can make estimates about the relative frequency of events for which no hard historical records exist or for which it would be exceedingly difficult or impossible to acquire actual frequency distributions. Subjective probabilities can lead the unwary knowledge engineer or statistical analyst into unwanted or unsupported conclusions. This was the case in the Challenger space shuttle explosion, where subjective estimates by NASA technical managers on the anticipated failure rate of the space shuttle were several orders of magnitude greater than warranted by any analysis of the physical systems themselves.

Yet, the area of subjective probabilities causes the most confusion in making a clear distinction between fuzzy logic and probability. To many untrained observers, fuzzy logic looks suspiciously like another form of Bayesian probability. This is especially difficult for those not versed in the fine details of fuzzy logic, since we say that fuzzy systems accumulate evidence for and against a solution state. However, we should bear in mind that probability accumulates evidence for or against the occurrence of an event, while fuzzy systems accumulate evidence for membership in a set of events. This is an extremely important and profound difference.

Mathematical Foundations (Briefly)

From a mathematical perspective, fuzzy membership values are commonly misunderstood to be probabilities, and fuzzy logic itself is often interpreted as a new way of manipulating probabilities. This, however, is not the case. A minimum requirement of probability is additivity, expressed in Equation 2.1.

$$P(s) = \sum_{i=1}^{N} p(x_i) = 1$$

Equation 2.1 The Additive Property of Probability

Additivity means that all the mutually independent probabilities of a particular system, P(s), must add together to one; or, in terms of the cumulative probabilities, the integral of their density curves must equal one. Fuzzy membership functions do not have this restriction. Although fuzzy memberships can be developed using probability density functions (or based heuristically on density functions), a wide spectrum of membership generating techniques exist that have nothing to do with probability or event frequencies.

Confusion of Aims

A great deal of confusion exists between the concepts and mathematical properties of probability and fuzzy logic. Unfamiliar with the actual mechanics and mathematics of both fuzzy set theory and approximate reasoning, a large audience of otherwise intelligent individuals have concluded, with little empirical evidence, that fuzzy logic is simply another form of subjective probability. Others have concluded that fuzzy logic, even if it is somehow different from probability, is an erroneous approach to uncertainty representation—that probability is the only correct method of expressing uncertainty.

Confusion of Methods

Much of this confusion stems both from the apparent external similarity between fuzzy logic and probability and from the sudden emergence of fuzzy logic from years of near obscurity. Both disciplines are concerned with measuring a form of uncertainty. Both use a metric scale between zero and one to encode the degree of uncertainty. And, both use Gaussian-like distribution functions to describe an event space. Furthermore, the long formal history of probability, coupled with the short and rather obscure history of fuzzy logic, leads many scientists and mathematicians to conclude that they must have a common root. As a cognitive scientist at IBM's research center in Thornwood, New York, remarked recently,

> Fuzzy logic *must* be a form of probability. It is just not possible for a new and important method of handling uncertainty to go unnoticed for thirty years.

This logic takes us back to that same intellectual tautology that lead ancient scientists to conclude that the earth was flat and sat at the center of the universe. If the earth was round, the ancient philosophers and scientists reasoned, how could it go undetected by Aristotle, the holy scriptures, and scientists of the new Renaissance?

Likelihood and Ambiguity

Fundamentally, the basic confusion between fuzzy logic and probability flows from the idea that they measure the same *kind* of uncertainty. In strictly semantic, as well as mechanistic, terms the two forms of uncertainty are distinct. Propositions in probability address the likelihood of an outcome for some discrete event. The event's outcome either happens or it does not happen. The propositions of fuzzy logic concern the degree or extent to which an event occurred. While a probability outcome happens unequivocally, whether a fuzzy event occurred may involve some degree of ambiguity or uncertainty.

The likelihood of an outcome in probability and the degree of ambiguity in an outcome in fuzzy logic are both measured on a [0,1] scale. But the interpretation of the values along this scale are different. In probability, a value of [0] means that the event cannot occur. A value of [1] means that an event is certain to happen. Intermediate values express the frequency of (or the odds for and against) a particular outcome. In fuzzy logic, however, the scale is a measure of ambiguity. The idea of ambiguity can also be expressed as a measure of compatibility or representation. An increase in ambiguity means that there is a corresponding (but not necessarily linear) decrease in the degree to which an event is compatible with or representative of a particular concept. In fuzzy logic, a membership value of [0] or a value of [1] means a complete lack of ambiguity in the outcome. When the membership is zero, the outcome is completely nonrepresentative of the event concept, and when the membership is one, the outcome is completely representative of the event concept. In simpler terms, a zero or a one means that no ambiguity exists regarding the description of an event or element. Intermediate values indicate various degrees of ambiguity in how we should interpret the outcome.

Also, probability deals with randomness or stochastic behavior in a large population. The uncertainty of probability concerns the occurrence of an event in this population given the distribution and density of all possible events. And, probabilities are time-dependent. Consider the toss of a die. We echo the words of Julius Caesar as he crossed the Rubicon and started the Civil War with Pompeius: *Alia iacta est—The die is cast*. When the die is moving through the air, each face has a one in six chance of appearing. This probability only exists while the die is in motion. When the die lands on the table top and stops rolling, the face is determined and all probability is gone.

But does it make any sense, knowing a probability distribution, to say that "Tom is probably tall" or "What is the probability that Tim is tall?" Probability tells us something about populations, not individual instances. Once we have an individual instance, probability evaporates. Before we select Tom from our population, we know the chances of Tom being tall, but, having selected Tom, this probability is gone. The process of choosing is an action involving time. This leads us to the conviction that probability is an uncertainty associated with time. Consider "There is a 50% chance of rain tomorrow." If we wait until tomorrow it will rain or it will not, thus the uncertainty associated with probability disappears. We take up this exact issue in Fuzzy Probabilities on page 62.

Fuzziness, on the other hand, concerns the ambiguity associated with the actual description of an event. These ambiguities are generally continuous valued phenomena where the boundary between different semantic groups is not precisely defined. Such phenomena are extremely common in the real world. When we speak about an individual as short, medium, or tall, we are applying labels to a group of heights over the continuum from roughly four feet to a little over six feet. Figure 2.14 shows the fuzzy sets for the ideas behind short and tall.

The boundary between *Short* and *Tall* is not only not precise, but it is also not unique. As an individual's height increases, they enter a world (a system space) where they have properties of shortness and tallness at the same time. Whether a particular individual near this boundary region is *Tall* can involve some undecidability or ambiguity. Fuzzy logic models the extent to which an individual height's is considered a member of the set *Tall* clients (or, perhaps, the elasticity of the concept *Tall* when applied to clients).

Thus, fuzzy logic is a calculus of compatibility. Unlike probability, which is based on frequency distributions in a random population (see the next section, Fuzzy Probabilities on page 62), fuzzy logic deals with describing the characteristics of properties. It describes properties that have continuously varying values by associating partitions of these values with a semantic label. Much of the descriptive power of fuzzy logic comes from the fact that these semantic partitions can overlap. This overlap corresponds to the transition from one state to the next. These transitions arise from the naturally occurring ambiguity associated with the intermediate states of the semantic labels.

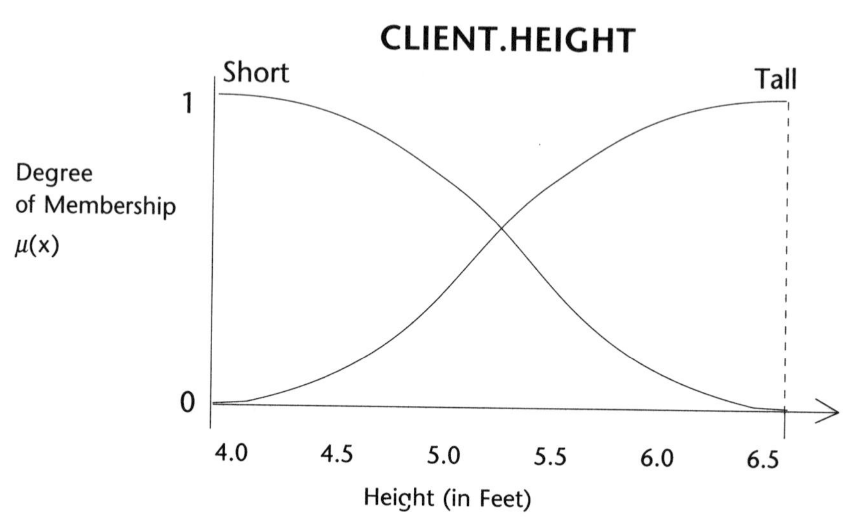

Figure 2.14 Fuzzy Sets for Short and Tall

FUZZY PROBABILITIES

Combining probability and fuzziness produces propositions that state the chance of an ambiguous outcome. They also highlight the distinct difference between probability and fuzzy logic. As a simple example, consider the following statement:

There is a fifty percent chance of a heavy rain tomorrow.

We have a statement of probability (50% chance) combined with a fuzzy description ("heavy rain"). As Figure 2.15 illustrates, there are a wide variety of intensities associated with a rain fall.

Figure 2.15 shows that the intensity associated with *Heavy* is not bounded by crisp rain fall quantities in inches per minute, but is an imprecise state bounded by *Moderate* rain and *Intense* rain. The fuzziness of the *kind* of rain is distinct from the chance that it will, in fact, rain. When we think of probability, we mean the chance that an event will happen in the future. The following statement makes very little sense in terms of probability:

There is a fifty percent chance of a heavy rain yesterday.

When an event has occurred, we know the outcome. No probability remains. If it rained yesterday, there is no probability that it will rain yesterday. However, the ambiguity associated with the fuzzy quantifier *Heavy* persists. You could still argue whether the rain was *Heavy* or *Moderate* or really *Intense*.

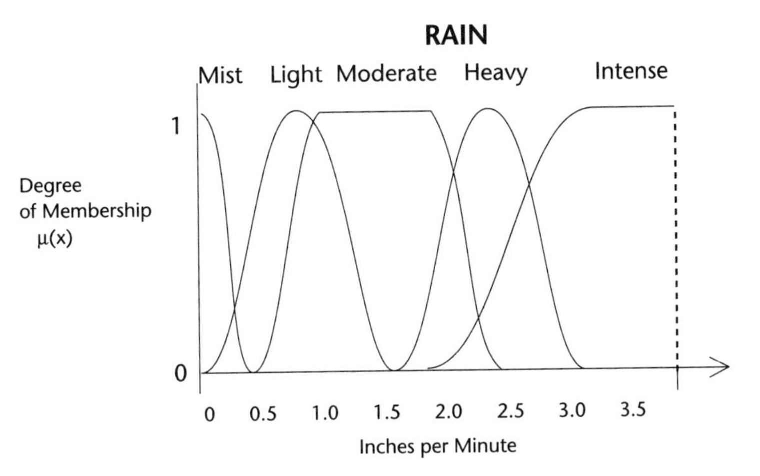

Figure 2.15 Fuzzy Sets Describing Rain Fall Intensity

In the first proposition, where probability is combined with fuzzy logic, we can see a difference between the two measurements by waiting one day. Either it will rain or it will not rain; in either case, the probability disappears. If it rains, however, we are still left with a descriptive ambiguity. Is the rain heavy, or is the rain not heavy? As Figure 2.15 shows, there is no fine boundary point between the idea of a heavy rain and a rain that is not heavy. We move in fine granulation from a light rain, through a moderate rain, into a definitely hard or heavy rain. Thus, for some intensity of rain, we cannot tell whether or not the fuzzy outcome occurred. We can only measure the extent to which the rain is heavy.

Bayes Theorem and Fuzzy Probability

Less rigorous statisticians abandon the frequentist notion of probability and assert that fuzzy logic is simply another method of encoding subjective or Bayesian probabilities. In this approach, the probability is taken as evidence for or against the occurrence of an event. Bayesian probabilities are often expressed as the odds for event **A** occurring given the accumulation of evidence for some conditional event **B**. This is based on Bayes Theorem. Equation 2.2 shows how we derive the probability of **A** given evidence **B**.

$$\frac{p(A \wedge B)}{p(A \wedge B) + p(\neg A \wedge B)} = \frac{p(A \wedge B)}{p(B)} = p(A \mid B)$$

Equation 2.2 Bayes Theorem (In a Variety of Forms)

From this equation, we might be tempted to construct statements such as, "The probability that **A** is *Tall* given his/her height is **B**." If we have a probability distribution of heights in a sample population, we can, of course, define *Tall* as a set of heights more than one (or two) standard deviations from the mean. If we now know (or can estimate or can speculate on) the probability that an individual "**X**" chosen at random from the population has a certain height, we can assign **X** into the class of *Tall* individuals by solving Bayes Theorem:

```
prob(X is Tall | X has height h)
```

While this is a simplification of both Bayes Theorem and the methods of accumulating evidence, the example does illustrate the fundamental difference between any form of probability and fuzziness. Even using Bayes Theorem, the knowledge engineer is postulating the occurrence of event *X* given prior evidence for events that ultimately lead to *X*. Once we have discovered that event *X* has occurred, all notions of probability must evaporate. Thus, picking *X*, our individual *X* with height "h" is assigned to a class of *Tall* individuals based not on the probability that he or she is *Tall*, but on the measurement of his/her height. In any case, referring to Figure 2.14, there remains some linguistic ambiguity about whether *X* is representative of *Short*, *Medium*, or *Tall* individuals (where we might consider *Medium* as located somewhere in the intersection of the *Short* and *Tall* regions).

Furthermore, unlike fuzziness, linguistic terms in probability have no underlying semantics. We simply and arbitrarily divide a probability distribution function (PDF) into named segments. Thus, given a normal distribution of heights, we can label the right-hand tail as *Tall*, the middle of the distribution as *Average* or *Medium* height, and the left-hand tail as *Short*. But these labels have no underlying organization. They represent a form of interval arithmetic. Thus we cannot determine a specific degree of *Tall* or *Short*. The transition across such a probability distribution function can, perhaps, yield something like a collection of fuzzy memberships. By labeling the right distribution region at two standard deviations or more as *Tall*, the left distribution region at two standard deviations as *Short*, and the central region as *Average*, the frequency distribution at any point could give us something similar to a membership function.

Fuzzy Logic

So we are left with a final, important question: Just what is "fuzziness"? Fuzziness is a measure of how well an instance (value) conforms to a semantic ideal or concept. Fuzziness describes the degree of membership in a fuzzy set. This degree of membership can be viewed as the level of compatibility between an instance from the set's domain and the concept overlying the set. That is, in the fuzzy set *Tall*, the value 5'2" has a degree [.38], meaning that it is only moderately compatible with *Tall*. We call this measure "fuzziness" because it is used to assess the degree of ambiguity or uncertainty attached to each set measurement. Some measurements have minimal fuzziness or ambiguity—those that fall at the extreme edges of a fuzzy region since they are highly compatible with the set's concept. In between, these properties have varying degrees of ambiguity. They can belong to different fuzzy sets simultaneously; in fact, a point can belong equally to neighboring fuzzy sets, thereby maximizing its ambiguity. Unlike probability, fuzziness does not dissipate with time. Fuzziness is an intrinsic property of an event or object.

The purpose of *The Fuzzy Systems Handbook* is to explore the ways in which fuzziness comes into play during the modeling of information decision systems across a wide spectrum of applications and industries. The handbook attempts to refine the definition of fuzziness so that knowledge engineers, systems analysts, and technical managers can evaluate its use in the design of expert and decision support systems (whether standalone or as an integral part of larger systems). The handbook also contains a set of software tools that can be used to build, test, and deploy fuzzy systems.

NOTES

1. In specifying measurements using scientific notation, we are indicating the number of significant digits. Significant digits reveal the level of granularity in the measuring system. This last digit is nearly always an estimate at one subdivision less than the finest division of the instrument.

2. An antique and unfortunately obsolete device used to perform complex mathematical calculations such as multiplication and division. I omit any more detailed descriptions since those who have never used or seen one would never believe how it worked!

3. Named after the mathematician, August F. Mobius, 1790–1868.

Three

Fuzzy Sets

Actioni contrarium semper et aequalem esse reactionem:
Sive corporum duorum actiones in se mutuo semper esse
Aequales et in partes contrarias dirigi

To every action there is always opposed an equal reaction:
Or, the mutual actions of two bodies upon each other are
Always equal, and directed to contrary parts.

> Sir Isaac Newton (1642–1727)
> *Principia Mathematica* (1687 ed.) Laws of Motion 3

Entia non sunt multiplicanda praeter necessitatem

Things should not be multiplied (added) without necessity.

> William of Occam (ca. 1285–1349)
> from *Quodlibeta* (ca. 1324) No. 5, Question 1, Art. 2
> Also known as "Occam's Razor," an ancient principle of reasoning usually attributed to Occam, but found earlier in origin (see M. T. Cicero, *Tusculan Disputations*).

In this chapter, we begin our exploration of fuzzy systems with the fundamentals: the meaning of fuzziness and its representation through fuzzy sets. We will examine the difference between fuzzy sets and classical or Boolean sets, determine the way fuzzy sets are constructed, define fundamental terms and concepts, and briefly look at some advanced fuzzy representations.

The Age of Science

Isaac Netwon ushered in the Age of Reason with his *Principia Mathematica*. Newton, the co-inventor of calculus and the discoverer of the laws of optics, motion, and gravitation, demonstrated that the deep laws of nature were both tractable to men's minds and expressible as mathematical equations. He elevated the differential equation to its exalted position as the linchpin of modern science. And, as the father of modern scientific thought, Newton also ushered in the Age of the Fine Grain Measurement. Scientific laws demand detailed knowledge of the fundamental constants.

And we are a society seemingly obsessed with precision. We see it as the core of Western scientific thought. Scientists from Robert Millikan to Murray Gell-Mann instituted a detailed search for the fundamental units of measurement, speed, elasticity, mass, and charge. In this quest, we have been amazingly successful. We have found the speed of light (to zero error):

$$2.99792458 \times 10^{10}\, cm \cdot \sec^{-1}$$

We have looked deep into the heart of matter to discover such strange values as Planck's Constant, measured to an accuracy of thirty-four decimal places and within .60 parts per million:

$$6.6260755 \times 10^{-27}\, erg \cdot \sec$$

We now know with extreme accuracy the exact number of molecules in a molecular weight's worth of pure chemical assay, the precise mass of a proton, the strict wavelength equivalent of any moving body, and the absolutely correct tug of gravity between any two bodies. From our extensive and precise knowledge of mathematics and physics, mankind has sent spacecraft to the outer planets, bounced a lander on the surface of Mars, mapped the genetic code of complex organisms, and reduced all the world's computing power of just thirty years ago into a few silicon chips the size of a dime. Mathematical precision sits at the heart of our fundamental understanding of the cosmos. It provides a common infrastructure for complex world views spanning the range from superstring vibrations and the infrared confinement of quarks, to the Carnot cycle describing our car's internal combustion engine, to the mechanics of stellar nucleosynthesis and the roiling virtual particles of the quantum vacuum that slowly dissipates black holes.

But despite our fascination with details and our love of mathematical precision (or, perhaps, mathematical purity), we live in a world where imprecision is an intrinsic feature of nearly all natural phenomena (see Imprecision in the Everyday World on page 70). It is a strange state of science that what we observe around us finds so little reflection in our theoretical descriptions of the cosmos. Scientists, engineers, and mathematicians act as though the imprecision of the real world is simply an inconvenient state that they can over-come by more extensive and exact theories. The universe offers so many precise constants, and mathematics seems to work so well, certainly the observed ambiguities and imprecisions of the world would be resolved if only we had better and more accurate measuring devices. But, alas, this just isn't so. Boyle's Law is a modest but fairly good example. This law says:

$$\left[V \ \frac{1}{\alpha} \ P \right]^{t}$$

Thus, the volume(V) of a given quantity of gas is inversely proportional to its pressure (P) at a constant temperature (t). But —and they don't tell you this in first-year chemistry — this "law" is only approximate at low and high pressures. In fact, there is a Boyle's Temperate at which the behavior coefficients of a gas change sign. Close to this temperature, Boyle's Law provides a fairly good approximation to the equation of state for the gas. The boundaries in which Boyle's Law accurately describes the behavior of a gas are defined by imprecise terms: *High*, *Low*, and *Close to*. As Figure 3.1 illustrates, the concepts of *Low* and *High* pressures do not have crisply defined boundaries. In fact, as we will see, these are examples of fuzzy sets.

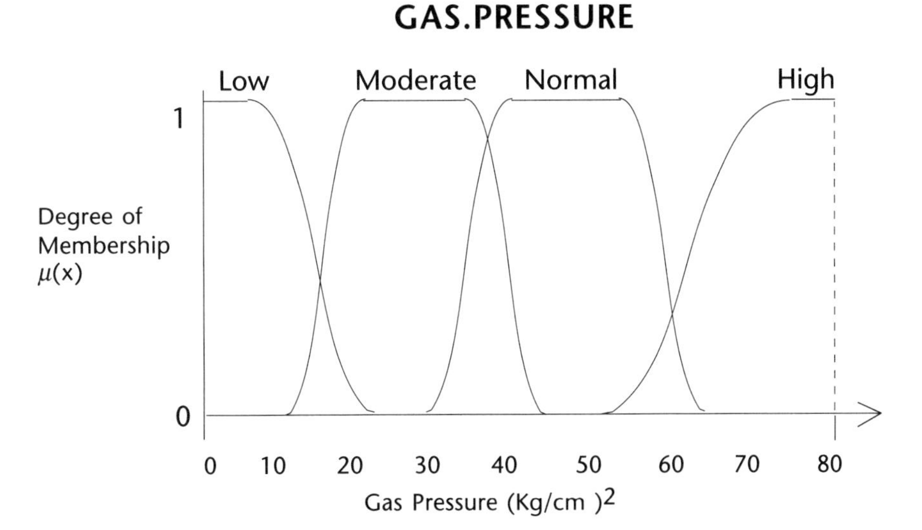

Figure 3.1 Imprecise Regions (Fuzzy Sets) for Gas Pressure

Educators seem to ignore the constraints on Boyle's Law; I suspect because the equation itself does not rigidly define the constraints . But the real-world limitations cannot be simply ignored. Nor can physicists eliminate the idea of *Low* or *High* pressure by any amount of increased pressure measurement. The indistinctness of a concept like *High* is an intrinsic property of the physical world. No amount of detailed measurement, other than declaring an arbitrary boundary, will find the exact point where *Normal* pressure suddenly becomes *High* pressure. This is part of the imprecision we find in the everyday world.

Imprecision in the Everyday World

Most of the phenomena we encounter every day are imprecise; that is, they carry a certain degree of fuzziness in the description of their nature. This imprecision may be associated with their shape, position, momentum, color, texture, or even the semantics that describe what they are. In many cases, the same concept will have different degrees of imprecision in different contexts or times. Thus, a hot day in winter is not exactly the same as a hot day in summer. The exact definition of when the temperature goes from warm to hot is imprecise—we cannot identify a single temperature point that is warm, and, when we move up a single degree, the temperature is now considered hot. This kind of imprecision or fuzziness associated with continuous phenomena is common in all fields of study: sociology, physics, biology, finance, marketing, engineering, oceanography, psychology, health management, and so forth.

Imprecise Concepts

Few of us pay any attention to the imprecision of our world. We accept it as a natural consequence of the way things work. The dichotomy between the rigor and precision of mathematical modeling in all fields of endeavor and the intrinsic uncertainty of the "real world" is generally not addressed by scientists, philosophers, and business analysts. We simply approximate these events as numerical functions and choose a result that either makes sense from the empirical data or satisfies our statistical filters. Yet we process and understand intrinsically imprecise data easily, from morning traffic reports to complex business analyses.

Intrinsic imprecision is an aid to understanding, not a hindrance. We are able to formulate plans, make decisions, and recognize compatible concepts at high levels of vagueness and ambiguity. Consider the following statements:

The engine temperature is hot.
Current inflation rates are rising rapidly.
Long projects are generally late.
A cruise missile has a long range at a high speed.
Our price should be around twice the competition's price.
The ratio of code to comment lines in program A is high.
Bill's cholesterol count is very high.
There is significant risk in dealing with Iranian moderates.
Firm commitments from upper management tend to weaken as costs increase.
Disposable incomes in the middle tax tier are adequate.
That project requires a large manpower commitment.
Plutonium is extremely radioactive.
Acme Corporation is a large and aggressive company.
Tom is rather tall, but Judy is short.
If you are tall, then you are quite likely heavy.

These propositions form the nucleus of our relationship with "the way things work" in the world. Yet they are incompatible with traditional modeling and information system designs. Current computer-based systems require much more precision on our part. Somehow we would like to bridge this gap between machine systems and human systems. If we can make decisions based on imprecise information, so should our machines. Thus, a heuristic such as

when engine temperature is hot, increase fan speed

can be readily understood, even if we do not have an exact threshold point where an engine is warm and then hot. Similarly, even more imprecise heuristics such as

if the current inflation rates are rapidly rising
and the job market is sluggish
then disposable incomes will be small

are easily understood by professionals versed in the jargon of econometric modeling. Even we can understand a concept like "current inflation" as the inflation rate of the recent past, without having a precise time boundary, or "sluggish" as a slow down in the rate of new hires, or "small" as a relative decrease in the amount of "after necessity" retained income. Of course, politicians are able to discuss the policies of "Iranian moderates" without understanding who or what they actually are, or even if they actually exist (undecidability and self-delusion are discussed later).

The Nature of Fuzziness

Is imprecision or fuzziness an artificial concept used to hide our inability to measure one or more of a phenomenon's properties, or is it an intrinsic part of the phenomenon itself? This is an important question since it strikes at the very heart of fuzzy set and fuzzy measurement theory. As we will see, fuzziness is independent of any measurement capability. While some forms of measurements will resolve stochastic uncertainty, fuzziness is a property of the phenomenon itself. Let's consider the following statement in terms of a precise understanding of heavy and hot:

A heavily stressed motor will run hot

The set of speeds at which a motor is heavily stressed is not distinct and bounded from the set of speeds at which the motor is not heavily stressed. Likewise, the set of temperatures considered hot is not discrete and bounded from those temperatures considered warm or slightly hot. In Figure 3.2, using a form that will become familiar, we can look at the idea of the concept *Hot* as a continuous curve connecting different temperatures.

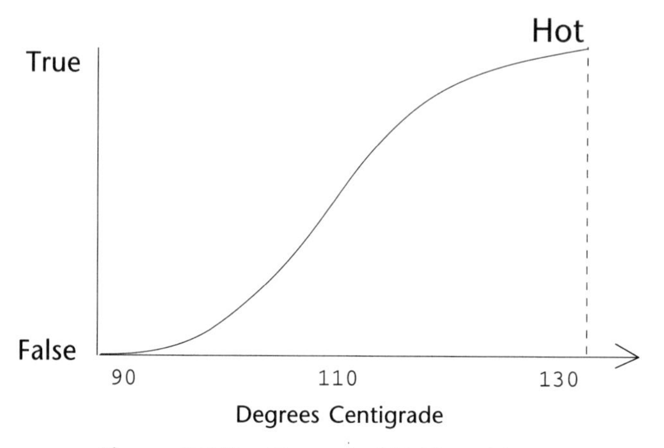

Figure 3.2 The Concept of *Hot* for a Motor

This means that 90˚C is absolutely **not** *Hot*, while everyone would agree that 180°C is definitely *Hot*. As Figure 3.3 illustrates, temperatures between these two extremes have various degrees of compatibility with the idea of *Hot* for a motor.

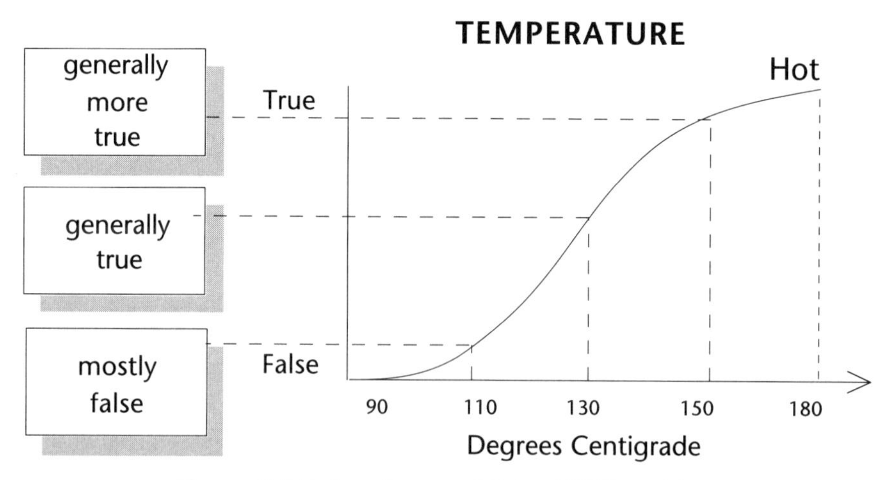

Figure 3.3 Various Degree of Being HOT for a Motor

Perhaps we want to think of the motor's temperature in terms other than simply *Hot* or *Not Hot*. Nothing prevents us from dividing up the entire operating range of the motor into several regions. We might, as an example, decide that a motor can be *Warm* over some range of temperatures and *Hot* over others. Since the idea of warm increases from *Cool* (as an example) and then decreases as the motor becomes *Hot*, Figure 3.4 shows one way of representing the idea of *Warm*.

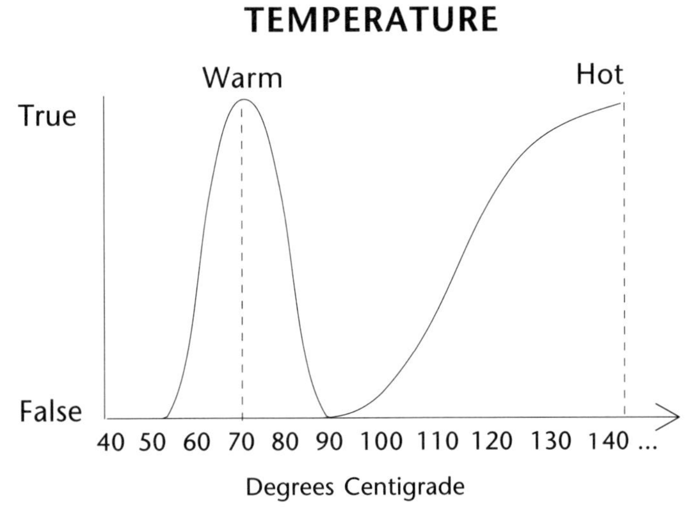

Figure 3.4 WARM and HOT as Distinct Regions

But this representation is at odds with our logical notion of the relationship between warm and hot. Surely as the temperature becomes less and less warm, it also becomes more and more hot. Thus, in Figure 3.4, the right-hand segment of the *Warm* region should overlap, to some degree, the left-hand part of the *Hot* region. Figure 3.5 illustrates a more logical and real-world representation of these two related concepts.

What do the graphs in Figure 3.5 represent? In particular the right-hand side of the `Warm` region needs some explanation. Intuitively we know that, as the temperature increases, the motor goes from *Warm* to *Hot*. So, first we might expect to see something like the graph in Figure 3.6, where *Warm* blends with various degrees of *Hot* (tepid, luke-warm, mildly hot, and so forth) until the curve crosses into the actual *Hot* temperatures.

But Figure 3.6 is the Boolean equivalent of the separation of *Warm* and *Hot*. At approximately 112°C, the curve reaches an inflection point. Figure 3.5, however, reflects the actuality of our experience. As the motor temperature increases, the curve associated with *Warm* moves toward zero membership, while the curve associated with *Hot* begins to gain increased truth. At this time, individual temperature measurements have some characteristics of both *Warm* and *Hot*. When the temperature reaches 112°C, the *Hot* curve crosses the *Warm* curve, and we would be more likely to call the motor temperature *Hot* rather than *Warm*. (Yet an increasingly smaller population of motor maintenance workers might still think the motor is warm or "slightly Hot.")

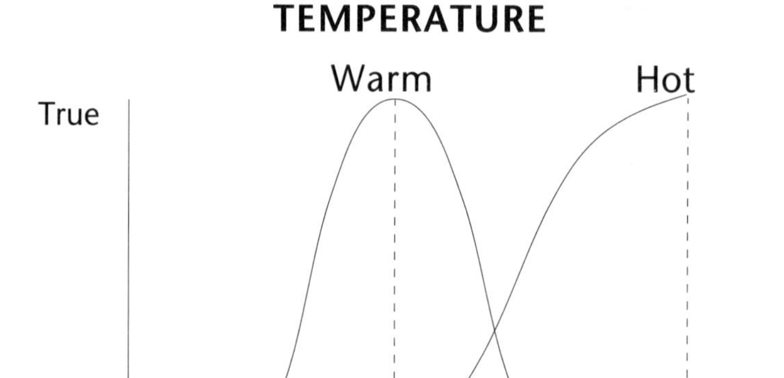

Figure 3.5 *Warm* and *Hot* as Regions in Motor Temperatures

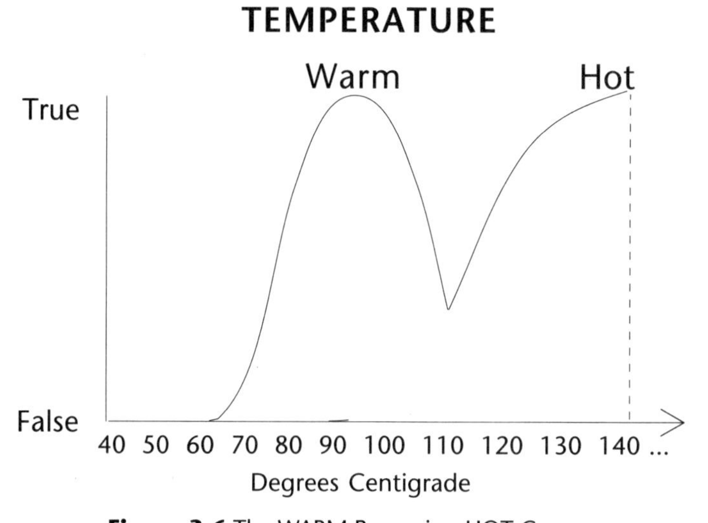

Figure 3.6 The WARM Becoming HOT Curve

Although this reasoning seems like common sense, it flies in the face of nearly all classical methods of data representation and system modeling. In such classical models, we must explicitly decide on a precise point where a temperature is *Hot* and where it is

Not Hot. In fact, there may not be any such point! To see this, we can break down the operating temperatures for the motor into just two regions, those that are *Hot* and those that are *Not Hot*. We define the idea of *Not Hot* as the complement of *Hot*. Figure 3.7 shows how these two regions might appear.

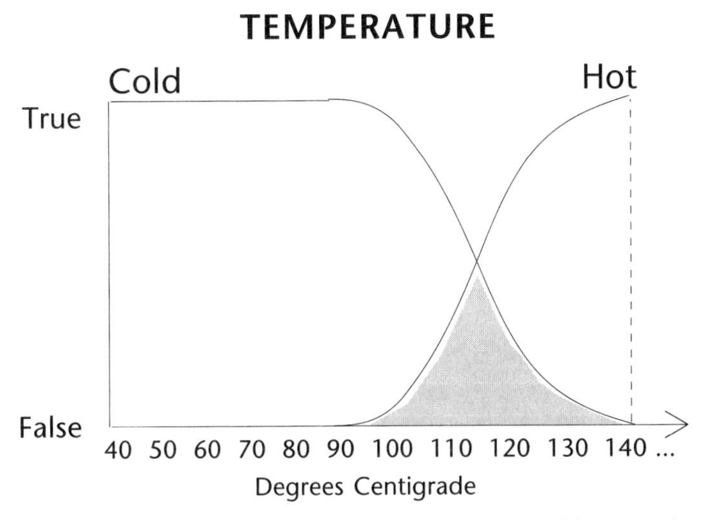

Figure 3.7 The Region *Hot* and Its Complement, *Cold (Not Hot)*

We immediately notice that there is an area shaded in Figure 3.7 where a temperature belongs both to the *Hot* and the *Not Hot* regions. This violates an Aristotelian principle called the Law of Noncontradiction (often confused with the Law of the Excluded Middle).

Fuzziness and Imprecision

This imprecision or ambiguity is related to the decoherence between our classical understanding of a phenomenon and its actual existence in the real world. How can this be? The answer lies in the tools that we use to construct mappings between the real world and our models. These tools are based on an ancient and pervasive view of the world, formalized since the time of Aristotle and refined by such thinkers as George Boole, Georg Cantor, David Hilbert, and Bertrand Russell. The reasoning process underlying this view, called *Boolean logic,* looks at the world in terms of well-structured categories (or sets). Each set has a well-defined membership roster (independent of whether or not we can actually count or conceive of the set members). An item is either a member of a set or it is not, and this membership is quantum mechanical in nature; that is, when we know the value of an object, it either jumps completely into or out of a set. There is no middle ground.

In the everyday world of our experience, however, such categorization is not so crisply defined. Consider the statement:

That car is going fast.

What do we mean by the concept fast? What is the set of *Fast* cars? In many ways the concept is context dependent, *Fast* on a highway is not the same as *Fast* in a residential neighborhood. But ponder the idea of fast on the Hollywood Freeway. (For the purposes of this discussion, assume you are driving at three o'clock in the morning so the overall car density is below $600/m^2$.) Obviously, driving along at 20 miles per hour is generally not going fast, but driving at 80 miles per hour is indeed traveling at a high rate of speed. How do we make a correspondence between the actuality of fast and the representation of fast in a sharply defined set?

The difficulty in deciding on a crisp boundary for intrinsically imprecise concepts can be illustrated by following Joe and Mary, two forensic epistemologists working for the California Highway Patrol and responsible for defining "how fast is fast" on the state's open highways.

Joe — Well, yes, but that's an extreme. A car traveling over 60 miles an hour is fast. I think we'd all agree to that. So, we can easily make a category for fast cars:

$$\mu_{FAST}[x] = \{\text{speed} \geq 60\}$$

Mary — Wait a minute, Joe. Do you mean that a car going 59 miles an hour is absolutely not fast, but, just by increasing its speed by one mile an hour, now it is? How about a car traveling 59.5 or 59.9 miles per hour?

Joe — (Giving this some thought) No, I guess not. Okay, Mary, that's easy enough to fix. We can just expand the category limits a little to include those cars going 59 miles per hour:

$$\mu_{FAST}[x] = \{\text{speed} \geq 59\}$$

There. Happy?

Mary — Not quite. What about someone driving along at 58 or 58.5 miles per hour—are they not fast simply because they're traveling at half a mile an hour slower?

Joe	(Seeing where this is going) Look, Mary, you've gotta draw the line somewhere!
Mary	Why?
Joe	Well, if we keep lowering the bottom rung of what we mean by fast, pretty soon we'll lose the whole concept of fast. Eventually the category will include cars that are moving quickly, slowly, or just rolling along. I mean, there has to be a point where we could both agree that a car just isn't fast.
Mary	You've definitely got a point, Joe. Okay, let's say that a car isn't fast if it's going less than 35 miles an hour, but it is definitely going fast if it's going over 60 miles an hour.
Joe	Now we're getting somewhere. I can describe the category for fast as:

$$\mu_{FAST}[x] = \{speed \geq 35 \ \& \ speed \geq 60\}\}$$

(shaking his head violently) No, that's not right. I can't give two limits on what we mean by fast! So the set membership function has to be:

$$\mu_{FAST}[x] = \{speed \geq 35\}$$

(throwing down his marker and collapsing in disgust) But that's not right either; 35 miles an hour is too slow to be considered fast.

Mary	No problem. We'll just say that 35 miles per hour is the minimum speed that some of us might consider fast—but not everyone, and 60 is the speed that we could all agree on as being fast. Anything in between just has varying degrees of fastness.
Joe	What good will that do us? Who ever heard of "varying degrees" of fastness? That would mean some cars might be more strongly associated with the category fast than others. Certainly, a car close to 60 miles an hour is, by consensus, more descriptive of fast than a car going along at 38 miles an hour.
Mary	Well, why not! That sounds good to me. Can't we just take a different view of categories? The membership in a category can take on a whole range of values, not just one and zero. That means, Joe, that 38 miles per hour might have a membership of [.15], 55 miles per hour might have a membership of [.87], and any value over 60 is [1]. (Mary walks to the chalkboard and draws.) What if we looked at "fast" as something like this (see Figure 3.8):
Joe	What's that?
Mary	A new way of representing categories, Joe. Now speeds have partial memberships in the set defined as Fast Cars.
Joe	That's silly.

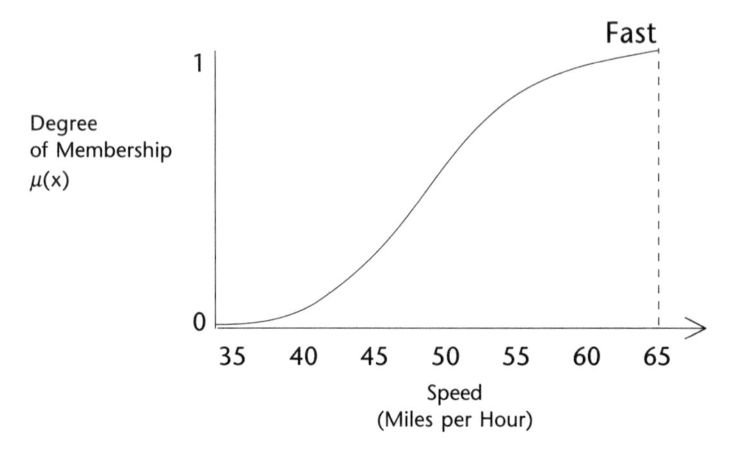

Figure 3.8 A Description of the Concept *Fast*

Mary has articulated the case for imprecision in our everyday experience, and her attempt at dealing with this imprecision leads to a formulation—as we will shortly see—of a proper fuzzy set for the concept *Fast*. We can see that this imprecision is an intrinsic property of the concept *Fast*. It is not related to any measurement of speed on our part, nor the precision of our equipment, nor the fine granularity of the outcome metric. This is the nature of fuzziness.

Representing Imprecision with Fuzzy Sets

Saying that some concept is imprecise and using that imprecision in a model are two different things. The bridge between them is the concept of a fuzzy set, an idea that Mary articulated in her attempt to understand the concept of *Fast*. We need to understand generally what a fuzzy set looks like, how it encodes imprecision, and how we specify its various components in order to discuss the allowable operations on fuzzy sets and how they are used in fuzzy system models.

To see how these properties are used in constructing an actual fuzzy set, consider the concept of *Tall* applied to American males.[2] Given a particular value for an individual's height, what values are considered *Tall*? In classical set theory, we are forced to chose an arbitrary cutoff point, say, six feet. Since the boundaries between what is in a set and what is outside a set are very sharp, these types of constructs are called *crisp* sets. A characteristic function for such a set appears as:

$$\mu_{TALL} = \{\text{height} \geq 6\}$$

Thus, anyone over six feet is tall. The membership graph for this set appears in Figure 3.9.

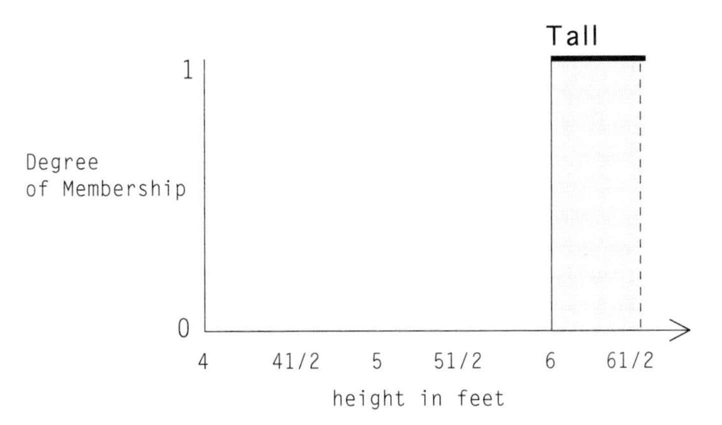

Figure 3.9 The Crisp Set for the Concept *Tall*

The discriminant or characteristic function for this set reflects its Boolean nature. As we move along the domain, the membership of heights in the set *Tall* remains false (zero) until we reach exactly six feet, when it jumps immediately to true (one). Note that in the membership transition graph for crisp sets the line connecting nonmembership and membership is dimensionless. All classical or crisp sets have this kind of membership function. Although this dichotomization of the sample space may work well for grainy, lumpy, and other noncontinuous collections, it generally fails when applied to phenomena that have a continuously (and monotonically) changing set of values.

FUZZY SETS

The idea of *Tall*, illustrated in Figure 3.10, is the classical example of a fuzzy set and illustrates the intrinsic properties of fuzzy spaces. The domain of this set, indicated along the horizontal axis, is the range of heights between four feet and six and a half feet. The degree of membership or truth function is indicated on the vertical axis to the far left. In general, the membership goes from zero (no membership) to one (complete membership). The membership function and the domain are connected, in this case, by a simple linear curve, which shows that tallness is directly proportional to height. Now, given a value for height, we can determine its degree of membership in the fuzzy set.

Thus, a height of five feet has a [.46] degree of membership. The interpretation of this value corresponds to the truth of the proposition:

Five feet is Tall

If the value for height is less than four feet, its membership is zero. If the height is greater than or equal to six and a half feet, then its membership value is one. This membership

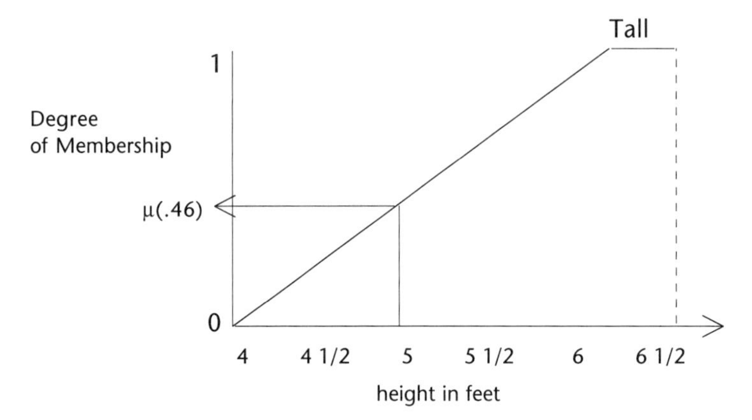

Figure 3.10 Determining the Degree of Membership in *Tall*

function, as we will see, is interpreted as a measure of the compatibility between a value from the domain and the idea underlying the fuzzy set. In the case of five feet, the membership of [.46] means that it is moderately but not strongly compatible with the notion of *Tall*. As the membership value moves toward zero the sample becomes less and less compatible with the semantics of the fuzzy set and, as the membership value moves toward one, the samples becomes more and more compatible with the fuzzy set's semantic property.

As Figure 3.11 illustrates, a fuzzy set consists of three components—a horizontal domain axis of monotonically increasing real numbers that constitute the population of the fuzzy set; a vertical membership axis between zero and one, indicating the degree of membership in the fuzzy set; and the surface of the fuzzy set itself that connects an element in the domain with a degree of membership in the set.

This degree of membership is also known as the membership, or *truth*, function (Equation 3.1) since it establishes a one-to-one correspondence between an element in the domain and a truth value indicating its degree of membership in the set.

$$\mu_A(x) \leftarrow f(x \in A)$$

Equation 3.1 The Fuzzy Truth-Generating Function

As we will soon see, the truth-generating function (Equation 3.1) is not symmetrical. Although each domain point has only a single truth function value, a truth function grade can be mapped to more than one domain point (for convex fuzzy sets such as those represented by triangles, beta, and Gaussian or other bell curves).

A fuzzy set encodes the imprecision or fuzziness associated with a phenomenon through its surface. The shape of the curve, in fact, represents the semantics of the actual

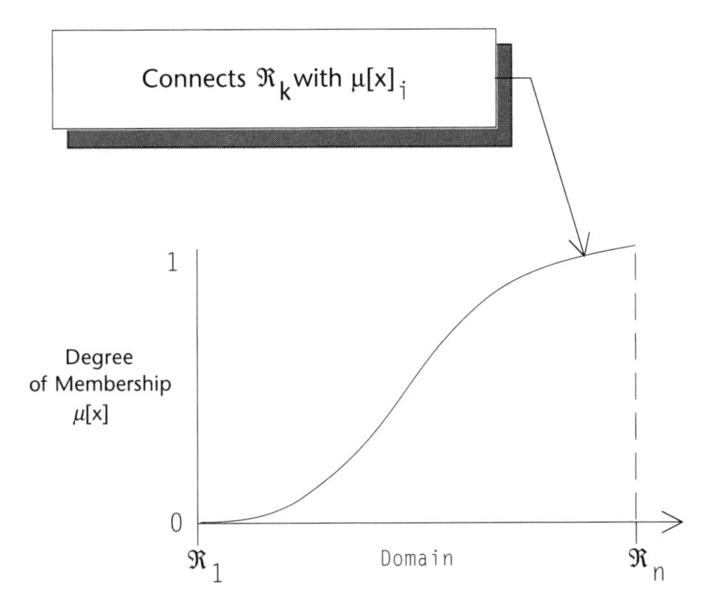

Figure 3.11 The General Structure of a Fuzzy Set

concept. Homophonic fuzzy sets are fuzzy sets with the same name but used to represent different concepts. This type of fuzzy set will reflect its own semantics not only by changes in its domain, but also in the shape of the fuzzy set curve connecting the domain to the membership function.

We can now revisit the idea of a motor's warm and hot operating temperatures. Figure 3.12 shows how these concepts are cast in terms of fuzzy sets.

The difference between the two representations is small yet important. By replacing the vertical axis with a calibrated metric measuring the truth of each membership point, we have combined imprecision with the ability to calculate a degree of membership at each point in the domain. This provides the basis for the calculus of fuzzy logic.

Representing Fuzzy Sets in Software

We begin the development of our fuzzy modeling system with the description of fuzzy sets—a fuzzy set descriptor block *(*FDB*)*—and the operations that allow us to find the degree of membership for a value and, conversely, the value associated with a particular degree of membership. A fuzzy set is essentially a look-up table containing a series of truth memberships. This stream of memberships is interpreted as the surface of the fuzzy set. We can probe this surface by knowing the domain of the set (its minimum and maximum values) and the maximum number of elements in the membership array.

Fuzzy sets are chained to a dictionary of active fuzzy sets. To accommodate this search facility, each FDB includes a pointer to the next FDB at a particular dictionary

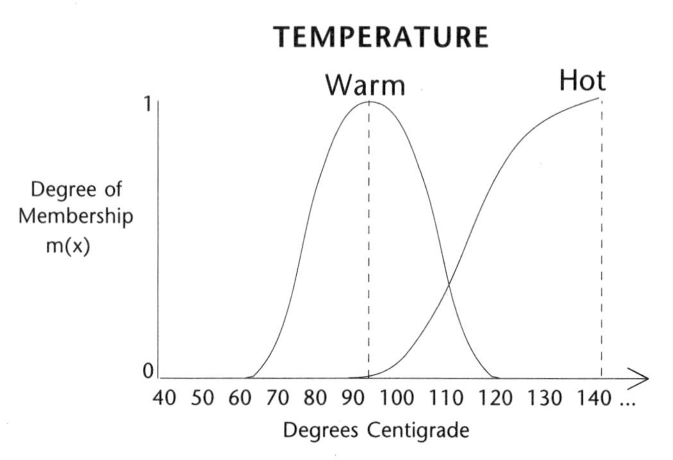

Figure 3.12 *Warm* and *Hot* as Fuzzy Sets

location. If you are using this fuzzy logic code without including the entire modeling infrastructure, these linkage pointers can be ignored.

Each fuzzy set in our modeling system is represented by a fuzzy set description block, called an FDB. The FDB describes the control parameters of the fuzzy set as well as a fixed-length floating-point vector containing the truth distribution of the membership function.

```
//--The Fuzzyset Descriptor Block. Each fuzzy set
//--is defined through one FDB. These sets are stored as
//--truth vectors with their associated bounding domains.
#include "mtypes.hpp"
struct FDB
  {
   char        FDBid[IDENLEN+1],     // Identifier name
               FDBdesc[DESCLEN+1];   // Description
   Ctlswitch   FDBgentype;           // Curve type
   bool        FDBempty;             // Set status
   int         FDBorder;             // Fuzzy degree
   double      FDBdomain[2],         // Lo and Hi edges
               FDBparms[4];          // Generator parameters
   float       FDBalfacut,           // AlfaCut
               FDBvector[VECMAX];    // Membership array
   FDB        *FDBnext;              // Link to next set
  };
```

Listing 3.1 The Fuzzy Set Descriptor Block [FDB]

The description of the fuzzy set is simple and straightforward. We make a trade-off between memory requirements and speed. This is a sensible exchange since today's powerful workstations have RAM memories in the multimegabyte range, with the new virtual memory-based operating systems promising flat memory models in the giga- and terabyte ranges.

The membership function of the fuzzy set is maintained as a floating-point array. The extent of this array, currently 256 cells, is held in the *VECMAX* system constant. The truth membership is proportionally divided between the low and high extents of the domains (*LoDomain* and *HiDomain.*) This is the *Range* of the fuzzy set. For any scalar value from the domain we can find its place in the membership function (*TVcell*) by means of the simple expression shown in Equation 3.2.

$$TVcell = \left(\frac{Scalar - LoDomain}{Range} \right) \times VECMAX$$

Equation 3.2 Locating a Scalar's Cell Position in a Fuzzy Set

On the other hand, finding a scalar associated with a particular truth value involves a two-step process of searching the membership function for the cell containing the truth value and then using this cell number to derive the scalar value. This process is used in fuzzy logic during the process of monotonic inference and defuzzification.

The function in Listing 3.2 finds a domain value corresponding to a given truth membership.

```
#include "FDB.hpp"
#include "mtypes.hpp"
#include "mtsptype.hpp"
double FzyEquivalentScalar(
   FDB *FDBptr,float Grade,int *statusPtr)
 {
  int    i;
  double scalar=0;

  *statusPtr=0;
  for(i=0;i<VECMAX;i++)
    {
     if(Grade>=FDBptr->FDBvector[i])
       {
        scalar=FzyGetScalar(FDBptr,i,statusPtr);
        return(scalar);
       }
    }
  *statusPtr=1;
  return(scalar);
 }
```

Listing 3.2 Finding the Scalar Associated with Truth Membership

The function scans the membership array of the incoming fuzzy set (whose address is in **FDBptr*) until it finds a membership that is equal to or less than the specified membership (*Grade*). The *FzyGetScalar* function, which uses the actual cell number of the array, is then called to find the underlying domain value. As we have already seen, however, a membership function can be associated with more than one scalar value. This function will find the first scalar closest to the lower edge of the set's domain.

Basic Properties and Characteristics of Fuzzy Sets

A fuzzy set has several intrinsic properties that affect the way the set is used and how it participates in a model. Before we turn to how fuzzy sets' curves are defined and what they tend to mean, we will examine a few of the important properties of any fuzzy set. These properties concern the vertical dimension of the fuzzy set (height and normalization) as well as the horizontal dimension of the fuzzy set (support sets and alpha cuts).

Fuzzy Set Height and Normalization

The height of a fuzzy set is its maximum degree of membership and is closely tied to the concept of normalization. As an example, consider the two bell-shaped fuzzy sets of Figure 3.13.

The height of the set *Around 4* is [1.0], while the height of the set *Around 50* is [0.82]. The *Around 4* fuzzy set is in normal form. Fuzzy sets are in *maximum normal form* when at least one element has a membership value of one [1.0] and one element has a value of zero [0]. A fuzzy set is in *minimum normal form* when at least one element has a membership value of [1.0]. In fuzzy system modeling, we are normally concerned only with minimum normal form.

Fuzzy sets are normalized by readjusting all the truth membership values proportionally around the maximum membership value, as shown in Figure 3.14.

The strength of a fuzzy model is tied to the process of normalization for all the variable fuzzy sets. As we will see, the output of a fuzzy system is correlated to the degrees of truth associated with the fuzzy sets used in the rules. If these fuzzy sets are not normal, then the truth of the fuzzy sets in a rule predicate will be arbitrarily truncated and thus will dampen the model's effect.

All the base fuzzy sets in a model must be in normal form; otherwise significant model processing problems will occur. Unnormalized sets also complicate the tasks of both verifying a model and applying model stability metrics. These three routines provide all the necessary functionality to maintain normal fuzzy sets.

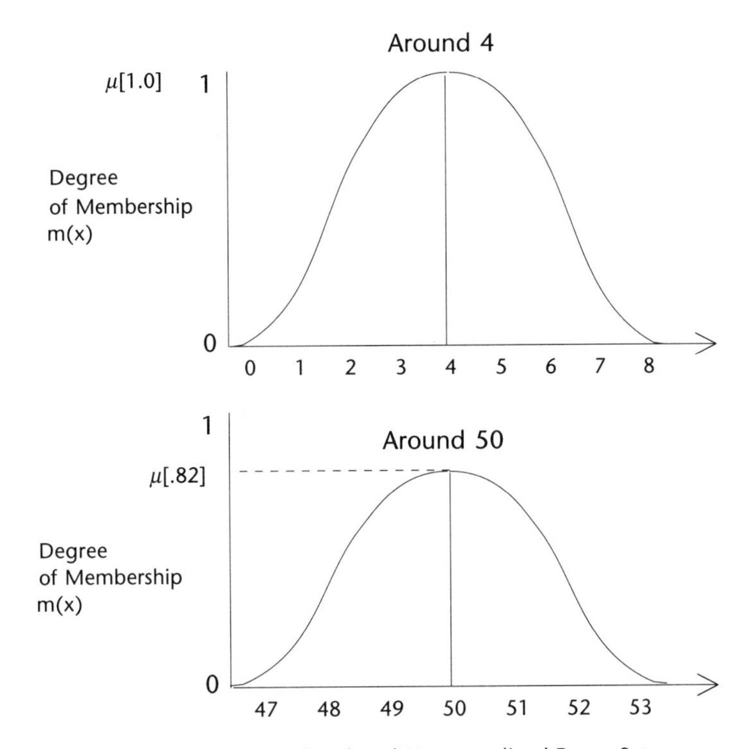

Figure 3.13 Normalized and Unnormalized Fuzzy Sets

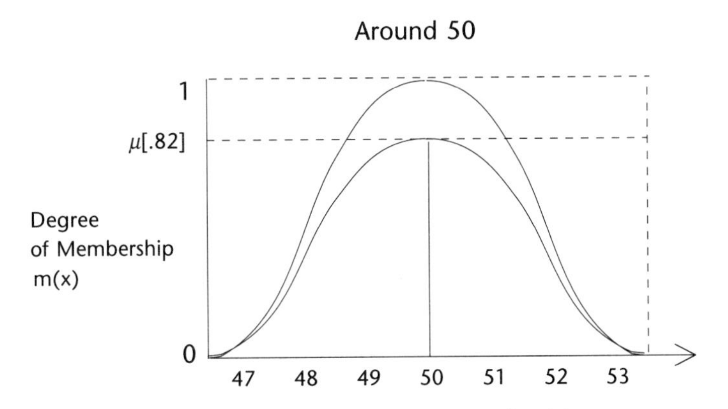

Figure 3.14 *Around 50* as a Normalized Set

```
#include "mtypes.hpp"
#include "fuzzy.hpp"
bool FzyIsNormal(FDB *FDBptr)
  {
   int i;
   for(i=0;i<VECMAX;i++)
      if(FDBptr->FDBvector[i]==1.0) return(TRUE);
   return(FALSE);
  }
```

Listing 3.3 Function to Detect a Normal Fuzzy Set

```
#include "fuzzy.hpp"
#include "FDB.hpp"
float FzyGetHeight(FDB* FDBptr)
  {
   int i;
   float max_memval=FDBptr->FDBvector[0];
   for(i=0;i<VECMAX;i++)
      if(FDBptr->FDBvector[i]>max_memval)
              max_memval=FDBptr->FDBvector[i];
   return(max_memval);
  }
```

Listing 3.4 Function to Return the Maximum Height of a Fuzzy Set

```
#include "FDB.hpp"
#include "fuzzy.hpp"
#include "mtypes.hpp"
void FzyNormalizeSet(FDB* FDBptr)
  {
   int i;
   float max_memval=FzyGetHeight(FDBptr);
   if(max_memval==0.0) return;
   for(i=0;i<VECMAX;i++)
      FDBptr->FDBvector[i]=FDBptr->FDBvector[i]/max_memval;
   return;
  }
```

Listing 3.5 Procedure to Normalize a Fuzzy Set

The functions in Listing 3.3 to 3.5 hardly need any explanation, but a quick review is in order. The *FzyIsNormal* function indicates whether or not the fuzzy set is already in normal form. The *FzyGetHeight* function returns the maximum membership value in the fuzzy set. This function will play an important role when we turn to measuring how well a model is constructed and its compatibility with its data. It is also used during the defuzzification process. Finally, the *FzyNormalizeSet* routine normalizes a fuzzy set by finding the maximum height and using this height to scale the membership array.

Domains, Alpha-level Sets, and Support Sets

We now take up several important properties of a fuzzy set that determine what we actually observe when using a fuzzy set in a model. Underlying the surface of the fuzzy region is the universe of values that we map back to this membership array. What we actually measure with a fuzzy set depends on an understanding of the range of possible values that the fuzzy set covers as well as another important factor called the *alpha threshold limit.*

The Fuzzy Set Domain

The total allowable universe of values is called the *domain* of the fuzzy set. The domain is a set of real numbers, increasing monotonically from left to right. The values can be both positive and negative. You select the domain to represent the complete operating range of values for the fuzzy set within the context of your model. As an example, the fuzzy set *Heavy* (for American males) has a domain that runs from 180 pounds to 280 pounds as graphed in Figure 3.15.

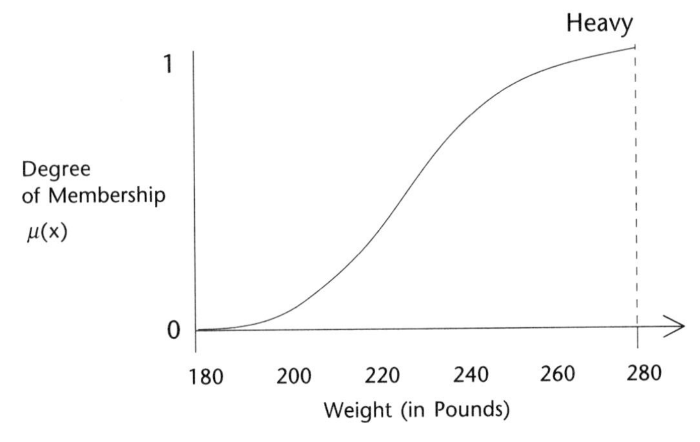

Figure 3.15 The Fuzzy Set *Heavy* Based on Weight in Pounds

Note that the domain has both an upper and a lower limit even though, in reality, the scope of values associated with the fuzzy concept may be open ended. In the *Heavy* fuzzy set, the possible upper weights may go beyond 280 pounds. (We could envision a person weighing, say, 380 or even 1500 pounds.) But since the fuzzy set reaches unity at 280 pounds (any weight above this is definitely *Heavy*), we terminate the domain at this point.

A fuzzy set *Stable* measures the stability of a rotor blade and has both positive and negative components. As the blade pitches off center (as measured in radians per second), the idea of stability decreases in both directions (Figure 3.16).

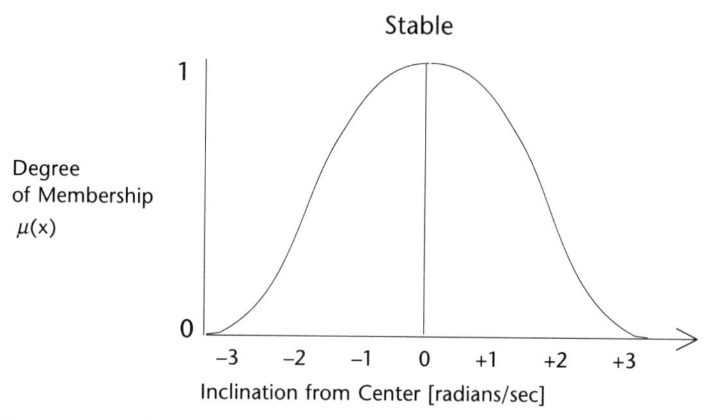

Figure 3.16 The Fuzzy Set *Stable* Measured as an Offset from Center

In designing and building fuzzy models, you must give some consideration to the relationship of domains between fuzzy sets. Domains for implicitly related fuzzy sets are conformal. That is, if a model uses a value of height with the fuzzy set *Tall* to infer a value for weight from another fuzzy set *Heavy*, then the domain of the two sets must be drawn from related populations (Figure 3.17).

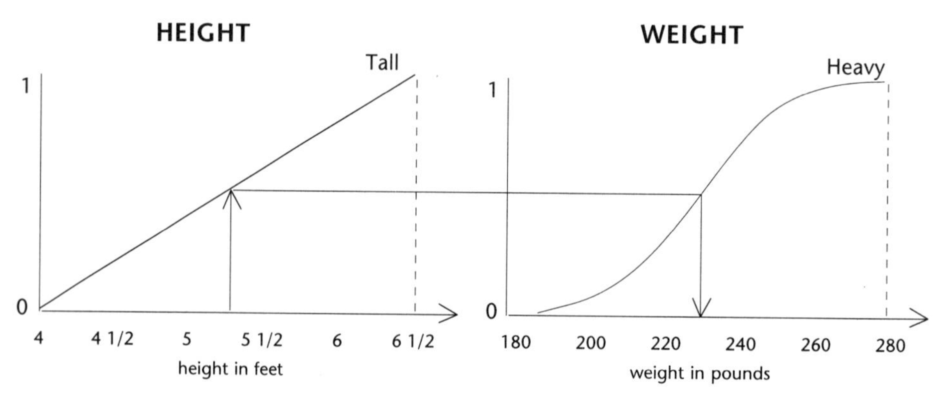

Figure 3.17 Mapping Conformal Domains in Two Related Fuzzy Regions

This process does not mean that both fuzzt sets have the same domain, only that the domain values are somewhat synchronized. This fact will become important when we turn to the concepts of monotonic reasoning and fuzzy implication functions.

The Universe of Discourse

A numeric model variable is generally described in terms of its fuzzy space. This space is composed of multiple, overlapping fuzzy sets, with each fuzzy set describing a semantic partition of the variable's allowable problem state. Figure 3.18 illustrates this concept. The model parameter *Temperature* is broken down into four fuzzy sets: *Cold*, *Cool*, *Warm*, and *Hot*.

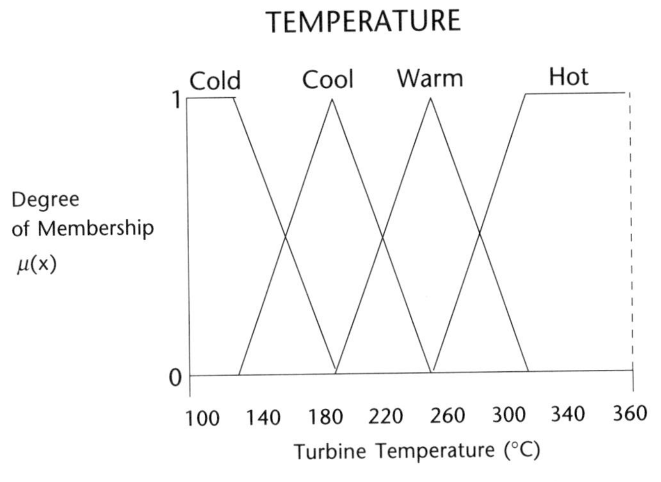

Figure 3.18 The Universe of Discourse for Turbine Temperature

This total problem space, from the smallest to the largest allowable value, is called the *universe of discourse* (UoD). The universe of discourse for the model variable *Temperature* is 100 to 360°C, while the domain for the fuzzy set *Warm* is 180 to 310°C. The fuzzy sets describing this universe of discourse need not be symmetric, but they will always overlap to some degree. The importance of this overlap and how it influences model behavior is an important topic considered in Chapter 7, *Fuzzy Models*.

Note that the universe of discourse is associated with a model *variable*, not with a particular fuzzy set (the range of an individual fuzzy set is the domain). The variable's allowable range of values constitutes its working problem space. This space is then decomposed into a number of overlapping fuzzy regions. Each region is assigned a term name so that it can be referenced in the model (you can only write rules associated with fuzzy regions in the variable's universe of discourse). This collection of fuzzy sets associated with a variable is often called the *term set*.

The Support Set

Sometimes the nonzero part of the fuzzy set does not extend across the entire domain. As an example, the formal domain for heavy might be 180 to 280 pounds; however, the actual contour curve starts at 190 and reaches full membership at 265 pounds (Figure 3.19).

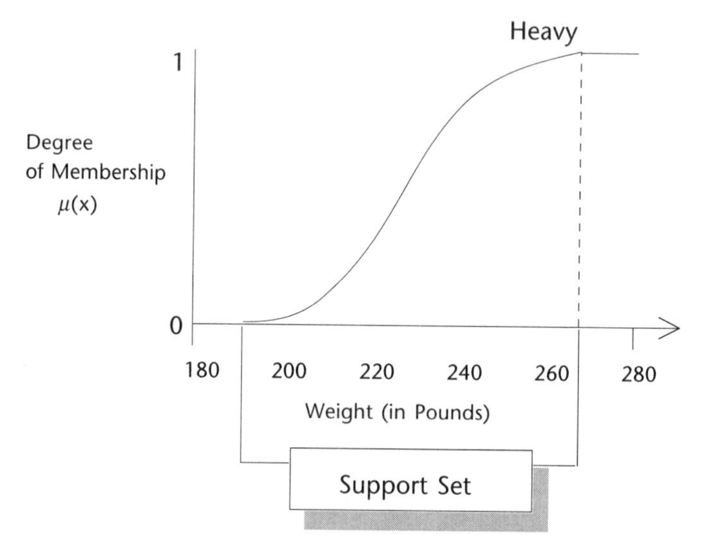

Figure 3.19 The Support Set for the Fuzzy Set `Heavy`

This area is called the *support set*. While support sets are not usually associated with base fuzzy sets (the sets that a designer creates and installs as part of a model's vocabulary), they are important facets in the interpretation and management of dynamic fuzzy regions.

Use of Psychometric Domains

There are some concepts whose domains are subjective and unitless rather than objective. An objective domain has some relationship to physical objects and units of measure such as feet, dollars, meters, item counts, ratios, and so forth. Other domains —those associated with concepts rather than things—reflect a subjective, if not quite arbitrary, scale against which the range of the fuzzy model is mapped. These domains are devised by the model builder, are unitless, and have no corresponding parallel in the real world. Examples of rules using unitless domains include:

```
if project.budget is high, then risk is increased.
if NumberofParts is large, then complexity is elevated.
if temperature is hot, then efficiency is reduced.
```

This kind of scale, since it comes from the designer's mind, is called a **psychometric scale**. A typical example of this scale is shown in Figure 3.20.

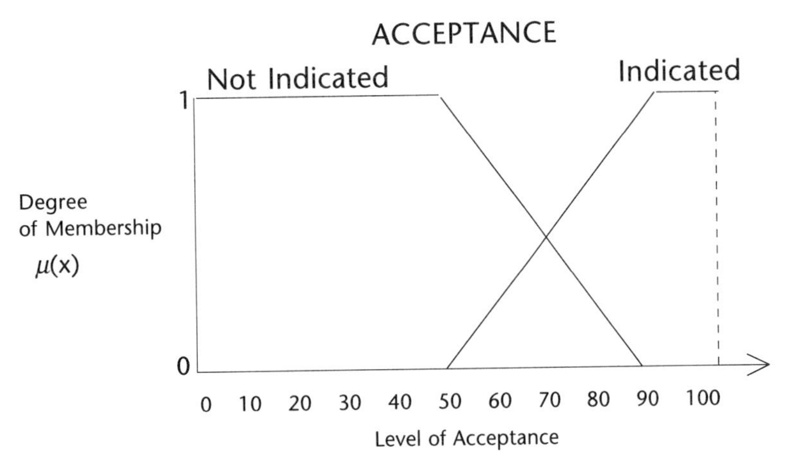

Figure 3.20 The *Acceptance* Psychometric Fuzzy Set

In Figure 3.20, the domain runs from 0 to 100 indicating the degree to which the concept *Acceptance* is indicated or not indicated. The range of the psychometric scale is determined by the level of granularity and fine detail in the model (an issue we will take up shortly). Both objective and psychometric domains can be combined in a fuzzy model. As an example, consider a small expert system that selects candidates for a basketball team based solely on height. Figure 3.21 shows the semantics for player height.

We can now map each player's height to the degree of acceptance through a set of rules using the psychometric scale. In its simplest form, the system might contain just two rules:

```
if Height is Tall, then Acceptance is Indicated;
if Height is Short, then Acceptance is Not Indicated;
```

The taller the individual, the more evidence exists that he or she should be accepted on the team; and, conversely, the shorter the individual, the less evidence exists. The degree of evidence is reflected in the psychometric scale underlying the *Acceptance* parameter. The higher the number, the more *Acceptance* is *Indicated.* As we evolve our expert system, we might begin to add rules that collect better evidence (or we might refine our judgment of who should be a candidate for the basketball team). As an example:

```
if Height is somewhat Tall
   then Acceptance is slightly Indicated
if Height is very Tall
   then Acceptance is positively Indicated
```

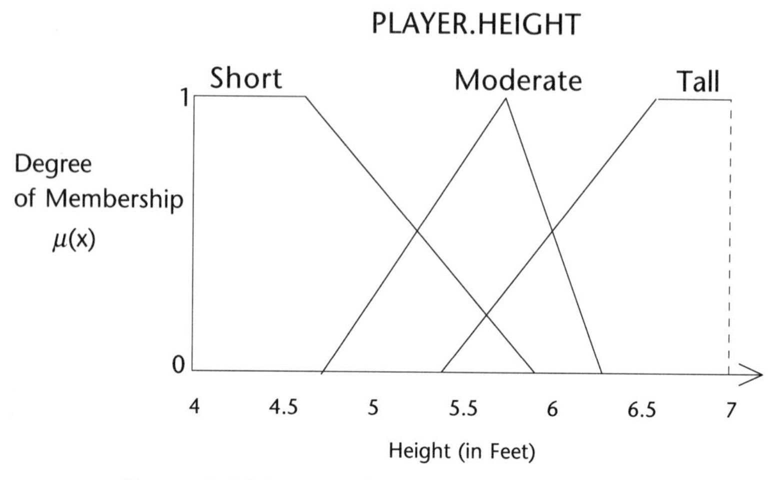

Figure 3.21 Fuzzy Set for Basketball Player Height

```
if Height is above Moderate
   then Acceptance is generally Indicated
```

(Refer to Chapter 5, *Fuzzy Set Hedges*, for a discussion of the qualifiers *slightly, positively,* and *generally.*) In each of these examples, we are connecting measurements in the real world into a conceptual framework that isolates a segment of our psychometric scale. While many models use the psychometric scaling in the consequent of the rule set, the same kind of unitless inferencing can be used on the antecedent side of the rule process. As a simple example, Figure 3.22 illustrates a collection of fuzzy sets used to evaluate the results from our basketball player selection model.

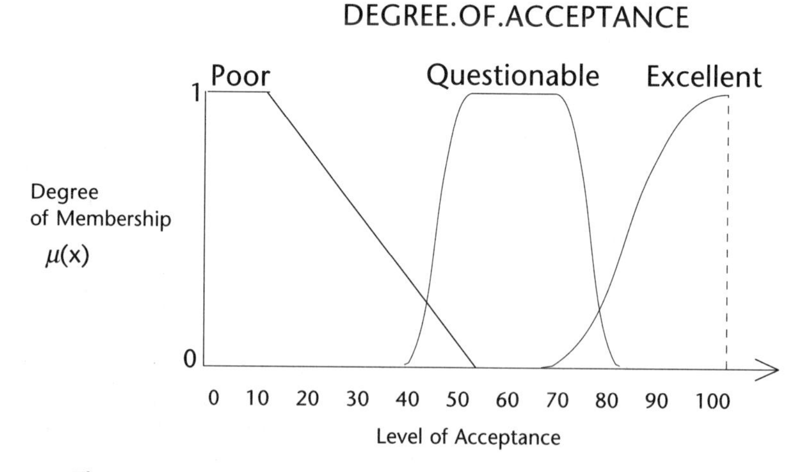

Figure 3.22 Antecedent Fuzzy Sets for *Acceptance* Fuzzy Model

The final result of the selection model is a scalar number drawn from the psychometric scale interval [0,100] (see the discussion of defuzzification in Chapter 6, *Fuzzy Reasoning*). The *Degree.of.Acceptance* then takes this scalar value and runs it through a post-processing fuzzy model to determine whether or not the candidate should be interviewed (or just signed to play). These rules (extremely simplified) might appear as:

```
Degree.of.Acceptance = Acceptance
if Degree.of.Acceptance is Questionable
   then Interview is Needed;
if Degree.of.Acceptance is Excellent
   then Interview is not Needed;
```

Here, *Needed* and *Not Needed* are, of course, fuzzy sets in the *Interview* variable overlaid with their own psychometric scale. This small set of rules is evaluated to produce another number that indicates the degree to which an interview should be conducted.

The range of the psychometric scale depends on the desired level of granularity. The smaller the overall range of values, the fewer values that will be generated. Figure 3.23 illustrates the same fuzzy sets mapped to different psychometric domain scales.

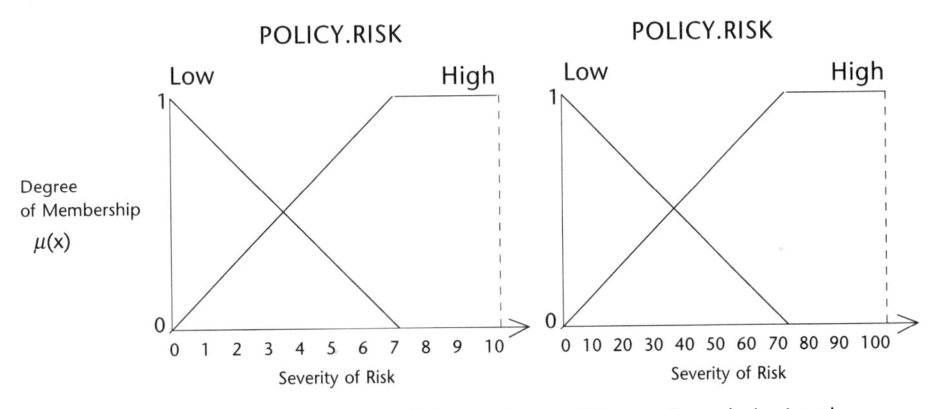

Figure 3.23 Psychometric Risk Fuzzy Sets at Different Granularity Levels

The granularity of the domain is especially important for calibrating the mechanics of the inference engine insofar as they affect the number of outcome states available for the solution variable. As an example, in a vehicle underwriting expert system, consider a straightforward rule such as:

```
if Client.Age is Young then Policy.Risk is High
```

A slight change in the antecedent's degree of membership in *Young* produces different results in the left or right *High* fuzzy set shown in Figure 3.23. The left *High* skips from one complete decimal class to the next (from 5 to 6), while the right *High* moves within the same decimal range (51 to 55, as an example). On a scale of [0,1000], the movement will be even smaller (relative to the range of the domain).

Psychometric scaling plays an important role when we are constructing fuzzy models in population dynamics, urban planning, econometrics, risk assessment, threat and threat classification analysis, social behavior models, and other areas where the model is resolving conceptual problems. The use of arbitrary scaling for conceptual domains has been only sparsely discussed in literature and is little understood by model designers working outside the control and physical problem arenas.

FUZZY ALPHA-CUT THRESHOLDS

A technical concept closely related to the support set is the alpha-level set or the "α-cut." An alpha level is a threshold restriction on the domain based on the membership grade of each domain value. This set contains all the domain values that are part of the fuzzy set at a minimum membership value of α. There are two kinds of α-cuts: weak and strong. The strong α-cut is defined as:

$$\mu_A(x) \geq \alpha$$

The weak alpha-cut is defined as:

$$\mu_A(x) > \alpha$$

This threshold cut restricts the considered domain of the fuzzy set. As an example, for an α-cut of [.2], the heavy fuzzy set is restricted to the domain of 210 to 280 pounds (Figure 3.24).

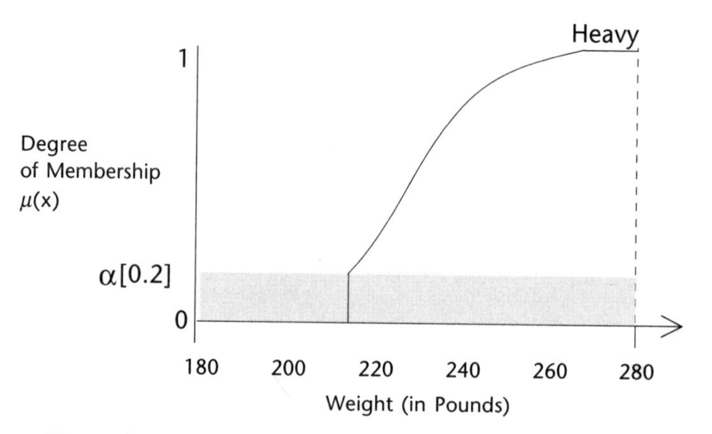

Figure 3.24 An Alpha-cut Threshold for the Fuzzy Set *Heavy*

But why should we care about α-cuts? There are two reasons. First, the strong α-cut at zero [0] defines the support set for a fuzzy set. You can see this easily by comparing the α-level set produced by this slicing function (Figure 3.19) and the support set for *Heavy* in Figure 3.25.

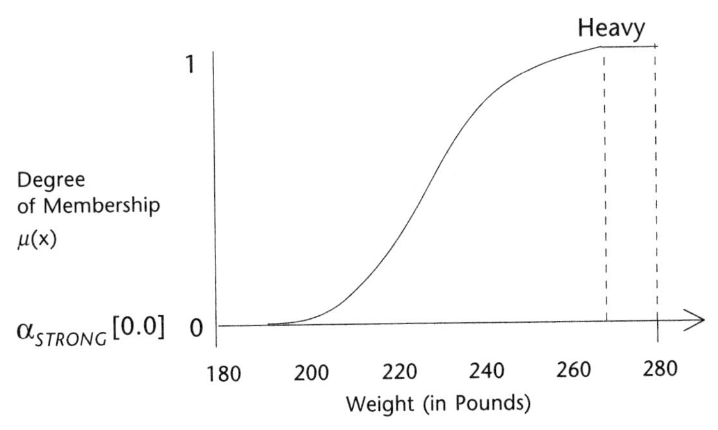

Figure 3.25 The Support Set for Heavy as an Alpha-cut Threshold

Second, the alpha-level set describes a power or strength function that is used by fuzzy models to decide whether or not a truth value should be considered equivalent to zero [0]. This is an important facility that controls the execution of fuzzy rules as well as the intersection of multiple fuzzy sets. This latter property of alpha cut brings us to another and often critical problem in fuzzy modeling.

Alpha Cuts, Transition Walls, and Control Voids

Moving the alpha cut upwards increases the hurdle threshold for a value's membership in a particular fuzzy set. This action profoundly affects the way rules are selected and fired in a fuzzy model. However, when dealing with a set of overlapping fuzzy sets that describe the semantics of a model variable, some care must be exercised in how high the alpha threshold is raised. Otherwise, we can introduce control voids in the model. As an example, consider Figure 3.26, which shows the term set for the variable *Temperature*.

In our previous discussion of fuzzy sets, we have made an unspoken assumption that the alpha threshold for the sets is essentially [0]. By moving the alpha threshold upward we begin to change the way fuzzy regions overlap, and, consequently, we sharpen the discontinuity between states in the model. This can, and usually does, lead to unrecognized transition walls in the variable. A transition wall is a region or place of abrupt discontinuity between adjoining fuzzy sets. As an example, Figure 3.27 shows the term set for *Temperature* when the alpha cut threshold is moved to [.40].

Raising the alpha threshold creates transition walls inside the collection of fuzzy sets. In essence, the alpha threshold becomes the floor of the fuzzy set, with a reduced overlap, and with a minimum truth across all the sets of [.40]. This has obvious effects on the performance of the fuzzy model. And, as Figure 3.28 shows, increasing the alpha threshold to the [.50] membership level, where the fuzzy sets intersect, creates a sharp bifurcation of the control regions.

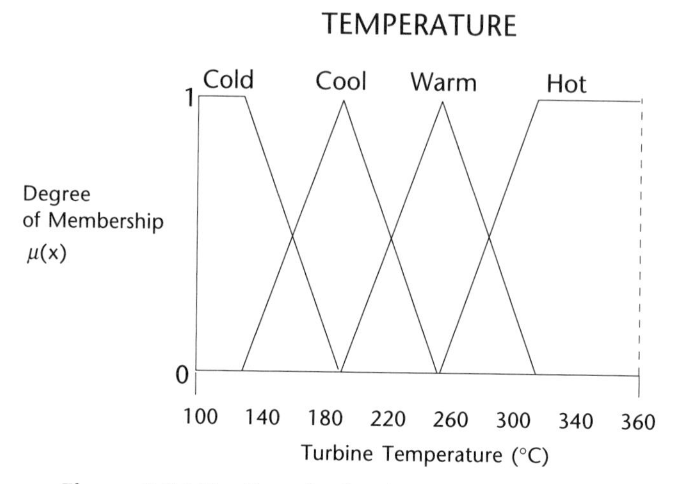

Figure 3.26 The Term Set for the Variable *Temperature*

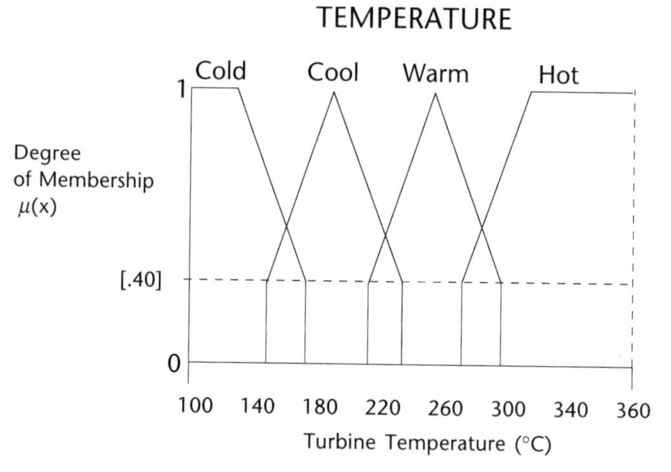

Figure 3.27 Transition Walls in the *Temperature* Variable

At this point, the overlap between fuzzy sets is eliminated. As a result of the missing overlap, control jumps from one fuzzy region to the next. The fuzzy sets then become simple interval spans with varying degrees of membership in the interval. Without the overlap of fuzzy sets, a single variable value cannot fire multiple rules associated with multiple model states.

An even more serious problem arises when the alpha threshold is moved above the intersection point of one or more fuzzy sets (in our example, all the underlying fuzzy sets

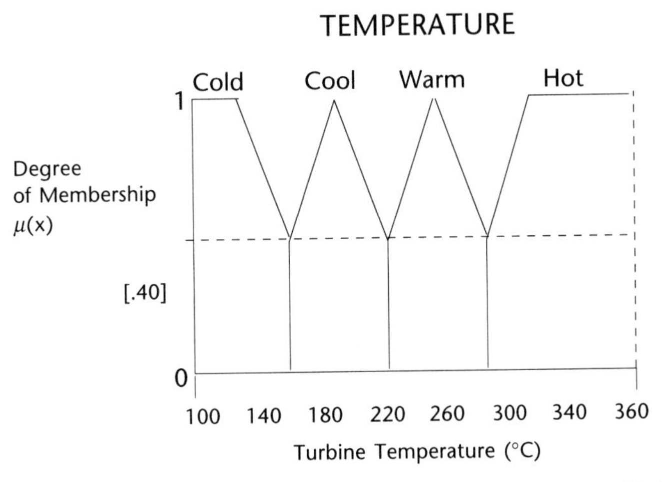

Figure 3.28 Bifurcation Transition Walls in the *Temperature* Variable

have the same intersection point, but this is not a requirement of fuzzy models; for example, refer back to Figure 3.21). As Figure 3.29 shows, we then introduce control voids or dead zones into the model.

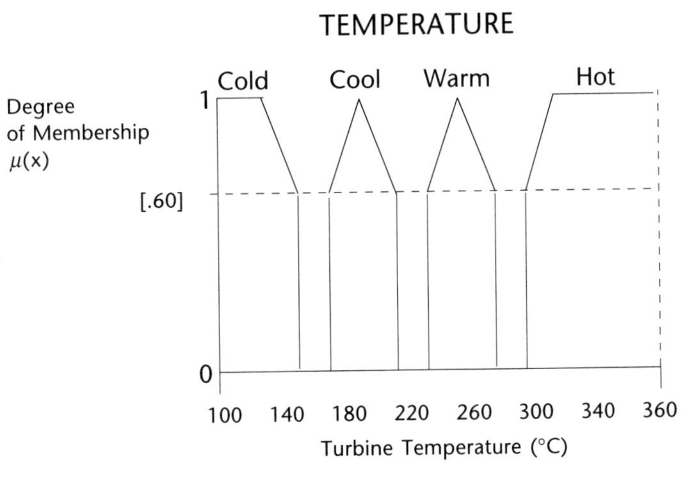

Figure 3.29 Control Voids in the *Temperature* Variable

Control voids introduce serious errors in a fuzzy model. Consider the following two rules from a fuzzy turbine control system:

```
if temp is Cold then ThrottleAction is PL
if temp is Cool then ThrottleAction is PM
```

PL is a large positive movement and *PM* is a moderate positive movement. When the temperature moves between 140 and 160 degrees Celsius, there are no rules that govern the turbine action. Both the *Cold* and *Cool* fuzzy sets have zero memberships during this interval. Thus, while we can use the alpha-cut threshold to focus a model on data that lies well within the control region of a fuzzy set, and hence modify the way it is executed, the same ability can, if used without care, severely disrupt the model's sensitive control space.

The functions used to find the support sets and the alpha-cut thresholds are closely related. We will use them often when applying operators to fuzzy sets, deciding whether or not to execute a rule, and in evaluating the final output from a fuzzy model.

```
#include "fuzzy.hpp"
#include "mtypes.hpp"
bool FzyAboveAlfa(float memval,float Alfa,int AlfaType)
  {
   switch(AlfaType)
    {
     case STRONG:
      if(memval>=Alfa)  return(TRUE);
        else            return(FALSE);
     case WEAK:
      if(memval>Alfa)   return(TRUE);
        else            return(FALSE);
     default:
      return(FAILED);
    }
  }
```

Listing 3.6 Find Memberships Above the Alpha Cut

```
#include "FDB.hpp"
#include "fuzzy.hpp"
#include "mtsptype.hpp"
void FzyApplyAlfa(
     FDB* FDBptr,const float Alfa,const int AlfaType)
  {
     int i;
     for(i=0;i<VECMAX;i++)
        if(!FzyAboveAlfa(FDBptr->FDBvector[i],Alfa,AlfaType))
            FDBptr->FDBvector[i]=0;
  }
```

Listing 3.7 Apply an Alpha Cut to a Fuzzy Set

```
#include "FDB.hpp"
#include "mtsptype.hpp"
void FzySupportSet(
      FDB* FDBptr,double Scalars[],int Truth[],int* statusPtr)
*statusPtr=0;
int i,j;
for(i=0;i<VECMAX;i++)
    {
      if(FDBptr->FDBvector[i]>0)
        {
         for(j=VECMAX-1;j>=0;j--)
          {
           if(FDBptr->FDBvector[j]>0)
             {
              Truth[0]  =i;
              Scalars[0]=FzyGetScalar(FDBptr,i,statusPtr);
              Truth[1]  =j;
              Scalars[1]=FzyGetScalar(FDBptr,j,statusPtr);
              return;
             }
          }
        }
    }
  }
```

Listing 3.8 Alpha Cut and Fuzzy Support Region Routines

The *FzyAboveAlfa* function examines a particular truth membership array element. If the element is above the specified type of alpha cut (either *STRONG* or *WEAK*) the function returns true; otherwise, it returns false. This function is used by the *FzyApplyAlfa* routine to process a fuzzy set's truth membership array. Any truth function array element that is not above the specified alpha-cut threshold is set to zero.

Finding the support region in a fuzzy set is done through the *FzySupportSet* function. Here we simply start from the left-hand side of the set and move to the right until a nonzero truth membership value is found. We then move from the extreme right of the set toward the left until a nonzero membership value is found. These two positions define the support set. The function returns the index of the left and right *FDBvector[]* positions as well as the actual domain values at these points.

Encoding Information with Fuzzy Sets

Now that we know what a fuzzy set is, we might reasonably want to know how it represents knowledge. That is, what does it look like, and how is the shape of a fuzzy set related to the knowledge in a fuzzy model? We have already encountered a wide spectrum of fuzzy sets in the previous chapter on fuzziness and uncertainty and also in the first part of this chapter. These fuzzy sets all seem to share a familiar set of shapes: bell curves, lines, S-curves, and an occasional trapezoid. But the surface of a fuzzy set, the part of the set that actually defines the membership function, can assume a wide range of shapes. In fact, although there are some standard fuzzy shapes, the actual contour of a set is based completely on the semantics of the concept it is intended to represent. Which brings us to an important point:

There are no universal or pre-defined fuzzy sets.

A fuzzy set has no meaning except in the context of the model. This does not mean that the knowledge engineer is free to chose a shape at random; it means that certain shapes are representative of particular classes of knowledge. Within each class, of course, the exact width, slope, plateau structure, and breadth of a fuzzy set must be adjusted to match the target concept. In some rare cases, when none of the standard shapes are adequate, a designer must create a new fuzzy surface by specifying either the membership of some data points in the domain or else a set of membership values that are spaced equidistantly across the domain. We now turn to the two basic kinds of fuzzy sets and then to an examination of the various standard fuzzy set shapes. We'll end this chapter with two methods of defining irregular fuzzy sets.

Expressing a Fuzzy Concept

Almost always (with a few minor exceptions) the surface of a fuzzy set is a continuous line from the left to the right edge of the set. The contours of a fuzzy set represent the semantic properties of the underlying concept, so the closer the set surface maps to the behavior of a physical or conceptual phenomenon, the better our model will reflect the real world. Understandably, the method of recognizing fuzzy attributes and drafting the fuzzy set is an important technique. As we will see, however, when we turn to building fuzzy models, fuzzy systems are tolerant of approximations not only in their problem spaces but also in the representation of fuzzy sets. This means that they perform well even when the fuzzy set does not map exactly with its model concept. There are, generally, two kinds of fuzzy sets: numbers and quantifiers. Both are used in the design of robust fuzzy models, but they have their own use and their own structure.

Fuzzy Numbers

Fuzzy numbers are sets that represent an approximate numeric quantity. These are convex fuzzy sets. Representing a fuzzy number can be done in a variety of ways. Figure 3.30 shows the concept *Around 5* modeled as a bell curve. Figure 3.31 shows the same concept modeled as a triangular space (a popular method of fuzzy representation used in the control engineering environment). When we overlay the two conceptual representa-

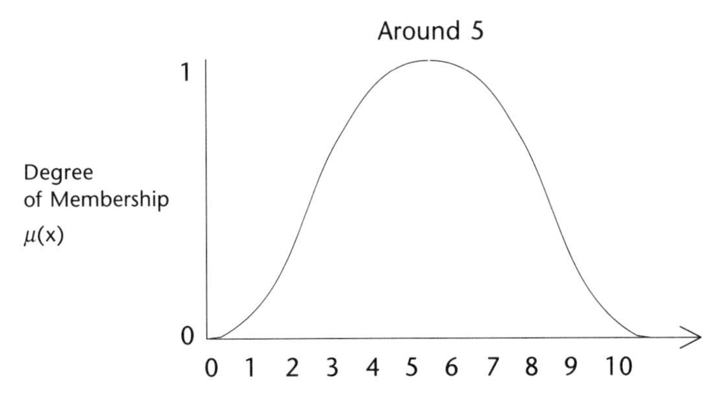

Figure 3.30 *AROUND 5* as a Bell Curve

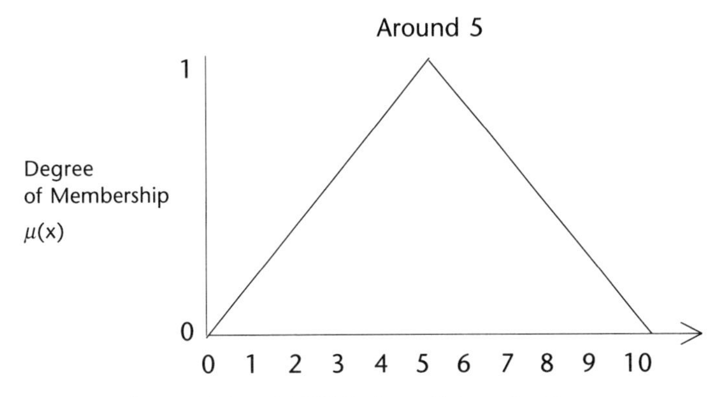

Figure 3.31 *AROUND 5* as a Triangular Curve

tions, as shown in Figure 3.32, the difference is apparent. Yet, in practice, fuzzy models are rarely sensitive to this kind of elasticity in the fuzzy set descriptions. Such insensitivity makes fuzzy models quite robust and resilient, which are important properties when models are initially prototyped.

Numbers on bell and triangular curves have a single point where their truth value is [1.0], generally at the domain point whose value the set is approximating. This kind of approximation is not always reasonable or appropriate. You can represent a broader class of fuzzy numbers as platykurtic ("fat") bell curves or trapezoids. An example from this class of numbers is illustrated in Figure 3.33.

Trapezoidal numbers play an important role in many business, public policy, and general decision models where both antecedent quantities and outcome quantities are grouped

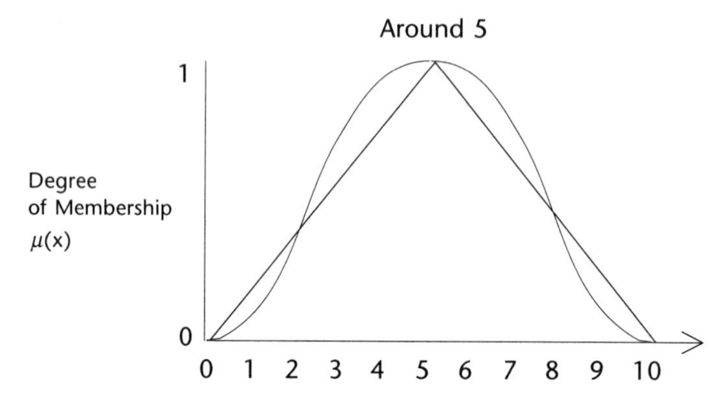

Figure 3.32 The Difference in *AROUND 5* Representations

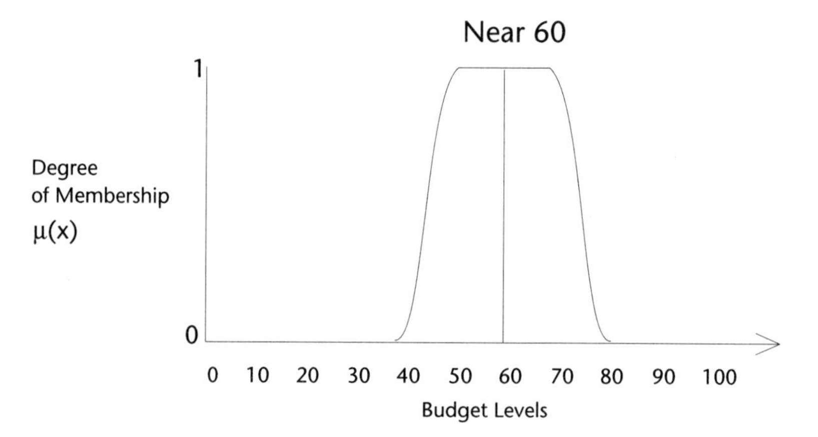

Figure 3.33 Near (Broadly Around) Sixty as a Trapezoidal Fuzzy Set

into intervals, clusters, or categories. A value within the category has complete membership in the category, but, as we move away from the category boundaries, the degree of membership drops off rapidly.

Fuzzy Qualifiers

Fuzzy qualifiers are fuzzy sets that are not fuzzy numbers. Qualifiers represent open-ended concepts and are expressed in fuzzy logic primarily through linear and S-curve sets. These sets provide the framework for describing generally unbounded concepts (or concepts that are theoretically unbounded). Figure 3.34 shows examples of linear fuzzy sets.

We now examine how fuzzy sets are constructed and how they encode a particular kind of information. First, we will look at the "standard" fuzzy shapes used most often in

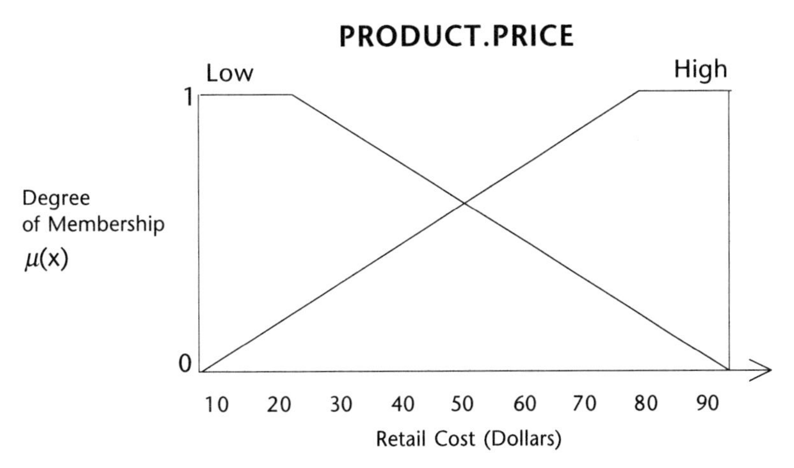

Figure 3.34 Low and High Product Price

fuzzy models—lines, S-curves, and bell shapes. Next, we will look at how more specialized fuzzy sets are constructed and what they mean. Bear in mind that this construction process encodes information about the underlying semantic concept, and the actual meaning of a fuzzy set is dependent on the model context.

Generating Fuzzy Membership Functions

> The fuzzy set membership functions described in this section change the fuzzy set curve surface; they do not create the actual `FDB`. We take this approach so that you can easily alter the set characteristics without affecting the storage allocated to an existing fuzzy set. Separating the truth membership generator from the memory manager for fuzzy sets provides a wealth of benefits, allowing us to fine-tune fuzzy sets, change the shape of a fuzzy set, and work with adaptive, genetically optimized, and self-organizing fuzzy systems (although these techniques themselves are not covered in this handbook).

We assume that memory is allocated for a fuzzy set and its address is stored in a pointer variable. Throughout this book, we use the naming convention **FDBptr* as a standard reference to a fuzzy set. Listing 3.9 illustrates a code fragment that allocates memory for a fuzzy set and calls the fuzzy set initializer to fill out each element in the `FDB` structure.

```
//--Create a new fuzzy set
//--and initialize it to the default state
```

```
if(!(FDBptr=new FDB))
  {
   *statusPtr=1;
   MtsSendError(2,PgmId,SetId);
   return(FDBptr);
  }
FzyInitFDB(FDBptr);
```

Listing 3.9 Allocating and Initializing Storage for a Fuzzy Set

We always protect the memory allocation process by checking for a *NULL* returned pointer value. If insufficient main memory exists, the error condition is raised and we take steps to terminate the process. When the new FDB structure is allocated, the *FzyInitFDB* procedure must be called to initialize all the fields and set the internal chain pointer.

Linear Representations

The linear proportionality surface is a straight line. This is perhaps the simplest fuzzy set and often a good choice when approximating an unknown or poorly understood concept that is not a fuzzy number. There are two states of a linear fuzzy set. The increasing set starts at a domain value that has zero membership in the set and moves to the right with values that have increasing set membership. The right-hand edge of the domain is the value with full membership (Figure 3.35).

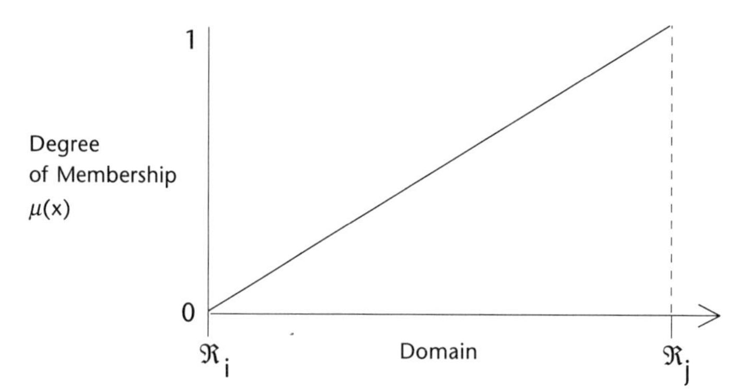

Figure 3.35 An Increasing Fuzzy Set

The decreasing fuzzy set, shown in Figure 3.36, is the opposite of the increasing set. The value at the left-hand edge of the domain has complete set membership while the value at the right-hand side of the edge has no membership.

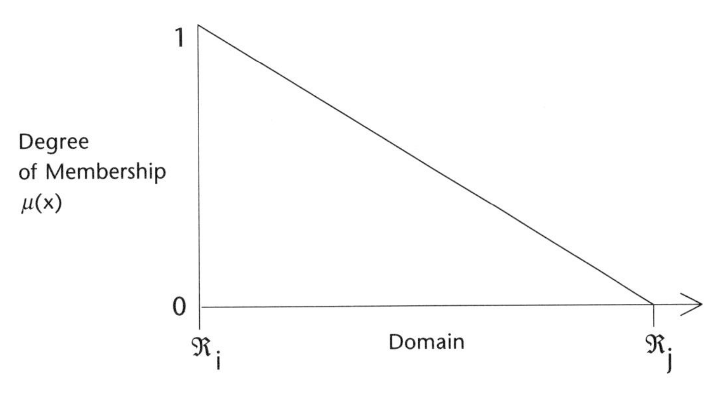

Figure 3.36 A Decreasing Fuzzy Set

Since the linear fuzzy shape connects a square fuzzy space at a 45-degree angle, the truth of any domain value is proportional to its distance (displacement) in the value axis.

The linear fuzzy set generator can produce both kinds of fuzzy sets. The parameter `INCREASE` produces a fuzzy set that goes from false to true, while the `DECREASE` parameter produces a fuzzy set that goes from true to false.

```
#include "fuzzy.hpp"
#include "FDB.hpp"
#include "mtsptype.hpp"
void FzyLinearCurve(
  FDB* FDBptr,
    double low,double high,int LineType,int *statusPtr)
  {
    int     i,j;
    double  swidth,
            dwidth,
            memval;
    char   *PgmId="mtfzlns";

    *statusPtr=0;
    if((low==high)||(low>high))
      {
       *statusPtr=1;
       MtsSendError(36,PgmId,FDBptr->FDBid);
       return;
      }
    swidth=high-low;
```

```
   dwidth=FDBptr->FDBdomain[1]-FDBptr->FDBdomain[0];
   for(i=0;i<VECMAX;i++)
      {
       memval=FDBptr->FDBdomain[0]+(float)i/VECMAX*dwidth;
       FDBptr->FDBvector[i]=0.0;
       if(memval>high) FDBptr->FDBvector[i]=1.00;
         else if(memval>low)
                FDBptr->FDBvector[i]=(memval-low)/swidth;
      }
   if(LineType==DECREASE)
      for(j=0;j<VECMAX;j++)
         FDBptr->FDBvector[j]=1-FDBptr->FDBvector[j];
   return;
 }
```

Listing 3.10 Creating a Linear Fuzzy Set

At the heart of this algorithm is the difference between the scope of the actual curve (held in the *swidth* variable) and the range of the fuzzy set's domain (held in the *dwidth* variable). The program creates a line that is proportional to the length of this actual curve but fills in the line with one and zero values above and below the line boundaries. As an example, the following figures (3.37, 3.38, and 3.39) are fuzzy sets created with a domain of [16,36] and the *low* and *high* parameters of *FzyLinearCurve* set to the same values.

```
double loEdge=16,HiEdge=36;
FDBptr->FDBdomain[0]=16;
FDBptr->FDBdomain[1]=36;
FzyLinearCurve(FDBptr,loEdge,HiEdge,INCREASE,&status)

FuzzySet:    Linear Up
       Description:
      1.00                                                                       .
      0.90                                                                 ......
      0.80                                                             .....
      0.70                                                        ......
      0.60                                                   ......
      0.50                                              ......
      0.40                                         ......
      0.30                                     .....
      0.20                                ......
      0.10                           ......
      0.00...........
         0---|---|---|---|---|---|---|---|---|---|---|---|---|---|---|---0
       16.00   18.50   21.00   23.50   26.00   28.50   31.00   33.50   36.00
       Domained:        16.00 to       36.00
```

Figure 3.37 A Linear Increasing Fuzzy Set

```
double loEdge=16,HiEdge=36;
FDBptr->FDBdomain[0]=16;
FDBptr->FDBdomain[1]=36;
FzyLinearCurve(FDBptr,loEdge,HiEdge,DECREASE,&status)
```

```
     FuzzySet:     Linear Down
     Description:
   1.00........
   0.90          .....
   0.80               ......
   0.70                     ......
   0.60                           ......
   0.50                                 .....
   0.40                                      ......
   0.30                                            ......
   0.20                                                  ......
   0.10                                                        ......
   0.00                                                              .....
       0---|---|---|---|---|---|---|---|---|---|---|---|---|---|---|---0
     16.00   18.50   21.00   23.50   26.00   28.50   31.00   33.50   36.00
     Domained:         16.00 to      36.00
```

Figure 3.38 A Linear Decreasing Fuzzy Set

```
double loEdge=16,HiEdge=36;
FDBptr->FDBdomain[0]=16;
FDBptr->FDBdomain[1]=42;
FzyLinearCurve(FDBptr,loEdge,HiEdge,INCREASE,&status)
```

```
     FuzzySet:     Linear Up
     Description:
   1.00                                                       ................
   0.90                                                   ....
   0.80                                               ....
   0.70                                          .....
   0.60                                      ....
   0.50                                 .....
   0.40                             ....
   0.30                        .....
   0.20                    ....
   0.10               .....
   0.00.........
       0---|---|---|---|---|---|---|---|---|---|---|---|---|---|---|---0
     16.00   19.25   22.50   25.75   29.00   32.25   35.50   38.75   42.00
     Domained:         16.00 to      42.00
```

Figure 3.39 Linear Fuzzy Set Within a Larger Domain

However, if the domain of the fuzzy set is set to [16,42] while the parameters *low* and *high* are set to [16,36], respectively, then the fuzzy set is drawn within the domain.

Many models uses a special linear fuzzy set called *True*. This fuzzy set provides an important calibration metric in, as an example, fuzzy database operations. This linear set (see the procedure in Listing 3.11) shows another alternative method of creating a proportionally spaced linear fuzzy set.

```
#include "FDB.hpp"
#include "mtsptype.hpp"
#include "mtypes.hpp"
void FzyTrueSet(FDB* FDBptr)
  {
   int i;
   FzyInitFDB(FDBptr);
   strcpy(FDBptr->FDBid,"TRUE");
   strcpy(FDBptr->FDBdesc,"The Orthogonal Truth Fuzzyset");
   FDBptr->FDBdomain[0]=0;
   FDBptr->FDBdomain[1]=1
   for(i=0;i<VECMAX;i++)
      FDBptr->FDBvector[i]=(((float)i)/VECMAX-1)*100;
   return;
  }
```

Listing 3.11 Creating the True Fuzzy Set

```
  FuzzySet:     TRUE
  Description: The Orthogonal Truth Fuzzyset
1.00                                                                    .
0.90                                                               ......
0.80                                                          .......
0.70                                                     ......
0.60                                               .......
0.50                                          ......
0.40                                     ......
0.30                               .......
0.20                          ......
0.10                    .......
0.00......
    0---|---|---|---|---|---|---|---|---|---|---|---|---|---|---|---|---0
  0.00     0.13     0.25     0.38     0.50     0.63     0.75     0.88     1.00
  Domained:          0.00 to          1.00
```

Figure 3.40 The Proportional *True* Fuzzy Set

The *True* fuzzy set (shown in Figure 3.40) is a square fuzzy space (N*N) giving the proportion of truth for each truth membership value. The *True* (truth proportional) fuzzy set permits us to reuse the truth of a fuzzy proposition in the decision-making process. Thus a propositional rule such as

if x is TRUE then s is W;

where x is the truth of a fuzzy proposition, can be used to ordinally rank the result of a fuzzy selection or to test for a minimum degree of truth.

S-Curve (Sigmoid/Logistic) Representations

The GROWTH and DECLINE curves are sigmoid curves or S-curves, that correspond to the increasing and decreasing nonlinear surfaces. A growth S-curve set moves from no membership at its extreme left-hand side to complete membership at its extreme right-hand side. The membership function is pivoted around its 50% membership point, called the *inflexion point*. Figure 3.41 illustrates the shape of a growth S-curve.

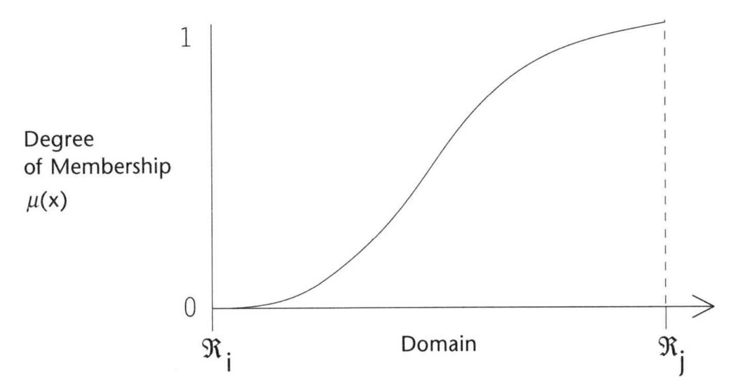

Figure 3.41 A Growth or Right-Facing S-Curve Fuzzy Set

The decline S-curve set moves from complete membership at its extreme left-hand side to zero membership at its extreme right-hand side. Figure 3.42 illustrates the shape of a declining S-curve.

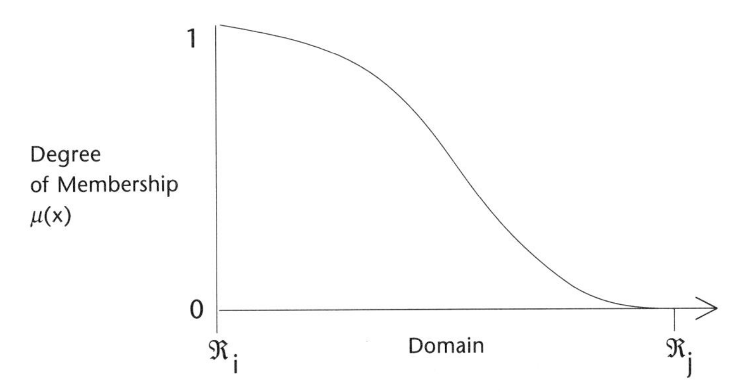

Figure 3.42 A Decline or Left-Facing S-Curve Fuzzy Set

An S-curve is defined using three parameters: its zero membership value (α), its complete membership value (γ), and a third piece of information—the inflection (or crossover) point (β). This is the point at which the domain value is 50% true. The value of the curve for domain point x is given in Equation 3.3.

$$S(x;\alpha,\beta,\gamma)=\begin{bmatrix}0 & \rightarrow x\leq\alpha \\ 2((x-\alpha)/(\gamma-\alpha))^2 & \rightarrow \alpha\leq x\leq\beta \\ 1-2((x-\gamma)/(\gamma-\alpha))^2 & \rightarrow \beta\leq x\leq\gamma \\ 1 & \rightarrow x\geq\gamma\end{bmatrix}$$

Equation 3.3 The S-Curve Formula

You can see clearly the characteristics of the S-curve in the schematic of Figure 3.43. The inflexion point is chosen by the analyst according to the apparent, suspected, or known distribution of the population.

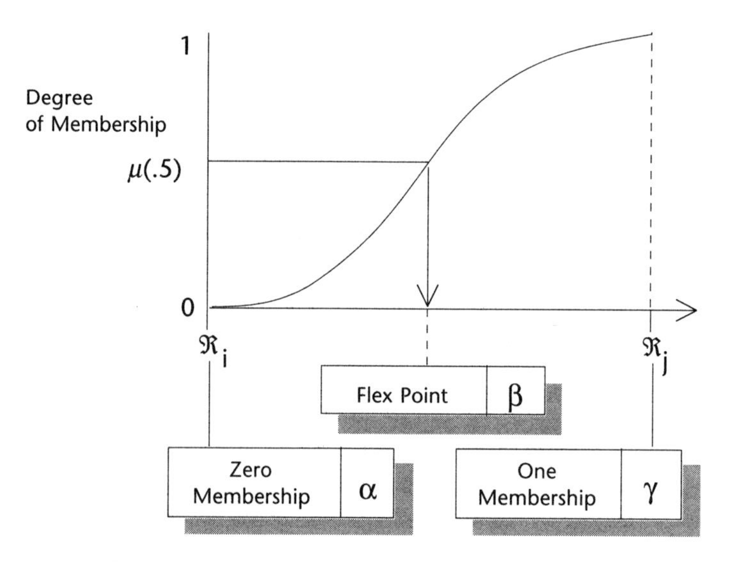

Figure 3.43 The S-Curve Function Characteristics

Growth curves play an important part in modeling population dynamics where the sampling of individual values approximates a continuous random variable. Examples of such systems and events include:

- the speed of a fighter jet
- the acceleration of a falling object
- the length of time between radioactive decays
- average income of executives on the East Coast
- the mean-time-between-failure (MTBF) of a hard disk drive
- an individual's height and weight

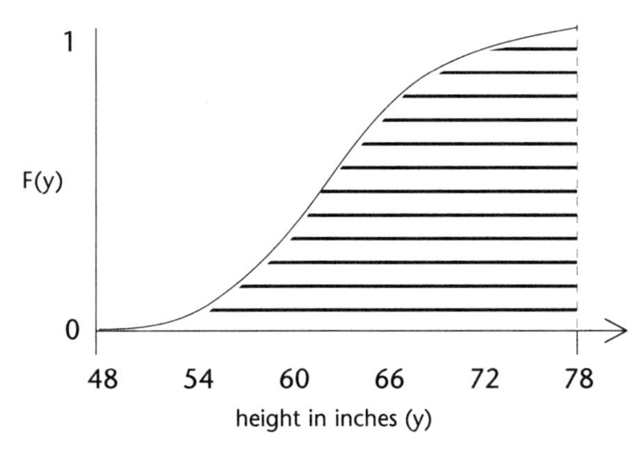

Figure 3.44 Cumulative Distribution Function for a Sample Population

- the interarrival time of the queue at a bank teller's window
- the point at which an artillery shell strikes its target
- the risk associated with morbidity underwriting

S-CURVES AND CUMULATIVE DISTRIBUTIONS

In the S-curve, as we shall see, the continuous cumulative distribution function [$F(y)$] for the underlying concept variable takes the form of an S-curve. Equation 3.4 is the definition for this function.

$$f(y) = prob(X \leq y)$$

Equation 3.4 The Probability Cumulative Distribution

In equation 3.4, y is a value from the underlying domain, and $\mathbf{X}$ is a randomly selected value from this domain. The function (Prob) returns the probability that $\mathbf{X}$ is less than or equal to y. Let's see how this works in practice by examining its relationship to the concept *Tall*. Figure 3.44 shows the cumulative distribution function for the heights of an arbitrary population sample.

This curve shows the cumulative probability of a sampled height being equal to or greater than any height selected from the domain. As an illustration, if we pick two individuals from our population sample—X_1 with a height of 57 inches and X_2 with a height of 69 inches—the cumulative distribution tells us that the first individual's height is greater than 38% of the population and the second individual's height is greater than 91% of the same population (Figure 3.45).

In effect then we are accumulating the probabilities as the heights move toward larger and larger numbers. The curve reaches one [1] at a point where the sampled height is equal to or greater than the tallest individual. The curve is at zero [0] when the height is less than the smallest individual. Of course, this is a behavior of cumulative distributions—as y becomes very large, $F(y)$ approaches one; and as y becomes increasingly small, $F(y)$ approaches zero.

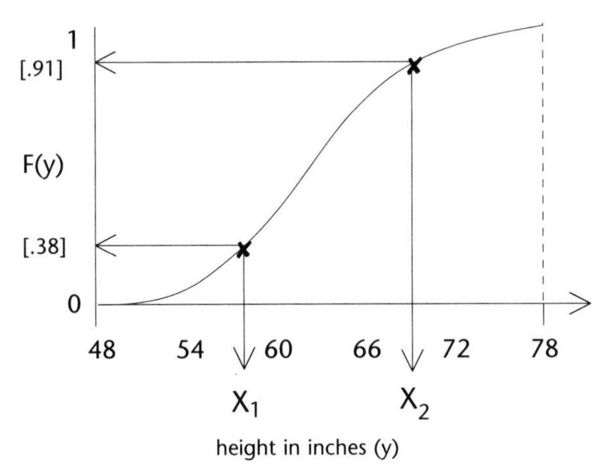

Figure 3.45 Probing a Cumulative Distribution Curve for X_1 and X_2

Let's take another look at the cumulative distribution function. Suppose we take a slightly different semantic track and redefine the function in this way:

$$f(R_i) = prob(X \leq R_i)$$

where we are now sampling any randomly chosen value from the domain. The function now returns the possibility that this value is regarded as *Tall*. This begins to look suspiciously like the membership function of a fuzzy set. Notice that we have defined *Tall* as a function of height drawn from the sample population. *Tall* is no longer a simple linear line, but reflects the underlying characteristics of the population. In this case, we take as absolutely *Tall* those heights whose cumulative frequency distribution places them above the .95 interval; all other values have some degree of tallness based on their cumulative distribution (Figure 3.46).

Understanding the characteristics of our base concept in the sample population yields a different representation for *Tall*. We have discarded our linear relationship and drawn a contour that reflects the distribution of heights in the sample. We still maintain a clear understanding of what we mean by *Tall*; only our mapping between the truth membership axis and the value of height has been changed.

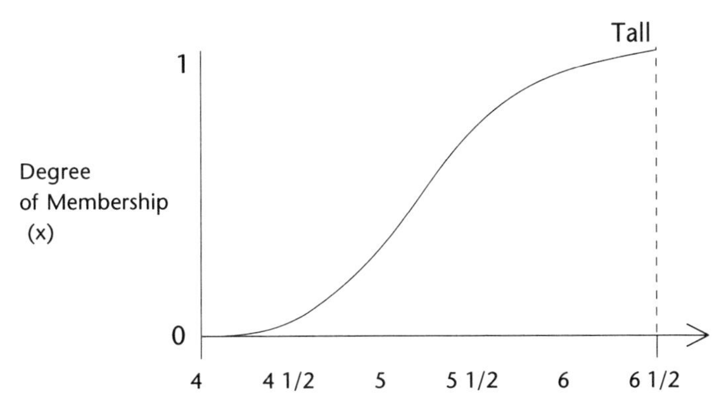

Figure 3.46 The *Tall* Fuzzy Set as an S-Curve

Proportional and Frequency Representations

Another common use of the S-curve is the representation of proportional dependencies and fuzzy sets, used in frequency representations. You can most easily see this occurrence in a fuzzy set like *usually* where the domain represents a fuzzy set of proportions (Figure 3.47).

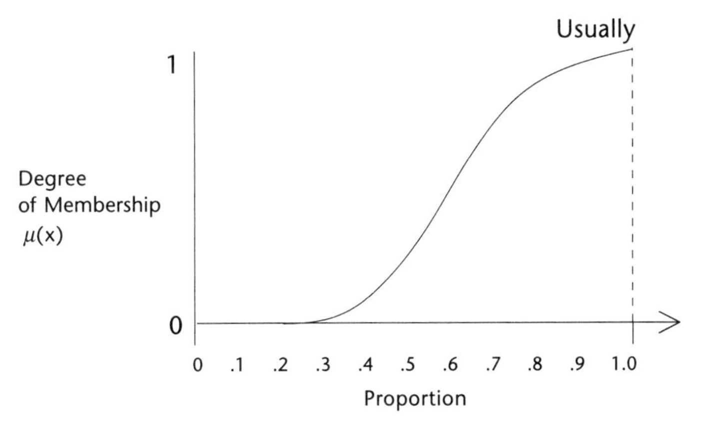

Figure 3.47 The Fuzzy Set *Usually*

Another qualifier used in the analysis of time series and groups is the fuzzy concept of *most*. This is also expressed as an S-curve, with its degree of proportionality skewed toward the right past the 50% population level (Figure 3.48).

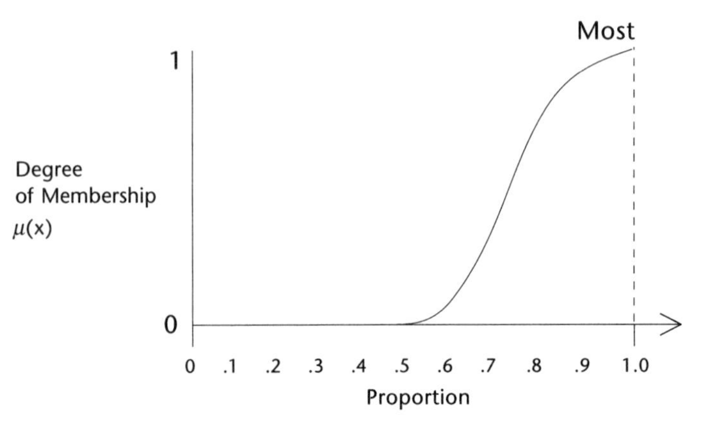

Figure 3.48 The Fuzzy Set *Most*

Other kinds of proportional sets are used to represent concepts like *nearly all* (a curve starting at the [.80] point) and *every* or *always* (a curve starting close to the [1.0] truth level). As we will see in the chapter on fuzzy modeling, these kinds of fuzzy sets provide the foundations for modeling knowledge that is imprecisely conditional. This knowledge is the kind that we use to express everyday thoughts, the basics of common sense. Some of the concepts—rules of thumb—that are usually but not always true include:

- Most projects are overdue.
- Long projects are always overdue.
- Most newly married couples are young.
- Costs usually exceed budgets.
- Snow is usually white.
- The sky is usually blue.
- Most adult Americans have a driver's license.
- Nearly all atomic nuclei contain neutrons.

This type of knowledge is often qualified with other conditionals:

Snow is usually white unless it falls in New York City.
The sky is usually blue unless you are in Great Britain.

As we shall see in the section on fuzzy inference and reasoning, conditional qualifiers and multiple fuzzy statements involving usuality terms lead to a class of ultra-fuzzy implications.

The S-curve fuzzy set generator can produce both left and right facing fuzzy sets. The parameter GROWTH produces a fuzzy set that goes from false to true, while the DECLINE parameter produces a fuzzy set that goes from true to false.

```
#include <math.h>
#include "FDB.hpp"
#include "fuzzy.hpp"
#include "mtsptype.hpp"
void FzySCurve(
      FDB* FDBptr,double left,double flexpoint,double right,
       int CurveType,int *statusPtr)
    {
      int     i;
      char    *PgmId="mtfzscs";
      double  swidth,dwidth,memval,temp1,temp2,temp3;

      *statusPtr=0;
      if((left>right)||(right<flexpoint))
{
          *statusPtr=1;
          MtsSendError(42,PgmId,FDBptr->FDBid);
          return;
         }
      swidth=right-left;
      dwidth=FDBptr->FDBdomain[1]-FDBptr->FDBdomain[0];
      for(i=0;i<VECMAX;i++)
         {
          memval=
            FDBptr->FDBdomain[0]+(float)i/(VECMAX-1)*dwidth;
          if(memval>=right) FDBptr->FDBvector[i]=1.00;
            else
          if(memval>flexpoint)
              {
               temp1=(memval-right)/swidth;
               FDBptr->FDBvector[i]=1-(2*(pow(temp1,2)));
              }
            else
          if(memval>left)
              {
               temp1=(memval-left)/swidth;
               FDBptr->FDBvector[i]=2*(pow(temp1,2));
              }
            else
```

```
            FDBptr->FDBvector[i]=0.0;
        }
    if(CurveType==DECLINE)
        for(i=0;i<VECMAX;i++)
            FDBptr->FDBvector[i]=1-FDBptr->FDBvector[i];
    return;
}
```

Listing 3.12 Creating an S-Curve Fuzzy Set

The core of this program in Listing 3.12 is located at the central *for* loop. Here we implement the decision tree specified in Equation 3.1. In a fashion similar to the linear fuzzy set curve, the S-curve is bounded between the actual domain of the fuzzy set (held in the *dwidth* variable) and the endpoints of the S-curve as it is placed inside the domain (held in the *swidth* variable.) After forming the basic right-facing S-curve, we examine the *CurveType* parameter. If this *CurveType* is *Decline*, we form the new curve by taking the complement of the generated curve.

Figures 3.49 and 3.50 are *Growth* and *Decline* S-curves that illustrate with a small code fragment how the S-curve is set within a wider domain.

```
FDBptr->FDBdomain[0]=16;
FDBptr->FDBdomain[1]=42;
LeftEdge=18;
Flexpoint=29;
RightEdge=40;
FzySCurve(FDBptr,LeftEdge,Flexpoint,RightEdge,GROWTH,&status);
```

```
  FuzzySet:    SCgrw[18,29,40]
  Description:
1.00                                                        .........
0.90                                                  .........
0.80                                             .....
0.70                                         ....
0.60                                      ...
0.50                                   ...
0.40                                ...
0.30                             ...
0.20                         ....
0.10                    .....
0.00.................
    0---|---|---|---|---|---|---|---|---|---|---|---|---|---|---|---0
  16.00   19.25   22.50   25.75   29.00   32.25   35.50   38.75   42.00
  Domained:         16.00 to       42.00
```

Figure 3.49 A *Growth* S-Curve Set Within a Wider Domain

```
FDBptr->FDBdomain[0]=16;
FDBptr->FDBdomain[1]=42;
LeftEdge=18;
Flexpoint=29;
RightEdge=40;
FzySCurve(FDBptr,LeftEdge,Flexpoint,RightEdge,DECLINE,&status);
```

```
  FuzzySet:    SCdcl[18,29,40]
  Description:
1.00.........
0.90         .........
0.80                  .....
0.70                       ....
0.60                           ...
0.50                              ...
0.40                                 ...
0.30                                    ...
0.20                                       ....
0.10                                           .....
0.00                                                .................
    0---|---|---|---|---|---|---|---|---|---|---|---|---|---|---|---0
  16.00   19.25   22.50   25.75   29.00   32.25   35.50   38.75   42.00
  Domained:         16.00 to       42.00
```

Figure 3.50 A *Decline* S-Curve Set Within a Wider Domain

The following S-curve fuzzy sets (as seen in Figures 3.51–3.54) illustrate how the proportionality sets appear. They are generated from the *PROPFZY.CPP* program. Each fuzzy set is positioned within the domain [0,1], with the inflexion point set at the center of the actual curve range. Each proportionality set returns a nonzero truth value with a range corresponding to the truth of another fuzzy proposition. Such composite fuzzy propositions are evaluated with a syntax similar to this code that preceds Figure 3.51:

```
if usually(x is Y) then s is W;
```

```
  FuzzySet:    Usually
  Description:
1.00                                                          .....
0.90                                                   ...........
0.80                                              .....
0.70                                         .....
0.60                                      ...
0.50                                  ....
0.40                               ...
0.30                           ....
0.20                      .....
0.10                ......
0.00..............
    0---|---|---|---|---|---|---|---|---|---|---|---|---|---|---|---0
  0.00    0.13    0.25    0.38    0.50    0.63    0.75    0.88    1.00
  Domained:          0.00 to        1.00
```

Figure 3.51 The *Usually* Proportional Fuzzy Set

This kind of evaluation interprets a higher truth value as a statement about the occurrence of an event in the underlying population. Thus we could use the fuzzy set *Mostly* (see Figure 3.52) to restrict propositions to truth values that fall above the zero truth level of the proportional fuzzy set. As with *Usually*, you can qualify *Mostly* fuzzy propositions:

```
if MOSTLY(x is Y) then s is W;
if MOSTLY(x is NOT Y) then s is W;
```

```
  FuzzySet:    Mostly
  Description:
1.00                                                              ...
0.90                                                         .....
0.80                                                      ...
0.70                                                    ..
0.60                                                  ..
0.50                                                ..
0.40                                               .
0.30                                             ..
0.20                                            .
0.10                                           .
0.00
                                             ..
    0---|---|---|---|---|---|---|---|---|---|---|---|---|---|---|---0
   0.00    0.13    0.25    0.38    0.50    0.63    0.75    0.88    1.00
  Domained:         0.00 to       1.00
```

Figure 3.52 The *Mostly* Proportional Fuzzy Set

```
  FuzzySet:    Nearly All
  Description:
1.00                                                                .
0.90                                                             ...
0.80                                                            .
0.70                                                           .
0.60
0.50                                                          .
0.40                                                         .
0.30
0.20                                                        .
0.10
0.00.......                                                .
    0---|---|---|---|---|---|---|---|---|---|---|---|---|---|---|---0
   0.00    0.13    0.25    0.38    0.50    0.63    0.75    0.88    1.00
  Domained:         0.00 to       1.00
```

Figure 3.53 The *Nearly All* Proportional Fuzzy Set

```
  FuzzySet:    Every
  Description:
1.00                                                                        .
0.90                                                                       .
0.80                                                                      .
0.70
0.60
0.50                                                                    .
0.40
0.30
0.20                                                                   .
0.10
0.00....
    0---|---|---|---|---|---|---|---|---|---|---|---|---|---|---|---0
   0.00    0.13    0.25    0.38    0.50    0.63    0.75    0.88    1.00
  Domained:         0.00 to        1.00
```

Figure 3.54 The *Every* Proportional Fuzzy Set

The fuzzy sets of Figures 3.53 and 3.54 illustrate the restriction proportionality sets for the concepts *nearly all* and *every*. They are used in the same fashion as the previously discussed *usually* and *mostly* fuzzy sets.

Fuzzy Numbers and "Around" Representations

This important class of fuzzy contours represents approximations of a central value and is graphically visualized as a class of "bell" curves. These are, in fact, fuzzy numbers. In general, there are three important classes of bell curves—the PI, the beta, and the Gaussian fuzzy sets. The difference between the three curve types has to do with the slope of the curve as well as the values at the curve endpoints. Fuzzy numbers are also represented with triangular fuzzy sets, especially in the process engineering field. These sets, along with "shouldered" and trapezoidal fuzzy sets, are covered in the next section as special cases of the general fuzzy number and fuzzy quantity sets.

Fuzzy Numbers

Bell-shaped fuzzy sets are used to represent fuzzy numbers as well as fuzzy spaces that are spread around a central value (which itself may be rather fuzzy). The generalized approximation curve indicated by *Around* or *Close to* produces a membership function for the fuzzy assertion representing a fuzzy space of all the numbers that are around or close to *Y*:

X is around Y

This space can be bounded, in the case of PI curves, or unbounded, in the case of beta and Gaussian curves. The width and slope of the bell curves indicate the degree of compactness associated with the fuzzy number.

An example of a fuzzy number is the concept *Approximately 5,* or *Around 5.* Figure 3.55 illustrates this fuzzy set. This set includes all the numbers that are close to the number five. Another way to approach *Around 5* is to think of a set whose numbers could be rounded off into a group of five, if we were separating items by fives.

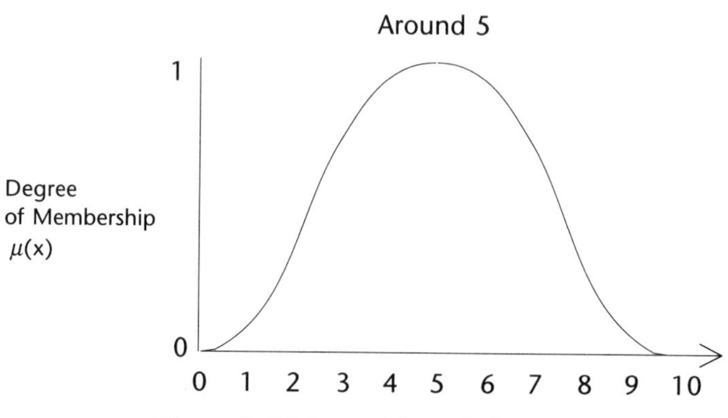

Figure 3.55 Around 5 as a Bell Curve

A fuzzy number represents an approximation of some value and usually provides a better working set than the corresponding crisp value. The concept of "middle aged" is a good example of this flexibility. In ordinary crisp terms, we are required to find some exact transition point where an individual's age moves him or her from the set *Young* into the set *MiddleAged* and then into the set *Old.* Middle age is defined, for our purposes, by the characteristic function in Equation 3.5.

$$m_{Tall} = \{Age \geq 40 \wedge Age \leq 55\}$$

Equation 3.5 Boolean Characteristic Function for MiddleAged

The crisp attributes for middle age define a plateau between ages 40 and 55. There is no gradualness associated with the set, simply a jump onto the plateau and a jump off. The crisp definition of MiddleAged is illustrated in Figure 3.56.

This, of course, means that $39\frac{2}{3}$ years old is not middle aged and that $55\frac{1}{4}$ years old is old. Figure 3.57 shows how the concept of middle age is easily represented as a fuzzy number.

In this fuzzy set we have decided that someone 40 years old represents the quintessence of middle-agedness. Everyone would vote that this individual is middle aged. We

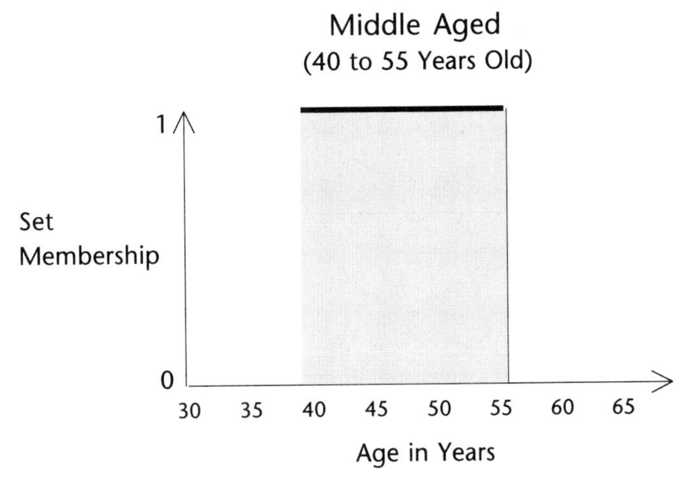

Figure 3.56 The Crisp Set for *MiddleAged*

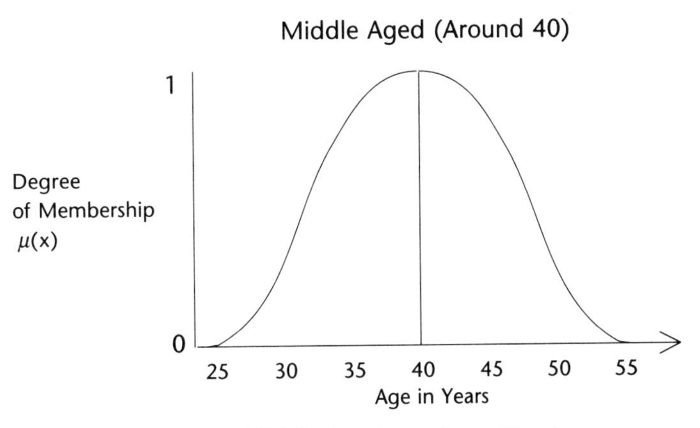

Figure 3.57 `MiddleAged` as a Fuzzy Number

use a bell-shaped curve to spread out the degree of middle agedness on both sides of the 40-years-old domain point. This corresponds to the general concept that someone 30 years old has a real, but not intense, degree of middle agedness. On the other side "of the hill," someone 50 years old is still considered, to some degree, middle aged.

Fuzzy Quantities and Counts

Fuzzy numbers also play an important role in the definition of weakly bounded concepts such as *Few* and *Some*. Obviously what we mean by Few or Some is context-dependent. For example, a jar with a few marbles is not the same as a jar with a few gas molecules. Nevertheless, the concepts are comparable. Taking the concept *Some*, as an example, we want to know when we have *Some* (but not many) marbles. Zero or one marble is obviously not *Some*. When we get to two or three marbles, the idea of having some but not many marbles becomes more real. Consequently, we might define the fuzzy set of Figure 3.58 to represent the concept *Some*.

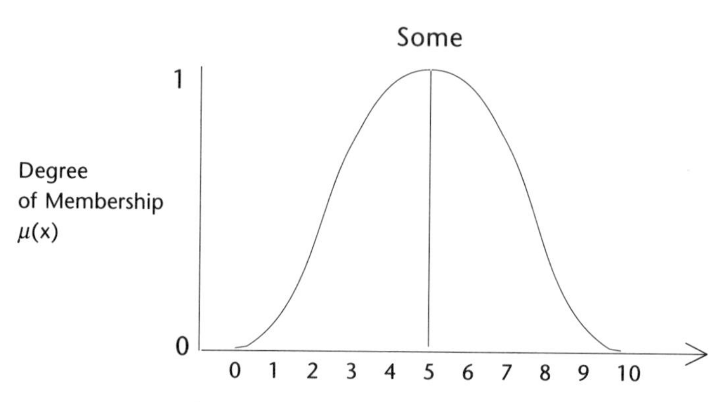

Figure 3.58 The Some Fuzzy Set

At five marbles, we are absolutely satisfied that we have *Some* marbles, but, as the number of marbles increases, the idea of *Some* gradually fades until it no longer even marginally describes the number of marbles. Now we don't have a *Some* anymore, we have something else—many, *a lot*, *a bunch*, *a pile*, etc. Consequently, the truth value begins to fall toward zero after reaching five.

The semantics of fuzzy quantities means that the fuzzy sets can overlap. As Figure 3.59 illustrates, the idea of *Few* is partially contained within the concept of *Some*.

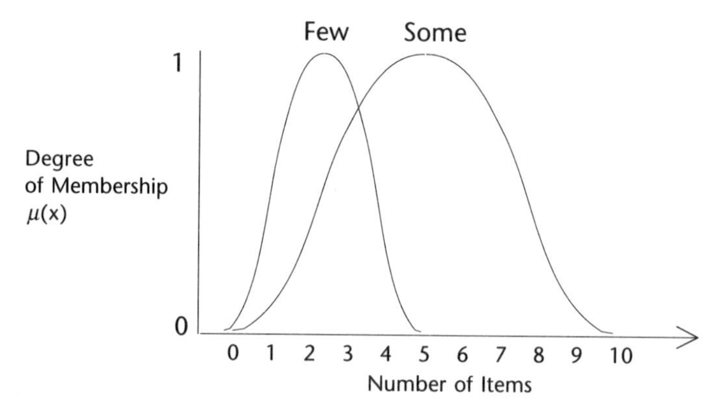

Figure 3.59 The Fuzzy Quantities Few and Some

The compactness of the *Few* curve in Figure 3.59 shows an important characteristic of representing fuzzy numbers and quantities: the width and slope of the curve determines much of the underlying semantics of the fuzzy set. Figure 3.60 shows another way that the concept *Some* might be defined.

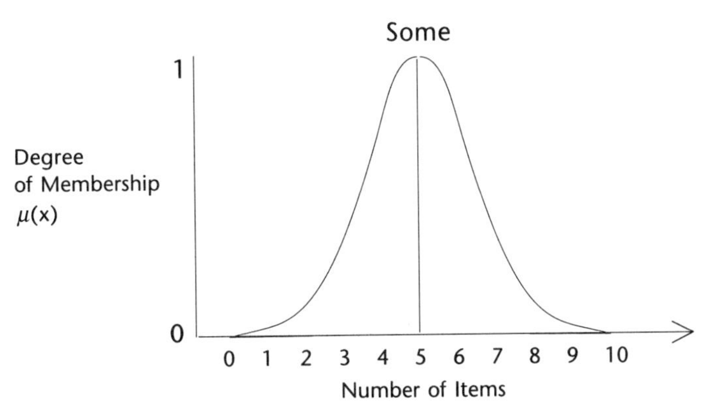

Figure 3.60 A Narrower Version of the Fuzzy Quantity Some

Naturally, as with all fuzzy sets, the exact shape of a fuzzy number or a fuzzy quantity depends on the model context as well as on the designer's understanding of the linguistic concept. With this knowledge, we now turn to the methods of representing fuzzy approximations.

PI CURVES

A PI curve is the preferred and default method of representing a fuzzy number. It provides a smooth descent gradient from the central value (the concept being approximated) to a zero membership point along the domain. The symmetric PI curve is centered on a single value from the domain (γ) with a single parameter that indicates the width of the curve's base (β). The value of the curve for domain point x is given in Equation 3.6.

$$\prod(x;\beta,\gamma) = \begin{bmatrix} S(x;\gamma-\beta,\gamma-\beta/2,\gamma) & \rightarrow x \le \gamma \\ 1-S(x;\gamma,\gamma+\beta/2,\gamma+\beta) & \rightarrow x > \gamma \end{bmatrix}$$

Equation 3.6 The Formula for the PI Curve

The left and right halves of the PI curve are defined as S-curve values, thereby using a function we already know. Figure 3.61 shows the principal characteristics of the PI curve. Note that the inflexion points are automatically determined. The PI curve has an important characteristic—its membership value becomes zero at a discrete and specified point. We say that it is not asymptotic.

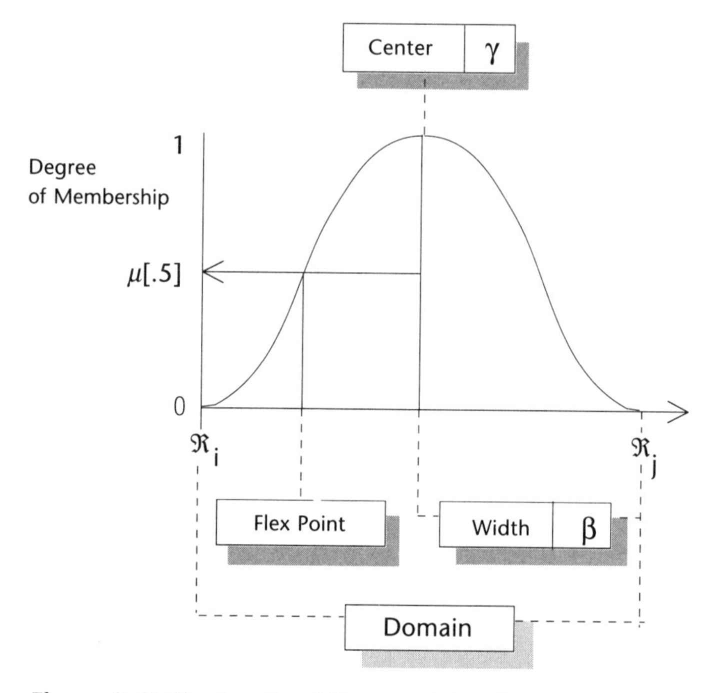

Figure 3.61 The Functional Characteristics of a PI Curve

The PI curve generator produces a bell-shaped fuzzy set centered around a single point in the domain. This is the point indicated by `Center`. The base of the PI curve—the displacement from `Center` to the zero membership point—is indicated by the `Width` parameter.

```
#include  <math.h>
#include "FDB.hpp"
#include "fuzzy.hpp"
#include "mtypes.hpp"
#include "mtsptype.hpp"
void FzyPiCurve(
   FDB* FDBptr,double Center,double Width,int *statusPtr)
  {
   int      i;
   float    tempVector[VECMAX];
   double   inflexpoint,boundarypoint;
```

```
    *statusPtr=0;
    FzyInitVector(tempVector,VECMAX,ZERO);
    inflexpoint=Center+(Width/2);
    boundarypoint=Center+Width;
    FzySCurve(
     FDBptr,Center,inflexpoint,boundarypoint,GROWTH,statusPtr);
    for(i=0;i<VECMAX;i++) tempVector[i]=1.0-FDBptr->FDBvector[i];
    inflexpoint=Center-Width;
    boundarypoint=Center-(Width/2);
    FzySCurve(
FDBptr,inflexpoint,boundarypoint,Center,GROWTH,statusPtr);
    for(i=0;i<VECMAX;i++)
    FDBptr->FDBvector[i]= min(FDBptr->FDBvector[i],tempVector[i]);
    return;
}
```

Listing 3.13 Creating a PI Bell-Shaped Fuzzy Set

The PI curve function checks that the specified parameters are correct and then initializes a working array (*tempVector*) to zero. The inflexion point and the domain point at the edge of the fuzzy set are then calculated. With this information in hand, we create two S-curves, then we reverse the right-hand S-curve. This creates the bell-shaped PI fuzzy set.

The actual width of the PI curve depends on the *Width* parameter, which is often calculated as a percentage of the fuzzy number itself. The PI fuzzy sets of Figures 3.62 through 3.66 illustrate how the shape of the fuzzy number is affected by the percentage. These fuzzy sets are generated by the *FZYSETS.CPP* program.

```
  FuzzySet:    PI Set[40%]
  Description:
1.00                                            ......
0.90                                      ......      ......
0.80                                   ...                  ...
0.70                                ...                        ...
0.60                              ..                              ..
0.50                            ..                                  ..
0.40                          ..                                      ..
0.30                        ..                                          ..
0.20                     ...
0.10                  ...
0.00.............
     0---|---|---|---|---|---|---|---|---|---|---|---|---|---|---|---0
  16.00    18.50    21.00    23.50    26.00    28.50    31.00    33.50    36.00
  Domained:         16.00 to         36.00
```

Figure 3.62 Fuzzy Number 30 at a 40% Width

```
FuzzySet:    PI Set[25%]
       Description:
     1.00                                        ....
     0.90                                     ...    ....
     0.80                                   ..           ..
     0.70                                 ..               .
     0.60                                .                  ..
     0.50                              ..                     .
     0.40                             .                        .
     0.30                            .                          .
     0.20                          ..                            ..
     0.10                        ..                                ..
     0.00........................                                    .....
         0---|---|---|---|---|---|---|---|---|---|---|---|---|---|---|---0
       16.00   18.50   21.00   23.50   26.00   28.50   31.00   33.50   36.00
       Domained:        16.00 to      36.00
```

Figure 3.63 Fuzzy Number 30 at a 25% Width

```
FuzzySet:    Pi Set[15%]
       Description:
     1.00                                         ..
     0.90                                       ..  ..
     0.80                                      .      ..
     0.70                                     .         .
     0.60                                    .
     0.50                                   .            .
     0.40                                  .              .
     0.30                                 .                .
     0.20                                .                  .
     0.10                               .                    .
     0.00..............................                       ............
         0---|---|---|---|---|---|---|---|---|---|---|---|---|---|---|---0
       16.00   18.50   21.00   23.50   26.00   28.50   31.00   33.50   36.00
       Domained:        16.00 to      36.00
```

Figure 3.64 Fuzzy Number 30 at a 15% Width

```
       FuzzySet:    Pi Set[10%]
       Description:
     1.00                                         ..
     0.90                                        .  .
     0.80                                       .    .
     0.70                                      .      .
     0.60
     0.50                                     .        .
     0.40
     0.30                                    .          .
     0.20                                                .
     0.10                                   .
     0.00..................................            ................
         0---|---|---|---|---|---|---|---|---|---|---|---|---|---|---|---0
       16.00   18.50   21.00   23.50   26.00   28.50   31.00   33.50   36.00
       Domained:        16.00 to      36.00
```

Figure 3.65 Fuzzy Number 30 at a 10% Width

```
  FuzzySet:     Pi Set[ 5%]
  Description:
1.00
0.90                                          ..
0.80                                            .
0.70                                         .
0.60
0.50
0.40                                             .
0.30                                        .
0.20
0.10                                       .      .
0.00.......................................        ..................
    0---|---|---|---|---|---|---|---|---|---|---|---|---|---|---|---0
  16.00   18.50   21.00   23.50   26.00   28.50   31.00   33.50   36.00
  Domained:         16.00 to      36.00
```

Figure 3.66 Fuzzy Number 30 at a 5% Width

BETA CURVES

The beta curve is a more tightly compacted bell-shaped curve than the PI curve. The beta fuzzy set is defined, like the PI curve, with two parameters: the single value from the domain around which the curve is built (γ) and a value that indicates the half-width of the curve at the inflection point (β). The value of the curve for domain point x is given in Equation 3.7.

$$B(x;\gamma,\beta) = \frac{1}{1+\left(\dfrac{x-\gamma}{\beta}\right)^2}$$

Equation 3.7 The Formula for the Beta Curve

The curve produced from this formula resembles the PI curve with one major difference—the degree of membership function goes to zero only at extremely large values of beta (β); that is, at infinity (see Figure 3.67).

Naturally the half-width parameter (β) plays a critical role in the shape and scope of the `About` fuzzy set. The larger the value for β, the wider the curve, and, conversely, the smaller the value for β, the narrower the curve. Since a curve with a large β is dispersed more widely across the domain, any individual number must be much closer to the central value (γ) in order to have a significant truth value.

The beta curve function is much more straightforward than the PI curve generator. First, we find the span of the domain. Then, for each point along the beta curve after calculating a domain value at the *i*'th membership array position and locating its distance from the center of the curve, we generate a membership grade by resolving the beta logistic function.

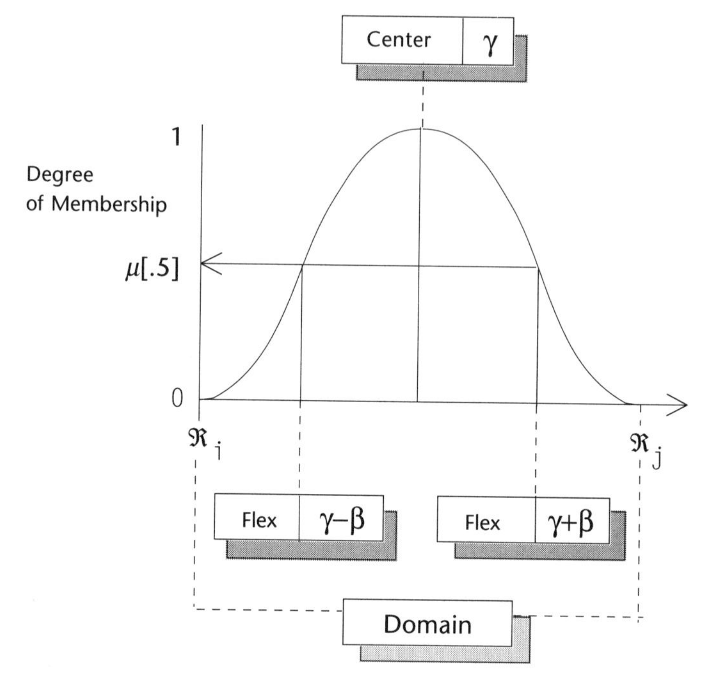

Figure 3.67 Beta Bell Curve Schematic

The beta curve generator produces a bell-shaped fuzzy set centered around a single point in the domain. This is the point indicated by `Center`. The edges of the beta curve are determined by the flexpoint parameter. There are two beta curve generators: *FzyBetaCurve* and *FzyWgtdBetaCurve*. The latter provides a weighting parameter that can attenuate (distort) the shape of the beta curve.

```
#include <stdio.h>
#include <math.h>
#include "FDB.hpp"
#include "fuzzy.hpp"
#include "mtsptype.hpp"
void FzyBetaCurve(
   FDB *FDBptr,double Center,double flexpoint,int *statusPtr)
 {
  int i;
  double  dwidth,memval,BetaPoint;

  *statusPtr=0;
  dwidth=FDBptr->FDBdomain[1]-FDBptr->FDBdomain[0];
```

```
   for(i=0;i<VECMAX;i++)
     {
      memval=FDBptr->FDBdomain[0]+(float)i/(VECMAX-1)*dwidth;
      BetaPoint=(memval-Center)/flexpoint;
      FDBptr->FDBvector[i]=1/(1+pow(BetaPoint,2));
     }
 }
```

Listing 3.14 Creating a Beta Bell-Shaped Fuzzy Set

The actual width of the beta curve depends on the *flexpoint* parameter. This width, like the previous PI curve example, is often calculated as a percentage of the fuzzy number itself. The beta fuzzy sets of Figures 3.68 through 3.72 illustrate how the shape of the fuzzy number is affected by a specified inflexion value. You should remember that, unlike the PI curve, the *flexpoint* parameter is a measure of the distance to the 50% inflexion point, essentially half the distance to the end of the beta curve. These fuzzy sets are generated by the FZYSETS.CPP program.

```
  FuzzySet:    Beta Set[   40%]
  Description:
1.00                                .....
0.90                        ........      ........
0.80                    ....                      ....
0.70               .....                              .....
0.60           ....                                        ....
0.50     ......                                                ......
0.40...                                                               ...
0.30
0.20
0.10
0.00
    0---|---|---|---|---|---|---|---|---|---|---|---|---|---|---|---0
  16.00   19.25   22.50   25.75   29.00   32.25   35.50   38.75   42.00
  Domained:         16.00 to       42.00
```

Figure 3.68 Fuzzy Number 30 at a 40% Inflexion Point

```
  FuzzySet:    Beta Set[   25%]
  Description:
1.00                                 ...
0.90                            .....   .....
0.80                         ...             ...
0.70                       ..                   ..
0.60                    ...                       ...
0.50                ....                             ....
0.40            ....                                     ....
0.30       .....                                             .....
0.20.....                                                         .....
0.10
0.00
    0---|---|---|---|---|---|---|---|---|---|---|---|---|---|---|---0
  16.00   19.25   22.50   25.75   29.00   32.25   35.50   38.75   42.00
  Domained:         16.00 to       42.00
```

Figure 3.69 Fuzzy Number 30 at a 25% Inflexion Point

```
   FuzzySet:     Beta Set[   15%]
   Description:
1.00                                    ...
0.90                                  ..   ..
0.80                                ..       ..
0.70                              ..           ..
0.60                             .               .
0.50                           ..                 ..
0.40                        ...                     ...
0.30                     ...                           ...
0.20               ......                                 ......
0.10..........                                                  ..........
0.00
    0---|---|---|---|---|---|---|---|---|---|---|---|---|---|---|---|---0
  16.00   19.25   22.50   25.75   29.00   32.25   35.50   38.75   42.00
  Domained:          16.00 to        42.00
```

Figure 3.70 Fuzzy Number 30 at a 15% Inflexion Point

```
   FuzzySet:     Beta Set[   10%]
   Description:
1.00                                     .
0.90                                   .. ..
0.80                                  .     .
0.70                                 .       .
0.60                                .         .
0.50                              ..           ..
0.40                             .               .
0.30                          ...                 ...
0.20                       ...                       ...
0.10               ........                             ........
0.00..........                                                  ..........
    0---|---|---|---|---|---|---|---|---|---|---|---|---|---|---|---|---0
  16.00   19.25   22.50   25.75   29.00   32.25   35.50   38.75   42.00
  Domained:          16.00 to        42.00
```

Figure 3.71 Fuzzy Number 30 at a 10% Inflexion Point

```
   FuzzySet:     Beta Set[    5%]
   Description:
1.00                                     .
0.90                                    . .
0.80
0.70                                   .   .
0.60
0.50                                  .     .
0.40                                 .       .
0.30                                .         .
0.20                              ..           ..
0.10                          ....               ....
0.00.....................                            .....................
    0---|---|---|---|---|---|---|---|---|---|---|---|---|---|---|---|---0
  16.00   19.25   22.50   25.75   29.00   32.25   35.50   38.75   42.00
  Domained:          16.00 to        42.00
```

Figure 3.72 Fuzzy Number 30 at a 5% Inflexion Point

The weighted beta curve generator takes, as an additional parameter, an attenuation weight (*Wght*) that is applied to the beta logistic function. This action has the effect of broadening or narrowing the shape of the curve for any particular inflexion point.

```
#include <stdio.h>
#include <math.h>
#include "FDB.hpp"
#include "fuzzy.hpp"
#include "mtsptype.hpp"
void FzyWgtdBetaCurve(
   FDB *FDBptr,
     double Center,double flexpoint,double Wght,int *statusPtr)
 {
  int i;
  double  dwidth,memval,BetaPoint;

  *statusPtr=0;
  dwidth=FDBptr->FDBdomain[1]-FDBptr->FDBdomain[0];
  for(i=0;i<VECMAX;i++)
    {
     memval=FDBptr->FDBdomain[0]+(float)i/(VECMAX-1)*dwidth;
     BetaPoint=(memval-Center)/flexpoint;
     FDBptr->FDBvector[i]=1/(1+(Wght*pow(BetaPoint,2)));
    }
 }
```

Listing 3.15 The Weighted Beta Curve Procedure

The following examples of Figures 3.73 through 3.77 illustrate the effects of various attenuating weights applied to the same beta curve. The beta curve is centered at 30 with a 25% inflexion point. Of course, a weight of one (1) generates a curve equivalent to the unweighted beta curve version.

```
  FuzzySet:    WBeta-    0.90
  Description:
1.00                                     ...
0.90                                .....   .....
0.80                             ...             ...
0.70                          ...                   ...
0.60                       ...                         ...
0.50                   ....                               ....
0.40               ....                                       ....
0.30         ......                                               ......
0.20...                                                                 ...
0.10
0.00
    0---|---|---|---|---|---|---|---|---|---|---|---|---|---|---|---0
  16.00   19.25   22.50   25.75   29.00   32.25   35.50   38.75   42.00
  Domained:        16.00 to        42.00
```

Figure 3.73 Weighted Beta (.90) Version of Fuzzy Number 30

```
   FuzzySet:    WBeta-   0.70
   Description:
 1.00                                        .....
 0.90                                  .....     .....
 0.80                              ...               ...
 0.70                          ....                     ....
 0.60                      ...                              ...
 0.50                  ....                                    ....
 0.40            .....                                             .....
 0.30......                                                             ......
 0.20
 0.10
 0.00
      0---|---|---|---|---|---|---|---|---|---|---|---|---|---|---|---0
   16.00   19.25   22.50   25.75   29.00   32.25   35.50   38.75   42.00
   Domained:        16.00 to       42.00
```

Figure 3.74 Weighted Beta (.70) Version of Fuzzy Number 30

```
   FuzzySet:    WBeta-   0.50
   Description:
 1.00                                        .....
 0.90                                 ......     ......
 0.80                            .....                 .....
 0.70                         ...                           ...
 0.60                    .....                                 .....
 0.50               ....                                            ....
 0.40 ......                                                            ......
 0.30.                                                                        .
 0.20
 0.10
 0.00
      0---|---|---|---|---|---|---|---|---|---|---|---|---|---|---|---0
   16.00   19.25   22.50   25.75   29.00   32.25   35.50   38.75   42.00
   Domained:        16.00 to       42.00
```

Figure 3.75 Weighted Beta (.50) Version of Fuzzy Number 30

```
   FuzzySet:    WBeta-   0.10
   Description:
 1.00                                    ...........
 0.90                  ..............               ..............
 0.80     ..........                                              ..........
 0.70...                                                                    ...
 0.60
 0.50
 0.40
 0.30
 0.20
 0.10
 0.00
      0---|---|---|---|---|---|---|---|---|---|---|---|---|---|---|---0
   16.00   19.25   22.50   25.75   29.00   32.25   35.50   38.75   42.00
   Domained:        16.00 to       42.00
```

Figure 3.76 Weighted Beta (.10) Version of Fuzzy Number 30

```
   FuzzySet:    WBeta-   1.00
   Description:
1.00                              ...
0.90                          .....   .....
0.80                       ...           ...
0.70                     ..                ..
0.60                  ...                    ...
0.50              ....                          ....
0.40          ....                                  ....
0.30     .....                                          .....
0.20.....                                                    .....
0.10
0.00
    0---|---|---|---|---|---|---|---|---|---|---|---|---|---|---|---0
   16.00   19.25   22.50   25.75   29.00   32.25   35.50   38.75   42.00
   Domained:        16.00 to        42.00
```

Figure 3.77 Weighted Beta (1.0) Version of Fuzzy Number 30

GAUSSIAN CURVES

The Gaussian, or exponential, curve is another, though less popular, method of representing fuzzy numbers. The Gaussian curve is defined, like the PI curve, with two parameters: the single value from the domain around which the curve is built (γ) and a value that indicates the width of the bell-shaped curve (k). The value of the curve for domain point x is given in Equation 3.8.

$$G(x;k;\gamma) = e^{-K(\gamma - x)^2}$$

Equation 3.8 The Gaussian Curve Formula

The curve produced from this formula resembles the beta curve with one difference—the slope of membership function goes to zero very quickly with a very short tail (Figure 3.78). In this regard, the Gaussian curve has some of the properties of the PI curve.

Naturally the width parameter (k) plays a critical role in the shape and scope of the About fuzzy set. The larger the value for k, the wider the curve, and, conversely, the smaller the value for k, the narrower the curve. The inability to predict exactly the overall shape of the resulting bell curve for a particular value of k makes the Gaussian function difficult to use.

The Gaussian curve generator produces a bell-shaped fuzzy set centered around a single point in the domain. This is the point indicated by *Center*. The width of the Gaussian curve is determined by the width factor (*WFactor*) parameter. The width parameter can be in the range [>0,∞], but useful values generally fall in the range [.9,5] inclusive.

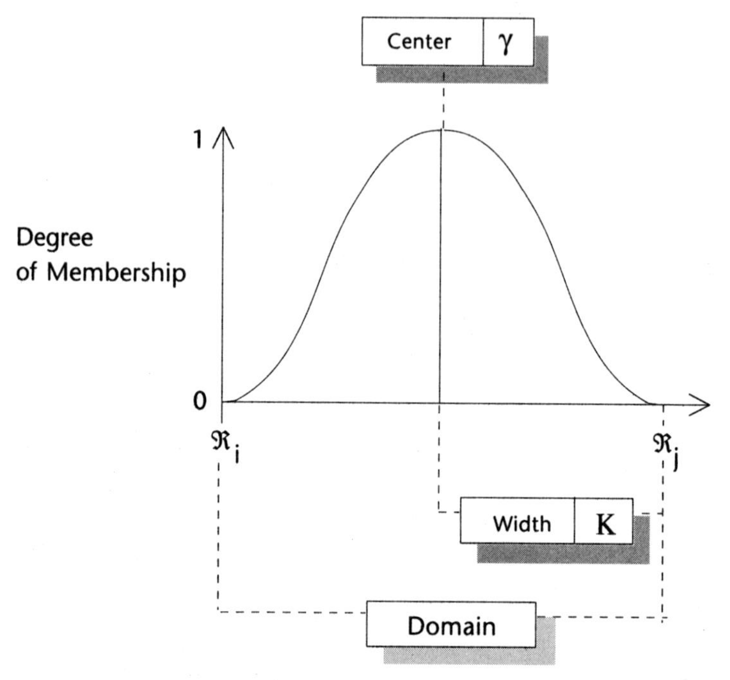

Figure 3.78 Functional Characteristics of a Gaussian Curve

```
#include <math.h>
#include "FDB.hpp"
#include "mtypes.hpp"
void FzyGaussianCurve(
  FDB *FDBptr,double Center,double WFactor,int *statusPtr)
 {
  int    i;
  double dwidth=FDBptr->FDBdomain[1]-FDBptr->FDBdomain[0];
  double thisScalar,GaussPt;

  *statusPtr=0;
  for(i=0;i<VECMAX;i++)
    {
     thisScalar=FDBptr->FDBdomain[0]+(float)i/(VECMAX-1)*dwidth;
     GaussPt=-WFactor*pow((Center-thisScalar),2);
     FDBptr->FDBvector[i]=pow(E,GaussPt);
    }
  return;
 }
```

The examples of Figures 3.79 through 3.83 illustrate the effects of various width factors applied to the same Gaussian curve centered at 30.

```
  FuzzySet:     Gauss-   0.90
  Description:
1.00                                 .
0.90
0.80                                . .
0.70
0.60
0.50                               .   .
0.40
0.30
0.20                              .     .
0.10                             .       .
0.00.............................         ............................
    0---|---|---|---|---|---|---|---|---|---|---|---|---|---|---|---0
  16.00   19.25   22.50   25.75   29.00   32.25   35.50   38.75   42.00
  Domained:         16.00 to      42.00
```

Figure 3.79 Fuzzy Number 30 with Gaussian Width Factor of .90

```
  FuzzySet:     Gauss-   0.50
  Description:
1.00                                 .
0.90                                . .
0.80
0.70                               .   .
0.60
0.50
0.40                              .     .
0.30
0.20                             .       .
0.10                            .         .
0.00............................           ...........................
    0---|---|---|---|---|---|---|---|---|---|---|---|---|---|---|---0
  16.00   19.25   22.50   25.75   29.00   32.25   35.50   38.75   42.00
  Domained:         16.00 to      42.00
```

Figure 3.80 Fuzzy Number 30 with Gaussian Width Factor of .50

```
FuzzySet:     Gauss-   0.10
        Description:
      1.00                                 .
      0.90                               .. ..
      0.80                              .     .
      0.70                             .       .
      0.60                            .         .
      0.50                           .           .
      0.40                          .             .
      0.30                         .               .
      0.20                       ..                 ..
      0.10                     ..                     ..
      0.00.....................                         ....................
          0---|---|---|---|---|---|---|---|---|---|---|---|---|---|---|---0
        16.00   19.25   22.50   25.75   29.00   32.25   35.50   38.75   42.00
        Domained:         16.00 to      42.00
```

Figure 3.81 Fuzzy Number 30 with Gaussian Width Factor of .10

```
   FuzzySet:    Gauss-   1.00
   Description:
1.00                                    .
0.90
0.80                                   . .
0.70
0.60
0.50                                  .   .
0.40
0.30
0.20                                 .     .
0.10
0.00............................       ............................
    0---|---|---|---|---|---|---|---|---|---|---|---|---|---|---|---0
   16.00   19.25   22.50   25.75   29.00   32.25   35.50   38.75   42.00
   Domained:        16.00 to      42.00
```

Figure 3.82 Fuzzy Number 30 with Gaussian Width Factor of 1.0

```
   FuzzySet:    Gauss-   5.00
   Description:
1.00                                    .
0.90
0.80
0.70
0.60
0.50
0.40                                   . .
0.30
0.20
0.10
0.00..............................   ...............................
    0---|---|---|---|---|---|---|---|---|---|---|---|---|---|---|---0
   16.00   19.25   22.50   25.75   29.00   32.25   35.50   38.75   42.00
   Domained:        16.00 to      42.00
```

Figure 3.83 Fuzzy Number 30 with Gaussian Width Factor of 5.0

Triangular, Trapezoidal, and Shouldered Fuzzy Sets

With the rise of eight- and sixteen-bit microprocessor control units (MCUs), the "standard" fuzzy set shape has become variations on the triangle. This shape has emerged from considerations of minimal representation and maximum exploitation. Both triangular and trapezoidal fuzzy sets can be represented using four bytes: two for points in the surface and two for the slope of the fuzzy surface. Figure 3.84 shows how a trapezoidal fuzzy set is defined as points and inflexion angles.

In process control systems, variables are decomposed into overlapping arrays of triangular regions. The endpoints of these variables are often expressed as "shouldered" sets (bisected trapezoids) or as linear fuzzy sets, which are actually bisected triangles. Figure 3.85 shows the motor temperature as a collection of fuzzy sets.

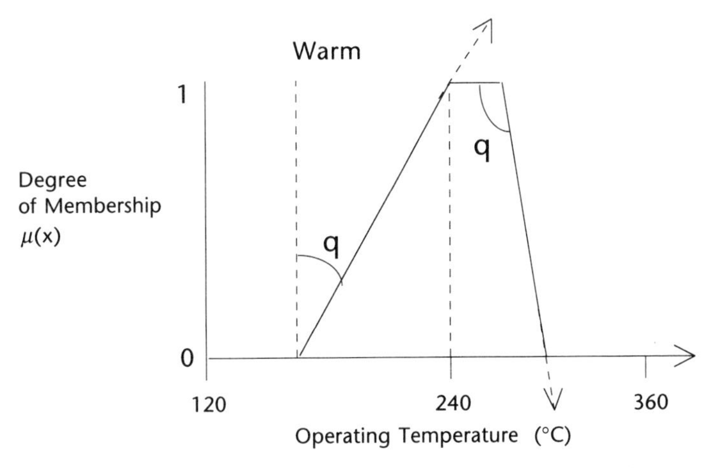

Figure 3.84 Components of a Trapezoidal Fuzzy Region

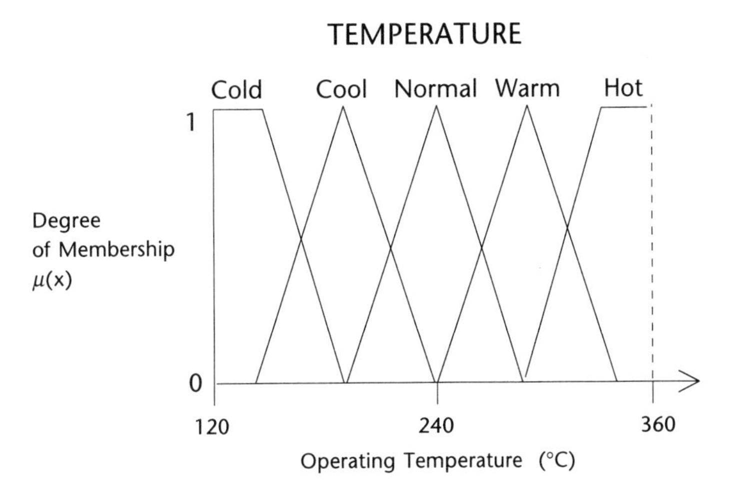

Figure 3.85 The Variable *Temperature* as Triangular Fuzzy Regions

In modeling dynamic systems, the use of triangular functions can approximate their behaviors to nearly any degree of precision. This often requires careful placement of triangles with intensive overlaps and widths. Generally, when we turn our attention to representing fuzzy variables in financial, marketing, transportation, and other business models, the bell-shaped curve provides a better way to represent individual fuzzy regions. Sigmoid curves are used to handle the variable endpoints.

Figure 3.86 illustrates the motor temperature variable represented as bell-shaped and S-curve fuzzy regions.

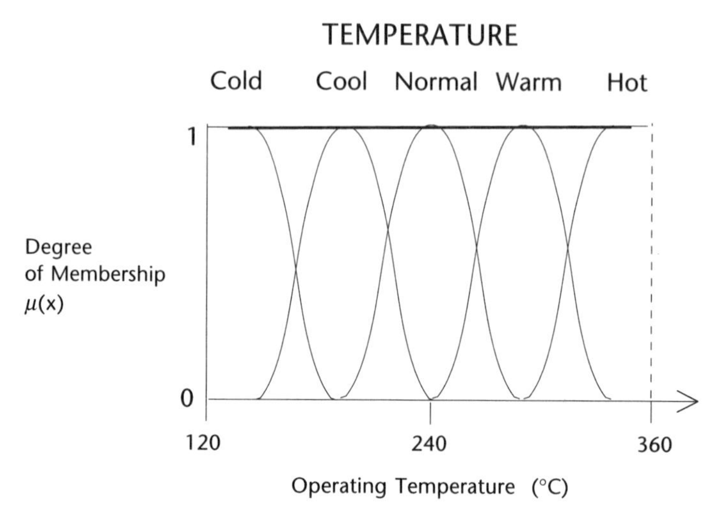

Figure 3.86 *Temperature* Composed of Bell-Shaped Fuzzy Sets

However, many business models continue to use triangular forms since they are simpler to specify, easier to visualize, and have the somewhat dubious advantage of mapping to fuzzy models in literature. Since most of the fuzzy proofs for such concepts as universal approximation, stability, and broad representation are cast in the form of triangular fuzzy sets, this provides a powerful motivation for using these fuzzy constructs. In this section, we explore how these forms are created. Triangular forms have been delayed to this section so that we can use the coordinate mapping function. (See Irregularly Shaped and Arbitrary Fuzzy Sets on page 149.)

TRIANGULAR FUZZY SETS

Triangular fuzzy sets, like PI and beta curves, represent fuzzy numbers. The apex of the triangle, its center, is the unity measure of the number. The left and right edges of the fuzzy region specify a linear decay from the center to the points where membership goes to zero. Figure 3.87 shows the schematic of a triangular region.

When you design a fuzzy variable partitioned into a collection of triangular fuzzy regions, triangles overlap to some extent. How this overlap works and why it is necessary is explored Chapter 7, *Fuzzy Models*. However, when you decompose a variable into fuzzy sets, the amount of overlap must vary between 10 to 50% (Figure 3.88).

The *FzyCoordSeries* function accepts two array parameters—the first contains the domain values and the second contains the corresponding truth membership values. Using these parameters, a new truth membership function for the fuzzy set is constructed.

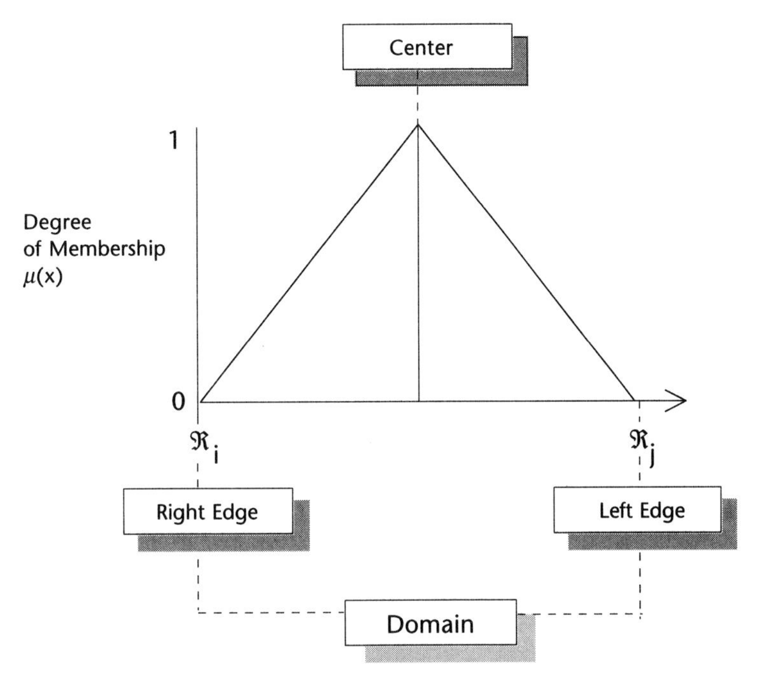

Figure 3.87 Characteristics of a Triangular Fuzzy Set

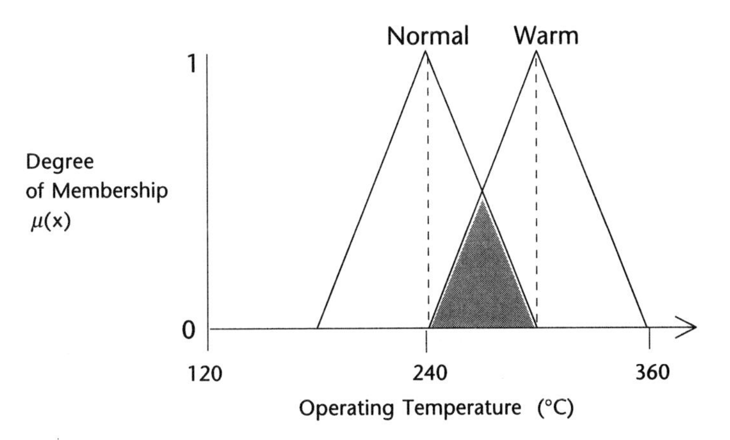

Figure 3.88 Neighboring Triangular Regions with 50% Overlap

```
#include <math.h>
#include "FDB.hpp"
#include "fuzzy.hpp"
#include "mtypes.hpp"
#include "mtsptype.hpp"
void FzyTriangleCurve(
   FDB* FDBptr,
     double Left,double Center,double Right,int *statusPtr)
  {
   int     i,TrgCnt=3;
   char   *PgmId="mtfztrs";
   double  TrgPoint[5];
   float   TrgGrade[5];

 if((Center<FDBptr->FDBdomain[0])||(Center>FDBptr->FDBdomain[1]))
     {
      *statusPtr=3;
      return;
     }
   if((Left<FDBptr->FDBdomain[0])||(Right>FDBptr->FDBdomain[1]))
     {
      *statusPtr=5;
      return;
     }
   TrgPoint[0]=Left;
   TrgGrade[0]=0.0;

   TrgPoint[1]=Center;
   TrgGrade[1]=1.0;

   TrgPoint[2]=Right;
   TrgGrade[2]=0.0;
   FzyCoordSeries(FDBptr,TrgPoint,TrgGrade,TrgCnt,statusPtr);
   return;
  }
```

Listing 3.16 Creating a Triangular Fuzzy Set (`mtfztrs`)

SHOULDERED FUZZY SETS

Regions in the middle of a variable are represented by triangles since their associated concept increases and then decreases (e.g., cold slowly gives way to warm and warm slowly gives way to hot). At the edges of the variable, however, the concept does not change. Once we reach *Hot*, as an example, all temperatures above this continue to be *Hot*. Shouldered fuzzy sets, not triangles, are used to bracket the endpoints of a variable's fuzzy region.

Shouldered Fuzzy Sets are essentially truncated trapezoidal fuzzy sets. The left shoulder goes from true to false across a small plateau. The right shoulder goes from false to true also across a small plateau. Figure 3.89 shows the variable *Temperature* with its shouldered regions.

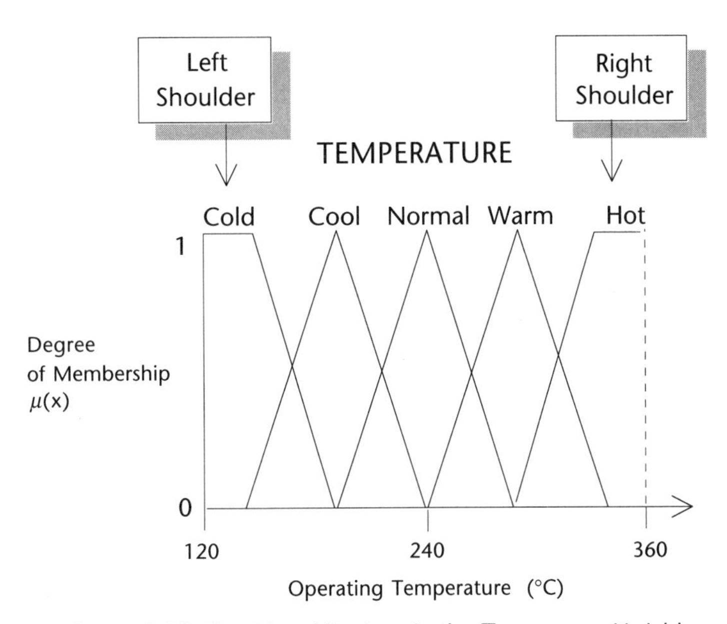

Figure 3.89 Shouldered Regions in the *Temperature* Variable

In many control applications, the width of the shoulder (plateau) regions provides an important surface. In these situations, the fuzzy sets that interact to provide the maximum degree of control are tightly clustered toward the center of the variable's scope. The shoulder region supports coarser grain control so that movement into them elicits a strong and general response. Figure 3.90 illustrates this concept for a steering controller.

The large shoulder plateaus provide a sensitive response surface for the steering controller, insuring that the steering state stays in the zero aperture zone. For details on shoulder controls and set overlap, see Chapter 7, *Fuzzy Models*.

By specifying a plateau of zero extent, we can construct apparently linear fuzzy sets that meet the endpoints of the variable's universe of discourse. When a variable's scope slices through a more extended fuzzy region, we can view this as the bisection of a triangular fuzzy set that overlaps the variable's endpoints (Figure 3.91).

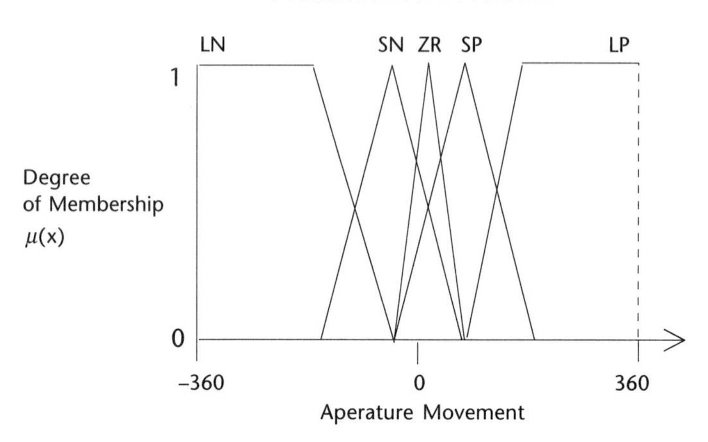

Figure 3.90 Extended Shoulder Regions in a Steering Control System

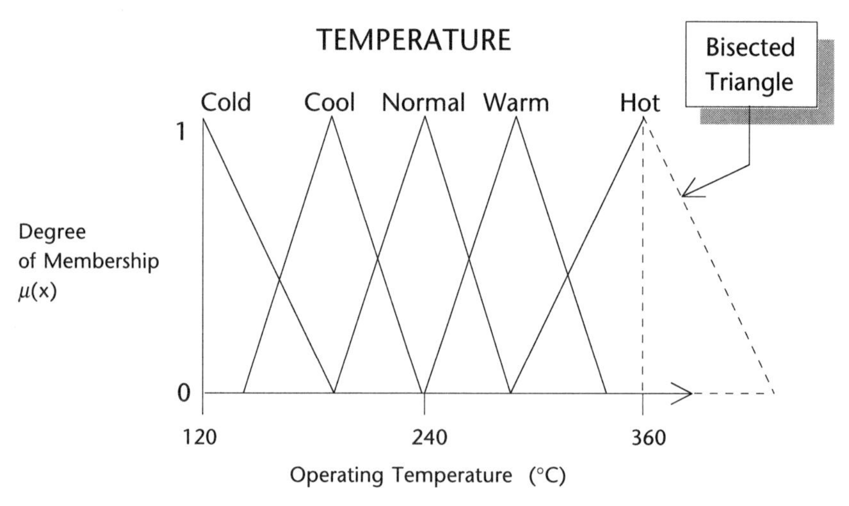

Figure 3.91 An Endpoint as a Bisected Triangular Fuzzy Set

The *FzyCoordSeries* function accepts two array parameters—the first contains the domain values and the second contains the corresponding truth membership values. Using these parameters, a new truth membership function for the fuzzy set is constructed. See "Domain-Based Coordinate Memberships" on page 159. for complete details on how the coordinate method of defining fuzzy sets is used.

```
#include "FDB.hpp"
#include "fuzzy.hpp"
#include "mtypes.hpp"
void FzyShoulderedCurve(
   FDB* FDBptr,int CurveType,double Edge,double Floor,int *sta-
tusPtr)
  {
   int     i,TrgCnt=3;
   char   *PgmId="mtfzsfs";
   double  TrgPoint[5];
   float   TrgGrade[5];

   if((Edge<FDBptr->FDBdomain[0])||(Edge>FDBptr->FDBdomain[1]))
     {
      *statusPtr=3;
      return;
     }
   if((Floor<FDBptr->FDBdomain[0])||(Floor>FDBptr->FDBdomain[1]))
     {
      *statusPtr=3;
      return;
     }
   if(CurveType==LEFTSHOULDER)
     {
      TrgPoint[0]=FDBptr->FDBdomain[0];
      TrgGrade[0]=1.0;
      TrgPoint[1]=Edge;
      TrgGrade[1]=1.0;
      TrgPoint[2]=Floor;
      TrgGrade[2]=0.0;
     }
   if(CurveType==RITESHOULDER)
     {
      TrgPoint[0]=Floor;
      TrgGrade[0]=0.0;
      TrgPoint[1]=Edge;
      TrgGrade[1]=1.0;
      TrgPoint[2]=FDBptr->FDBdomain[1];
      TrgGrade[2]=1.0;
     }
   FzyCoordSeries(FDBptr,TrgPoint,TrgGrade,TrgCnt,statusPtr);
   return;
  }
```

Listing 3.17 Creating Shouldered Fuzzy Sets (mtfzsfs)

Note that if we did not want to use shouldered fuzzy sets on the endpoints in our fuzzy variable, the linear *Increase* and linear *Decrease* sets can also be used to create the effect of bisected triangles. Bounding your variables without shoulders is generally fine since all values to the left and right of the endpoints have either zero [0] or one [1] values, respectively. However, as previously mentioned, in some control applications the shoulder extents are used to cover a large area lying outside the required operating range.

We conclude this section of the design and implementation of fuzzy sets with an example of how the triangular and shoulder fuzzy set generators are used to partition a model variable into a set of individual fuzzy regions.

The following program (part of the *FZYTRBN.CPP*—the fuzzy steam turbine—program) generates a set of fuzzy regions that partition the `Temperature` variable. The triangular and shoulder fuzzy set generators provide the tools necessary to construct the kinds of fuzzy representations common in control engineering. These are mainly arrays of overlapping triangular fuzzy regions.

```
char *FDBnames[]=
  {
   "COLD",
   "COOL",
   "NORMAL",
   "WARM",
   "HOT"
  };
const int FzyTermMax=5;
double Point[]={100,120,180,240,300,360,380};
void main(void)
 {
  FDB   *tempFDBptr[FzyTermMax];
  int     i,status,FDBcnt=5;
  status=0;
  MdlConnecttoFMS(&status);
  for(i=0;i<FzyTermMax;i++)
    {
     tempFDBptr[i]=new FDB;
     FzyInitFDB(tempFDBptr[i]);
     strcpy(tempFDBptr[i]->FDBid,FDBnames[i]);
     tempFDBptr[i]->FDBdomain[0]=Point[i];
     tempFDBptr[i]->FDBdomain[1]=Point[i+2];
    }
  FzyShoulderedCurve(tempFDBptr[0],LEFTSHOULDER,120,180,&status);
  for(i=1;i<4;i++)
     {
      FzyTriangleCurve(
```

```
        tempFDBptr[i],Point[i],Point[i+1],Point[i+2],&status);
     }
   FzyShoulderedCurve(tempFDBptr[4],RITESHOULDER,360,300,&status);
   FzyPlotVar("TEMPERATURE",tempFDBptr,FDBcnt,SYSLOGFILE,&status);
   return;
  }
```

Listing 3.18 Section of FZYTRBN.CPP that Produces the Temperature Fuzzy Sets

The fuzzy steam turbine program creates a series of empty, initialized fuzzy sets and establishes the proper domain for each individual partition (in the first *for* block). The fuzzy term partitions are created by simply creating the left and right shoulders and then building the three overlapping fuzzy sets in the second *for* loop. The fuzzy set partitions of Figures 3.92 through 3.96 are created using the shouldered and triangular generators. [The individual fuzzy set drawing routines (*FzyDrawSet*) appear in the original but have been deleted from this example; otherwise, this code has not been simplified in any way.]

```
  FuzzySet:    COLD
  Description:
1.00.................
0.90                 .....
0.80                      .....
0.70                           ....
0.60                               .....
0.50                                    .....
0.40                                         .....
0.30                                              .....
0.20                                                   .....
0.10                                                        ....
0.00                                                            .....
    0---|---|---|---|---|---|---|---|---|---|---|---|---|---|---|---0
 100.00  110.00  120.00  130.00  140.00  150.00  160.00  170.00  180.00
  Domained:       100.00 to     180.00
```

Figure 3.92 The Left-Shouldered *Cold* Fuzzy Set Partition of *Temperature*

```
  FuzzySet:    COOL
  Description:
1.00                                .
0.90                             ... ...
0.80                          ...       ...
0.70                       ...             ...
0.60                   ....                   ....
0.50                ...                           ...
0.40             ...                                 ...
0.30          ...                                       ...
0.20       ...                                             ....
0.10   ....                                                    ...
0.00...                                                           ...
    0---|---|---|---|---|---|---|---|---|---|---|---|---|---|---|---0
 120.00  135.00  150.00  165.00  180.00  195.00  210.00  225.00  240.00
  Domained:       120.00 to     240.00
```

Figure 3.93 The Triangular *Cool* Fuzzy Set Partition of *Temperature*

```
FuzzySet:    NORMAL
        Description:
      1.00                                .
      0.90                            ... ...
      0.80                          ...      ...
      0.70                       ...            ...
      0.60                   ....                  ....
      0.50                ...                          ...
      0.40             ...                                ...
      0.30          ...                                      ...
      0.20       ...                                            ....
      0.10   ....                                                   ...
      0.00...                                                          ...
          0---|---|---|---|---|---|---|---|---|---|---|---|---|---|---|---0
       180.00  195.00  210.00  225.00  240.00  255.00  270.00  285.00  300.00
        Domained:      180.00 to     300.00
```

Figure 3.94 The Triangular *Normal* Fuzzy Set Partition of *Temperature*

```
        FuzzySet:    WARM
        Description:
      1.00                                .
      0.90                            ... ...
      0.80                          ...      ...
      0.70                       ...            ...
      0.60                   ....                  ....
      0.50                ...                          ...
      0.40             ...                                ...
      0.30          ...                                      ...
      0.20       ...                                            ....
      0.10   ....                                                   ...
      0.00...                                                          ...
          0---|---|---|---|---|---|---|---|---|---|---|---|---|---|---|---0
       240.00  255.00  270.00  285.00  300.00  315.00  330.00  345.00  360.00
        Domained:      240.00 to     360.00
```

Figure 3.95 The Triangular *Warm* Fuzzy Set Partition of *Temperature*

```
        FuzzySet:    HOT
        Description:
      1.00                                               .................
      0.90                                          .....
      0.80                                     .....
      0.70                                 ....
      0.60                            .....
      0.50                       .....
      0.40                  .....
      0.30             .....
      0.20         ....
      0.10    .....
      0.00.....
          0---|---|---|---|---|---|---|---|---|---|---|---|---|---|---|---0
       300.00  310.00  320.00  330.00  340.00  350.00  360.00  370.00  380.00
        Domained:      300.00 to     380.00
```

Figure 3.96 The Right-Shouldered *Hot* Fuzzy Set Partition of *Temperature*

Once the complete collection of overlapping fuzzy sets has been created we have effectively partitioned the variable *Temperature* into a collection of regions that can be addressed by the propositions of our fuzzy model. Such a production rule might appear as

```
if CurTemp is WARM then ThrottleAction is LargeNegative;
```

where the term *LargeNegative* is a fuzzy partition of the variable *ThrottleAction*. Fuzzy propositions (rules) always address linguistic variables, which, in this instance, happen to coincide with the names of the variable's underlying fuzzy sets. Chapter 7, *Fuzzy Models*, addresses the issues associated with decomposing model variables into the correct number of fuzzy partitions with the right shape and proper amount of overlap.

The *FzyPlotVar* program is called to display the complete universe of discourse for the variable and the fuzzy sets. When drawn, we can see how the partitioning corresponds to the semantic nature of each region in the variable *Temperature*. (If it doesn't, then you must go back and restructure your fuzzy sets.)

The notation "*[UofD]*" beneath the variable name refers to the implicit universe of discourse for the variable. This range is calculated from the individual domains of each fuzzy set passed to the plotting program.

If you refer back to Figure 3.86, on page 138, we indicated that bell-shaped representations provide a more natural method of decomposing fuzzy variables, but for reasons associated with memory and processing speed control applications, we generally use triangular, trapezoidal, and shouldered fuzzy sets. The following is the same temperature variable (produced by the *FZYIRD.CPP* program) decomposed using PI curves and S-curves (at the shoulders).

```
TEMPERATURE
Domain [UofD]:     100.00 to     380.00

      1.00.....             *             +            :             ------
      0.90    ...          ***          ++++          ::::          --
      0.80      ..       **   *        ++  ++        ::  ::       --
      0.70       ..     **     **     ++     +      :     ::     --
      0.60        ... ***       **  ++        ++  ::        ::  --
      0.50          .**           *++           +:           :--
      0.40          **.          ++**          ::++           --::
      0.30        **   ..       ++  **        ::  ++       ---  ::
      0.20       **     ..     ++     **     :      ++    --     :::
      0.10     ***        .  ++        **  ::        ++  --        ::
      0.00    **           .++           *:            ---          ::
          0---|---|---|---|---|---|---|---|---|---|---|---|---|---|---|---0
       100.00  135.00  170.00  205.00  240.00  275.00  310.00  345.00  380.00
      .   FuzzySet:    COLD
          Domained:       100.00 to     180.00
      *   FuzzySet:    COOL
          Domained:       120.00 to     240.00
      +   FuzzySet:    NORMAL
          Domained:       180.00 to     300.00
      :   FuzzySet:    WARM
          Domained:       240.00 to     360.00
      -   FuzzySet:    HOT
          Domained:       300.00 to     380.00
```

Figure 3.97 The Variable *Temperature* Decomposed into Triangular Fuzzy Regions

```
FzySCurve(tempFDBptr[0],100,140,180,DECLINE,&status);
k=2;
for(i=1;i<4;i++)
   {
    FzyPiCurve(
      tempFDBptr[i],Point[k],Point[k+1]-Point[k],&status);
    k++;
   }
 FzySCurve(tempFDBptr[4],300,340,380,GROWTH,&status);
 FzyPlotVar("TEMPERATURE",tempFDBptr,FDBcnt,SYSLOGFILE,&status);
```

Listing 3.19 Segment of *FZYIRD.CPP* that Used PI and S-Curve Representations

Notice that this code fragment is remarkably close to the triangular representations. The only significant change is the way the center and width of the PI curves are specified. To get a complete 50% overlap, we set the PI curve *Width* parameter to the distance from the current *Center* point to the next *Center* point. Figure 3.98 shows the bell-shaped representation for *Temperature*.

```
TEMPERATURE
Domain [UofD]:     100.00 to     380.00

      1.00..               ***           ++            ::                --
      0.90 ....          *** ***       ++  +++      :::  ::          ----
      0.80    ..        *      **    ++      ++    ::      ::       --
      0.70      ..     *         *  ++        ++  ::        ::     --
      0.60       ..   *          ** +          ++::          :    -
      0.50        .. *            *+            +:            :  -
      0.40         ..*            +*            :+             :-
      0.30          *.           ++ *          ::++            -:
      0.20        ** ..         ++  **        ::  ++         -- ::
      0.10       **   ...      ++     **     :      +      ---   ::
      0.00    ****      ....+++        ***:::        ++-----       :::
          0---|---|---|---|---|---|---|---|---|---|---|---|---|---|---|---0
       100.00  135.00  170.00  205.00  240.00  275.00  310.00  345.00  380.00
     .   FuzzySet:     COLD
        Description:
        Domained:        100.00 to      180.00
     *   FuzzySet:     COOL
        Description:
        Domained:        120.00 to      240.00
     +   FuzzySet:     NORMAL
        Description:
        Domained:        180.00 to      300.00
     :   FuzzySet:     WARM
        Description:
        Domained:        240.00 to      360.00
     -   FuzzySet:     HOT
        Description:
        Domained:        300.00 to      380.00
```

Figure 3.98 The Variable *Temperature* Decomposed into Bell-Shaped Fuzzy Regions

Irregularly Shaped and Arbitrary Fuzzy Sets

Thus far we have examined some of the basic fuzzy sets used in a wide variety of fuzzy models. The linear, S-curve, bell-shaped, triangular, and trapezoidal sets form the basic vocabulary of a fuzzy modeling system. Occasionally, however, the standard fuzzy set shapes fail to capture the semantics of a particular model variable, or the representation using these standard forms becomes unwieldy and cumbersome. This case is especially true when the fuzzy set, and perhaps its complement, is used to independently represent the state of a variable. In this case, the designer must turn to a representation that more closely approximates the behavior of the model variable. This means defining an arbitrarily shaped fuzzy set.

CAUTION
Irregularly defined fuzzy sets can introduce behaviors into your model that may be difficult to evaluate. These sets can have unpredictable effects on the model performance and can complicate model verification and validation. This problem is especially true for customized fuzzy sets that might be multimodal, have multiple plateaus, have severely irregular surfaces, or have extended skew morphologies. You should be an experienced knowledge engineer familiar with conventional fuzzy modeling techniques before introducing irregularly shaped fuzzy sets into your system.

Variations on the sigmoid and distribution curves are common in most models. But some fuzzy shapes are much more "wavy," matching behavior patterns that undulate over the domain. Remembering that fuzzy memberships reflect subjective judgment, the fuzzy set that determines *Tall* for a building might appear something like that of Figure 3.99.

As the number of floors above 20 increases, our perception that this is a tall building rises rapidly. But after 50 or so stories, all the buildings begin to look pretty much the same. As we add more floors, the perception that one building is taller than the next is not sharply increased. That is, after a certain height, all the buildings begin to look pretty much the same. We can confirm this by watching tourists standing at 34th Street and Fifth Avenue trying desperately to find the Empire State building.

The point we have been making is simply that fuzzy sets reflect the membership patterns of the underlying concept, and these patterns are often not directly adaptable to the standard curve architectures. Some fuzzy sets, in fact, are pretty wavy. As an example, consider a fuzzy set that represents high traffic volumes on Eighth Avenue in New York City. Figure 3.100 illustrates how the time of day can be mapped to a fuzzy definition of heavy traffic.

Right after the morning rush hour, the traffic is rather light. Around noon and a little after, it begins to build again, and after four o'clock it is definitely, positively, absolutely, unbearably heavy. From this observation, we can see that the *Heavy* fuzzy set in Figure 3.101 is derived from a three-dimensional fuzzy region that maps vehicle volume per minute to the time of day to a sense of *Heavy* for traffic.

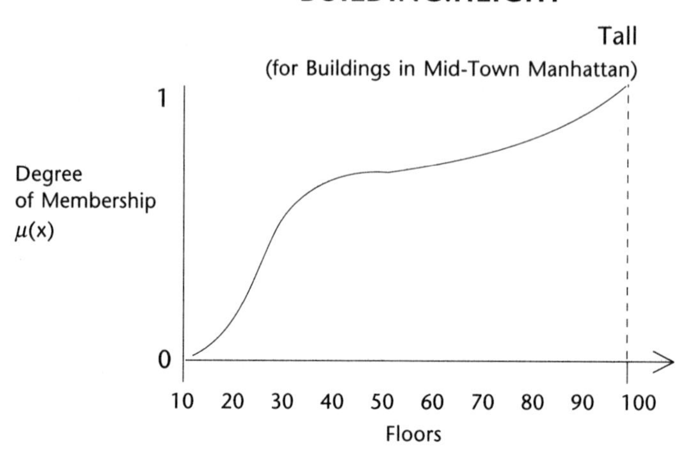

Figure 3.99 Fuzzy Set for `Tall Building` as a Nonlinear Approximation

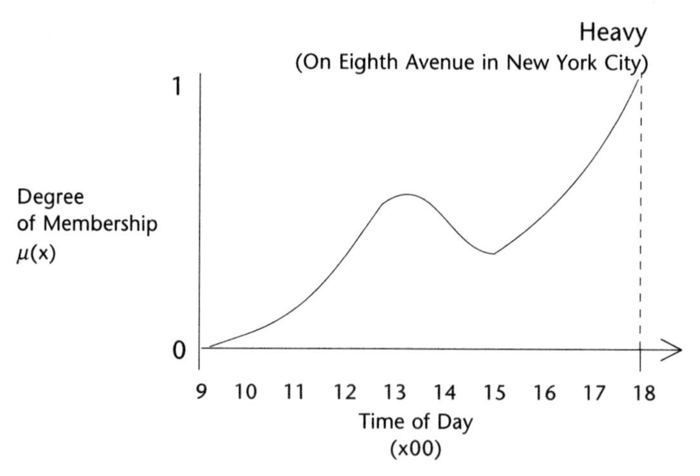

Figure 3.100 Heavy Traffic Fuzzy Representation as a Convex Fuzzy Set

Other fuzzy sets may be discontinuous. Consider a fuzzy set that relates the concept of alloy elasticity to the current temperature and pressure (Figure 3.102). As the ratio increases, the elasticity falls off (due to quantum mechanical "freezing" of the metallic surface). But, at a certain point, the kinetic energy of the heat is stronger than the frozen electronic gas. At this point, elasticity once more increases.

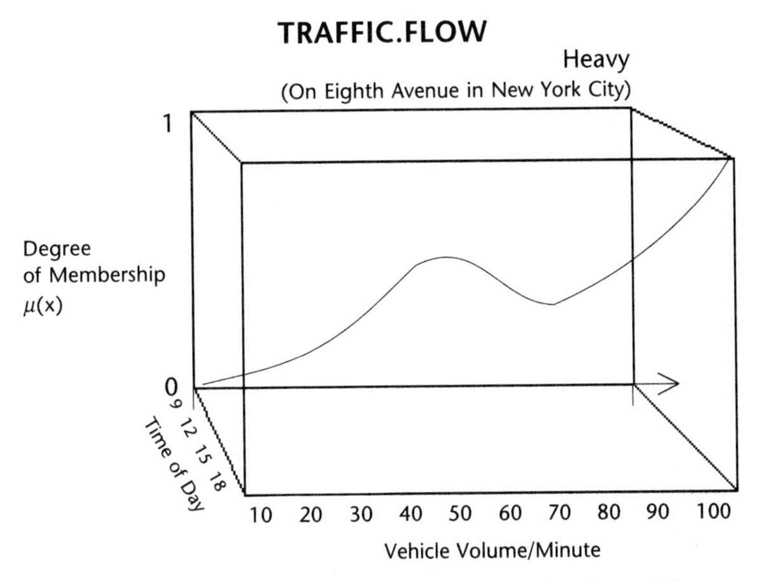

Figure 3.101 Heavy Traffic as a Set of Fuzzy Sets Across Time

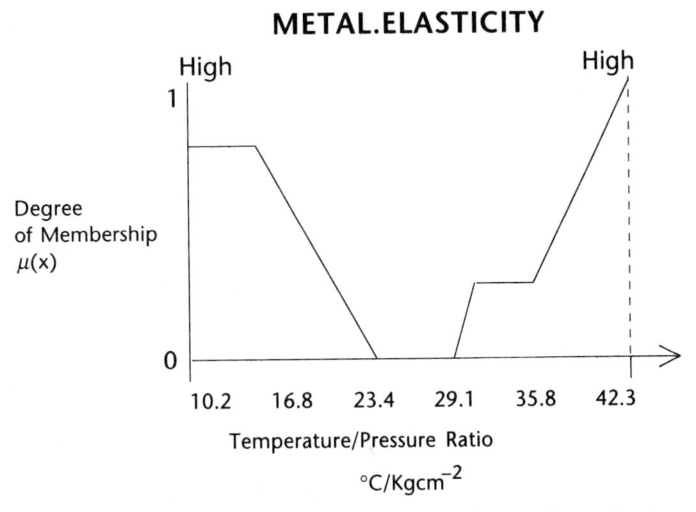

Figure 3.102 Alloy Elasticity as a Noncontinuous Fuzzy Region

Although this fuzzy set looks a little odd, it is really no different than the *Heavy Traffic* set—except that the membership curve falls to zero for some part of the set.

In an automobile underwriting application, the age of the driver affects the risk of the underwriting. Younger drivers and very old drivers both constitute high risks. A fuzzy set—*High Driving Risk*—associated with the driver's age might appear as given in Figure 3.103.

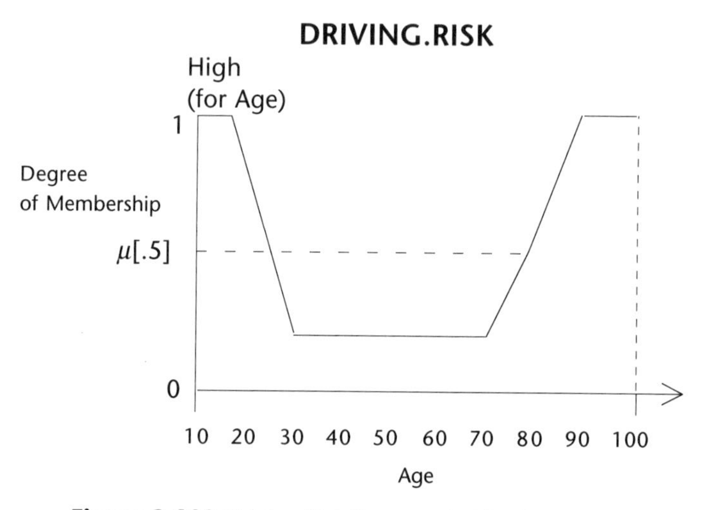

Figure 3.103 Driving Risk Represented by Segments

The degree of risk drops quickly from 16, the minimum legal driving age, and then flattens out into a plateau from about 28 years old until 68 years old. After the individual's 68th birthday, the risk factor gradually increases until it becomes very high after the age of 80. This fuzzy set contour resembles the inverted PI curve, and, in fact, when we overlay the *High Driving Risk* set with the standard bell curve, the approximation is evident (Figure 3.104).

Yet, the flanged edges of the fuzzy set supported by the PI curve will distort the actual membership function associated with the risk. In this instance, we must design a fuzzy contour that more accurately reflects the precise and observed nature of risk (as determined by the underwriters).

In this section, then, we take up the idea of defining arbitrarily shaped fuzzy sets. Through the use of a domain-to-truth membership mapping function and the use of equidistantly spaced truth membership values, we can define fuzzy sets that have any surface. The techniques developed in this section are used to describe more regular fuzzy sets in the next section (and the reason for this digression into irregular set shapes will be apparent).

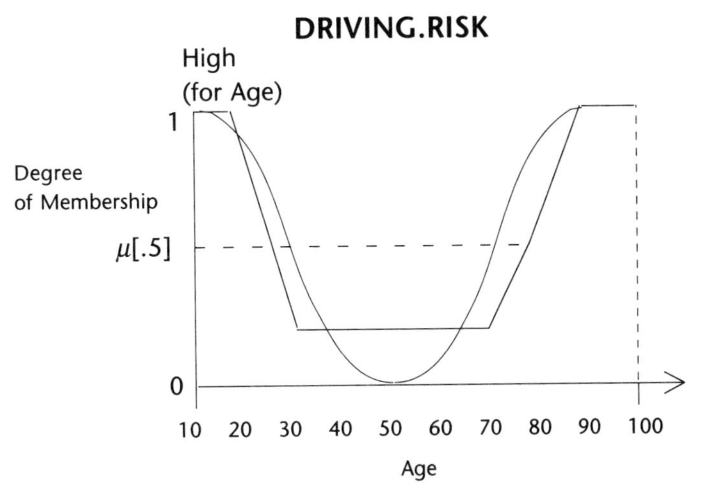

Figure 3.104 Driving Risk as an Approximate PI Curve

This is the linear interpolation function. A fuzzy surface is constructed by linearly interpolating between truth membership points across the domain. We need at least two points in the domain. The following algorithm fills in the space between domain points with proportional truth values.

```
#include <stdio.h>
#include "fuzzy.hpp"
#include "mtypes.hpp"
#include "mtsptype.hpp"
void FzyInterpVec(float TempVector[],int *statusPtr)
   {
    int    i,j,k,Pnt1,Pnt2;
    float  y,Seg1,Seg2;
*statusPtr=0;
    i=0;
    j=0;
//--This loop moves across the vector until it finds
//-- a slot that does not contain a -1. We then linearly
//--interpolate between the previous boundary and this cell.
    interpolate:
     j++;
     if(j>VECMAX)
       {
```

```
          *statusPtr=1;
           return;
          }
        if(TempVector[j]==-1) goto interpolate;

//--OKay, let's interpolate. The idea here is very simple,
//--we calculate "y" where is the proportion of the distance
//--from the anchor point to the current point (k). We then
//--multiply this with the linear distance between the two
//--points.
         Pnt1=i+1;
         Pnt2=j;
         for(k=Pnt1;k<Pnt2+1;k++)
            {
             Seg1=k-i;
             Seg2=j-i;
             y=Seg1/Seg2;
             if(k>VECMAX) {*statusPtr=77;return;}
             TempVector[k]=
             TempVector[i]+y*(TempVector[j]-TempVector[i]);
            }
         i=j;
         if(i>=VECMAX) return;
       goto interpolate;
    }
```

Listing 3.20 Linear Interpolation Procedure (`mtfzipv`)

This function needs a little explanation. In order for the interpolation function to work, the truth membership vector must be properly constructed. Every cell of the truth membership array is initialized to "–1." The domain truth values are then inserted into the vector at their corresponding cell locations along the truth function. We linearly interpolate by finding the segments between runs of "–1" values and then construct, for each cell, a truth value equal to its proportion of the line.

Truth Series Descriptions

A fuzzy set can be specified by simply listing a series of membership truth values between [0,1]. These values are spread out evenly across the domain, and linear interpolation is used to produce a final fuzzy contour. There is a certain elegance and advantage to this method—the designer need only sketch out the truth of the surface; the underlying domain points are picked up automatically. A triangular fuzzy set can be easily created:

```
//--Define the WARM fuzzy set
memSeries[]={0,  1,  0};
memCnt=3;
FzyMemSeries(WarmFDBptr,memSeries,memCnt,&status);
```

The truth values are spread across the domain. The membership function associated with "absolutely true" will be placed at the midpoint of the domain. Such a fuzzy set appears as shown in Figure 3.105.

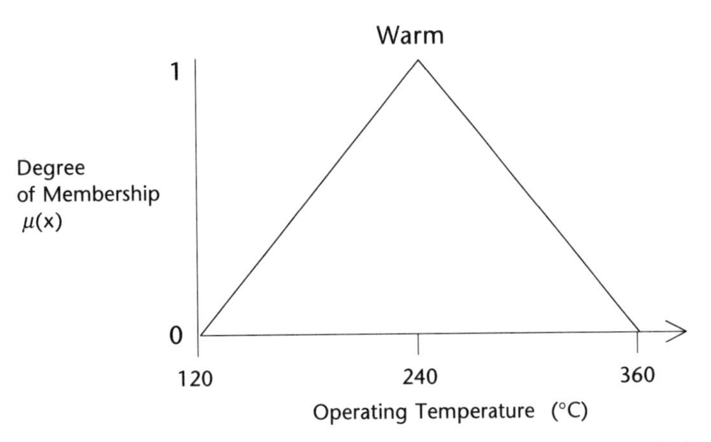

Figure 3.105 Generating a Triangular Function by Linear Interpolation

Creating a truncated triangular fuzzy set with a crossover plateau (also called a trapezoidal set function) is simplified with the truth series approach to set description. For example:

```
//--Create the WARM fuzzy set
memSeries[]={0,1,1,1,1,0};
memCnt=6;
FzyMemSeries(WarmFDBptr,memSeries,memCnt,&status);
```

The width of the plateau is automatically determined from the proportional number of true [1] membership points. This produces the fuzzy set of Figure 3.106.

We will discuss the details on triangular, trapezoidal, and shouldered fuzzy sets, using the coordinate mapping approach, in the next section. The truth series specification provides an alternative method of specifying similar curves across the fuzzy set's domain without, however, the same precise positioning available when membership grades are coupled with actual domain elements.

The membership series method of specifying a fuzzy set is implemented in the following function. The truth series must be normal, and you must include at least two membership points.

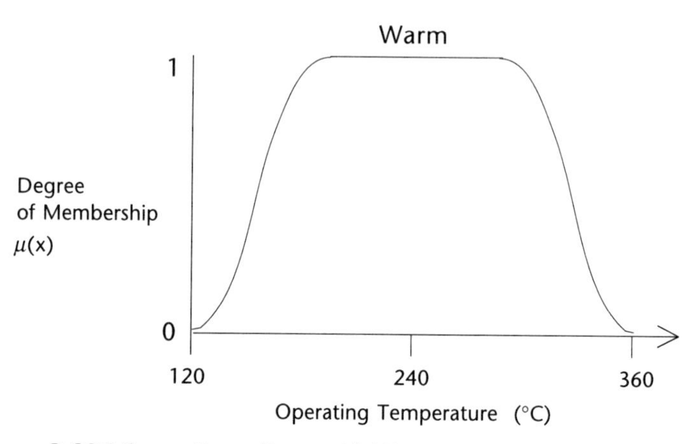

Figure 3.106 Generating a Trapezoidal Function by Linear Interpolation

```
#include <stdio.h>
#include "FDB.hpp"
#include "fuzzy.hpp"
#include "mtypes.hpp"
#include "mtsptype.hpp"
void FzyMemSeries(
      FDB* FDBptr,float TruthSeries[],const int Scnt,int *sta-
tusPtr)
   {
    bool Normal;
    int  i,
         j,
         Serlen,
         remaining;
    float       TempVector[VECMAX];
    const float NUDGE=0.99999;
    *statusPtr=0;
    Normal=FALSE;
    for(i=0;i<Scnt;i++)
      if(TruthSeries[i]==1.0) Normal=TRUE;
    if(!Normal) {*statusPtr=1;return;}
    FzyInitVector(TempVector,VECMAX,MINUSONE);
    TempVector[0]=TruthSeries[0];
    Serlen=Scnt-1;
    for(remaining=1;remaining<Serlen+1;remaining++)
       {
       j=(int)
          (NUDGE+((float)remaining/(float)Serlen)*(VECMAX-1));
```

```
        if(j>VECMAX) {*statusPtr=66; return;}
        TempVector[j]=TruthSeries[remaining];
       }
    FzyInterpVec(TempVector,statusPtr);
    if(*statusPtr!=0) {*statusPtr=88;return;}
    FzyCopyVector(FDBptr->FDBvector,TempVector,VECMAX);
    return;
   }
```

Listing 3.21 Creating a Fuzzy Set Through a Truth Membership Series

This is a simple function. A temporary working vector is initialized to "–1" (required for the use of the vector interpolation function, *FzyInterpVec*). We place the first membership value from the incoming *TruthSeries[]* array into the first position of the temporary working vector, *TempVector[]*. We then calculate each new array position so that the remaining membership values are spread equidistantly across the membership vector.

In the following example (from the *FZYCOORD.CPP* program), we create a truth series fuzzy set with a specific plateau at the [.5] grade level. We assume that the fuzzy set has already been created and that **FDBptr* points to its address (Figure 3.107).

```
FzyInitFDB(FDBptr);
FDBptr->FDBdomain[0]=16;
FDBptr->FDBdomain[1]=36;
//
//---Fuzzy Set via a Membership Function
//
strcpy(FDBptr->FDBid,"MemSeries");
float memvalues[]={0,.1,.1,.1,.3,.5,.5,.5,.5,.6,.8,1};
int   memcnt=12;
FzyMemSeries(FDBptr,memvalues,memcnt,&status);
FzyDrawSet(FDBptr,SYSLOGFILE,&status);
```

```
  FuzzySet:    MemSeries
  Description:
1.00                                                                .
0.90                                                             ...
0.80                                                          ...
0.70                                                        ..
0.60                                                    ....
0.50                             .......................
0.40                           ..
0.30                        ...
0.20                     ...
0.10      ...............
0.00......
    0---|---|---|---|---|---|---|---|---|---|---|---|---|---|---|---0
  16.00   18.50   21.00   23.50   26.00   28.50   31.00   33.50   36.00
  Domained:        16.00 to     36.00
```

Figure 3.107 Fuzzy Set from an Arbitrary Truth Membership Series

The triangular fuzzy set (see Figure 3.105 and associated commentary) is easily produced with the truth series technique. Figure 3.108 shows part of the output from the *FZYCOORD.CPP* program.

```
FzyInitFDB(FDBptr);
FDBptr->FDBdomain[0]=120;
FDBptr->FDBdomain[1]=360;
strcpy(FDBptr->FDBid,"Triangle");
float mem2values[]={0,1,0};
memcnt=3;
FzyMemSeries(FDBptr,mem2values,memcnt,&status);
FzyDrawSet(FDBptr,1,&status);
```

```
    FuzzySet:    Triangle
    Description:
  1.00                                      .
  0.90                                  ... ...
  0.80                                ...     ...
  0.70                             ...           ...
  0.60                         ....                 ....
  0.50                      ...                         ...
  0.40                   ...                               ...
  0.30                ...                                     ...
  0.20             ...                                           ...
  0.10         ....                                                 ....
  0.00...                                                                ...
     0---|---|---|---|---|---|---|---|---|---|---|---|---|---|---|---|---0
  120.00  150.00  180.00  210.00  240.00  270.00  300.00  330.00  360.00
   Domained:        120.00 to      360.00
```

Figure 3.108 Triangular Fuzzy Set from a Three-Point Truth Membership Series

Plateau or trapezoid fuzzy sets (see Figure 3.106 and associated commentary) can also be easily produced by specifying the number of points across the plateau. As you can see, this is simply a special case of the triangular generating technique (with the apex consisting of multiple μ [1.0] points instead of one). Figure 3.109 is also part of the output from the *FZYCOORD.CPP* program.

```
strcpy(FDBptr->FDBid,"Trapezoid");
float mem3values[]={0,1,1,1,1,0};
memcnt=6;
FzyMemSeries(FDBptr,mem3values,memcnt,&status);
FzyDrawSet(FDBptr,1,&status);
```

```
       FuzzySet:    Trapezoid
       Description:
     1.00                 ...................................
     0.90                .                                   .
     0.80               .                                     .
     0.70             ..                                       ..
     0.60            .                                           .
     0.50           .                                             .
     0.40          .                                               ..
     0.30        ..                                                  .
     0.20       .                                                     .
     0.10      .                                                       .
     0.00..                                                             ..
         0---|---|---|---|---|---|---|---|---|---|---|---|---|---|---|---0
      120.00  150.00  180.00  210.00  240.00  270.00  300.00  330.00  360.00
       Domained:        120.00 to      360.00
```

Figure 3.109 Trapezoid Fuzzy Set from a Six-Point Truth Membership Series

Domain-Based Coordinate Memberships

A straightforward method for defining any fuzzy set is simply to specify a value from the domain and its membership in the set. This description consists of a series of number pairs—the domain value and a truth membership value. The components are separated by a slash (/). This has the form of:

$Scalar_i/Grade_i$

"Scalar" is a number drawn from the fuzzy set's domain, and "Grade" is the scalar's degree of membership in the fuzzy set. The scalars can be specified in any order. As an example, in an insurance underwriting system, the fuzzy set that describes the risk associated with a driver's age is "U" shaped. The coordinate set could be described as the following seven pairs:

16/1 21/.6 28/.3 68/.3 76/.5 80/.7 96/1

Figure 3.110 shows how these coordinates specify points along the domain of a fuzzy set. All the points must lie within the domain of the fuzzy set, and at least one of these points should have a truth value of [1.0]. Once these points have been positioned, linear interpolation is used to create the fuzzy surface. Figure 3.111 shows how this final curve will appear.

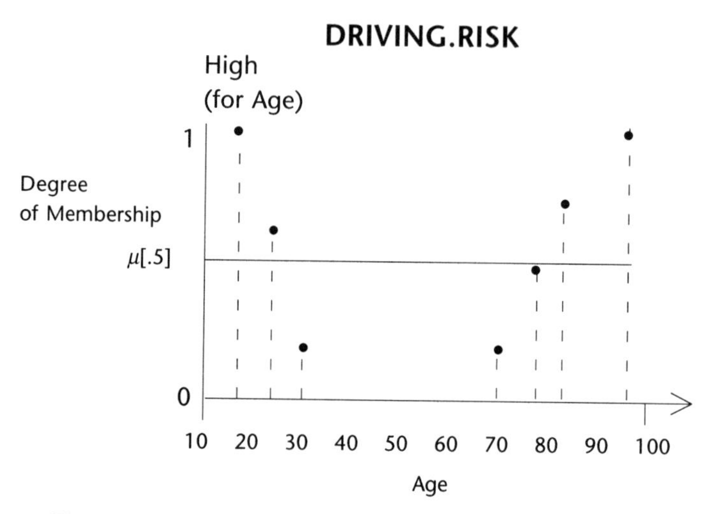

Figure 3.110 The Coordinate Points for High Driving Risk

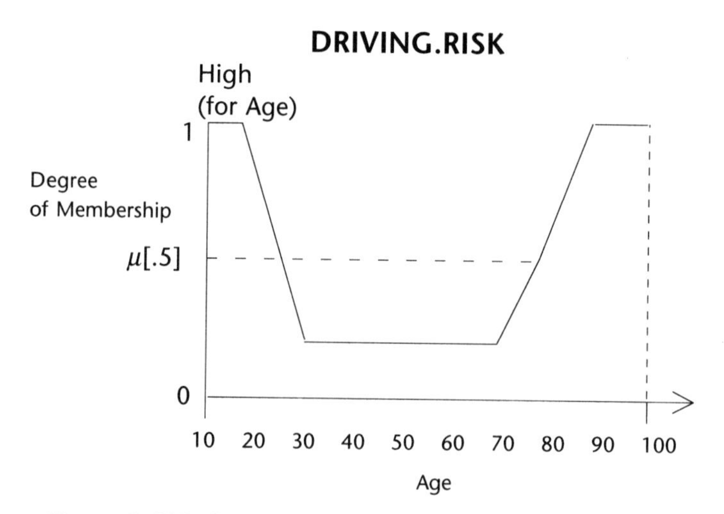

Figure 3.111 The Connected Curve for High Driving Risk

You must provide at least two points for the coordinate fuzzy set process, but naturally the more points you provide, the finer the curve detail.

The *FzyCoordSeries* function accepts two array parameters—the first contains the domain values and the second contains the corresponding truth membership values. Using these parameters, a new truth membership function for the fuzzy set is constructed.

```
#include "FDB.hpp"
#include "mtypes.hpp"
#include "mtsptype.hpp"
void FzyCoordSeries(
  FDB* FDBptr,
   double Scalars[],
   float Grades[],
   const int ValCnt,
   int *statusPtr)
  {
   int     i,j,Cellpos;
   double  Hi,Lo,Range;
   float   tempVector[VECMAX+1];
   bool    Normalset;

   *statusPtr=0;
   Hi=FDBptr->FDBdomain[1];
   Lo=FDBptr->FDBdomain[0];
   Range=Hi-Lo;
   Normalset=FALSE;
   FzyInitVector(tempVector,VECMAX,MINUSONE);
   for(i=0;i<ValCnt;i++)
    {
     Cellpos=(int)((Scalars[i]-Lo)/Range)*VECMAX);
     tempVector[Cellpos]=Grades[i];
     if(Grades[i]==1.0) Normalset=TRUE;
    }
   if(tempVector[0]==-1) tempVector[0]=0.0;
   FzyInterpVec(tempVector,statusPtr);
   FzyCopyVector(FDBptr->FDBvector,tempVector,VECMAX);
   if(Normalset==FALSE) FzyNormalizeSet(FDBptr);
   return;
  }
```

Listing 3.22 Creating a Fuzzy Set from Coordinate Pairs (`mtfzcds`)

The *FzyCoordSeries* is simple and straightforward. After initializing a temporary working vector to "–1," we examine each of the domain values. For each domain value (*Scalars[i]*), its cell position in the truth membership array (*tempVector[Cellpos]*) is calculated. This position is filled with the corresponding truth value (*Grades[i]*). After completing the temporary vector, the linear interpolation routine is called to complete the fuzzy set surface. The coordinate routine keeps track of the maximum membership value in the *Grades[*]* array. If the fuzzy set is not normal, it is automatically normalized.

Figure 3.112 (from the *FZYCOORD.CPP* program) shows the *High Driving Risk* fuzzy set created by specifying the domain value and the associated truth membership grade. See Figure 3.110 and associated commentary for additional detail.

```
FzyInitFDB(FDBptr);
FDBptr->FDBdomain[0]=16;
FDBptr->FDBdomain[1]=100;
strcpy(FDBptr->FDBid,"High Driving Risk");
fprintf(stdout,"%s\n","A coordinate-curve fuzzyset");
Values[0]=16; Members[0]=1.0;
Values[1]=21; Members[1]=.6;
Values[2]=28; Members[2]=.3;
Values[3]=68; Members[3]=.3;
Values[4]=76; Members[4]=.5;
Values[5]=80; Members[5]=.7;
Values[6]=96; Members[6]=1.0;
Values[7]=100;Members[7]=1.0;
Vcnt=8;
FzyCoordSeries(FDBptr,Values,Members,Vcnt,&status);
FzyDrawSet(FDBptr,SYSLOGFILE,&status);
```

```
  FuzzySet:    High Risk
  Description:
1.00.                                                          ....
0.90 .                                                     ....
0.80                                                   ....
0.70  .                                            ....
0.60   .                                          .
0.50    ..                                      ..
0.40      ..                                 ...
0.30        ..................................
0.20
0.10
0.00
     0---|---|---|---|---|---|---|---|---|---|---|---|---|---|---|---0
  16.00   26.50   37.00   47.50   58.00   68.50   79.00   89.50  100.00
  Domained:          16.00 to      100.00
```

Figure 3.112 The *High Driving Risk* for Age Coordinate-Based Fuzzy Set

The *FzyGetCoordinates* function provides a shell that accepts a character string containing the pairs of domain and membership grade values and returns the arrays that can be used by *FzyCoordSeries*. A count of the extracted pairs is also returned. As an example:

```
//--Specify the High Risk Fuzzy Region
char *CoordSeries=
"16/1 21/.6 28/.3 68/.3 76/.5 80/.7 96/1";
FzyGetCoordinates(
CoordSeries,Scalars,Grades,&Paircnt,&status);
```

This allows us to specify the shape of a fuzzy region in a buffer and then quickly convert this to the coordinate arrays. Listing 3.23 is a nearly complete listing of the coordinate processing function.

```
#include <string.h>
#include <stdlib.h>
#include "mtsptype.hpp"
void FzyGetCoordinates(
  char   *Coords,
  double  Values[],
  float   Members[],
  int    *VcntPtr,
  int    *statusPtr)
  {
    char   ScalarPart[16],MemberPart[16],thisToken[16];
    char  *Tokens[128];
    char  *PgmId="mtfzgcd";
    int    i,j,k,n,NumType,TokCnt;
    int    PartLength=16;
    int    MaxTokens=128;
    double Scalar,MFvalue;

    *VcntPtr=0;
    *statusPtr=0;
//--Step 1. Entoken the character string into a set
//--of individual value+membership elements. This simplifies
//--our processing of each element.
    TokCnt=0;
    MtsEntoken(Coords,Tokens,MaxTokens,&TokCnt,statusPtr);
    if(*statusPtr!=0) return;
```

```
    if(TokCnt==0)      return;
//--Step 2. For each token, extract the scalar value and the
//--membership. We convert the scalar to a double precision
//--number and the membership to the single precision
//--floating-point number.
    for(i=0;i<TokCnt;i++)
     {
       strcpy(thisToken,Tokens[i]);
       j=MtsStrIndex(thisToken,'/');
       memset(ScalarPart,'\0',PartLength);
       for(n=0;n<j;n++) ScalarPart[n]=thisToken[n];
       if(!(MtsIsNumeric(ScalarPart,&NumType)))
         {
          *statusPtr=1;
          return;
         }
       Scalar=atof(ScalarPart);
       memset(MemberPart,'\0',PartLength);
       for(n=j+1,k=0;thisToken[n];n++,k++)
                   MemberPart[k]=thisToken[n];
       if(!(MtsIsNumeric(MemberPart,&NumType)))
         {
          *statusPtr=3;
          return;
         }
       MFvalue=atof(MemberPart);
       Values[*VcntPtr]=Scalar;
       Members[*VcntPtr]=MFvalue;
       (*VcntPtr)++;
     }
    return;
 }
```

Listing 3.23 Building the Coordinate Arrays from Domain and Grade Specifications

This program calls the utility entokening routine (*MtsEntoken*) to break the string down into a table of coordinate pairs. The general entokener recognizes both blanks and commas as pair delimiters. Each pair is then decomposed into its parts, which are converted to actual numeric representations. The number of points you can process is limited by the size of *Tokens[]*, which is currently set to 128 elements. The *Values[]* and *Members[]* arrays must be dimensioned to handle the maximum possible number of coordinate points.

NOTES

1. Using the Boolean convention of equating one (1) with *True* and zero (0) with *False*, our partial degrees of truth must lie on a scale somewhere between zero and one. The distance used to measure the complement of *Hot* is simply 1 minus x, where x is the partial degree of membership for the idea of *Hot*. Later, when we begin to define formally the ideas underlying fuzzy set theory, this concept will play an important part.

2. Note that we qualified the scope of the fuzzy set. As we will soon see, all fuzzy sets are context-dependent. In fuzzy models, a concept such as *High* can qualify such concepts as budgets, project priorities, staffing levels, capital expenditures, etc.

3. In fact, a small jar with ten marbles might be considered largely full, but a jar with a few gas molecules would probably be considered a perfect vacuum. Again, we must remember that the meaning of a fuzzy set is context-dependent.

Four

Fuzzy Set Operators

Se non e vero, e molto ben trovato
If it is not true, it is a happy invention

Giordano Bruno (1585)

All great truths begin as blasphemies.

George Bernard Shaw (1856-1950)
Annajanska (1919), p. 262

Having examined the architecture of a fuzzy set, we are now ready to look at the various operations that can be reasonably applied to fuzzy sets and general fuzzy regions. We begin with a brief review of the operations normally associated with conventional or Boolean sets and then move quickly into the set theoretic operations associated with fuzzy logic. We are also going to look at some of the operations not associated nor allowed with Boolean sets, in particular, alternative interpretations of conjunctive and disjunctive operators, and the meaning of a fuzzy complement. It is in this area, the interpretation of the complement in fuzzy logic, that we will encounter one of the striking logical principles that separates fuzzy logic from conventional (Boolean) logic.

Conventional (Crisp) Set Operations

In traditional Boolean logic, sets are considered as bivalent systems with their states alternating between inclusion and exclusion. The characteristic or discriminant function reflects this two-valued space:

$$\mu_A[x] = \begin{matrix} 0 & x \notin A \\ 1 & x \in A \end{matrix}$$

This equation says that the membership function for the set A is zero [0] if x **is not** an element in A, and the membership function is one [1] if x *is* an element in A. Since there are only two states, the transition between these states is always immediate and crisp (thus, the term "crisp" sets). Members of crisp sets are always fully categorized, and there is no ambiguity or dichotomy about the membership. As Figure 4.1 illustrates, there are four basic set operations that can be performed on crisp sets.

The union of sets S_1 and S_2 ($S_1 \cup S_2$) contains all the elements that appear in either set S_1 or in set S_2. The intersection of sets S_1 and S_2 ($S_1 \cap S_2$) contains all the elements that appear in both set S_1 and in set S_2. The complement of a set, such as S_1 (and indicated by $\sim S_1$), consists of all the elements, drawn from the possible universe of the set (in this case, the domain of all real positive numbers), that are not in the set.

Another kind of union, called the *exclusive-OR*, and represented by $S_1 \oplus S_2$, contains all the elements that are in S_1 or in S_2, but not in both. Exclusive-OR operations are important in Boolean systems (especially in electronic gate switching, computer instruction operations, and signal process filtering). As we shall see, fuzzy exclusive-OR operations combine the idea of partial membership with the α-cut threshold partition discussed in the previous chapter.

Basic Zadeh-Type Operations on Fuzzy Sets

Like conventional sets, there are specifically defined operations for combining and modifying fuzzy sets. These set theoretic functions provide the fundamental tools of the logic. Following the conventional fuzzy logic operations initially defined by Zadeh, the basic operations are:

Intersection	$A \cap B$	$\min(\mu_A[x], \mu_B[y])$
Union	$A \cup B$	$\max(\mu_A[x], \mu_B[y])$
Complement	$\sim A$	$1 - \mu_A[x]$

Since fuzzy sets are not crisply partitioned in the same sense as Boolean sets, these operations are applied at the truth membership level. As a consequence of a fuzzy set's somewhat fluid characteristic function, deciding whether or not a value is a member of any particular set requires some notion about how the set is constructed, the manifold of the connecting surface, and the current α-threshold limits.

Fuzzy Set Membership and Elements

An element is a member of a fuzzy set if (1) it falls within the underlying domain of the set, (2) its membership truth value is greater than zero, and (3) it lies above the current α-cut threshold. While considerations of the α-cut threshold are not usually important at

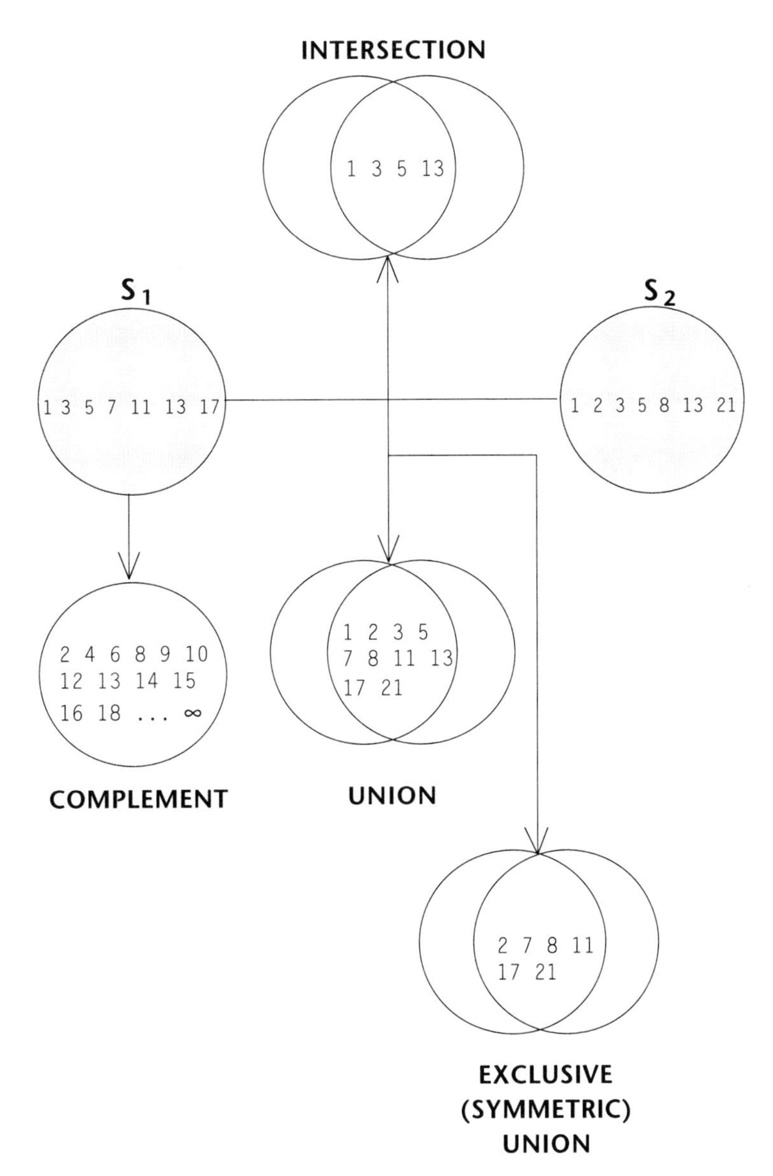

Figure 4.1 Crisp Set Operations

the static (permanent) fuzzy set level, they are essential when evaluating the active fuzzy region produced by fuzzy set theoretic and aggregation processes. Listing 4.1 is a Boolean function indicating whether or not an arbitrary value (`domainval`) is a member of the fuzzy set.

The Rules of Noncontradiction

While I am assuming a familiarity with the basic notions and operations of classical set theory, there is an important distinction between fuzzy set logic and crisp set logic that is, in principle, an abrupt change in the way we view the relationship of things in the real world. This characteristic of the classical set universe deals with the way in which the intersection and union of complementary sets are handled. Going back to the time of Aristotle, classical symbolic logic has always maintained two inviolate laws: The Law of Noncontradiction and the Law of the Excluded Middle.

Noncontradiction

The Law of Noncontradiction,

$$S_1 \cap \sim S_1 \equiv \varnothing$$

says that "the intersection of a set with its complement results in an empty or null set."

That is, there are no elements that exist in the set S_1 and that also exist in its complement $\sim S_1$. This would be, in the eyes of Aristotle (and also according to the everyday laws of Boolean logic), a contradiction in the logical process (hence the name). In the highly dichotomous world of Boolean logic, this law makes complete sense. If we consider the set $P_{expensive}$ of expensive projects, where expensive, in the crisp sense, is defined as

$$\mu_{\text{expensive}}[x] = \{\text{budget} \geq \$1{,}500{,}000\}$$

then, by definition, the complement of this set—P_{cheap}

$$\mu_{\sim\text{expensive}}[x] = \{\text{budget} < \$1{,}500{,}000\}$$

contains all the projects whose authorized budgets are less than $1,500,000. The intersection of the sets $P_{expensive}$ and P_{cheap} is empty since no project can be common to both sets. In fuzzy logic, however, we will shortly find that this convention—an atomic principle of Boolean logic—is routinely broken. In fact, some of the strongest and deepest capabilities in fuzzy reasoning stem from its failure to obey this dictum of Boolean logic.

Excluded Middle

The Law of the Excluded Middle,

$$S_1 \cup \sim S_1 \equiv X$$

says that "the union of a set with its complement results in the universal set of the underlying domain."

This law of Boolean logic, if you give it a moment's thought, is a direct outgrowth of the Law of Noncontradiction since its assumes that the sets in question contain completely separate and disjoint elements (the results of applying the negation operator to the initial set in the first place). In this way, the Law of the Excluded Middle is symmetrical with the Law of Noncontradiction—one operation insures the disjointedness of sets along an arbitrarily crisp boundary (the set definitional edges) and the other operation combines the disjoint sets to produce the conjoint universal domain. Since fuzzy logic does not adhere to the Law of Noncontradiction, it follows that it will not obey the Law of the Excluded Middle. These aspects of fuzzy logic are discussed in detail when we turn to the idea of fuzzy complements.

In deciding whether of not a number (scalar) is currently a member of a fuzzy set, we get its truth function value (grade of membership) and see if it is either zero or falls below the current alpha-cut threshold established in the model's system control block. Notice that we are interrogating its current membership—changing the alpha-cut threshold may alter this membership judgment.

```
bool FzyIsMemberof(
  FDB *FDBptr,double Scalar,float *grade,int *statusPtr)
 {
  int  i;
  *statusPtr=0;
  *grade=FzyGetMembership(FDBptr,Scalar,&i,statusPtr);
  if(*grade==0)                return(FALSE);
  if(*grade<XSYSctl.XSYSalfacut) return(FALSE);
  return(TRUE);
 }
```

Listing 4.1 Determine Whether a Number Is in a Fuzzy Set (`mtfziss`)

There may be some occasions when we want to relax the truth membership restriction on fuzzy set membership. For sets that have discontinuous surfaces (they are zero at some points between the low and high end of the explicit domain), a value contained within the domain, but falling within these empty or cusp sectors, might still be considered within the proper fuzzy set. Such a relaxation of fuzzy set membership is occasionally important in determining whether or not the domains of two fuzzy sets are comparable or compatible.

The Intersection of Fuzzy Sets

In a crisp system the intersection of two sets contains the elements that are common to both sets. This is equivalent to the arithmetic or logical AND operation. In conventional fuzzy logic, the AND operator is supported by taking the minimum of the truth membership grades. Table 4.1 (produced by the ZADEHAND.CPP program) gives the fuzzy AND value for representative *x* and *y* memberships.

ZADEH 'AND' OPERATOR

	0.0000	0.2500	0.5000	0.7500	1.0000
0.00	0.0000	0.0000	0.0000	0.0000	0.0000
0.25	0.0000	0.2500	0.2500	0.2500	0.2500
0.50	0.0000	0.2500	0.5000	0.5000	0.5000
0.75	0.0000	0.2500	0.5000	0.7500	0.7500
1.00	0.0000	0.2500	0.5000	0.7500	1.0000

Table 4.1 The Zadeh AND (Intersection) Truth Table

The intersection operator is the most common form of restriction used in fuzzy rule antecedents. When we evaluate rules in a form such as

```
if x is Y and z is W then m is P;
```

the elastic membership strength between the consequent (outcome) *m* and the fuzzy region *P* is determined by the strength of the premise or antecedent. The truth of this antecedent is determined by taking the min(μ[x is Y],μ[z is W]).

To see how fuzzy intersections differ from their classical counterparts, we can examine the effects of intersection operations on both crisp and fuzzy sets. Consider the crisp representations for *MiddleAged* and *Tall* shown in Figures 4.2 and 4.3. A characteristic function for MiddleAged is given by:

$$\mu_{middleaged}[x] = \{Age \geq 35 \wedge Age \leq 45\}$$

Thus, the members of this set are all individuals between the ages of 35 and 45, inclusive. The membership graph for this crisp set appears in Figure 4.2.

Figure 4.2 Crisp Set Operations

The characteristic function for *Tall* is given by,

$$\mu tall[x] = \{Height \geq 6'\}$$

specifying all individuals over six feet in height. The membership graph for this set appears in Figure 4.3.

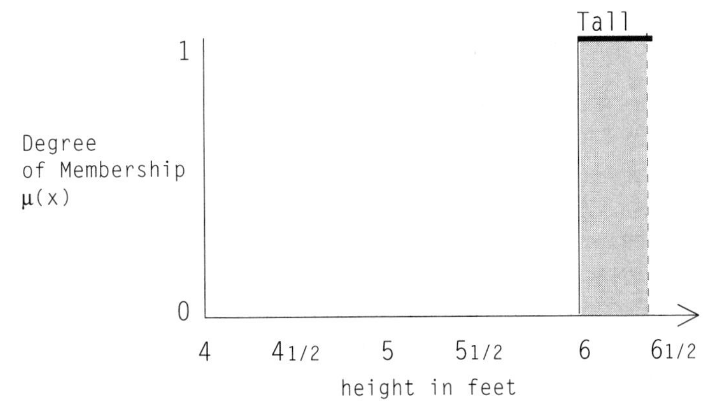

Figure 4.3 Crisp Representation of *Tall*

We want to answer a simple question: *What members of a sample population are both MiddleAged and tall?* To answer this question, we select a small sample of project managers (for a list of names, see Table 4.2, which we will use again throughout this chapter).

NAME	AGE	HEIGHT
ABLE	36	5'9"
COLLINS	58	6'0"
FRANKLE	64	5'7"
JONES	32	6'2"
MILLER	40	6'1"
PETERS	22	5'3"
SMITH	47	5'11"
WILLIAMS	25	6'0"

Table 4.2 Project Manager Profiles Showing Age and Height

In Boolean logic, those individuals that are both MiddleAged and tall are found with the AND operator. We can visualize this process as the bitwise ANDing of Boolean vectors representing the truth of the set characteristic expression for each category. Since these are Boolean discriminants, the results can only be true [1] or false [0]. Table 4.3 shows the result of applying the Boolean AND operator to this sample population.

Only a single instance of this table meets the strict dichotomous edge enforced by the Boolean AND operator. In the intersection operation, both characteristic function values must be true.

What is the intersection of *Tall* and *MiddleAged* people in a fuzzy system? Before we can answer that, we need to define these concepts in terms of fuzzy spaces. In a fuzzy set the membership edge is not abruptly disjoint at a particular domain value, but reflects the degree of membership for values that lie at the extremes of our reasonable expectations for set membership. Figure 4.4 is a fuzzy set representing *MiddleAged*.

This set starts at 25 as the youngest age that someone might have the smallest degree of middle-agedness. The membership curve climbs steadily until it reaches 40 years old—the absolute quintessence of MiddleAgedness. After 40, the curve drops off again, so that someone around 50 is only moderately MiddleAged, and someone over 55 has no membership in the concept MiddleAged.

The same kind of analysis is applied in Figure 4.5 to the concept of *Tall*. For simplicity, we take a linear view of tallness—as your height increases, your membership in *Tall*

	MiddleAged age≥35^ age≤45	**tall height≥6**	**middleaged ∩ tall**
ABLE	1	0	0
COLLINS	0	1	0
FRANKLE	0	0	0
JONES	0	1	0
MILLER	1	1	1
PETERS	0	0	0
SMITH	0	0	0
WILLIAMS	0	1	0

Table 4.3 The AND Bit Vectors for MiddleAged and Tall

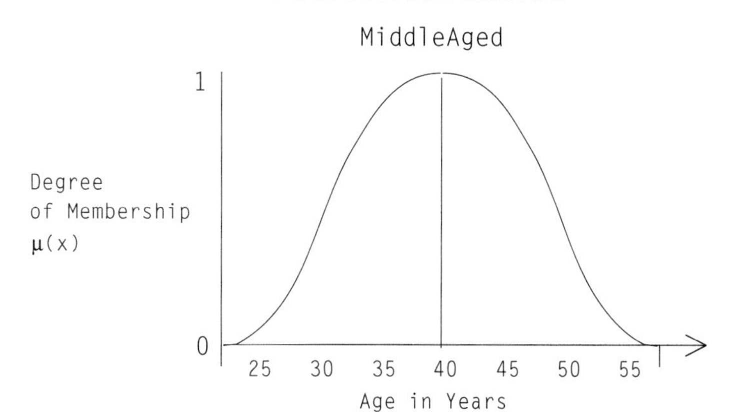

Figure 4.4 Fuzzy Representation of *MiddleAged*

increases proportionally. The fuzzy set starts at 4 feet as the value with no membership in the set and increases linearly until we reach unity at 6.5 feet. All heights below 4 feet are absolutely *Not Tall* and all heights above 6.5 are absolutely *Tall.*

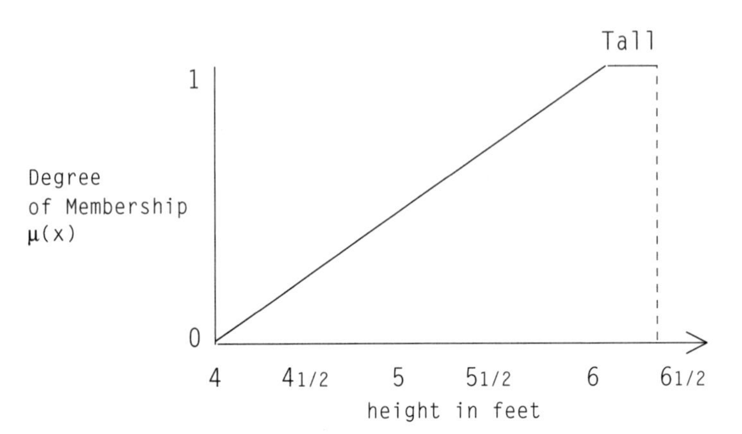

Figure 4.5 Fuzzy Representation of *Tall*

Following the basic Zadeh rules for fuzzy intersection, the common space between the two sets is found by applying the operation:

$$A \cap B = \min(\mu_A[x], \mu_B[y])$$

That is, you take the minimum of the two membership values when the fuzzy truth function (T) is applied to the implicit fuzzy assertions:

$$\begin{bmatrix} Age \in MiddleAged & T \leftarrow \mu_A[x] \\ Height \in Tall & T \leftarrow \mu_B[y] \end{bmatrix}$$

Table 4.4 shows the results of intersecting the fuzzy sets *MiddleAged* and *Tall* for each instance in the sample population. The final, right-most, column shows the truth value resulting from the intersection or *AND*ing of the two truth values for *MiddleAged* and *Tall*.

The intersection determines the compatibility between each fuzzy set and its domain value. This produces a truth membership value. The minimum of these truth values is used to determine the degree to which the sets are intersected. In contrast to the Boolean intersection, the fuzzy operation identifies five instances. In particular, we can see that both Able and Smith have been included in the selection—corresponding to the common sense view that, although they fall outside a strict Boolean partition, they are semantically close to the concept we want enforced (as opposed to enforcing a rigid arithmetic rule).

	Age is MiddleAged	Height is Tall	MiddleAged and Tall
ABLE	0.92	0.84	0.84
COLLINS	0.00	0.92	0.00
FRANKLE	0.00	0.68	0.00
JONES	0.47	0.96	0.47
MILLER	1.00	0.94	0.94
PETERS	0.00	0.39	0.00
SMITH	0.74	0.90	0.74
WILLIAMS	0.10	0.92	0.10

Table 4.4 The AND Truth Vectors for MiddleAged and Tall

The basic Zadeh AND operator is implemented by taking the minimum of the membership values in two fuzzy sets. Since the intersection of two fuzzy sets is often used as a new fuzzy region in a model, we produce the AND by creating and populating a new fuzzy set.

```
#include <math.h>
float min(float,float);
float FzyAND(float truth1,float truth2)
  {
    if(truth1<0||truth1>1) {return(-1);}
    if(truth2<0||truth2>1) {return(-2);}
    return(min(truth1,truth2));
  }
```

Listing 4.2 Performing the Simple Zadeh AND on Two Truth Values

```
#include "FDB.hpp"
#include "mtypes.hpp"
#include "mtsptype.hpp"
FDB *FzyApplyAND(FDB *FDBptr1,FDB *FDBptr2,int *statusPtr)
 {
```

```
    FDB  *FDBoutptr,*FDBnull=NULL;
    int   i;
    *statusPtr=0;
    if(!(FDBoutptr=new FDB))
      {
       *statusPtr=1;
       return(FDBnull);
      }
    FzyInitFDB(FDBoutptr);
    FzyCopySet(FDBptr1,FDBoutptr,statusPtr);
    for(i=0;i<VECMAX;i++)
       {
        FDBoutptr->FDBvector[i]=
          min(FDBptr1->FDBvector[i],FDBptr2->FDBvector[i]);
       }
    return(FDBoutptr);
  }
```

Listing 4.3 Taking the Intersection (AND) of Two Fuzzy Sets

The Union of Fuzzy Sets

The union of two sets—finding those elements that are in either of the sets—is performed with the OR operator. Using conventional fuzzy logic, the OR operator is supported by taking the maximum of the truth membership grades. Table 4.5 (produced by the ZADEHOR.CPP program) gives the fuzzy OR value for representative x and y memberships.

ZADEH 'OR' OPERATORS

	0.0000	0.2500	0.5000	0.7500	1.0000
0.00	0.0000	0.2500	0.5000	0.7500	1.0000
0.25	0.2500	0.2500	0.5000	0.7500	1.0000
0.50	0.5000	0.5000	0.5000	0.7500	1.0000
0.75	0.7500	0.7500	0.7500	0.7500	1.0000
1.00	1.0000	1.0000	1.0000	1.0000	1.0000

Table 4.5 The Zadeh OR (Union) Truth Table

The fuzzy OR operator is less frequently used in system modeling since its effect can, for relatively simple predicates, be decomposed into multiple fuzzy propositions.[1] As an example, the rule

```
if x is Y or z is W then m is P;
```

can be rewritten as two separate fuzzy rules

if x is Y then m is P
if z is W then m is P

In both cases, the elastic membership strength between the consequent *m* and the fuzzy region *P* is determined by taking the `max(μ[x is Y],μ[z is W])`. Naturally, for more complex antecedent expressions, this restructuring may be impossible or very difficult. Even in cases where the restructuring is possible, we might not want to sacrifice logical clarity and long-term maintenance simply to avoid using the `OR` operator.

Using the same problem of finding tall and MiddleAged project managers, we can examine the effects of the union operator on crisp and fuzzy sets. In this case, we want to answer the question: *What members of the population are either MiddleAged or tall?*

Like the `AND` process, we can visualize this process as the `OR`ing of bit vectors representing the truth of the characteristic function at each instance in the population. Table 4.6 shows the results of applying the Boolean `OR` operator to the population of project managers.

	MiddleAged age≥34^ age≤45	**tall height≥6'**	**middleaged ∪tall**
ABLE	1	0	1
COLLINS	0	1	1
FRANKLE	0	0	0
JONES	0	1	1
MILLER	1	1	1
PETERS	0	0	0
SMITH	0	0	0
WILLIAMS	0	1	1

Table 4.6 The *OR* Bit Vectors for *MiddleAged* and *Tall*

The `OR` operator provides the least amount of discrimination in the sets and selects any case where either of the discriminant functions is true. Instead of a single instance chosen by the `AND` process, the `OR` operator selects five of the eight (or 62.5%).

Building on the fuzzy sets used in the intersection example, we now ask: *What is the union of Tall and MiddleAged in a fuzzy system?* Following the Zadeh rules for fuzzy union, the common conjoint space between the two sets is found by applying the following operation across the membership spaces of each set:

$$A \cup B = \max(\mu_A[x], \mu_B[y])$$

We take the maximum of the two membership values when the fuzzy truth function (T) is applied to the implicit fuzzy assertions:

$$\begin{bmatrix} Age \in MiddleAged & T \leftarrow \mu_A[x] \\ Height \in Tall & T \leftarrow \mu_B[y] \end{bmatrix}$$

Table 4.7 shows the results of union of the fuzzy sets *MiddleAged* and *Tall* for each instance in our sample population.

	Age is MiddleAged	**Height is Tall**	**MiddleAged or Tall**
ABLE	0.92	0.84	0.92
COLLINS	0.00	0.92	0.92
FRANKLE	0.00	0.68	0.68
JONES	0.47	0.96	0.96
MILLER	1.00	0.94	1.00
PETERS	0.00	0.39	0.39
SMITH	0.74	0.90	0.90
WILLIAMS	0.10	0.92	0.92

Table 4.7 The AND Truth Vectors for *MiddleAged* and *Tall*

The union determines the compatibility between each fuzzy set and its domain value. This process then produces a truth membership value. The maximum of these truth values is used to determine the degree to which the sets are combined.

The basic Zadeh OR operator is implemented by taking the maximum of the membership values in two fuzzy sets. Since the intersection of two fuzzy sets is often used as a new fuzzy region in a model, we produce the OR by creating and populating a new fuzzy set.

```
#include <math.h>
float max(float,float);
float FzyOR (float truth1,float truth2)
  {
    if(truth1<0||truth1>1) {return(-1);}
    if(truth2<0||truth2>1) {return(-2);}
    return(max(truth1,truth2));
  }
```

Listing 4.4 Performing the Simple Zadeh OR on Two Truth Values

```
#include "FDB.hpp"
#include "mtypes.hpp"
#include "mtsptype.hpp"
FDB *FzyApplyOR(FDB *FDBptr1,FDB *FDBptr2,int *statusPtr)
 {
   FDB  *FDBoutptr,*FDBnull=NULL;
   int   i;

   *statusPtr=0;
   if(!(FDBoutptr=new FDB))
     {
      *statusPtr=1;
      return(FDBnull);
     }
   FzyInitFDB(FDBoutptr);
   FzyCopySet(FDBptr1,FDBoutptr,statusPtr);
   for(i=0;i<VECMAX;i++)
      {
       FDBoutptr->FDBvector[i]=
         max(FDBptr1->FDBvector[i],FDBptr2->FDBvector[i]);
      }
   return(FDBoutptr);
 }
```

Listing 4.5 Taking the Union (OR) of Two Fuzzy Sets

The Complement (Negation) of Fuzzy Sets

A *complement*, or negation, of set S_1 contains all the elements that are not in S_1 and is represented by the characteristic:

$$\begin{bmatrix} \mu_{\sim S_1}[x] = 0 & x \in S_1 \\ \mu_{\sim S_1}[x] = 1 & x \notin S_1 \end{bmatrix}$$

To see how the complement of a fuzzy region differs significantly from its Boolean counterpart, we begin by examining the complements of the MiddleAged and Tall crisp sets. The shape and characteristics for these sets appear as Figures 4.6 and 4.7:

$$\mu_{\sim MiddleAged}[x] = \{Age < 35 \vee Age > 45\}$$

Figure 4.6 The Complement of the Crisp Set *MiddleAged*

$$\mu_{Tall}[x] = \{Height < 6'\}$$

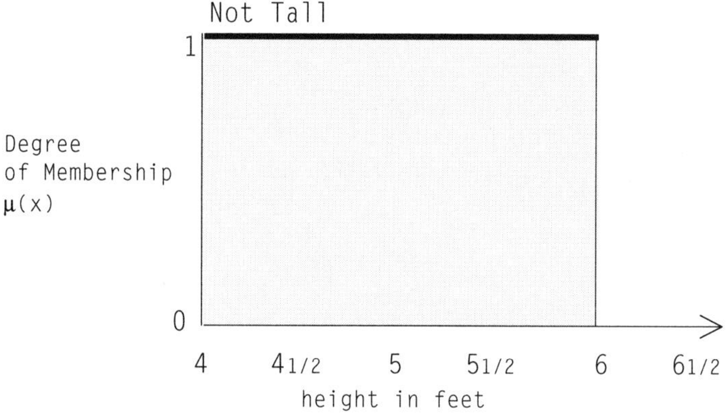

Figure 4.7 The Complement of the Crisp Set *Tall*

We now ask the question: *What members of the project manager population are not tall and not MiddleAged?* This is the intersection of the two complement sets and also produces a bit vector projection along the population table. Table 4.8 shows the results of applying the intersection operator to the complements of the *MiddleAged* and *Tall* sets.

	~middleaged age<35^ age>45	**~tall height<6'**	**~middleaged ∩~tall**
ABLE	0	1	0
COLLINS	1	0	0
FRANKLE	1	1	1
JONES	1	0	0
MILLER	0	0	0
PETERS	1	1	1
SMITH	1	1	1
WILLIAMS	1	0	0

Table 4.8 The *AND* Bit Vectors for the Crisp *~MiddleAged* and *~Tall* Sets

In fuzzy set logic, the complement is produced by inverting the truth function along each point of the fuzzy set. This process is done by applying the following transforming operation to the fuzzy set:

$$\sim \mu_A[x] = 1 - \mu_A[x]$$

Figures 4.8 and 4.9 show the *Tall* and *MiddleAged* fuzzy set complements.

Like the intersection and union operations, a fuzzy complement does not dichotomize the set into crisply defined partitions. The complement registers the degree to which an element is complementary to the underlying fuzzy set concept. That is, how compatible an element's value [*x*] is with the assertion, *x is NOT Y*, where *x* is an element from the domain, and *Y* is a fuzzy region. We can see this clearly by restating the same intersection of *~MiddleAged* and *~Tall* in their fuzzy form. The membership is generated by applying the discriminant function

$$\mu_{\sim MiddleAged} \cap \mu_{\sim Tall} = \min(\mu_{\sim MiddleAged}[x], \mu_{\sim Tall}[y])$$

and produces the results in Table 4.9.

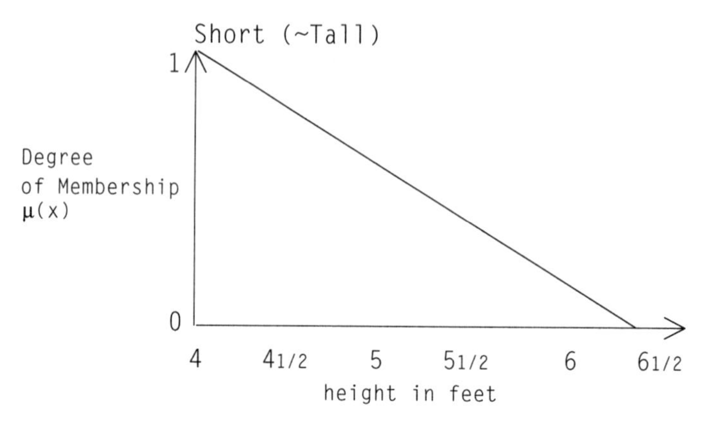

Figure 4.8 The Fuzzy Complement of the Fuzzy Set *Tall*

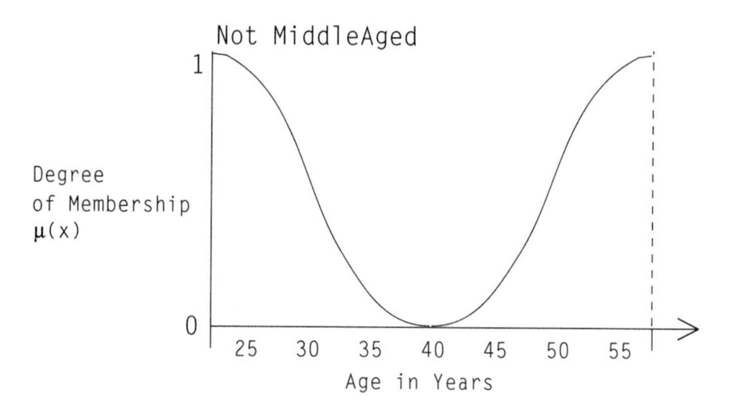

Figure 4.9 The Fuzzy Complement of the Fuzzy Set *MiddleAged*

The basic Zadeh NOT operator is implemented by taking one minus the membership value at each point along the truth function. Since the complement of a fuzzy set is often used as a new fuzzy region in a model, we produce the NOT by creating and populating a new fuzzy set.

	Age is Not MiddleAged	Height is Not Tall	Not MiddleAged and Not Tall
ABLE	0.08	0.16	0.08
COLLINS	1.00	0.08	0.08
FRANKLE	1.00	0.32	0.32
JONES	0.53	0.04	0.04
MILLER	0.00	0.06	0.00
PETERS	1.00	0.61	0.61
SMITH	0.26	0.10	0.10
WILLIAMS	0.90	0.08	0.08

Table 4.9 The *AND* Truth Vectors for *MiddleAged* and *Tall*

```
#include "FDB.hpp"
#include "mtypes.hpp"
#include "mtsptype.hpp"
FDB *FzyApplyNOT(FDB *inFDBptr,int *statusPtr)
 {
   FDB  *outFDBptr,*FDBnull=NULL;
   int   i;

   *statusPtr=0;
   if(!(outFDBptr=new FDB))
     {
      *statusPtr=1;
      return(FDBnull);
     }
   FzyInitFDB(outFDBptr);
   FzyCopySet(inFDBptr,outFDBptr,statusPtr);
   for(i=0;i<VECMAX;i++)
      {
       outFDBptr->FDBvector[i]=1-inFDBptr1->FDBvector[i];
      }
   return(outFDBptr);
  }
```

Listing 4.6 Creating the Complement (Negation) of a Fuzzy Set

If you refer to Table 4.9, you will notice that the complement memberships are indeed reflective around the base fuzzy set memberships. You cannot, however, take the result of the joint set operations and produce the complement by applying the fuzzy transformation operator.

Counterintuitives and the Law of Noncontradiction

An intuitive and, for some, an obvious observation is that the intersection of a set, *S1*, with its complement, $\sim S_1$, is empty. This phenomenon is called the *Law of Noncontradiction* and traces its roots back to the Organon of Aristotle. We can see this process clearly in Figure 4.10 where the membership space for *Short* (that is, ~*Tall*) and *Tall* are shown. Note that no overlap exists between the two regions.

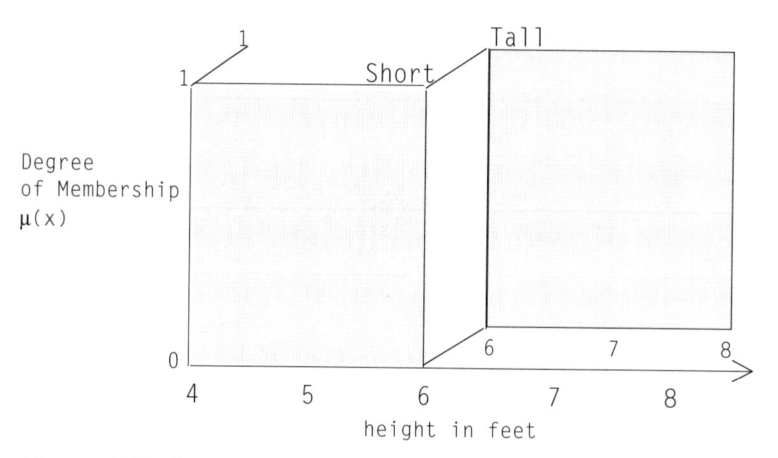

Figure 4.10 The Intersection of the Crisp Complements *Short* and *Tall*

Does the same kind of partitioning exist in the complement of fuzzy sets? The answer is no. Because a fuzzy complement measures the degree to which a value from the domain is compatible with the concept of the fuzzy complement, we might expect, and in fact find, that a certain amount of ambiguity exists when we attempt to partition a fuzzy set crisply. That is, at some region in the set, values of the domain exist in both the base fuzzy set and its complement.

In this regard, a fuzzy complement is actually a metric. It measures the distance between two points in the fuzzy regions at the same domain. The linear displacement between the complementary regions of the fuzzy regions determines the degree to which one set is a counterexample of the other set. We can also view this as a measure of the fuzziness or information entropy in the set.

To see this, consider the intersection of *Tall* and *Short* (~*Tall*), illustrated in Figure 4.11. Instead of two disjoint and clearly separable regions, the partial membership organization of the fuzzy sets creates an area (shaded) where values are considered both *Tall* and *Short* (~*Tall*).

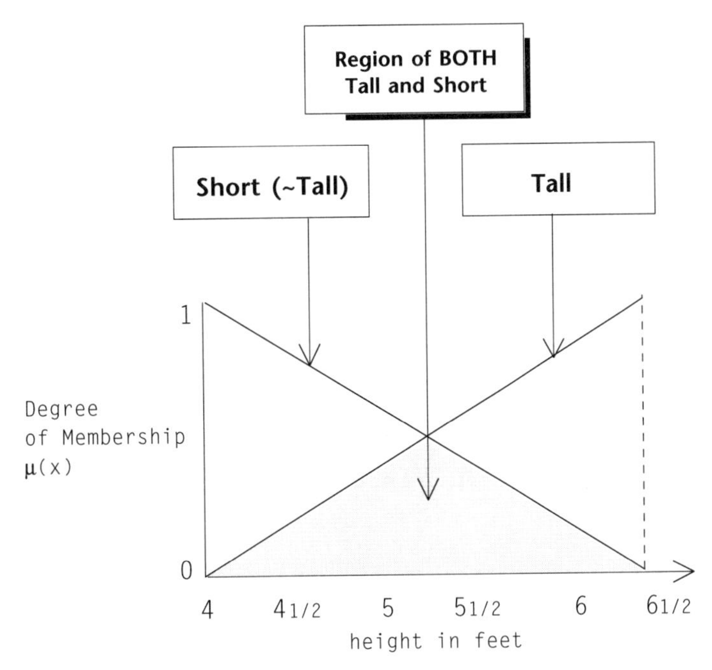

Figure 4.11 The Intersection of the Fuzzy Complements *Short* and *Tall*

This idea corresponds to our everyday experience that some individuals are neither absolutely tall nor short. Some individuals exist that have a degree of both shortness and tallness. In our fuzzy modeling language we could construct a perfectly acceptable rule such as:

```
if height is tall and height is not tall
   then weight is about medium;
```

This is the kind of reality modeling that is an intrinsic feature of fuzzy logic. In Boolean representations, no such partitioning can exist since it would violate the Law of Noncontradiction. (*"Something cannot be Tall and ~Tall at the same time."*) In this respect, then, fuzzy set logic has an ability to easily represent concepts that are derived or inherited from a vocabulary of base fuzzy sets.

The same ambiguity about set membership and categorization carries over into fuzzy sets that share parts of the same domain. As an example, consider the concept of Young and Old as complementary ideas to MiddleAged. The fuzzy sets for these two concepts might be constructed as shown in Figure 4.12.

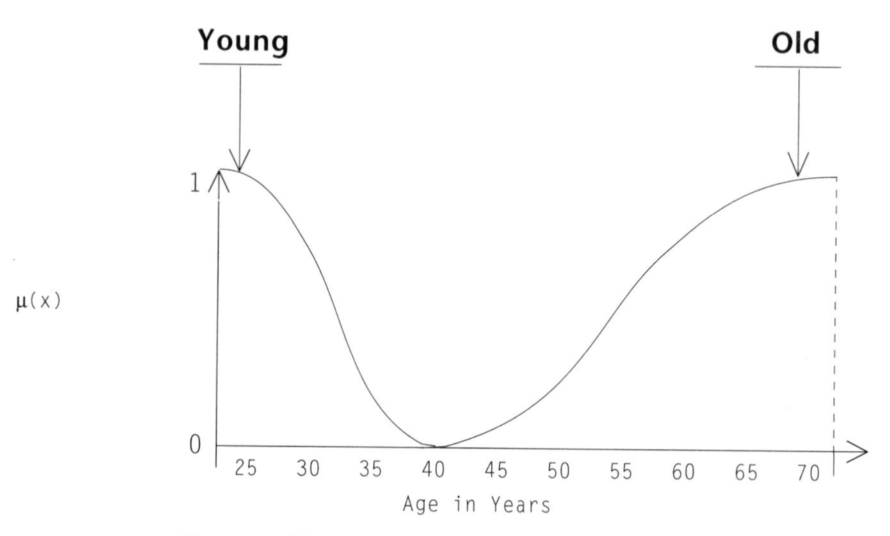

Figure 4.12 The Fuzzy Concepts Young and Old

From this graph we can see that anyone less than 25 years old is definitely *Young*, but his/her compatibility with the concept young (or the truth of the assertion *x is young*) declines until it becomes zero at 40 years old. The concept old is similarly defined: people less than 40 are definitely not *old* (they are incompatible with the concept of *Old*), but as their age increases, their compatibility with the concept *old* also increases until, at 70, they are definitely *Old*. (The truth of the assertion *x is old* is [1].)

The ambiguity of our definitions appears when we go back and consider the definition of *MiddleAged*. Figure 4.13 shows how *MiddleAged* intersects with the fuzzy space for both *Young* and *Old*. There are some regions where an individual is both *Young* and *MiddleAged*, and other regions where an individual is both *MiddleAged* and *Old*. Once more, this kind of undecidability or ambiguity corresponds to our experience. At 33 years old, as an example, an individual is strongly equidistant into both the *Young* and *MiddleAged* sets—we have a hard time deciding whether this person is young or MiddleAged. The same is true for 54 years old—the individual is weakly equidistant into the *MiddleAged* and *Old* sets: we have a hard time calling this person either MiddleAged or old.

This kind of ambiguity in partitioning concept spaces has applicability to real-world modeling. Unless we understand how related fuzzy concepts overlap and the degree of their truth membership functions, we cannot build models with adequate support regions, nor can we tune a model to ensure that rules are fired for regions with sufficient membership strength. Figure 4.14 illustrates how a fuzzy vocabulary for project costing might be constructed.

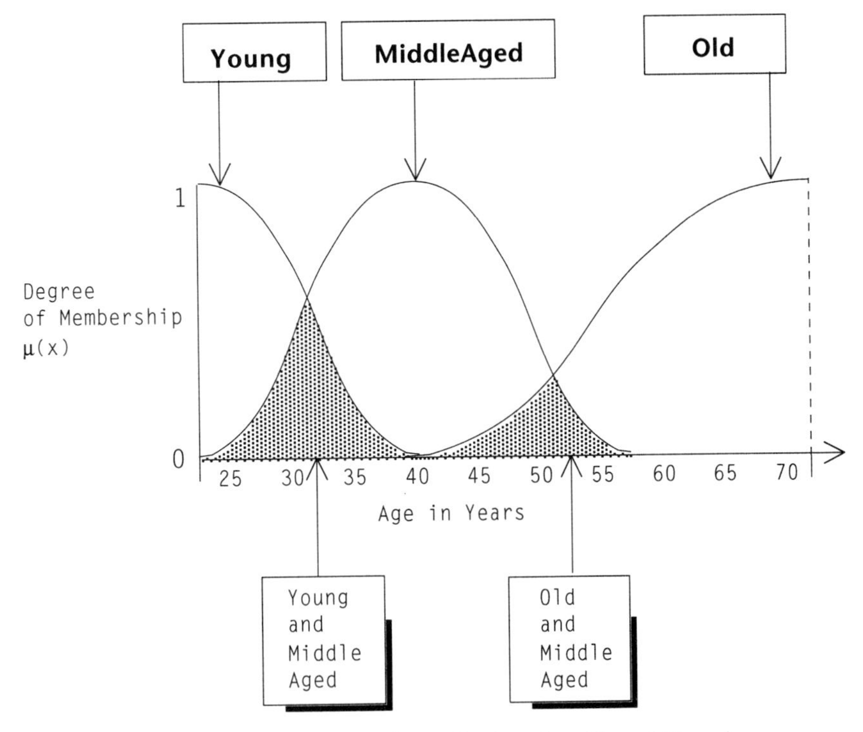

Figure 4.13 Ambiguous Fuzzy Regions in a Shared Domain

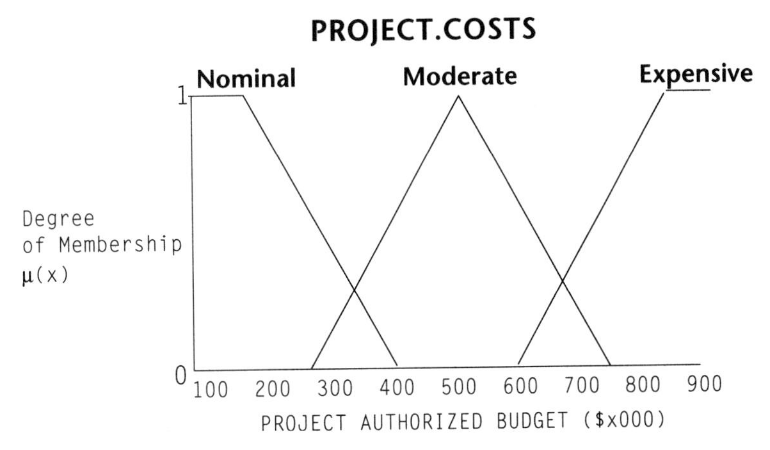

Figure 4.14 Project Cost Vocabulary Fuzzy Sets

In this model vocabulary, costs are partitioned according to the builder's perceptions of nominal, moderate, or expensive projects (these are obviously government-related). Yet between around \$300K and \$450K, projects are both nominal and moderate; and, between roughly \$600K and \$800K, projects are both moderate and expensive. In our rule language, this means that the following two rules can both execute for the same data point:

```
if project.budget is nominal
   then acceptance is increased;

if project.budget is moderate
   then acceptance is somewhat increased;
```

At some point between \$300K and \$450K, the fuzzy predicate assertions

```
project.budget is nominal
project.budget is moderate
```

will have a nonzero degree of truth. Both rules will fire and the *acceptance* consequent fuzzy region will be updated by both rules. Now, in practical fuzzy system modeling, as we will soon discover, the model vocabulary must be adequately partitioned to avoid extreme degrees of correlated fuzzy ambiguity. As an example, Figure 4.15 shows a possible fuzzy representation for budget costs—*Expensive*, and its complement, ~*Expensive*. At nearly every point in the fuzzy spaces, both spaces have some degree of truth.

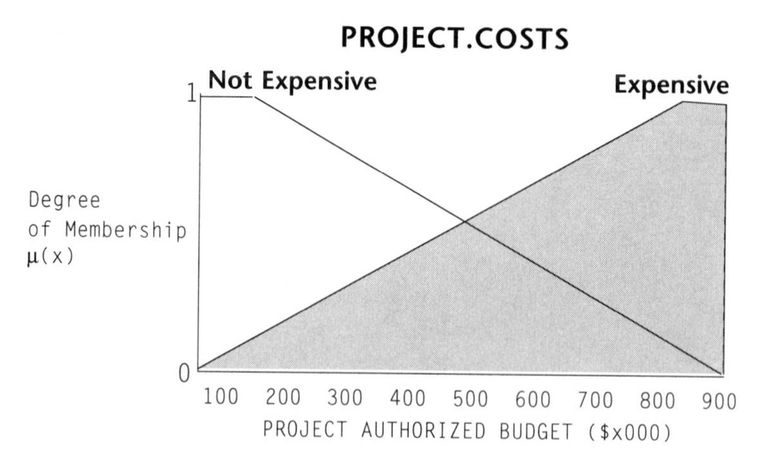

Figure 4.15 A Highly Ambiguous Fuzzy Vocabulary

In such a model, the following rules could fire in unpredictable ways:

```
if project.budget is expensive
   then risk is elevated;
if project.budget is not expensive
   then risk is reduced;
```

This apparently unpredictable behavior is directly related to the ambiguity in the partitioning of the two complementary fuzzy regions. At each point in the model, for any values above $100K and below $900K, the fuzzy predicates will simultaneously have a significant degree of truth.

This does not mean that fuzzy region overlap and ambiguity must be avoided. Indeed, fuzzy models depend on the overlap of neighboring fuzzy sets with related semantics. The stability and performance of a fuzzy model depends on the continuous mapping of domain spaces in the Cartesian space associated with all active control and solution variables. By automatically formulating a fuzzy space that represents all *n* degrees of freedom in the model, a fuzzy system can run smoothly and predictably.

Non-Zadeh and Compensatory Fuzzy Set Operations

The fundamental ways in which fuzzy sets are combined define the limits and mechanics of fuzzy logic. In the first part of this chapter, we looked at the union, intersection, and complement of fuzzy sets according to the rules originally defined by Lotfi Zadeh. A large body of work exists, however, that defines alternative set theoretic operations; and practical experience in constructing fuzzy models shows that such alternative forms are required. This is especially true in the case of the fuzzy intersection and, to a slightly lesser degree, for the fuzzy complement (if we want to maintain the law of noncontradiction).

The alternative forms of `AND`, `OR`, and `NOT` are called *compensatory operators* since they act to compensate for the strict minimum, maximum, and complement of the Zadeh operators. The compensatory operators generally provide a weaker or less sensitive relationship among propositions when their truth values are widely separated. In a few cases, the strength of the operators can be controlled so that a single operator can take on, to some degree, the properties of both an intersection and a union.

In this section we examine a few (but not all) of the important alternative operators whose characteristics are determined by functions of the form:

$$\begin{bmatrix} Intersection & A \cap B & A \cdot B = f_{AND}(\mu_A[x], \mu_B[y], k) \\ Union & A \cup B & A \cdot B = f_{OR}(\mu_A[x], \mu_B[y], k) \\ Complement & \sim A & nA = f_{NOT}(\mu_A[x], k) \end{bmatrix}$$

The function *f(x,k)* is often called the *class operator* (with *k* standing for the transformation algebra type) and represents a family or class of related operations expressed through a similarly structured set of transformers. There are two general types of alternative operators: those based on simple arithmetic transformations and those based on more complex functional transformations. The bounded sum, product, and average are examples of the first type. The Yager class is a good example of the second type.

Selecting an appropriate compensatory operator for a particular model proposition requires some experience and is generally done heuristically by observing the effect of the proposition (rule) on the overall model problem state. You should always obey the *Law of Least Complications* (an offshoot of Ockham's razor) by starting with the conventional Zadeh operators. The use of the product, average, or bounded sum operators should be investigated before moving on to more complex compensatory functions (such as the Yager class).

The necessity for alternative ways of interpreting the union and intersection becomes evident in complex fuzzy models. Since a model often combines control variables in an *n*-dimensional Cartesian space through the intersection of their respective fuzzy regions, the `AND` (and, less often, the `OR`) operator's behavior directly affects the model performance. In a rule such as

```
if costs are high
   and market share is low
   and margins are less than adequate
       then inventory should be
              just below around the reorder_point;
```

a single low truth value in any of the predicate propositions will ripple through the antecedent and suppress the truth function of the consequent fuzzy region. Figure 4.16 illustrates the effect of a single proposition with a low truth value.

If the current value for margins is sufficiently high so that the proposition *margins are less than adequate* has a low truth value, the entire antecedent truth value is affected. Applying the Zadeh intersection operation to the rule premise yields the following truth value:

$$\mu_{predicate} = \min(\mu_{Costs}[x], \mu_{Marketshare}[x], \mu_{LessThanAdequate}[x])$$
$$\mu_{predicate} = \min(.94, .58, .23) = .23$$

We can clearly see that the minimum of any predicate expression will control the truth of the entire expression (for expressions containing all `AND` connectors). Thus, if the

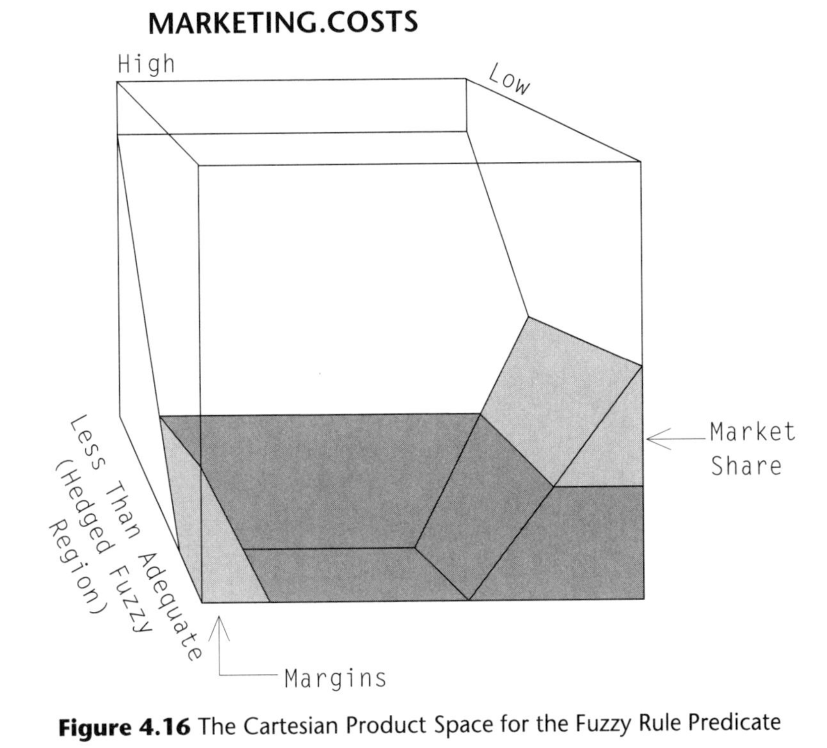

Figure 4.16 The Cartesian Product Space for the Fuzzy Rule Predicate

proposition margins are less than adequate, have a zero truth value, or fall below the current α-threshold cut for the rule, the expression reduces to

$$\mu_{predicate} = \min(.94, .58, 0) = 0$$

This seems a little extreme both in terms of its effect on the rule and its implication for a consistent epistemology for handling truth in a modeling system. The effects of the strict minimum operation for the fuzzy AND are further exacerbated if the rule premise involves semi-fuzzy[2] and crisp control variables.

The following modification to the previous rule illustrates this problem:

```
if costs > average(product.costs,prev(year))
   and market share is low
   and margins are less than adequate
       then inventory should be
       just below around the reorder_point;
```

In this case, costs is used in a crisp expression and will produce a Boolean result. If the answer is true [1], the entire predicate will have a value of [1]; if the answer is false [0], then the entire predicate will have a value of [0]. We could modify this predicate by turning it into a fuzzy proposition

```
if costs are
above about the average(product.costs,prev(year))
```

but such conversions are not always feasible nor desirable. In modeling the behavior of real world systems, a more robust and tolerant approach to rule truth management is required. In response to this requirement, a family of alternative operators, basically concerned with the operation of the fuzzy intersection and complement, has been developed.

General Algebraic Operations

The initial class of alternative operators, shown in Table 4.10, changes the union and intersection through relatively simple algebraic transformations. For reference, the standard Zadeh min/max operation is included as the first type.

	Intersection	Union
Zadeh	$\min(\mu_A[x],\mu_B[y])$	$\max(\mu_A[x],\mu_B[y])$
Mean	$\frac{\mu_A[x],\mu_B[y]}{2}$	$\frac{2\times\min(\mu_A[x],\mu_B[y])+4\times\min(\mu_A[x],\mu_B[y])}{6}$
Product	$\mu_A[x]\times\mu_B[y]$	$(\mu_A[x]+\mu_B[y])-(\mu_A[x]\times\mu_B[y])$
Bounded Sum	$\max(0,(\mu_A[x]+\mu_B[y])-1$	$\min(1,\mu_A[x]+\mu_B[y])$

Table 4.10 Algebraic Intersection and Union Compensatory Operators

The Mean and Weighted Mean Operators

In interpreting the action of the intersection (AND) operators, the first algebraically convincing operation involves taking the mean or average of the two truth member functions. The mean compensatory operator satisfies two important criteria: first, it's not overly sensitive to minimum and maximum values, and second, it is distributive—the order in which the two truth values are evaluated does not affect the operation. It also behaves well at the truth function extremes:

$$f_{MEAN}(0,0) = 0$$
$$f_{MEAN}(1,1) = 1$$

This is a property not shared by all compensatory operators, including those outside the algebraic class. Table 4.11 (from the ALTANDS.CPP program) shows the mean AND truth table for representative truth membership functions.

MEAN (AVERAGE) FUZZY 'AND' OPERATOR

	0.0000	0.2500	0.5000	0.7500	1.0000
0.00	0.0000	0.1250	0.2500	0.3750	0.5000
0.25	0.1250	0.2500	0.3750	0.5000	0.6250
0.50	0.2500	0.3750	0.5000	0.6250	0.7500
0.75	0.3750	0.5000	0.6250	0.7500	0.8750
1.00	0.5000	0.6250	0.7500	0.8750	1.0000

Table 4.11 A Truth Table for the Mean AND operator $(\mu_A[x]+\mu_B[y])/2$

The mean AND operator compensates for the minimal truth values of fuzzy antecedent propositions. Returning to our previous rule,

```
if costs are high
   mean.and market share is low
   mean.and margins are less than adequate
       then inventory should be
              just below around the reorder_point;
```

the rule's fuzzy predicate is now evaluated differently,

$$\mu_{predicate}[] = \frac{\left(\frac{\mu_{Costs}[x] + \mu_{Marketshare}[y]}{2}\right) + \mu_{LessThanAdequate}[z])}{2}$$

$$\mu_{predicate}[] = \frac{\left(\frac{.94 + .58}{2}\right) + .23}{2} = .495$$

but notice that the order of evaluation is significant. Reordering the rule so that the predicate propositions appear in reverse order changes the composite truth of the premise. Here is the new rule:

```
if margins are less than adequate
   mean.and market share is low
   mean.and costs are high
       then inventory should be
              just below around the reorder_point;
```

The evaluation averages the two weaker truth functions and then averages this result with the strongest truth function. The result is materially larger.

$$\mu_{predicate}[] = \frac{(\frac{\mu_{LessThanAdequate}[x] + \mu_{Marketshare}[y]}{2}) + \mu_{Costs}[z])}{2}$$

$$\mu_{predicate}[] = \frac{(\frac{.23 + .58}{2}) + .94}{2} = .673$$

This process is not necessarily a shortcoming of the mean AND operator, but it applies to all compensatory, dyadic operators that are not commutative.[3] We can see from the previous example that the general truth management function is sensitive to the specification order, the final expression providing the largest contribution. As an example, consider the general rule:

```
if p1 mean.and p2 mean.and p3 mean.and p4
```

This is evaluated according to the following scheme:

$$\frac{\frac{\frac{\mu_{p1}[w] + \mu_{p2}[x]}{2} + \mu_{p3}[y]}{2} + \mu_{p4}[z]}{2}$$

We note that the truth function is a series that converges on the final value as the number of predicate propositions increases. Listing 4.7 is a brief program that generates an average series.

What we want in a modeling system is an alternative way of applying compensatory operators to a complex rule predicate. The best way to accomplish this is through a function rather than an operator approach. A compensatory union or intersection function takes, as its argument, a series of fuzzy propositions and returns the truth of the compensatory operator applied to all the statements. A revision of the rule using a compensatory function might appear as:

```
if mean.and(costs are high,
   market share is low,
   margins are less than adequate)
      then inventory should be
            just below around the reorder_point;
```

where T_e is the estimated time. Setting a=b, the expression resolves to

$$T_e = \frac{2a + 4m}{6}$$

```
void MeanSeries(double StartValue, double *AvgVal)
{
    double const SeriesMax = 50;
    double PrevVal,NextVal;
    int i;
    PrevVal=StartValue;
    for(i=0;i<SeriesMax;i++)
    {
    *AvgVal=(PrevVal+i)/2;
    PrevVal=AvgVal;
    }
```

Listing 4.7 A Program to Compute the Cumulative Series Mean

This arithmetic sum series takes a starting value and then runs through fifty consecutive real numbers, computing the running series average. As the series progresses, the value of *AvgVal* begins to center around the current end-of-series value.

and is the syntax for the mean compensatory OR operation. The operation is weighted toward the maximum of the two membership functions.

Table 4.12 (from the ALTORS.CPP program) shows the mean OR truth table for representative truth membership functions.

WEIGHTED MEAN (AVERAGE) FUZZY 'OR' OPERATOR

	0.0000	0.2500	0.5000	0.7500	1.0000
0.00	0.0000	0.1667	0.3333	0.5000	0.6667
0.25	0.1667	0.2500	0.4167	0.5833	0.7500
0.50	0.3333	0.4167	0.5000	0.6667	0.8333
0.75	0.5000	0.5833	0.6667	0.7500	0.9167
1.00	0.6667	0.7500	0.8333	0.9167	1.0000

Table 4.12 A Truth Table for the Mean OR Operator
$(2*\min(\mu_A[x],\mu_B[y])+4*(\max(\mu_A[x],\mu_B[y]))/6$

Since we are manipulating truth membership functions in the range [0,1], a subfamily of extended operators based on Zadeh's concept of *very* and *somewhat* can be defined on any compensatory operation. These operators either intensify or dilute the membership value elicited from the algebraic operation. Table 4.13 shows the truth table for the intensified mean AND, and Table 4.14 shows the corresponding truth table for the diluted AND operator.

INTENSIFIED MEAN (AVERAGE) FUZZY 'AND' OPERATOR

	0.0000	0.2500	0.5000	0.7500	1.0000
0.00	0.0000	0.0156	0.0625	0.1406	0.2500
0.25	0.0156	0.0625	0.1406	0.2500	0.3906
0.50	0.0625	0.1406	0.2500	0.3906	0.5625
0.75	0.1406	0.2500	0.3906	0.5625	0.7656
1.00	0.2500	0.3906	0.5625	0.7656	1.0000

Table 4.13 A Truth Table for the Intensified Mean AND Operator (mean^2)

DILUTED (AVERAGE) FUZZY 'AND' OPERATOR

	0.0000	0.2500	0.5000	0.7500	1.0000
	0.0000	0.2500	0.5000	0.7500	1.0000
0.00	0.0000	0.3536	0.5000	0.6124	0.7071
0.25	0.3536	0.5000	0.6124	0.7071	0.7906
0.50	0.5000	0.6124	0.7071	0.7906	0.8660
0.75	0.6124	0.7071	0.7906	0.8660	0.9354
1.00	0.7071	0.7906	0.8660	0.9354	1.0000

Table 4.14 A Truth Table for the Diluted Mean AND Operator (mean$^{1/2}$)

The same semantic intentionality is applied to the intensification and dilution compensatory operator as we applied to the corresponding hedges *very* and *somewhat*. By intensification, we dampen the membership function and reduce the contribution to the consequent fuzzy region in the same manner as the *very* hedge dampens a fuzzy set and reduces the possible candidate population. The dilation operator, like the *somewhat* hedge, increases the contribution to the consequent fuzzy region and also, in a manner similar to the *somewhat* hedge, expands the candidate population space.

THE PRODUCT OPERATOR

The product operator provides another algebraic transformer that is well-behaved. The product operator also has the property of respecting the min/max characteristics of the Zadeh fuzzy intersection:

$$f_{PRODUCT}(0, \forall(\mu_A[x]) = 0$$
$$f_{PRODUCT}(1, \mu_A[x]) = \mu_A[x]$$

Table 4.15 (from the ALTANDS.CPP program) shows the product AND truth table for representative truth membership functions.

The product AND has another attribute not associated with Zadeh's minimum operator. This attribute is potential interactively, that is, the product value changes for each (μ_B[x],μ_B[y]) value pair. This is not necessarily true for the minimum rule. The responsiveness of the product

PRODUCT FUZZY 'AND' OPERATOR

	0.0000	0.2500	0.5000	0.7500	1.0000
0.00	0.0000	0.0000	0.0000	0.0000	0.0000
0.25	0.0000	0.0625	0.1250	0.1875	0.2500
0.50	0.0000	0.1250	0.2500	0.3750	0.5000
0.75	0.0000	0.1875	0.3750	0.5625	0.7500
1.00	0.0000	0.2500	0.5000	0.7500	1.0000

Table 4.15 A Truth Table for the Compensatory Product Intersection Operator ($\mu_B[x]*\mu_B[y]$)

operator to continuous changes in the fuzzy space provides a basis for solving some problems associated with the nature of fuzziness.

In particular, the product operator satisfactorily describes the behavior of time-varying fuzzy states, such as changes in sales, the slippage of a project, the credit worthiness of a customer, the behavior of fraudulent service providers, and the speed of noncatalytic chemical reactions. Model problem states that inherit the current state (S_i) from a previous state (S_{i-n}) through some consistent transformation function

$$S_i = S_{i-n} \cdot f(S_i)$$

can become "stuck" in a minimum truth plateau. If $f(S_i)$ reduces the problem state truth by the same amount in each period, the truth stabilizes at the region

$$\min(f(S_i), f(S_{i-1}))$$

that is a constant. We need some way to escape from this plateau and make the model state more responsive to the change in fuzzy truth values.

The Heap Metaphor

The difficulty with incremental fuzzy changes follows from the paradox of the dialectician, Eubulides of Miletus, an opponent of Aristotle. The argument runs as follows:

```
Given a pile of stones that makes a heap
Removing a single stone does not destroy the heap
```

It follows from this logic that we can never unmake a heap. The minimum operator fails to resolve this paradox. Let's follow the logic of this situation. We start with a truth value for the fuzzy proposition:

$$\text{The Pile is a Heap} \quad \mu_{\text{heap}}[1.0]$$

Removing a stone reduces by some truth value (say, [.05]) the compatibility between the concept of a heap and the physical existence of the heap so that the proposition now assumes a new truth value:

$$\text{The Pile is a Heap} \quad \mu_{\text{heap}}[.95]$$

Taking away additional stones results in the same truth reduction. If we repeat this action continuously, however, the truth of the proposition never changes:

$$\text{The Pile is a Heap} \quad \min(\mu_{\text{heap}}[1.0], \mu_{\text{heap}}[.95], \mu_{\text{heap}}[.95], \mu_{\text{heap}}[.95], \ldots) = \mu_{\text{heap}}[.95]$$

This is equivalent to the extended fuzzy statement

> *"removing a single stone does not destroy a heap and removing a single stone does not destroy a heap and removing a single stone does not destroy a heap, and…" ad infinitum.*

The minimum of the current truth position and the position after removing the stone ($min(\mu[.95],\mu[.95])$) is exactly the same. Substituting the product for the minimum operator resolves this paradox. Each successive evaluation of the statement reduces the truth of the proposition that the pile is a heap.

Table 4.16 shows the compatibility between the size of the pile and the concept of a heap after removing a number of stones.[4]

Stones Removed	Truth of ***Pile is a Heap***
2	.902
5	.773
10	.598
15	.463
20	.358
50	.076
100	.00592
500	.000000000000346

Table 4.16 Reduction in Truth of the Fuzzy Proposition *Pile is a Heap*

The product operator is responsive to just this kind of incremental change in the fuzziness of a statement. This is an important facility in fuzzy models that measure changes over time or incrementally across process states. The product AND operator appears to function as an approximation of the way we conceptualize such changes across time or from one system state to the next.

The product union operator (also called the probabilistic or algebraic sum) is also well behaved and follows, at the extremes, the maximum values in a fashion like Zadeh's fuzzy union. Table 4.17 (from the ALTORS.CPP program) is a truth table for representative truth membership functions.

PROBABILISTIC SUM FUZZY 'OR' OPERATOR

	0.0000	0.2500	0.5000	0.7500	1.0000
0.00	0.0000	0.2500	0.5000	0.7500	1.0000
0.25	0.2500	0.4375	0.6250	0.8125	1.0000
0.50	0.5000	0.6250	0.7500	0.8750	1.0000
0.75	0.7500	0.8125	0.8750	0.9375	1.0000
1.00	1.0000	1.0000	1.0000	1.0000	1.0000

Table 4.17 A Truth Table for the Compensatory Product Union Operator
$(\mu_A[x]+\mu_B[y])-(\mu_A[x]*\mu_B[y])$

THE BOUNDED DIFFERENCE AND SUM OPERATORS

The final algebraic compensatory operation is often called the bounded sum or the restricted sum operator. The intersection form of the bounded sum is sometimes called the bounded difference. Table 4.18 is a truth table for representative truth membership functions in the intersection bounded-difference space.

BOUNDED DIFFERENCE 'AND' OPERATOR

	0.0000	0.2500	0.5000	0.7500	1.0000
0.00					0.0000
0.25					0.2500
0.50				0.2500	0.5000
0.75			0.2500	0.5000	0.7500
1.00	0.0000	0.2500	0.5000	0.7500	1.0000

Table 4.18 A Truth Table for the Bounded-Difference Intersection Operator
$\max(0,\mu_A[x]+\mu_B[y]-1)$

We note immediately that the bounded difference operation is highly restrictive. The intersection membership functions are subject to a higher thresholding than we have found in any of the previous compensatory operators. As Table 4.19 shows, the truth table is partitioned diagonally between the nonmembership and membership space.

BOUNDED DIFFERENCE 'AND' OPERATOR

	0.0000	0.2500	0.5000	0.7500	1.0000
0.00					0.0000
0.25					0.2500
0.50				0.2500	0.5000
0.75			0.2500	0.5000	0.7500
1.00	0.0000	0.2500	0.5000	0.7500	1.0000

Table 4.19 The Nonzero Partitioning of the Bounded-Sum Intersection Space

We can view the bounded-sum intersection as a selective filter that admits only members that individually have high truth membership values. Only the unity membership value is sufficient to define a nonzero space if the complementary membership is weak, but nonzero. The bounded sum (or difference) acts as a hurdle barrier in the intersection plane.

Just as the bounded-sum AND operator is highly restrictive; the bounded-sum union (OR) operation is highly open. Table 4.20 (from the ALTORS.CPP program) is a truth table for representative truth membership functions in the union bounded-sum space.

BOUNDED SUM 'OR' OPERATOR

	0.0000	0.2500	0.5000	0.7500	1.0000
0.00	0.0000	0.2500	0.5000	0.7500	1.0000
0.25	0.2500	0.5000	0.7500	1.0000	1.0000
0.50	0.5000	0.7500	1.0000	1.0000	1.0000
0.75	0.7500	1.0000	1.0000	1.0000	1.0000
1.00	1.0000	1.0000	1.0000	1.0000	1.0000

Table 4.20 A Truth Table for the Compensatory Bounded-Sum Union Operator $\min(1,\mu_A[x]+\mu_B[y])$

A bounded-sum OR operator establishes a relatively low membership hurdle and then creates a membership space when either of the fuzzy regions has a significant degree of truth. The bounded sum moves to a unity value very quickly. This can be viewed as a measure of the compatibility between the conceptual spaces in the two fuzzy regions. It presupposes that the OR connected propositions can be decoupled without a significant loss of membership reinforcement.

Functional Compensatory Classes

These broad ranges of basic algebraic operators are based on mathematical operations drawn from the algebra of real numbers and groups. In this section, we explore some of

The bounded-sum intersection operator has the surprising property of respecting the Law of Noncontradiction and the Law of the Excluded Middle. This is a consequence of its algebraic formulation. Since the complement of a fuzzy region is produced by the transformation

$$\mu_{\sim A}[x] = 1 - \mu_A[x]$$

the intersection of a fuzzy set *A* with its complete ~*A* at each isographic location is given by:

$$\mu_A[x] + \mu_{\sim A}[x] = 1$$

But the bounded difference operation reduces this back to its zero membership level as follows:

$$(\mu_A[x] + \mu_{\sim A}[x]) - 1 = 0$$

In this manner, the bounded sum intersection is closer to Boolean and forms of Godelian logic than to fuzzy logic. We can see that this is also reflected in the polarity of the truth tables associated with the fuzzy intersection since there appears to be fairly crisp delineation between the concepts of membership and nonmembership.

the alternative functional approaches to fuzzy set operations developed by fuzzy logic researchers. These operations are not quite as intuitive and most of them depend on *k*, a class transformation parameter. These operators, in particular the Yager class, are important in constructing verifiable and correct fuzzy models. Just as the crispness of Boolean algebra is restrictive in representing the real world, the dichotomous partitioning of membership values through the minimum and maximum often results in a model that behaves erratically when data fall at the extremes of its support set. In many cases we can use the class parameter *k* as a measure of fuzziness in the model, thus forcing the union or intersection to reflect a degree of softness or hardness proportional to the model problem state.

The Yager Compensatory Operators

First proposed by Ron Yager of Iona College's Institute for Machine Intelligence, the Yager class of compensatory operators has a class parameter *k* representing the strength or weight of the connection. The Yager union and intersection operator functions converge on the Zadeh min/max representation as the parameter *k* grows very large (that is, as $k \rightarrow \infty$). Such a behavior means that we can interpret the class parameter as a filter on the importance of the set membership truth functions.

The Yager AND Operator

The Yager compensatory *AND* operator is based on the k'th root of the sum of the unit differences between the set membership truths. Equation 4.1 shows the method of calculating this Yager class operator.

$$\mu_{A \cap B}[x] = 1 - \min(1, ((1 - \mu_A[x])^k + (1 - \mu_B[y])^k)^{\frac{1}{k}}$$

Equation 4.1 The Yager Compensatory AND

When k is small, the intersection is responsive to high truth functions. That is, it imposes a hard intersection. When k is very large, the intersection is responsive to smaller and smaller degrees of truth until it stabilizes at the minimum of the truth functions. This means that a single truth table can represent the effect of a Yager union or intersection. The output from the program YAGERAND.CPP in Table 4.21 is the truth tables for representative intersection memberships at increasing class strength parameters.

YAGER COMPENSATORY FUZZY 'AND' OPERATORS

Class Intensity Value: 1.00

	0.0000	0.2500	0.5000	0.7500	1.0000
0.00	0.0000	0.0000	0.0000	0.0000	0.0000
0.25	0.0000	0.0000	0.0000	0.0000	0.2500
0.50	0.0000	0.0000	0.0000	0.2500	0.5000
0.75	0.0000	0.0000	0.2500	0.5000	0.7500
1.00	0.0000	0.2500	0.5000	0.7500	1.0000

Class Intensity Value: 16.00

	0.0000	0.2500	0.5000	0.7500	1.0000
0.00	0.0000	0.0000	0.0000	0.0000	0.0000
0.25	0.0000	0.2168	0.2499	0.2500	0.2500
0.50	0.0000	0.2499	0.4779	0.5000	0.5000
0.75	0.0000	0.2500	0.5000	0.7389	0.7500
1.00	0.0000	0.2500	0.5000	0.7500	1.0000

Class Intensity Value: 48.00

	0.0000	0.2500	0.5000	0.7500	1.0000
0.00	0.0000	0.0000	0.0000	0.0000	0.0000
0.25	0.0000	0.2391	0.2500	0.2500	0.2500
0.50	0.0000	0.2500	0.4927	0.5000	0.5000
0.75	0.0000	0.2500	0.5000	0.7464	0.7500
1.00	0.0000	0.2500	0.5000	0.7500	1.0000

Table 4.21 The Yager AND Values at Various Strength Levels

Class Intensity Value: 64.00

	0.0000	0.2500	0.5000	0.7500	1.0000
0.00	0.0000	0.0000	0.0000	0.0000	0.0000
0.25	0.0000	0.2418	0.2500	0.2500	0.2500
0.50	0.0000	0.2500	0.4946	0.5000	0.5000
0.75	0.0000	0.2500	0.5000	0.7473	0.7500
1.00	0.0000	0.2500	0.5000	0.7500	1.0000

Class Intensity Value: 96.00

	0.0000	0.2500	0.5000	0.7500	1.0000
0.00	0.0000	0.0000	0.0000	0.0000	0.0000
0.25	0.0000	0.2446	0.2500	0.2500	0.2500
0.50	0.0000	0.2500	0.4964	0.5000	0.5000
0.75	0.0000	0.2500	0.5000	0.7482	0.7500
1.00	0.0000	0.2500	0.5000	0.7500	1.0000

Class Intensity Value: 128.00

	0.0000	0.2500	0.5000	0.7500	1.0000
0.00	0.0000	0.0000	0.0000	0.0000	0.0000
0.25	0.0000	0.2459	0.2500	0.2500	0.2500
0.50	0.0000	0.2500	0.4973	0.5000	0.5000
0.75	0.0000	0.2500	0.5000	0.7486	0.7500
1.00	0.0000	0.2500	0.5000	0.7500	1.0000

Table 4.21 The Yager AND Values at Various Strength Levels

The final table shows the truth table when $k \approx \infty$ (since an exponent value 128 is the maximum value we can use without triggering an overflow interrupt on a 32-bit machine).

The Yager OR Operator

Like the compensatory AND operator, the Yager compensatory OR operator is based on the k'th root of the sum of the unit differences between the set membership truths. Equation 4.1 shows the method of calculating this Yager class operator.

$$\mu_{A \cup B}[x] = \min(1, ((1-\mu_A[x])^k + (1-\mu_B[y])^k)^{\frac{1}{k}}$$

Equation 4.1 The Yager Compensatory `OR`

When k is small, the union is responsive to low truth functions. That is, it imposes a weaker form of union. When k is very large, the union is responsive to larger and larger degrees of truth until it stabilizes at the maximum of the truth functions. The output (from the program YAGEROR.CPP) from Table 4.22 is the truth tables for representative union memberships at increasing class strength parameters.

YAGER COMPENSATORY FUZZY 'OR' OPERATORS

Class Intensity Value: 1.00

	0.0000	0.2500	0.5000	0.7500	1.0000
0.00	0.0000	0.2500	0.5000	0.7500	1.0000
0.25	0.2500	0.5000	0.7500	1.0000	1.0000
0.50	0.5000	0.7500	1.0000	1.0000	1.0000
0.75	0.7500	1.0000	1.0000	1.0000	1.0000
1.00	1.0000	1.0000	1.0000	1.0000	1.0000

Class Intensity Value: 16.00

	0.0000	0.2500	0.5000	0.7500	1.0000
0.00	0.0000	0.2500	0.5000	0.7500	1.0000
0.25	0.2500	0.2611	0.5000	0.7500	1.0000
0.50	0.5000	0.5000	0.5221	0.7501	1.0000
0.75	0.7500	0.7500	0.7501	0.7832	1.0000
1.00	1.0000	1.0000	1.0000	1.0000	1.0000

Class Intensity Value: 48.00

	0.0000	0.2500	0.5000	0.7500	1.0000
0.00	0.0000	0.2500	0.5000	0.7500	1.0000
0.25	0.2500	0.2500	0.5000	0.7500	1.0000
0.50	0.5000	0.5000	0.5000	0.7500	1.0000
0.75	0.7500	0.7500	0.7500	0.7500	1.0000
1.00	1.0000	1.0000	1.0000	1.0000	1.0000

Class Intensity Value: 64.00

	0.0000	0.2500	0.5000	0.7500	1.0000
0.00	0.0000	0.2500	0.5000	0.7500	1.0000
0.25	0.2500	0.2500	0.5000	0.7500	1.0000
0.50	0.5000	0.5000	0.5000	0.7500	1.0000
0.75	0.7500	0.7500	0.7500	0.7582	1.0000
1.00	1.0000	1.0000	1.0000	1.0000	1.0000

Class Intensity Value: 96.00

	0.0000	0.2500	0.5000	0.7500	1.0000
0.00	0.0000	0.2500	0.5000	0.7500	1.0000
0.25	0.2500	0.2500	0.5000	0.7500	1.0000
0.50	0.5000	0.5000	0.5000	0.7500	1.0000
0.75	0.7500	0.7500	0.7500	0.7500	1.0000
1.00	1.0000	1.0000	1.0000	1.0000	1.0000

Class Intensity Value: 128.00

	0.0000	0.2500	0.5000	0.7500	1.0000
0.00	0.0000	0.2500	0.5000	0.7500	1.0000
0.25	0.2500	0.2500	0.5000	0.7500	1.0000
0.50	0.5000	0.5000	0.5000	0.7500	1.0000
0.75	0.7500	0.7500	0.7500	0.7500	1.0000
1.00	1.0000	1.0000	1.0000	1.0000	1.0000

Table 4.22 The Yager *OR* Values at Various Strength Levels

The wide variety of compensatory operators for fuzzy set unions and intersections are collected into two functions: *FzyCompAND* and *FzyCompOR*. The functions support the basic algebraic operator types as well as the Yager compensatory class. Note that (1) the intensified and the diluted average *AND* are not supported and (2) the standard Zadeh operator is included for completeness and ease of use.

```
#include <math.h>
#include "mtypes.hpp"
#include "fuzzy.hpp"
#include "mtsptype.hpp"

float FzyCompAND(
  int ANDClass,
    double ANDCoeff,float truth1,float truth2,int *statusPtr)
  {
    float tempTruth;
    *statusPtr=0;
    switch(ANDClass)
     {
      case MEANAND:
       {
        tempTruth=(truth1+truth2)/2;
        if(ANDCoeff!=0&&ANDCoeff<2) tempTruth=tempTruth*ANDCoeff;
        return(tempTruth);
       }
      case PRODUCTAND:
       {
        tempTruth=(truth1*truth2);
        if(ANDCoeff!=0&&ANDCoeff<2) tempTruth=tempTruth*ANDCoeff;
        return(tempTruth);
       }
      case BOUNDEDAND:
       {
        tempTruth=max(0,(truth1+truth2)-1);
        if(ANDCoeff!=0&&ANDCoeff<2) tempTruth=tempTruth*ANDCoeff;
        return(tempTruth);
       }
      case YAGERAND:
       {
        double Exponent=(1/ANDCoeff);
        double Part1=pow((1-truth1),ANDCoeff);
        double Part2=pow((1-truth2),ANDCoeff);
        tempTruth=1-min(1,pow(Part1+Part2,Exponent));
        return(tempTruth);
```

```
        }
       case ZADEHAND:
        {
         return(min(truth1,truth2));
        }
       default:
        *statusPtr=1;
        return((float)-1);
      }
  }
```

Listing 4.8 The Compensatory AND Function (*FzyCompAND*)

The Yager compensatory intersection and union operators provide a flexible and easily tunable method of adjusting the strengths of the fuzzy AND and OR connectors. The single class strength attribute can be set on a model-by-model or on a proposition-by-proposition basis. This tunability is important in complex fuzzy models that involve many antecedent conditions, and the strength of a single antecedent proposition is not as important as the aggregate ("best likely fit," so to speak) of the entire predicate.

```
#include <math.h>
#include "mtypes.hpp"
#include "fuzzy.hpp"
#include "mtsptype.hpp"

float FzyCompOR(
  int ORClass,
     double ORCoeff,float truth1,float truth2,int *statusPtr)
  {
    float tempTruth,t1,t2;
    *statusPtr=0;
    switch(ORClass)
     {
      case MEANOR:
       {
        t1=2*min(truth1,truth2);
        t2=4*max(truth1,truth2);
        tempTruth=(t1+t2)/2;
        if(ORCoeff!=0&&ORCoeff<2) tempTruth=tempTruth*ORCoeff;
        return(tempTruth);
       }
      case PRODUCTOR:
       {
        tempTruth=(truth1*truth2);
        if(ORCoeff!=0&&ORCoeff<2) tempTruth=tempTruth*ORCoeff;
        return(tempTruth);
```

```
        }
       case BOUNDEDOR:
        {
         tempTruth=max(0,(truth1+truth2)-1);
         if(ORCoeff!=0&&ORCoeff<2) tempTruth=tempTruth*ORCoeff;
         return(tempTruth);
        }
       case YAGEROR:
        {
         double Exponent=(1/ORCoeff);
         double Part1=pow(truth1,ORCoeff);
         double Part2=pow(truth2,ORCoeff);
         tempTruth=min(1,pow(Part1+Part2,Exponent));
         return(tempTruth);
        }
       case ZADEHOR:
        {
         return(max(truth1,truth2));
        }
       default:
        *statusPtr=1;
        return((float)-1);
      }
 }
```

Listing 4.9 The Compensatory OR Function (*FzyCompAND*)

Both the compensatory AND and OR functions take a strength coefficient parameter (*ANDCoeff* and *ORCoeff*, respectively). This value is required for the Yager class function. For the other compensatory operators, the coefficient is used as a strength attenuation factor. The derived truth value will be multiplied by the coefficient parameter value before the final truth value is returned.

The Yager NOT Operator

Yager also defines an alternative form of the fuzzy complement having a power function form similar to that of the union and intersection—the NOT operator (as defined in Equation 4.3).

$$\mu_{\sim A}[x] = (1 - \mu_A[x])^k)^{\frac{1}{k}}$$

Equation 4.2 The Yager Compensatory *NOT* Operator

Here, the class function k is generally in the range [>0,<5]. The class function deforms the standard Zadeh complement (which is found when k=1). The following output (from the program YAGERNOT.CPP) of Table 4.23 is the truth tables for representative complement memberships at increasing class strength parameters.

YAGER COMPENSATORY FUZZY 'NOT' OPERATORS

Class Intensity Value: 0.70

0.0000	0.2500	0.5000	0.7500	1.0000
1.0000	0.5064	0.2552	0.0880	0.0000

Class Intensity Value: 0.50

0.0000	0.2500	0.5000	0.7500	1.0000
1.0000	0.2500	0.0858	0.0179	0.0000

Class Intensity Value: 1.00

0.0000	0.2500	0.5000	0.7500	1.0000
1.0000	0.7500	0.5000	0.2500	0.0000

Class Intensity Value: 2.00

0.0000	0.2500	0.5000	0.7500	1.0000
1.0000	0.9682	0.8660	0.6614	0.0000

Class Intensity Value: 3.00

0.0000	0.2500	0.5000	0.7500	1.0000
1.0000	0.9948	0.9565	0.8331	0.0000

Class Intensity Value: 4.00

0.0000	0.2500	0.5000	0.7500	1.0000
1.0000	0.9990	0.9840	0.9093	0.0000

Class Intensity Value: 5.00

0.0000	0.2500	0.5000	0.7500	1.0000
1.0000	0.9998	0.9937	0.9473	0.0000

Table 4.23 Yager NOT Values for Various Class Strengths

The class membership control in the Yager complement, like the union and intersection, provides a convenient and flexible method of adjusting the strength of the fuzzy NOT operator. For the endpoint conditions of zero and one, the Yager complement, regardless of the class strength parameter, always acts like the standard Zadeh complement. Between these points, however, the complement can be some degree of hard or soft.

THE SUGENO CLASS AND OTHER ALTERNATIVE NOT OPERATORS

We conclude this section on complementary operators by examining a few alternative negation functions: the Sugeno complement, the idea of a threshold NOT, and a continuous trigonometric NOT function based on the cosine function. Of these, only the Sugeno class is a true functional class as defined in this chapter. Like the Yager complement, the Sugeno complement takes a class parameter that determines the strength of the negation. The Sugeno class is defined according to Equation 4.3:

$$\mu_{\sim A}[x] = \frac{1-\mu_A[x]}{1+k\mu_A[x]}$$

Equation 4.3 The Sugeno Complementary NOT Function

In this case, the class parameter is in the range [-1,∞]. When $k=0$, the Sugeno complement has the desirable property of becoming the standard Zadeh complement. The output (from the program SUGENNOT.CPP) of Table 4.24 is the truth tables for representative complement memberships at increasing class strength parameters.

```
SUGENO COMPENSATORY FUZZY 'NOT' OPERATOR
        Class Intensity Value: -0.90
0.0000 0.2500 0.5000 0.7500 1.0000
------ ------ ------ ------ ------
1.0000 0.9677 0.9091 0.7692 0.0000
        Class Intensity Value: -0.50
0.0000 0.2500 0.5000 0.7500 1.0000
------ ------ ------ ------ ------
1.0000 0.8571 0.6667 0.4000 0.0000
        Class Intensity Value: 0.00
0.0000 0.2500 0.5000 0.7500 1.0000
------ ------ ------ ------ ------
1.0000 0.7500 0.5000 0.2500 0.0000
        Class Intensity Value: 1.00
0.0000 0.2500 0.5000 0.7500 1.0000
------ ------ ------ ------ ------
1.0000 0.6000 0.3333 0.1429 0.0000
        Class Intensity Value: 5.00
0.0000 0.2500 0.5000 0.7500 1.0000
------ ------ ------ ------ ------
1.0000 0.3333 0.1429 0.0526 0.0000
```

Table 4.24 Sugeno NOT Value at Various Class Strengths

```
Class Intensity Value: 9.00

0.0000 0.2500 0.5000 0.7500 1.0000
------ ------ ------ ------ ------
1.0000 0.2308 0.0909 0.0323 0.0000
```

Table 4.24 Sugeno NOT Value at Various Class Strengths

Threshold NOT Operator

The simplest alternate fuzzy complement is the threshold NOT. In this form, we pick an arbitrary truth membership value [k] (arbitrary in the sense that it is dictated by the model context and not by any considerations of performance remote from its utility within the model) and make this a division between membership and nonmembership. The threshold function is shown in Equation 4.4.

$$\mu_{\sim A}[x] = \begin{bmatrix} 1 & \mu_A[x] < k \\ 0 & \mu_A[x] \geq k \end{bmatrix}$$

Equation 4.4 The Threshold NOT Function

Table 4.25 output (from the program THLDNOT.CPP) is the truth tables for representative complement memberships at increasing truth membership values.

The threshold NOT should not be confused with the alpha-cut threshold (see Domains, Alpha-level Sets, and Support Sets on page 87) that can be applied to any fuzzy truth membership functions. The alpha threshold establishes an equivalence between a truth level and zero. The threshold NOT treats this boundary as a Boolean functions, dichotomously dividing the fuzzy set into zero and one members.

The Cosine NOT Function

We can use trigonometric functions to implement a wide variety of fuzzy complements. The negation based on the cosine function is illustrated in Equation 4.5:

$$\mu_{\sim A}[x] = \frac{1}{2}(1 + \cos(\pi \times \mu_A[x]))$$

Equation 4.5 The Cosine NOT Trigonometric Complement

The output (from the program COSNNOT.CPP) of Table 4.26 is the truth table for representative complement memberships.

Note that the S-curve passes through the [.5] membership point so that the complements of truth memberships close to the midpoint of the fuzzy set are very close to the truth values themselves. The values at the set extremes, [0] and [1], behave like conventional complements.

THRESHOLD COMPENSATORY FUZZY 'NOT' OPERATORS

Threshold Truth Value: 0.00

```
0.0000 0.2500 0.5000 0.7500 1.0000
------ ------ ------ ------ ------
1.0000 1.0000 1.0000 1.0000 1.0000
```

Threshold Truth Value: 0.25

```
0.0000 0.2500 0.5000 0.7500 1.0000
------ ------ ------ ------ ------
0.0000 1.0000 1.0000 1.0000 1.0000
```

Threshold Truth Value: 0.50

```
0.0000 0.2500 0.5000 0.7500 1.0000
------ ------ ------ ------ ------
0.0000 0.0000 1.0000 1.0000 1.0000
```

Threshold Truth Value: 0.75

```
0.0000 0.2500 0.5000 0.7500 1.0000
------ ------ ------ ------ ------
0.0000 0.0000 0.0000 1.0000 1.0000
```

Threshold Truth Value: 1.00

```
0.0000 0.2500 0.5000 0.7500 1.0000
------ ------ ------ ------ ------
0.0000 0.0000 0.0000 0.0000 1.0000
```

Table 4.25 Threshold NOT Values at Various Truth Levels

COSINE COMPENSATORY FUZZY 'NOT' OPERATORS

```
0.0000 0.2500 0.5000 0.7500 1.0000
------ ------ ------ ------ ------
1.0000 0.8536 0.5000 0.1464 0.0000
```

Table 4.26 Cosine NOT Values at Various Truth Memberships

The various compensatory operators for the fuzzy set complement are collected into a single function: FzyApplyNOT. The standard Zadeh complement is also included for completeness and general ease of use.

```
#include <math.h>
#include <stdio.h>
#include <string.h>
#include "FDB.hpp"
#include "fuzzy.hpp"
#include "mtypes.hpp"
#include "mtsptype.hpp"

void FzyApplyNOT(
   int NotClass,float NotWeight,FDB *FDBptr,int *statusPtr)

  {
   int        i;
   double     Power,InversePower;
   double     Pi=3.14159;
   float      thisMV;
   const int  MaxWeight = 128;

   *statusPtr=0;
   switch(NotClass)
    {
     case ZADEHNOT:
      for(i=0;i<VECMAX;i++)
         FDBptr->FDBvector[i]=1-FDBptr->FDBvector[i];
      CompleteNot("ZADEH",FDBptr);
      return;
     case YAGERNOT:
      if(NotWeight<1||NotWeight>MaxWeight)
        {
         *statusPtr=1;
         return;
        }
      Power=NotWeight;
      InversePower=(1/Power);
      for(i=0;i<VECMAX;i++)
          FDBptr->FDBvector[i]=
           pow(1-pow(FDBptr->FDBvector[i],Power),InversePower);
      CompleteNot("YAGER",FDBptr);
      return;
     case SUGENONOT:
      if(NotWeight<-1||NotWeight>MaxWeight)
        {
         *statusPtr=3;
         return;
        }
      for(i=0;i<VECMAX;i++)
         {
```

```
          thisMV=FDBptr->FDBvector[i];
          FDBptr->FDBvector[i]=(1-thisMV)/(1+(NotWeight*thisMV));
         }
       CompleteNot("SUGENO",FDBptr);
       return;
      case THRESHOLDNOT:
       if(NotWeight<0||NotWeight>VECMAX)
         {
          *statusPtr=5;
          return;
         }
       for(i=0;i<VECMAX;i++)
          {
           thisMV=FDBptr->FDBvector[i];
           if(thisMV<=NotWeight)
               FDBptr->FDBvector[i]=1;
              else
               FDBptr->FDBvector[i]=0;
          }
       CompleteNot("THRESHOLD",FDBptr);
       return;
      case COSINENOT:
       for(i=0;i<VECMAX;i++)
          {
           thisMV=FDBptr->FDBvector[i];
           FDBptr->FDBvector[i]=.5*(1+cos(Pi*thisMV));
          }
       CompleteNot("COSINE",FDBptr);
       return;
      default:
       *statusPtr=7;
       return;
     }
   }
//--Update the fuzzy set to indicate the the NOT hedge has been
//--applied. The NOT is a fuzzy operator but treated like a hedge.
  void CompleteNot(char *TypeofNot,FDB *FDBptr)
    {
     int        x,status;
     char       NameBuf[DESCLEN+1];
     char       wrkBuff[80];

     sprintf(wrkBuff,"%s%s%s%s%s",
      "Hedge '",TypeofNot,"' NOT applied to Fuzzy Set \"",
        FDBptr->FDBid,"\"");
     MtsWritetoLog(SYSMODFILE,wrkBuff,&status);
    //--Now update the name and the description of the new
```

```
//--hedged fuzzy set.
 strcpy(NameBuf,TypeofNot);
 strcat(NameBuf," NOT ");
 strcat(NameBuf,FDBptr->FDBid);
 strcpy(FDBptr->FDBdesc,NameBuf);
 return;
}
```

Listing 4.10 Performing Various NOT Operations on a Fuzzy Set

Important. Unlike the compensatory AND and OR functions, the FzyApplyNOT function modifies the incoming fuzzy set. This protocol change reflects the way the complement is usually used—as a function applied to intermediate fuzzy sets.

Notes

1. Under the assumptions covered in Chapter 6, the min/max implication method is used to combine the fuzzy consequents of all the rules that specify the same fuzzy output variable.

2. A variable is semi-fuzzy when it retains many of its original crisp behaviors. This is generally the result of such factors as being narrowly approximated, having a fuzzy surface topology that is roughly dichotomous (such as a step-like surface with many plateaus), or the use of built-in model functions that produce continuous value spaces based on probabilities, certainty, factors, or stochastic interpolation (random dispersion of the variable space). Some fuzzy variables behave like semi-fuzzy or crisp parameters if the compatibility between the expected domains in the model and their actual domains cause them to fall consistently at the extremes of the support fuzzy domains, thus yielding a high percentage of zero [0] or one [1] truth values.

3. A commutative operation is interchangeable, so that $x \bullet y = y \bullet x$, where "•" denotes some transitive operation of x and y. The min/max operations for the Zadeh AND and OR operators are commutative since the order in which the predicate propositions appear does not affect the final truth function.

4. A reduction from [1.0] to [.95] speaks to the granularity of the heap. If the heap, for instance, was a pile of gravel, then the truth change might be on the order of [.999]. In this case removing five stones produces a new truth of [.995], removing one hundred stones produces a new truth of [.9047]—still a highly compatible heap!

Five

Fuzzy Set Hedges

παντα ρει, ουδεν μενει.
All is flux, nothing is stationary.

Heracleitus (fl. 513 BC)
in Aristotle, *De Caelo*, 3.1.18

In this chapter we look at various kinds of fuzzy set shape transformers and generators known as hedges. Hedges play the same role in the fuzzy modeling system as adverbs and adjectives do in English: they modify the nature of a fuzzy set. Another class of hedges, the approximators, not only can diffuse a fuzzy set but also, when applied to a scalar, can convert the number to a fuzzy set (this creates a class of sets called "fuzzy numbers"). Hedges are important components of the fuzzy system, allowing us to model closely the semantics of the underlying knowledge as well as maintain a symmetry between fuzzy sets and numbers.

Hedges and Fuzzy Surface Transformers

A hedge modifies the shape of a fuzzy set's surface causing a change in the membership truth function. Thus, a hedge transforms one fuzzy set into another, new fuzzy set. In a fuzzy reasoning system, there are several different classes of hedge operations, each represented by a linguistic-like construct. There are hedges that intensify the characteristic of a fuzzy set (*very*, *extremely*), that dilute the membership curve (*somewhat*, *rather*, *quite*), that form the complement of a set (*not*), and that approximate a fuzzy region or convert a scalar to a fuzzy set (*about*, *near*, *close to*, *approximately*).

Hedges play the same role in fuzzy production rules that adjectives and adverbs play in English sentences, and they are processed in a similar manner. The application of a hedge changes the behavior of the logic, in the same way that adjectives change the meaning and intention of an English sentence. Accordingly, the number and order of hedges are significant, so that the expressions

not very high
very not high

are subject to two different interpretations—just as we would interpret them differently in English or other natural languages. Since the hedge and its fuzzy set constitute a single semantic entity in the modeling language, they are often collectively called a *linguistic variable* (refer back to "Fuzzy System Models" at the beginning of Chapter 1). This idea of a linguistic variable leads to another important property of a hedge called *closure*—that is, when the application of a hedge to a fuzzy set produces another fuzzy set.

Table 5.1 contains a representative collection of conventional fuzzy hedges with their effects.

HEDGE	MEANING
`about,around,near,roughly`	Approximate a scalar
`above, more than`	Restrict a fuzzy region
`almost,definitely,positively`	Contrast intensification
`below, less than`	Restrict a fuzzy region
`vicinity of`	Approximate broadly
`generally,usually`	Contrast diffusion
`neighboring,close to`	Approximate narrowly
`not`	Negate or complement
`quite,rather,somewhat`	Dilute a fuzzy region
`very,extremely`	Intensify a fuzzy region

Table 5.1 Fuzzy Linguistic Hedges and Their Approximate Meanings

THE MEANING AND INTERPRETATION OF HEDGES

The mechanics underlying a hedge operation are generally heuristic in nature. That is, the degree to which a fuzzy surface is transformed and the nature of the transformation are not based on a mathematical theory of fuzzy surface topology operations. Instead, they are associated with the perceived "fit" of the transformation and the psychological "goodness" of the resulting fuzzy region. Lotfi Zadeh's original definition of the hedge *very*, for instance, intensifies the fuzzy space by squaring the membership function at each point in the set ($\mu_A^2[x]$). Zadeh's definition of the hedge *somewhat* or *rather*, on the other hand, dilutes the fuzzy space by taking the square root of the membership function at each point along the set ($\mu_A^{1/2}[x]$).

Why do we square the membership function for the *very* hedge, and take the square root for the *somewhat* hedge? For no really good reason except that these operations seem to give a pretty good approximation of these concepts in a wide range of fuzzy spaces and operations, and both operations are domain independent (that is, they are model-free transformers). The way a hedge modifies the shape of a fuzzy set is, in most cases, an arbitrary but nevertheless informed choice. The choice is based on the desire to approximate the set's linguistic characteristics. In the case of *very*, as an example, we might choose to cube the membership (although this definition is usually reserved for *extremely*) or take some intermediate exponent value such as 1.3 or 1.8. In fact, the hedge function

$$\mu_A^{1.3}[x]$$

has been used frequently to implement the hedge *slightly* in such rules as

```
if costs are slightly high
   then margins should be increased
```

The same kind of model dependency exists for hedges that approximate scalars or broaden existing bell-shaped fuzzy regions. The PI curve is usually employed because it has the desirable properties of rotational symmetry and zero membership values at its left and right edges. Other curves—such as a triangular space, normal, Weibul, or Erlang distributions—can be used to represent the fuzzy space around a scalar.

Importance of Hedges in Fuzzy Modeling

Because hedges are heuristic in nature, several well-known academic experts in fuzzy logic have rejected the concept of hedges. According to their judgment, any hedged fuzzy set can (and should) be constructed as a normal part of the model vocabulary. By this assertion, they reveal not only their control engineering background, but also their lack of experience in building real-world fuzzy models in information decision support. There are two reasons why hedges are an integral part of fuzzy expert and decision support systems: dynamically created fuzzy sets and reduced rule complexity.

Dynamically Created Fuzzy Sets

Perhaps the most important use for fuzzy set hedges is the *around* approximation hedges, which convert scalar numbers into fuzzy sets. In a large number of fuzzy business models, it is simply not possible to create vocabulary fuzzy sets describing all possible fuzzy sets for a model input parameter. As an example, consider the following rule from a new product pricing model:

```
for each products
   if our.price is near 2*MfgCosts
      then suggested.price is increased
```

The *for each* rule statement reads a new candidate product from the *products* database table. The rule tests whether our current price is close to twice the manufacturing costs (if so, then the suggested price is updated from the *increased* outcome fuzzy set). Proposed products can have a wide range of manufacturing costs. As Figure 5.1 illustrates, these costs produce their own fuzzy numbers.

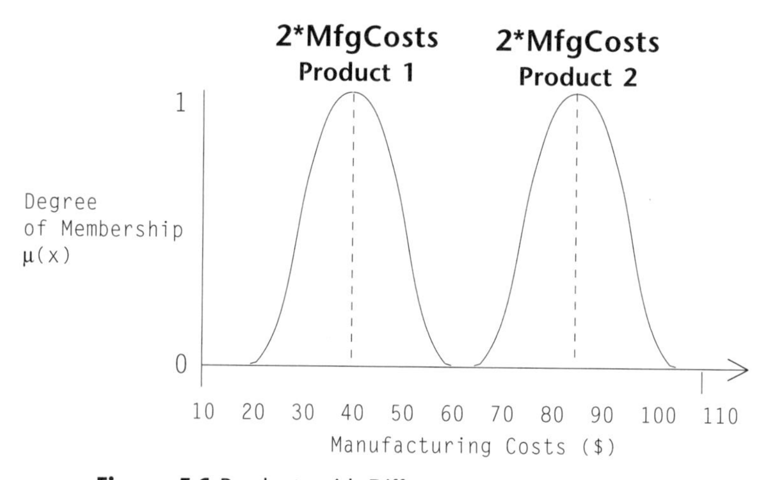

Figure 5.1 Products with Different Manufacturing Costs

Product 1 has a base manufacturing cost of 20 while product 2 has a manufacturing cost of 42.5. If the *products* table contains several hundred instances, we would need to know, *a priori*, the manufacturing costs for each product and then we could create a fuzzy set representing the bell- or triangular-shaped number around its value. Real-world fuzzy models often have even more complex decision spaces. Consider the following rule:

```
for each products
   suggested.price must be above around 2*MfgCosts
```

This pricing model rule uses the dynamically constructed manufacturing costs fuzzy set to create the fuzzy set actually used in the rule: *above (around 2*MfgCosts)*. So, for our hypothetical base vocabulary, you would need to create each separate fuzzy set representing twice the manufacturing costs, then create the sigmoid growth curve representing the set that is above the set's right surface. Figure 5.2 illustrates what this dynamically created, doubly hedged set looks like.

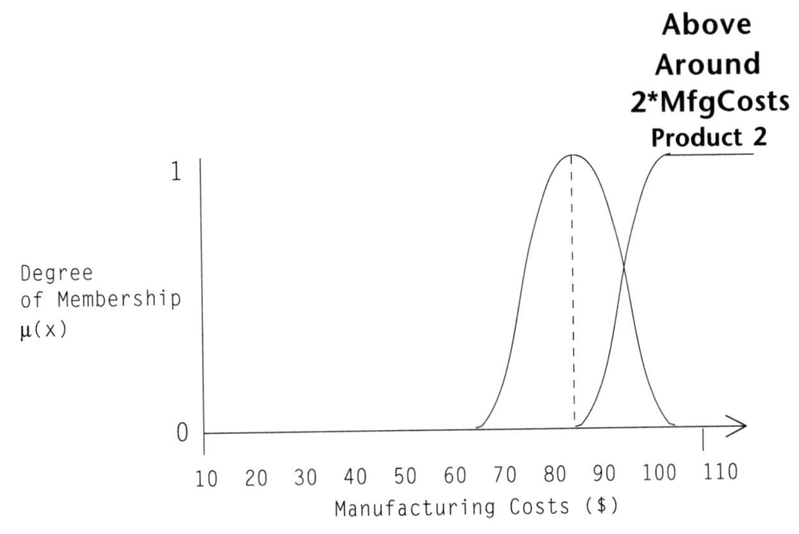

Figure 5.2 A Dynamic, Doubly Hedged Fuzzy Set

Even if we had the resources to profile all the possible manufacturing costs in the database, the fuzzy model must have the necessary logic to select the proper fuzzy set for the incoming manufacturing costs. Any reasonably complex information model requires the dynamic generation of fuzzy sets. Often these fuzzy sets are needed to compute the compatibility between model parameters that are not known or are only partially known beforehand. As an example, a rule from a financial forecasting model compares last period's income to this period's expenses:

```
if income[t-1] is near expenses[t] then margins[t] are reduced
```

If the two figures are close, then margins in this period are reduced. Such models involving lead/lag relationships are common in business and in areas such as urban planning. Typically, the lead and lag windows are adjusted dynamically by the model, based on the current context. Thus, there is no practical method of creating fuzzy sets as part of the static vocabulary. The approximation edge is the only satisfactory method of converting model numbers into fuzzy sets.

Reducing Rule Complexity

Hedges can reduce both the computational as well as psychological complexities of the rule. Computational complexity starts at both the base vocabulary and the rule specification levels. At the vocabulary level, hedges eliminate the need to create and maintain a large number of fuzzy sets. Predicating or generating one fuzzy set by dynamically hedging another keeps complementary concepts synchronized. As an example, the rule

```
if height is very Tall then weight is quite heavy
```

applies the hedge *very* to the fuzzy set *Tall* and the hedge *quite* to the outcome fuzzy set *Heavy*. If we create separate fuzzy sets in the model for *Tall*, *very Tall*, *Heavy*, and *quite Heavy*, then we run the real risk of changing the definition for *Tall* without changing the definition for *very Tall*. This can introduce subtle and hard-to-detect errors into the model.

Hedges also reduce the complexity of a rule by introducing English-like linguistic variables. These act much like nouns and adjectives or verbs and adverbs in English. Consequently, they enhance rule readability and improve knowledge base maintenance. As an example, these rules are part of an expert system that analyzes potential cardiac risk:

```
[R1]:
if height is rather Short
    and weight is very Heavy
   and serumCholesterol is positively elevated
       then CardiacRisk is generally increased;

[R2]:
if weight is heavy
   and age is around MiddleAge
       then CardiacRisk is elevated;
```

The examples use hedges to intensify or dilute the meaning of base fuzzy sets, thus allowing a more precise statement of the conditions under which cardiac risk is increased. We can see in rule [*R2*] how an approximation hedge (*around*) is used to dilute the width of an existing bell-shaped fuzzy set (*MiddleAge*).

Applying Hedges

Since a hedge is linguistic in nature, multiple hedges can be applied to a single fuzzy region in a manner that corresponds to a restriction and a fine definition of the region's semantic characteristics. As an example

```
very very high
positively not very high
generally around the median costs
```

are hedged fuzzy expressions. As we previously noted, hedges are processed in a manner analogous to English adjectives. Thus, the fuzzy proposition

```
positively not very high
```

is interpreted as:

```
positively(not(very high))
```

First, a new fuzzy region *very high* is created. The negation hedge, *not*, is applied to form a fuzzy set containing the complement of *very high*. Finally, the hedge *positively* is applied to this final fuzzy set. From this approach you can see that a statement such as

```
not positively very high
```

is interpreted quite differently than the previous example. This close affinity between the way hedges work and how we think about them in our models is clearly advantageous. By following the general rules for adjectives and adverbs, hedged statements help reduce cognitive dissonance—the difference between the model's representation and our way of thinking about the model.

Predicate and Consequent Hedges

Hedge operations can be used in both the predicate (antecedent) and the consequent (action) of a fuzzy rule as the following examples illustrate:[3]

```
if costs are very high then margins are small;

if inflation(t-1) is much greater than inflation(t)
   then sales are positively reduced;
```

When hedges are used in the consequent or outcome portion of a rule, the effects are often counter-intuitive. This results from the way hedged fuzzy sets are used on both sides of the **then** keyword. In the antecedent or premise of a rule, the hedged fuzzy set generates a truth function value. On the consequent side of the rule, the hedged fuzzy set itself participates in the actual outcome space. Often the semantics of *somewhat* and *very* are reversed when used in the outcome fuzzy set.

Fuzzy Region Approximation

The approximation hedges are an important class of surface transformers. They not only broaden or restrict existing bell-shaped fuzzy regions, but they convert scalar values into bell-shaped fuzzy regions.[4] The ability to convert any arbitrary scalar into a fuzzy region provides the fundamental symmetry between fuzzy sets and scalar control variables that is an essential characteristic of fuzzy models.

The basic approximation hedges—*about, around, near,* and *roughly*—all produce the same bell distribution called a PI curve. The hedge takes a scalar expression and produces a fuzzy region. We can see this by looking at a simple model rule that makes an assertion about the relation between product price and the costs of maintaining the distribution channel:

```
our price must be around 2*Distribution_Costs
```

Figure 5.3 shows the bell-curve created by approximating the expression "2*Distribution_Costs."

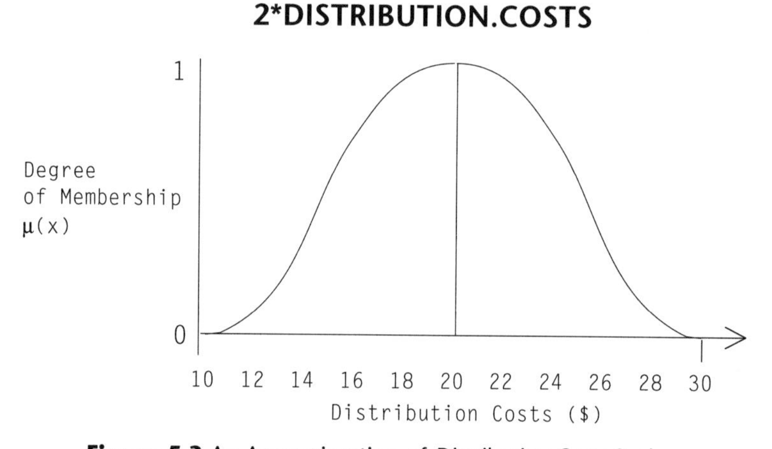

Figure 5.3 An Approximation of *Distribution Costs* Scalar

How does the approximation hedge know how wide to make the base of the PI curve? This is an important consideration in approximating a scalar since the domain of the generated fuzzy region must match the domain of the consequent fuzzy space or the comparator fuzzy region (in the case of union or intersection operations with other fuzzy regions). Without any other information, the modeling system selects the domain of the consequent fuzzy region. When the rule is a fuzzy assertion, the domain is implicitly transferred into the approximation space. Figure 5.4 illustrates this process.

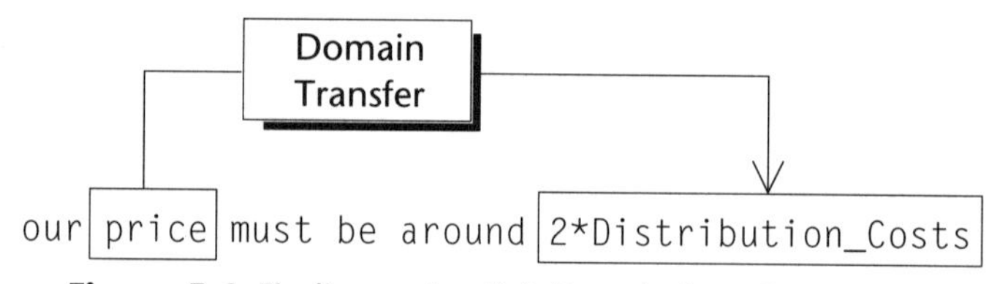

Figure 5.4 Finding an Implicit Domain for a Fuzzy Space

When a domain cannot be determined from context, then an explicit domain statement must be issued. By explicitly stating a domain, the approximation hedge varies the breadth of the PI curve. Figure 5.5 shows the results of the following fuzzy rule statements:

```
domain of Distribution_Costs is 14 to 26;
our price must be around 2*Distribution_Costs;
```

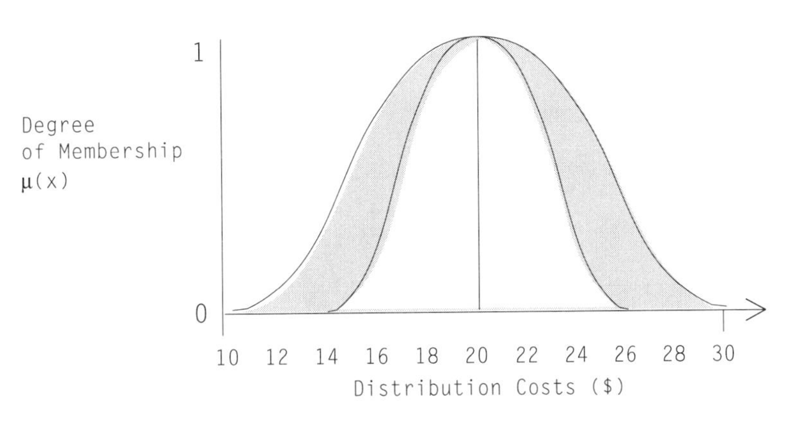

Figure 5.5 A Re-Domained Approximation of *Distribution Costs*

The approximation hedge can also be applied to existing bell-shaped fuzzy regions. The *about* hedge (and its synonyms) slightly broadens the fuzzy region. Applying around to the base *MiddleAge* fuzzy region produces a slightly wider fuzzy region as illustrated in Figure 5.6.

The standard approximation hedge (*about*) creates a space that is proportional to the height and width of the generated fuzzy space. While an explicit domain can attenuate the shape of the region, the *about* hedge still keeps the central measure of the fuzzy region. There are two other approximate hedges used to convert scalars to fuzzy sets and to reapproximate the space of existing bell-shaped fuzzy sets. The first hedge, {*in the*} *vicinity of*, constructs a very wide fuzzy region—much broader than the region created by about. Figure 5.7 shows the effect of applying *in the vicinity of* to the previous distribution costs expression.

Note that a broadly approximated space may exceed the general domain of the working fuzzy sets. If several closely allied sets share the same underlying domain, this approximation may lead to an unforeseen degree of ambiguity.

Complementing the *in the vicinity of* hedge are two surface transformers that produce narrowly approximated regions—*close to* and *neighboring*—which are synonyms. When *neighboring* is applied to a scalar or another bell-shaped fuzzy region it produces a spike-like region closely clustered around the central value (see Figure 5.8).

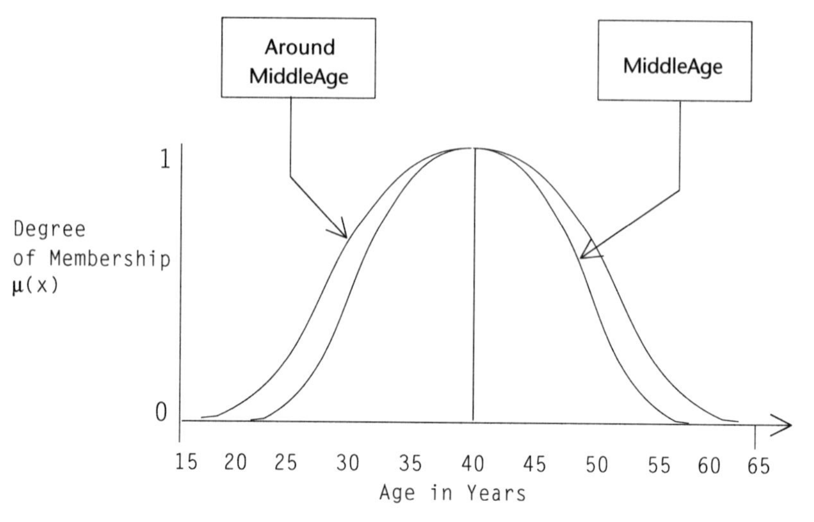

Figure 5.6 Approximating the *MiddleAge* Fuzzy Set

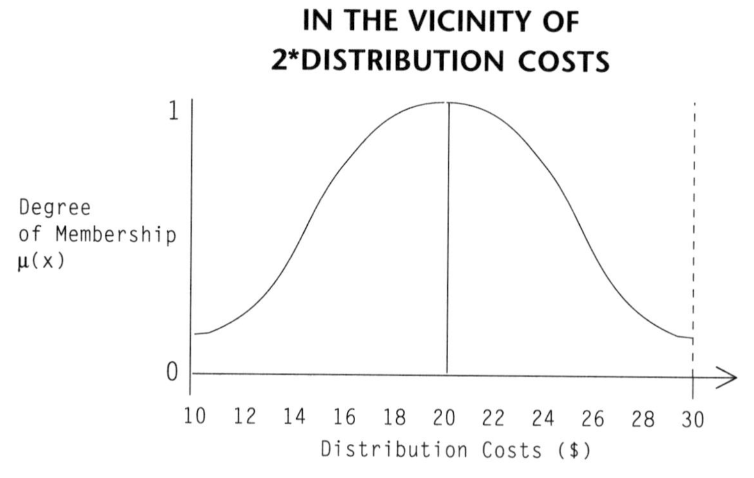

Figure 5.7 The Approximation Space for *in the vicinity of 2*Distribution Costs*

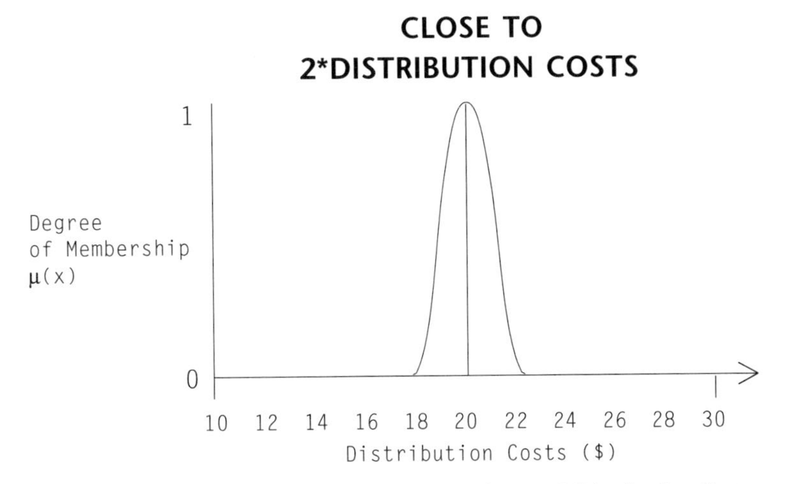

Figure 5.8 The Approximation Space for *close to 2*Distribution Costs*

Restricting a Fuzzy Region

Two special hedges are reserved for modifying the shape of directional and PI or bell-shaped fuzzy regions—these are *above* and *below*. The natural effect of applying an *above* or *below* hedge is the restricting of the fuzzy membership scope. The *below* hedge can only be applied to a fuzzy membership function that increases as the domain moves from left to right. Figure 5.9 shows the fuzzy region created when *below* is applied to the bell-shaped *MiddleAge* fuzzy set. The *below* fuzzy region declines to zero membership when the *MiddleAge* fuzzy set reaches unity.

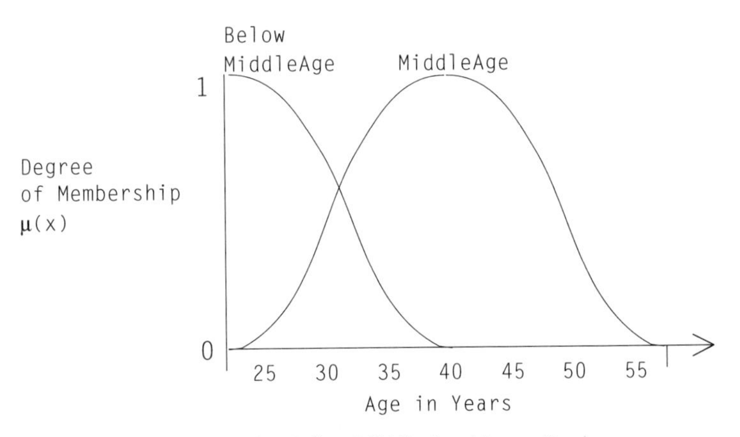

Figure 5.9 The *Below Middle Aged* Fuzzy Region

We can see that the effect of this edge is restrictive by looking at the following equivalent rules:

```
if Age is below MiddleAge
   then CardiacRisk is decreased;

if Age is less than MiddleAge
   then CardiacRisk is decreased;
```

The further to the left of the fuzzy region the value moves (moving toward the zero membership edge of *MiddleAge*), the higher degree of truth the *below* region assumes.

Figure 5.10 illustrates how *below* modifies a linear fuzzy region, in this case the fuzzy set *Tall*. The fuzzy region created by *below* is indicated by the shaded area. Note that the truth membership function must fall as the domain moves to the right. This hedge could not be used on ~*Tall* (*Short*).

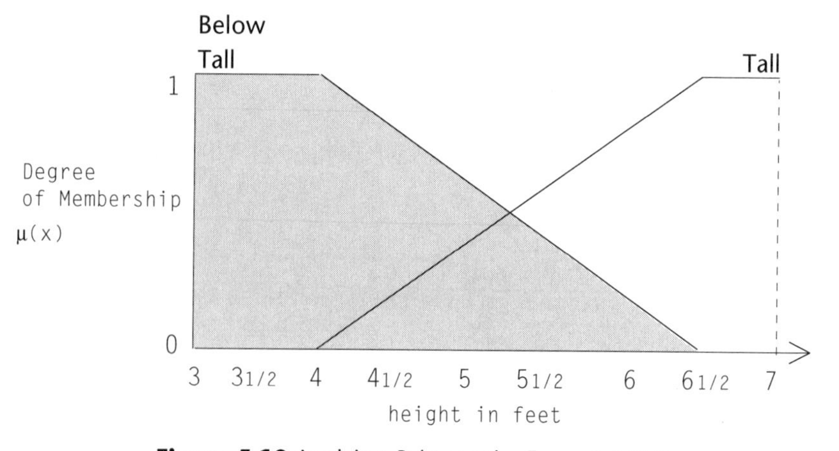

Figure 5.10 Applying *Below* to the Fuzzy Set *Tall*

The *above* hedge can only be applied to a fuzzy membership function that declines as the domain moves from left to right. *Above* is a symmetrical reflection of the *below* fuzzy region, and it restricts the fuzzy qualifier space to the region where the truth membership function is falling to zero [0]. Figure 5.11 shows the application of the hedge *above* to the *MiddleAge* fuzzy set.

We can see that the effect of this hedge is restrictive by looking at the following equivalent rules:

```
if Age is above MiddleAge
   then CardiacRisk is increased;
```

```
if Age is more than MiddleAge
   then CardiacRisk is increased;
```

The further to the right of the fuzzy region the value moves (moving toward the zero membership edge of *MiddleAge*), the higher degree of truth the *above* region assumes.

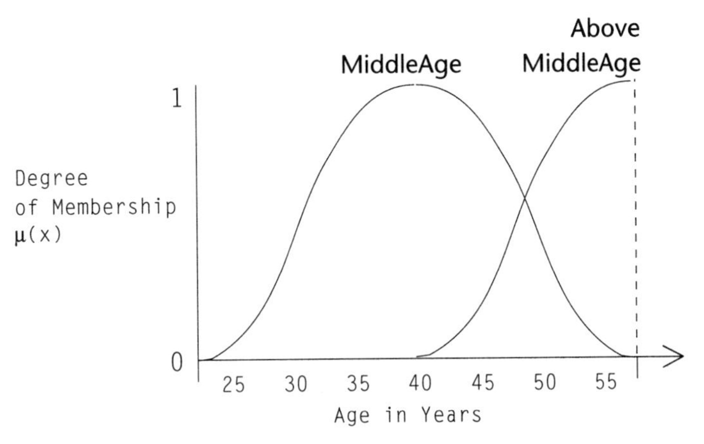

Figure 5.11 The Above *MiddleAge* Fuzzy Region

The *above* and *below* hedges can also be applied to unhedged scalars. In that case, the transformation produces a classical step-function space (see Figure 5.12), becoming fully true immediately after the scalar's value in the underlying domain.

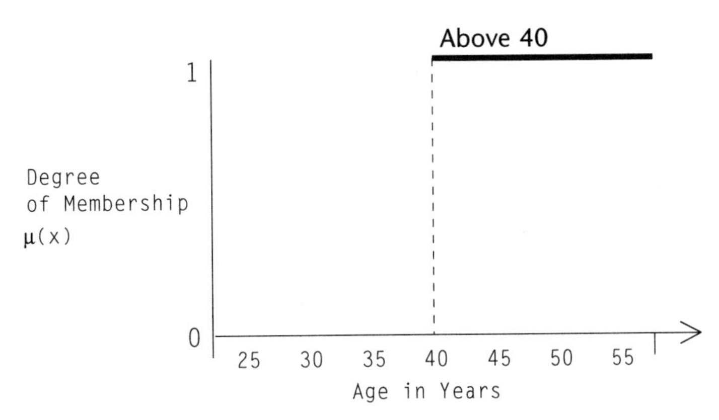

Figure 5.12 Applying *Above* to an Unapproximated Scalar

To prevent this rather crisp representation of what should intuitively be a fuzzy space, the scalar should be approximated as the restriction hedge is applied. Linguistic statements such as

```
above around 40
below close to our direct_costs
below the neighboring inventory levels
```

will first create the appropriate bell-shaped approximate fuzzy region and then apply the restriction hedge.

Important. You should be aware that the restriction operator can act on scalars as well as on fuzzy sets. If you want to restrict a scalar, and at the same time treat it as a fuzzy region, you must approximate the scalar before restricting it. On the other hand, directly restricting a scalar produces an interesting fuzzy space—a fuzzy set with Boolean properties (see Figure 5.12). Such sets are often useful when noncontinuous or step-behavior functions are being modeled.

Intensifying and Diluting Fuzzy Regions

The intensification transformers—represented by *very* and *extremely*—and the dilution transformers—represented by *somewhat*, *quite*, *rather*, and *sort of*—are an important class of hedges. With the exception of hedges used for scalar approximations, these are the most frequently used hedges in fuzzy system modeling.

Very is the most commonly discussed and used hedge. Zadeh's original definition made note of the fact that, when we consider the idea underlying *very*, we would expect a representative example from the intensified fuzzy set to be rated at least as true as the same example in the base fuzzy set, but the reciprocal is not true. In considering the fuzzy set *Tall*, we can see that a person who is *very Tall* would also generally be voted as simply *Tall*, but a simply *Tall* person would not be judged as *very Tall*. From this observation, we see that the effect of *very* is to reduce the candidate space within the fuzzy region so that the characteristic function relationship is:

$$\mu_{Tall}[x] \geq \mu_{VeryTall}[x]$$

We implement the *very* hedge (Zadeh calls this a *concentrator*) by squaring the membership function at each point in the fuzzy region. The transformational operation is:

$$\mu_{veryA}[x] = \mu_{A^2}[x]$$

Squaring any value in the interval [0,1] reduces its value. An important characteristic of *very* (and *somewhat*) is domain independence. The hedge rescales the fuzzy set surface but maintains the normalization space for the set. That is, the points representing absolute set membership [1] and set exclusion [0] are not changed.

This is not the case in alternative forms of the intensification operator where the membership function is shifted to the right by an arbitrary transforming factor. Consider one version of the transformational scheme:

$$\mu_{veryA}[x] = \mu_{\min(1, A+k)}[x]$$

the membership is skewed by an amount proportional to k in a manner that works like a the bounded sum operation. In similar transformations, the selection of k is usually domain and context dependent. Note that shifting intensification hedges are also restricted to fuzzy sets with monotonic membership functions rising from left to right (in keeping with the need to maintain the "less than" relationship between the base and the new membership space). Care must also be taken that shift transformers maintain, in some way, the normalization characteristics of the fuzzy region.

By observing that the standard Zadeh *very* hedge is equivalent to

$$\mu_{veryA}[x] = \mu_{A \times A}[x]$$

we can often replace it with alternative product operators. The point transformation parameter k in this scheme

$$\mu_{veryA}[x] = \mu_{A \times k}[x]$$

also occurs in the interval [0,1], exclusive of the endpoints. As noted, an advantage of the bounded-sum and product hedge transformations is that k can be selected as a context responding parameter. When $k=\mu_A[x]$ for the product or $k=0$ for the bounded sum, then the operation is equivalent to Zadeh's original definition for *very*.

THE VERY HEDGE

The idea behind the *very* hedge is simple: we depress the surface of the fuzzy set so that an element from the domain that had x degree of truth, now has $y \bullet x$ degree of truth, where y is the proportional reduction at domain point S in the fuzzy set. The effect of applying the hedge *very* to the fuzzy set *Tall* is illustrated in Figure 5.13 (the degree of surface shifting has been slightly increased for clarity).

Bowing the fuzzy set surface reduces the membership truth functions for each value of the domain except at the set extremes. We can observe how such a reduction affects the

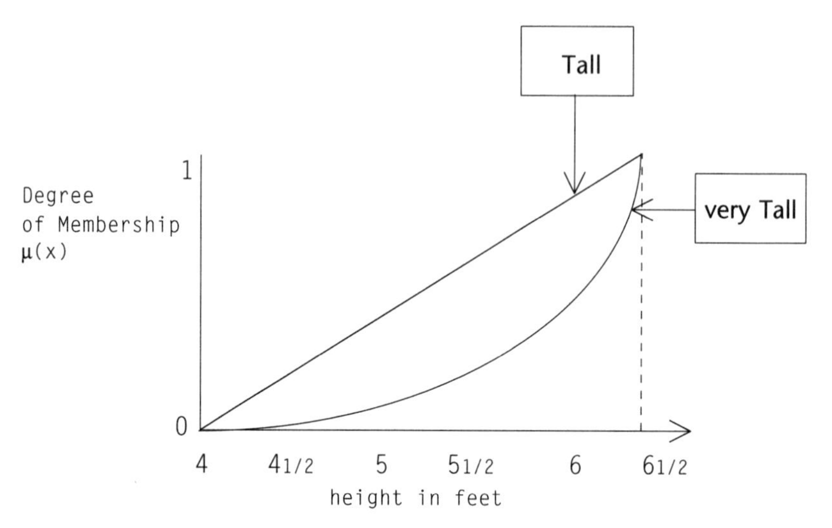

Figure 5.13 Applying the Hedge *very* to *Tall*

truth of fuzzy propositions by looking at two rules which estimate an individual's weight from their height:

```
if height is tall then weight is heavy;
if height is very tall then weight is heavy;
```

Figure 5.14 illustrates how this change in the characteristics of the fuzzy set topology produces a change in the truth function at each point along the line.

For a height of 5.5 feet, the predicate proposition, *height is tall*, has a truth of [.56] and will produce a significant contribution to the consequent fuzzy region *Heavy.* For the same height, the proposition *height is very tall* has a truth of [.28] and induces a weak truth function in the consequent fuzzy set *Heavy.* In effect, as Figure 5.15 shows, a *very* hedge means that, for equivalent truth membership values, a domain element must occur further to the right. In order for a *very tall* predicate to have a consequent influence of [.56], the height must be a little over 6'4".

Figures 5.17 shows that the intensification transformer works on a variety of curves, including those that are not uniform or truth-function monotonic.

While the *very* hedge provides a proportional decrease in the membership function for all uniformly orthogonal sets, including bell-shaped curves, it does have some anomalous behavior when applied to fuzzy regions that are multimodal or have large plateau areas. As Figure 5.17 shows, the membership function is depressed, but moving to the right does not necessarily increase the truth function.

The *very* hedge also concentrates or intensifies the truth membership function for bell-shaped fuzzy regions. This moves the curve closer to the center value of the fuzzy set.

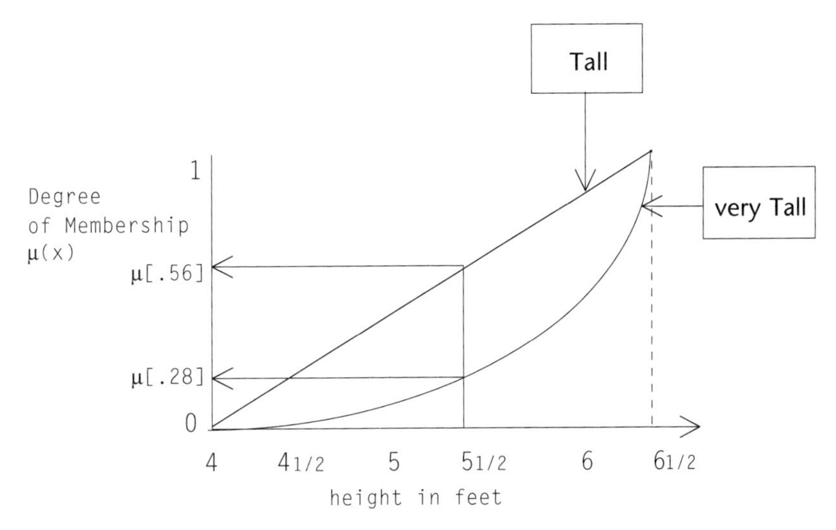

Figure 5.14 Comparing *Tall* and very *Tall* at the 5.5' Domain Point

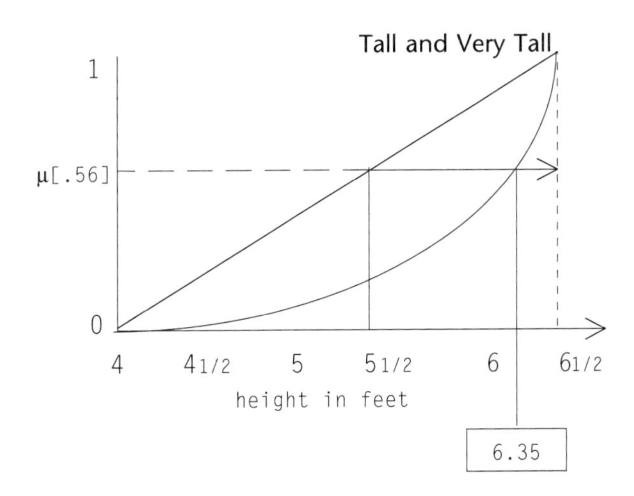

Figure 5.15 Domains with Equal Truth Membership Values in *Tall* and *very Tall*

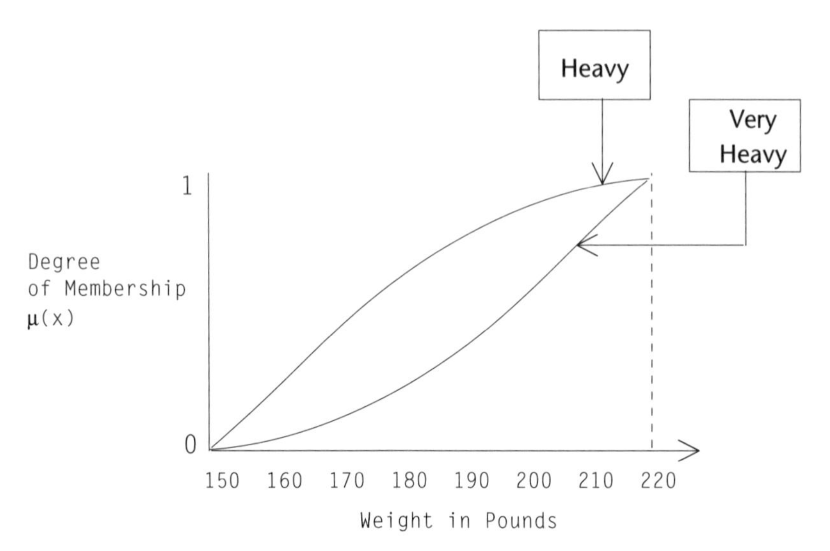

Figure 5.16 Applying the Hedge *very* to *heavy*

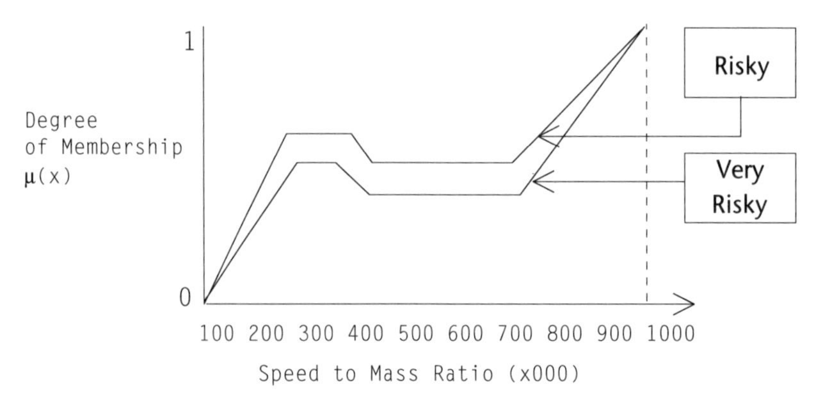

Figure 5.17 Applying the Hedge *very* to an Irregular Fuzzy Region

(Remember that bell- and triangular-shaped fuzzy sets are actually fuzzy numbers.) Figure 5.18 shows how the fuzzy set *MiddleAge* is modified.

A generalization of the Zadeh concentrator hedge simply replaces the exponent of the intensification function with a real positive number greater than one [1]. The concentrator has the construct:

$$\mu_{con(A)}[x] \;=\mu_{A^n}[x] \quad n>1$$

In practical use, n will fall in the interval [1,4]. Hyperquadratic hedges of the Zadeh concentrator type simply push the membership function toward zero (or, at the very least,

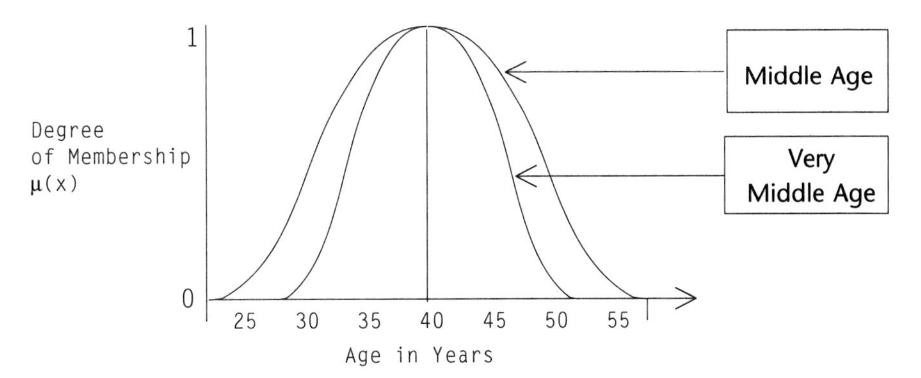

Figure 5.18 Applying the Hedge *very* to *MiddleAge*

below the α-cut threshold in the fuzzy model). The hedge *Extremely*, shown in Figure 5.19, is an example of a hedge that uses the general concentrator form with $n=3$.

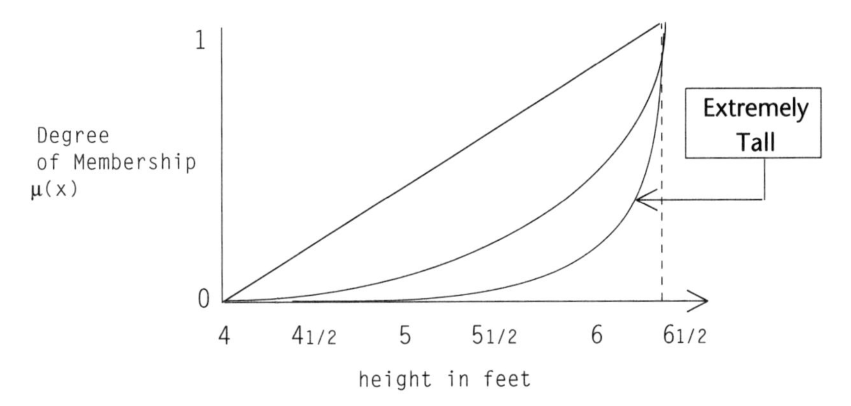

Figure 5.19 Applying the Hedge *extremely* to *tall*

The membership intensification function for *extremely* is given by:

$$\mu_{extremely(A)}[x] = \mu_{A^3}[x]$$

Cubing the membership sharply attenuates the truth function so that the candidate space is greatly reduced. Varying the strength of exponent, especially using fractional numbers, can produce a related family of concentrators. Hedges such as *slightly* or *a little* can be created with a fractional exponent:

$$\mu_{slightly(A)}[x] = \mu_{A^{1.7}}[x]$$
$$\mu_{ALittle(A)}[x] = \mu_{A^{1.3}}[x]$$

The use of fractional powers allows a finer resolution of the semantic meaning underlying the hedge, although the actual predictor between anticipated behavior and actual performance, as with all hedges, rests with the empirical evaluation of the model results. Fractional scaling, however, does place the truth function—depending on the exponent value—somewhere intermediate between the base set, *very*, and *extremely*.

The effects of various power values on surface intensification can be seen in the modifications of Figures 5.20 and 5.21 to a bell-shaped and linear fuzzy set (from the `INTHEDGE.CPP` program). To improve clarity and reduce clutter in Figures 5.22 through 5.26, the fuzzy set graph legends have been omitted. The base fuzzy set is indicated by a dot (.) and the modified fuzzy surface is represented by an asterisk (*).

The bell-shaped fuzzy set is compressed toward its central value. On the other hand, the linear fuzzy set is "bent" toward the domain axis (Figures 5.26 through 5.30). The higher the exponent, the greater this bending. Generally, power functions greater than four (4) produce hedged sets that are too skewed for effective use in a model.

```
  FuzzySet:    PI.Curve
  Description:
1.00                              ...
0.90                            ..   ..
0.80                          ..       ..
0.70                         .           .
0.60                        .             .
0.50                       .               .
0.40
0.30                      .                 .
0.20                    ..                   ..
0.10                   .                       .
0.00....................                        ....................
    0---|---|---|---|---|---|---|---|---|---|---|---|---|---|---|---0
  16.00   20.00   24.00   28.00   32.00   36.00   40.00   44.00   48.00
  Domained:         16.00 to      48.00
```

Figure 5.20 The Bell-shaped PI Fuzzy Set (Fuzzy Number 32 at a 25% Width)

```
  FuzzySet:    LINE.Curve
  Description:
1.00                                                                .
0.90                                                          ......
0.80                                                   .......
0.70                                             ......
0.60                                      .......
0.50                                ......
0.40                          ......
0.30                   .......
0.20             ......
0.10      .......
0.00......
    0---|---|---|---|---|---|---|---|---|---|---|---|---|---|---|---0
  16.00   20.00   24.00   28.00   32.00   36.00   40.00   44.00   48.00
  Domained:         16.00 to      48.00
```

Figure 5.21 The Linear Fuzzy Set Between 16 and 48

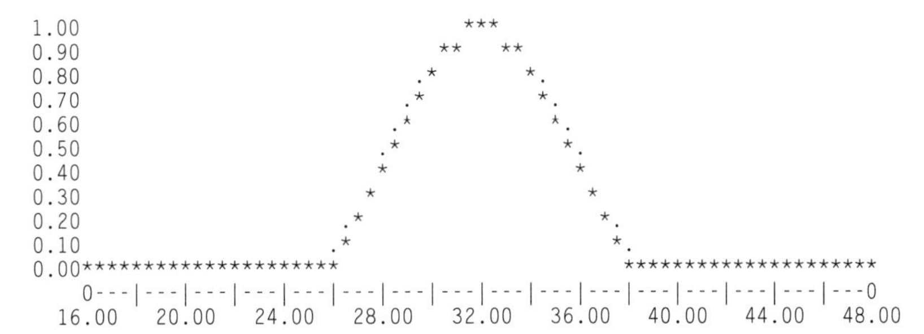

Figure 5.22 Bell Fuzzy Set Intensified with Exponent Power of 1.20

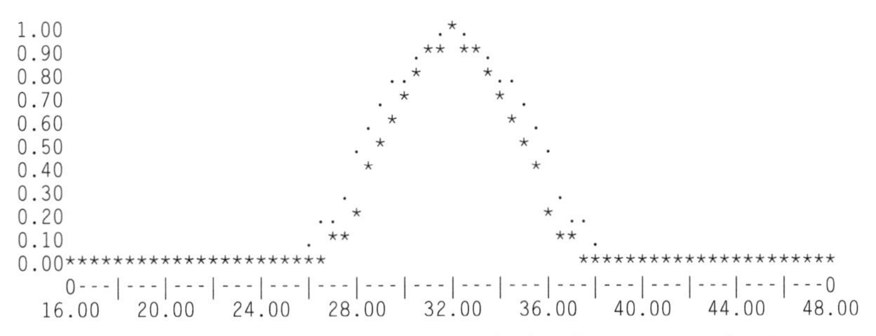

Figure 5.23 Bell Fuzzy Set Intensified with Exponent of 1.80

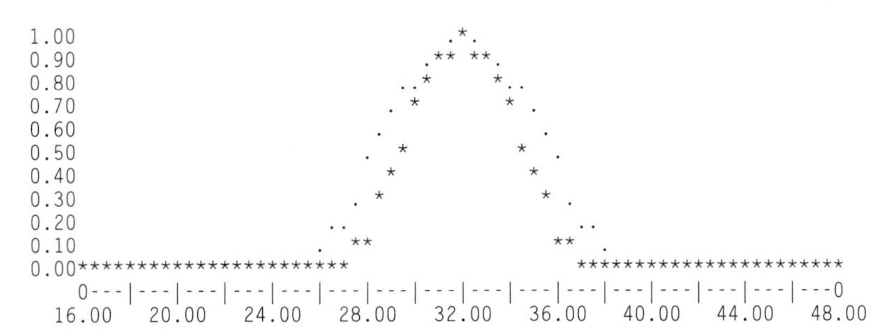

Figure 5.24 Bell Fuzzy Set Intensified with Exponent of 2.00

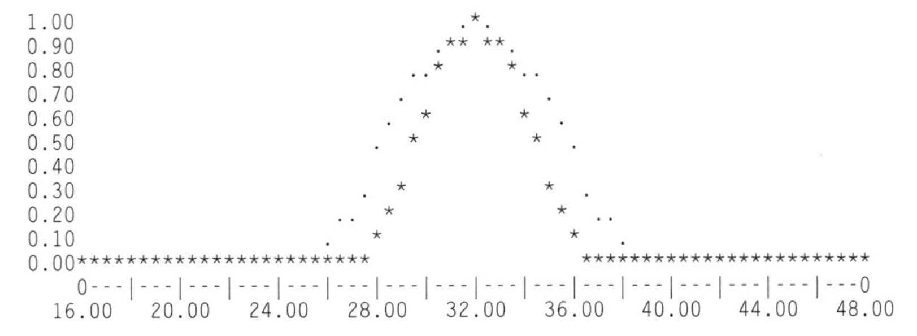

Figure 5.25 Bell Fuzzy Set Intensified with Exponent of 3.00

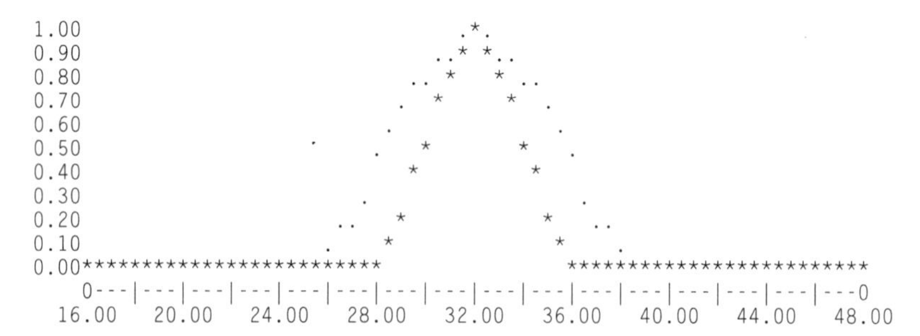

Figure 5.26 Bell Fuzzy Set Intensified with Exponent of 4.00

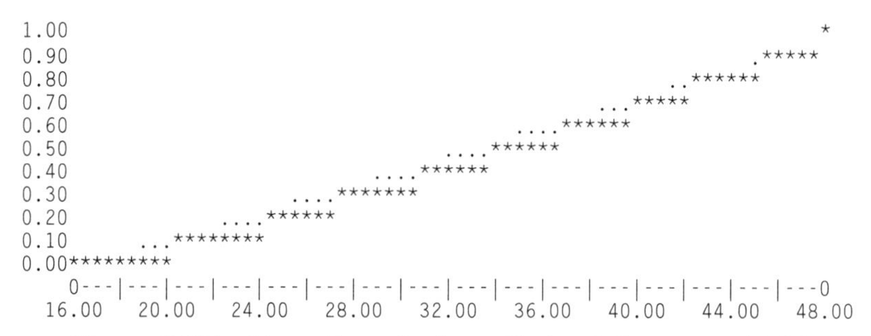

Figure 5.27 Linear Fuzzy Set Intensified with Exponent of 1.20

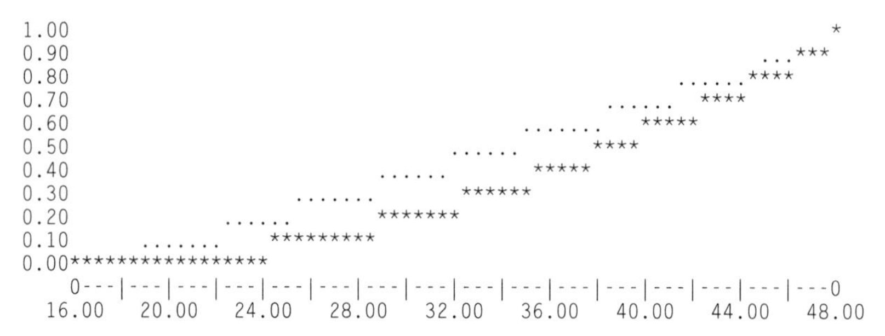

Figure 5.28 Linear Fuzzy Set Intensified with Exponent of 1.80

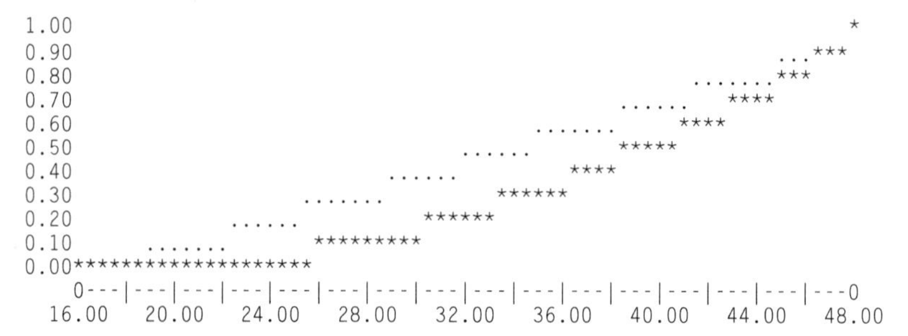

Figure 5.29 Linear Fuzzy Set Intensified with Exponent of 2.00

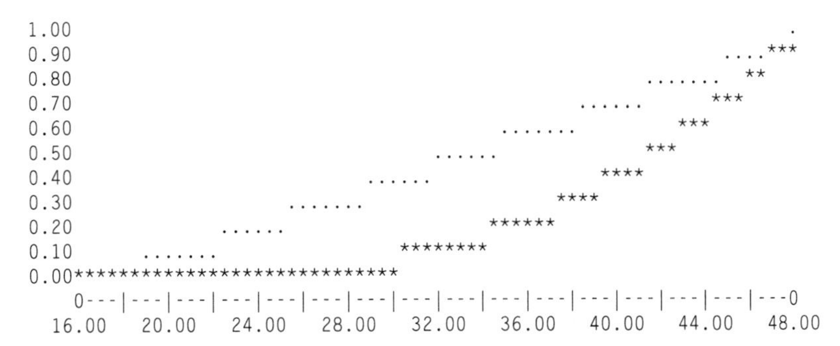

Figure 5.30 Linear Fuzzy Set Intensified with Exponent of 3.00

THE SOMEWHAT HEDGE

The complement of *very* is the hedge group represented by *somewhat*, *rather*, and *quite*. These hedges, basic synonyms for each other, dilute the force of the membership function in a particular fuzzy region. In a philosophical approach corresponding to the rational behind *very*, Zadeh notes that *somewhat* or *sort of* should have a membership value that is greater than the base truth function. This comes from the observation that someone that is judged tall would always be judged somewhat tall. The effect of *somewhat*, then, is to increase the candidate space within the fuzzy region with a characteristic function relationship of

$$\mu_{Tall}[x] \leq \mu_{SomewhatTall}[x]$$

We implement the *somewhat* hedge (Zadeh calls this a *dilator*) by taking the square root of the membership function at each point in the fuzzy region. The transformational operation is:

$$\mu_{somewhat}[x] = \mu_{A^{.5}}[x]$$

Taking the square root of any value in the interval [0,1] increases its value. Like *very*, *somewhat* is also domain-independent and maintains the normalization characteristics of the fuzzy set. Dilution hedges can also be represented by alternative forms in a manner analogous to the *very* hedge. As an example:

$$\mu_{somewhatA}[x] = \mu_{\max(0, A-k)}[x]$$

This transformation is also bounded at the zero domain edge. A bounded product version of the dilator is also possible in the form:

$$\mu_{somewhatA}[x] = \mu_{\max(1, A \times k)}[x]$$

The parameter k provides a context-sensitive adjusting control on the strength of the dilator. This is important in shift-type hedges since they are dependent on the edges of the underlying domain. A dilator hedge is easily converted to its power function counterpart by the appropriate substitution of the value for k.

The idea behind the *somewhat* hedge is simple: the surface of the fuzzy set is inflated so that an element from the domain that had x degrees of truth, now has $x+y$ degrees of truth, where y is the proportional increase at domain point S in the fuzzy set. Figure 5.31 show the effect of applying the hedge *somewhat* to the fuzzy set *Tall.*

The *somewhat* hedge increases the truth membership function for each domain value except at the fuzzy set extremes. Again (see the *very* hedge for a counterexample), we can see how this affects the truth of the fuzzy propositions by examining two rules that estimate weight from height:

```
if height is tall then weight is heavy
if height is somewhat tall then weight is heavy
```

Figure 5.31 illustrates how this change in the characteristics of the fuzzy set topology produces a change in the truth function at each point along the line. For a height of 5.5 feet, the predicate proposition, *height is tall*, has a truth of [.56] and will produce a significant contribution to the consequent fuzzy region *Heavy.* For the same height, the proposition *height is somewhat tall* has a truth of [.97] and induces a weak truth function in the consequent fuzzy set *Heavy.*

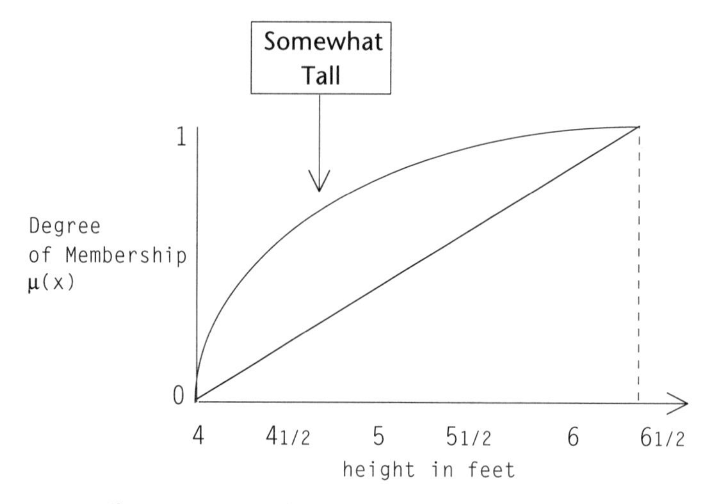

Figure 5.31 Applying the Hedge *Somewhat* to *Tall*

As Figure 5.32 shows, a *somewhat* hedge means that, for equivalent truth functions, a domain element must be located further to the left. In order for a *somewhat tall* predicate to have a consequent influence of [.56], the height need only be 4.5 feet.

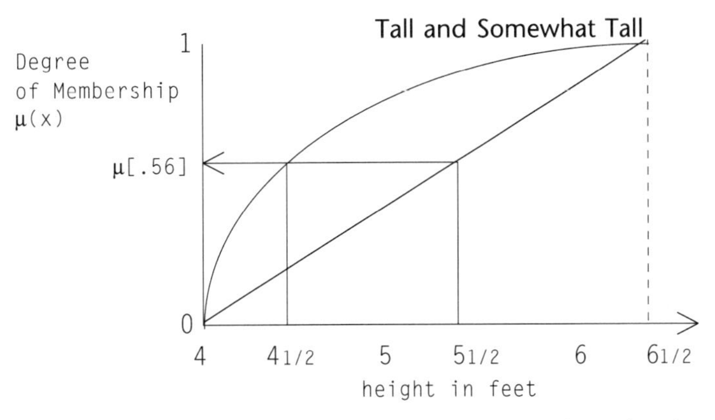

Figure 5.32 Domains with Equal Truth Membership Values in *Tall* to *Somewhat Tall*

As Figure 5.33 illustrates, while *somewhat*, like *very*, provides a proportional increase in the truth function for all uniformly orthogonal sets including bell-shaped curves, it also has some anomalous behavior when applied to fuzzy sets that are multi-modal or have large plateau areas (see commentary for Figure 5.17).

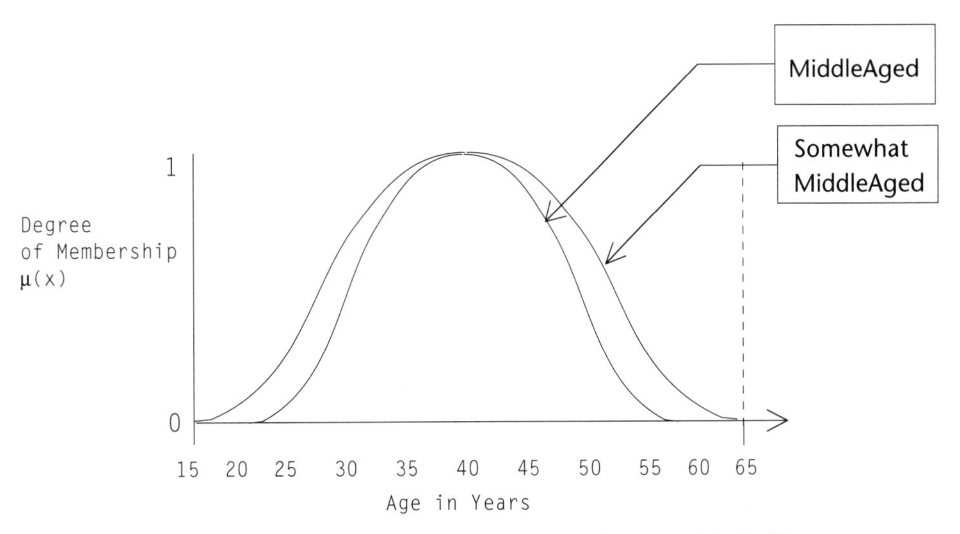

Figure 5.33 Applying the Hedge *Somewhat* to *MIDDLEAGE*

A generalization of the Zadeh dilator hedge simply replaces the exponent of the intensification function with a real positive number that is less than one [1] and is expressed as a fraction (*1/n*). The dilator has the construct:

$$\mu_{dil(A)}[x] = \mu_{A^{\frac{1}{n}}}[x] \quad n > 1$$

In practical use, n will fall in the interval [1,8]. Intensely root-functional hedges of the Zadeh dilator type simply push the membership function toward one [1]. A fuzzy region with its truth function at or near the unity [1] level will behave as a crisp variable and will disrupt the partitioning of the fuzzy model control spaces. A hedge such as *greatly* is represented by the dilator as

$$\mu_{greatlyA}[x] = \mu_{A^{.7}}[x]$$

and used in rules such as

```
if costs are greatly increased,
   then product.risks are somewhat high;
```

You can use the effects of various power values on surface dilution in the following modifications (Figures 5.34 and 5.35) to a bell-shaped and linear fuzzy set (from the `DILHEDGE.CPP` program.)

```
  FuzzySet:    PI.Curve
  Description:
1.00                               ...
0.90                             ..    ..
0.80                           ..       ..
0.70                          .           .
0.60                         .             .
0.50                        .               .
0.40
0.30                       .                 .
0.20                     ..                   ..
0.10                    .                       .
0.00....................                         ....................
    0---|---|---|---|---|---|---|---|---|---|---|---|---|---|---|---0
  16.00   20.00   24.00   28.00   32.00   36.00   40.00   44.00   48.00
  Domained:       16.00 to      48.00
```

Figure 5.34 The Bell-Shaped PI Fuzzy Set (Fuzzy Number 32 at a 25% Width)

To improve clarity and reduce clutter, the fuzzy set graph legends have been omitted in Figures 5.36 through 5.39. The base fuzzy set is indicated by a dot (.), and the modified fuzzy surface is represented by an asterisk (*).

```
 FuzzySet:     LINE.Curve
 Description:
1.00                                                                   .
0.90                                                             ......
0.80                                                     .......
0.70                                              ......
0.60                                        .......
0.50                                  ......
0.40                            ......
0.30                     .......
0.20              ......
0.10       .......
0.00......
    0---|---|---|---|---|---|---|---|---|---|---|---|---|---|---|---0
  16.00    20.00    24.00    28.00    32.00    36.00    40.00    44.00    48.00
 Domained:         16.00 to         48.00
```

Figure 5.35 The Linear Fuzzy Set Between 16 and 48

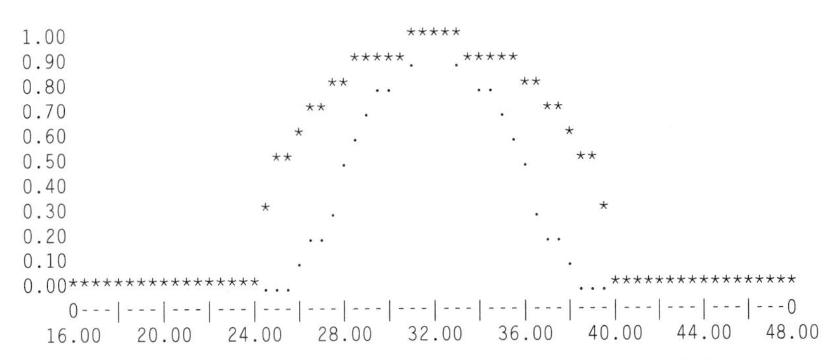

Figure 5.36 Bell Fuzzy Set Diluted with Exponent of .20

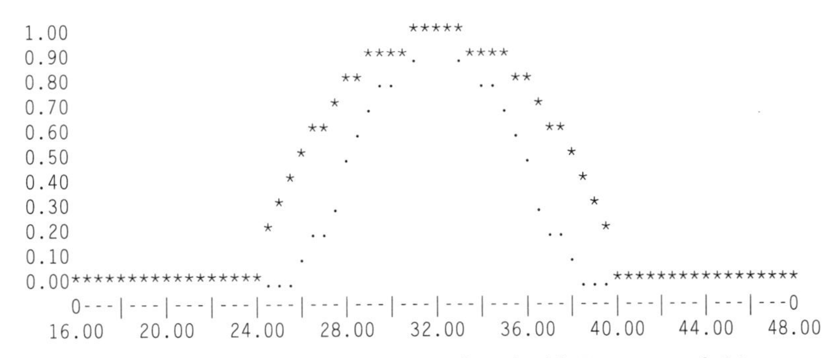

Figure 5.37 Bell Fuzzy Set Diluted with Exponent of .30

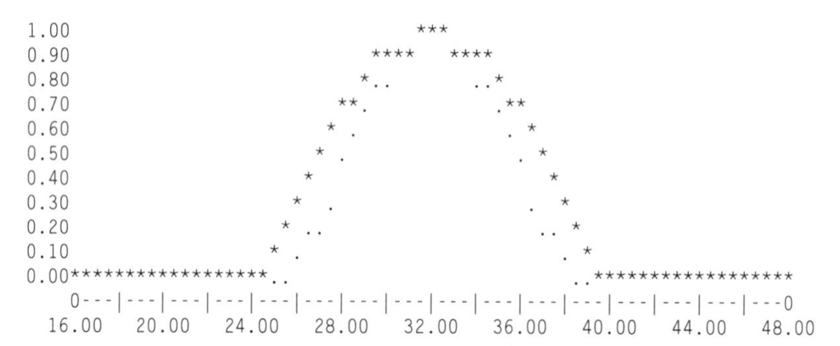

Figure 5.38 Bell Fuzzy Set Diluted with Exponent of .50

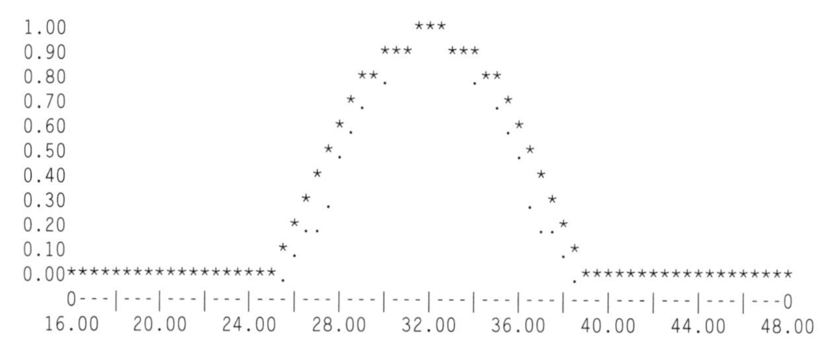

Figure 5.39 Bell Fuzzy Set Diluted with Exponent of .70

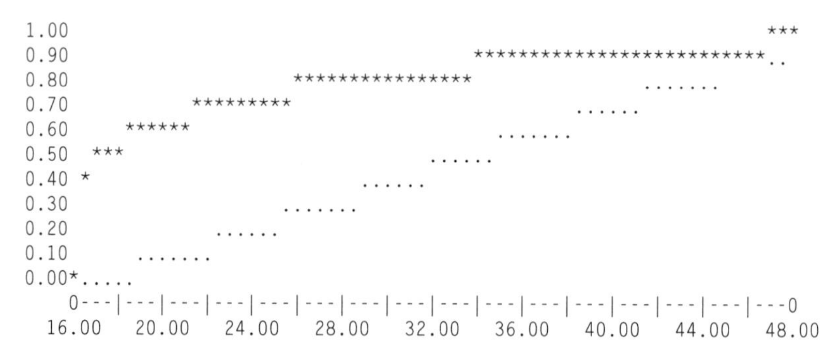

Figure 5.40 Linear Fuzzy Set Diluted with Exponent of .20

The bell-shaped fuzzy set is further fuzzified away from its central value. The linear fuzzy set, on the other hand, is "bent" outward, away from the domain axis (Figures 5.40 through 5.43). The smaller the exponent, the greater this bending. Generally, power functions greater than four (.7) produce hedged sets that are too close to the base set for effective use in a model.

```
1.00                                                         **
0.90                                     *******************.
0.80                      **************        .......
0.70             ***********                 ......
0.60     *******                      .......
0.50  ******                    ......
0.40   ***                ......
0.30  *            .......
0.20 *       ......
0.10     .......
0.00*.....
    0---|---|---|---|---|---|---|---|---|---|---|---|---|---|---|---0
  16.00   20.00   24.00   28.00   32.00   36.00   40.00   44.00   48.00
```

Figure 5.41 Linear Fuzzy Set Diluted with Exponent of .30

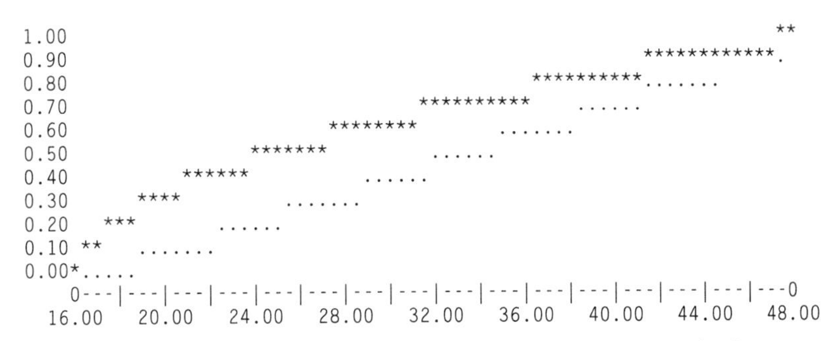

Figure 5.42 Linear Fuzzy Set Diluted with Exponent of .50

```
1.00                                                            *
0.90                                                    *********
0.80                                           *********...
0.70                                   ********.....
0.60                           *******.......
0.50                   *******  ......
0.40           *******  ......
0.30      ******  .......
0.20   *****  ......
0.10  ***.......
0.00***...
    0---|---|---|---|---|---|---|---|---|---|---|---|---|---|---|---0
  16.00   20.00   24.00   28.00   32.00   36.00   40.00   44.00   48.00
```

Figure 5.43 Linear Fuzzy Set Diluted with Exponent of .70

Reciprocal Nature of Very and Somewhat

The power function hedges, *very* and *somewhat*, are commutative and obey the rules of set idempotency. A commutative function follows the mathematically transitive mechanism

$$f(x, y) = f(y, x)$$

meaning that the linguistic statements

```
somewhat very tall
very somewhat tall
```

represent the same semantic space. This seems to contradict the principle that hedges follow the same linguistic rules as English adjectives and adverbs, but it is really a consequence of their reciprocal functional representations. This means that two different orders in which they are applied provide identical results:

$$\mu_{(A^2)^{\frac{1}{2}}}[x] \equiv \mu_{(A^{\frac{1}{2}})^2}[x]$$

These are the only naturally idempotent members of the hedge set, and they are the only commutative hedges (that is, where *positively somewhat tall* and *somewhat positively tall* are distinct linguistic variables).

Contrast Intensification and Diffusion

The contrast hedges change the nature of fuzzy regions by either making the region less fuzzy (*intensification*) or more fuzzy (*diffusion*). The idea behind contrast hedging is related to the concept of fuzzy entropy and intrinsic ambiguity. Between the two domain points of a fuzzy set, values have varying degrees of membership. We can view this membership as a degree of compatibility between a value (x) and a fuzzy set (Y) measured by the fuzzy proposition, *x is Y.*

At the extremes of the fuzzy set, membership is less ambiguous. The fuzziness associated with extreme values is minimized—we have less difficulty saying "*x is (is not) a member of Y.*" On the other hand, for values that are centered around the midpoint in the fuzzy truth function, we have increased difficulty in saying that a value (x) is compatible with the fuzzy concept (Y). This area, illustrated in Figure 5.44, is an area of maximum fuzziness.

In order to see that this area of increased fuzziness represents a region of heightened ambiguity, we must look at what happens when we form the intersection of a fuzzy set (s) with its own complement ($\sim S$). As Figure 5.45 shows, there is a region around the [.5] membership area that can belong to either set simultaneously. This is a region of increased fuzziness, expressed as ambiguity or undecidability.

The Positively Hedge

Represented by *positively*, *absolutely*, or *definitely*, the contrast intensification modifier changes the fuzzy surface by raising all the truth function values to above [.5] and

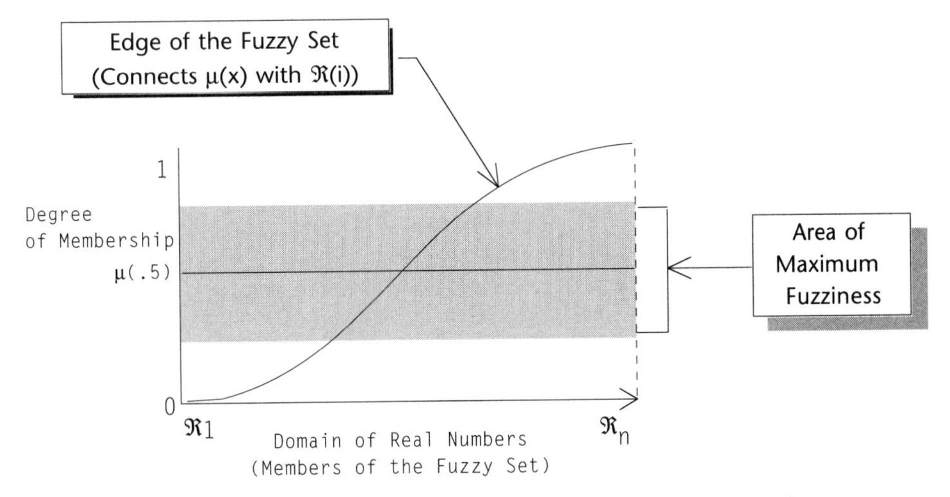

Figure 5.44 The Maximum Fuzzy Space in a Normal Fuzzy Region

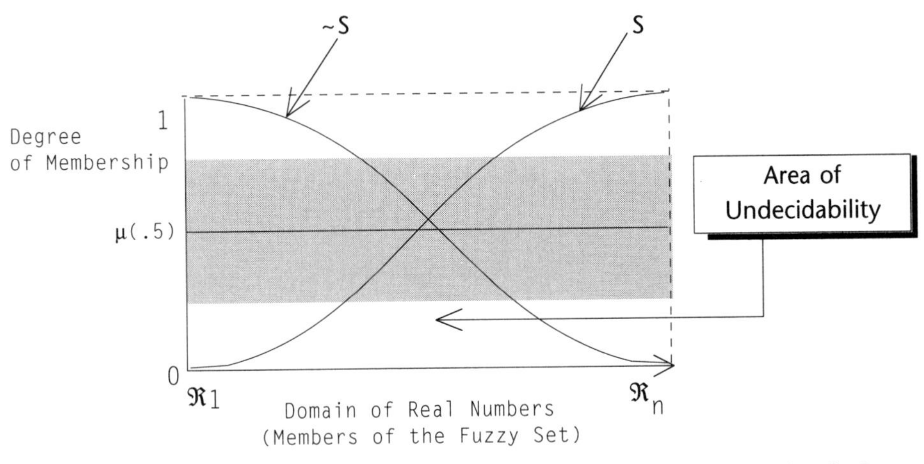

Figure 5.45 The Undecidable Region at the Intersection of Fuzzy Sets *S* and ~*S*

decreasing all the truth function values to below [.5]. This process moves the membership values closer to [1] or [0], and thus reduces the overall fuzziness of the region.

Zadeh's proposal for the contrast intensification follows the overall power function of the *very* hedge for memberships above [.5] and the bounded product for membership's below the [.5] threshold. The characteristic function for an intensified fuzzy region appears as:

$$\mu_{\text{int}(A)}[x] = \begin{cases} 2\times(\mu_{A^2}[x]) & \mu_A[x] \geq 0.5 \\ 1-2\times(\mu_{A^2}[x]) & \mu_A[x] < 0.5 \end{cases} \quad \text{when}$$

The middle axis of fuzzy set at the μ[.5] level acts as an inflection point. Figure 5.46 shows how the fuzzy set *Tall* is modified by intensification.

$$\mu_{\text{int}(A)}[x] = \begin{cases} 2\times(\mu_{A^2}[x]) & \mu_A[x] \geq 0.5 \\ 1-2\times(\mu_{A^2}[x]) & \mu_A[x] < 0.5 \end{cases} \quad \text{when}$$

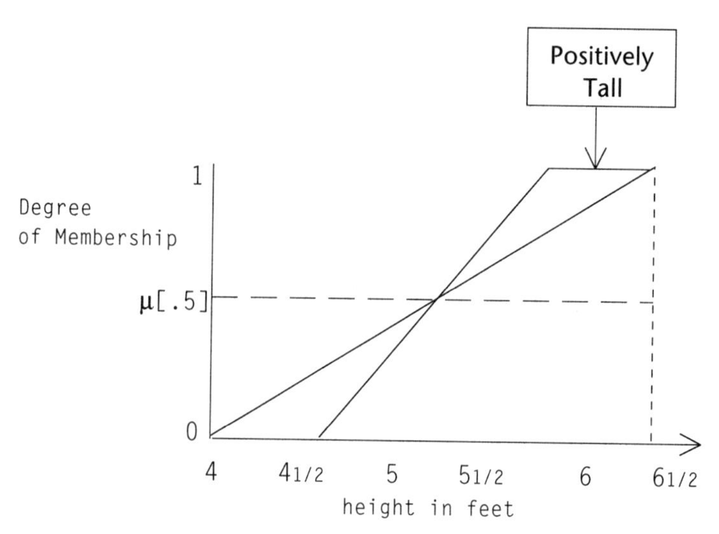

Figure 5.46 Applying the Hedge *positively* to the Fuzzy Set *TALL*

This inflection point reduces the overall fuzziness of the region. We can see how this affects the truth of the fuzzy proposition by examining rules that estimate weight from height:

```
if height is tall then weight is heavy;
if height is positively tall then weight is heavy
```

For a height of nearly 5 feet (Figure 5.47), the predicate proposition *height is tall* has a truth of [.42] and produces a moderate contribution to the consequent heavy fuzzy set.

For the same height, the proposition *height is positively tall* has a truth of [.28] and thus makes a significantly lesser contribution to the consequent heavy fuzzy set. If the height falls below 4.5 feet, then the membership in *positively Tall* becomes zero.

In a fashion similar to the *very* hedge, the intensification function can be generalized by changing the value of the truth membership exponent (*1/n*) and the displacement constant (*n*). This has a form of:

$$\mu_{\text{int}(A)}[x] = \begin{cases} n \times (\mu_{A^n}[x]) & \mu_A[x] \geq 0.5 \\ 1 - n \times (\mu_{A^n}[x]) & \mu_A[x] < 0.5 \end{cases} \quad \text{when}$$

The original Zadeh definition of intensification maintains fuzzy set normalization and respects the set's zero and unity boundaries. For general forms, any algorithm must maintain the normal and extreme characteristics of the fuzzy set.

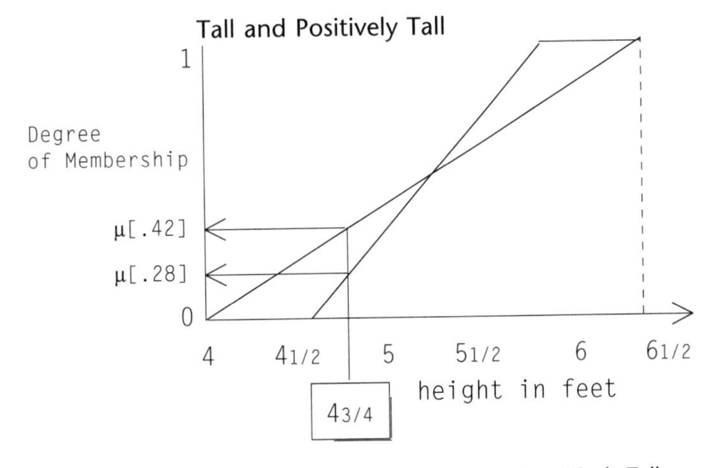

Figure 5.47 Differences Between *Tall* and *Positively Tall*

The Generally Hedge

Represented by *generally*, the diffusion modifier changes the fuzzy surface by reducing all the truth function values above [.5] and increasing all the truth function values below [.5]. This process moves the membership values closer to [.5], thus increasing the overall fuzziness of the region.

Zadeh's proposal for the contrast diffusion follows the overall square root function of the *somewhat* hedge for memberships. The characteristic function for a diffuse fuzzy region appears as:

$$\mu_{\text{int}(A)}[x] = \begin{cases} .5\times(\mu_{A^{.5}}[x]) & \mu_A[x]\geq 0.5 \\ 1-.5\times(\mu_{A^{.5}}[x]) & \mu_A[x]<0.5 \end{cases} \quad \text{when}$$

The middle axis of the fuzzy set at the μ[.5] level acts as an inflexion point. Figure 5.48 shows how the fuzzy set *Tall* is modified by diffusion.

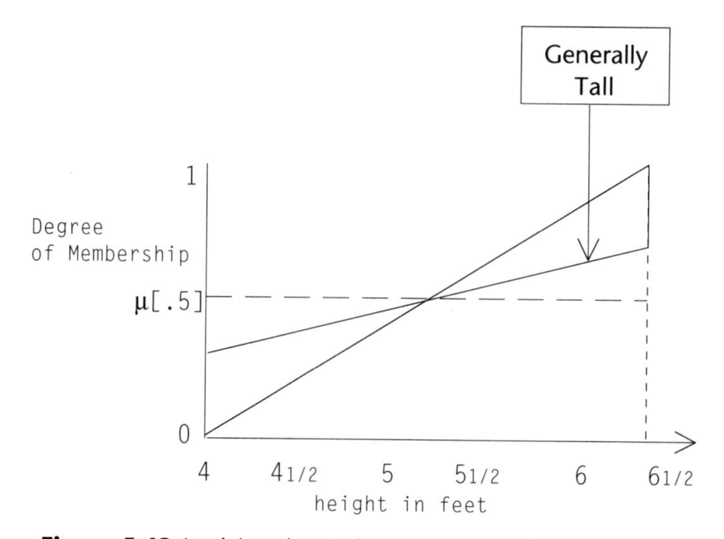

Figure 5.48 Applying the Hedge *Generally* to the Fuzzy Set *Tall*

This axis increases the overall fuzziness of the region. We can see how this affects the truth of the fuzzy proposition by examining rules that estimate weight from height:

```
if height is tall then weight is heavy;
if height is generally tall then weight is heavy
```

For a height of nearly 5 feet (see Figure 5.49), the predicate proposition *height is Tall* has a truth of [.42] and produces a moderate contribution to the consequent *Heavy* fuzzy set. For the same height, the proposition *height is generally Tall* has a truth of [.48] and thus makes a significantly larger contribution to the consequent *Heavy* fuzzy set.

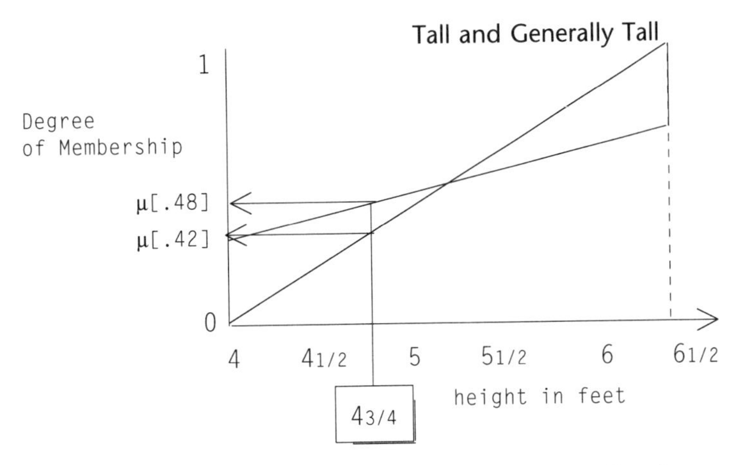

Figure 5.49 Difference Between *Tall* and *Generally Tall*

In a fashion similar to the `somewhat` hedge, the diffusion function can be generalized by changing the value of the truth membership exponent (`1/n`) and the displacement constant (`n`). This has a form of:

$$\mu_{\text{int}(A)}[x] = \begin{cases} \frac{1}{n} \times (\mu_{A^{\frac{1}{n}}}[x]) & \text{when } \mu_A[x] \geq 0.5 \\ 1 - \frac{1}{n} \times (\mu_{A^{\frac{1}{n}}}[x]) & \mu_A[x] < 0.5 \end{cases}$$

The original Zadeh definition of diffusion maintains fuzzy set normalization and respects the set's zero and unity boundaries. For general forms, any algorithm must maintain the normal and extreme characteristics of the fuzzy set.

The *positively* and *generally* hedges are important surface transformers. They allow us to make decisions based on whether or not a truth value is either broadly within the fuzzy set (generally inside the set) or definitely within the fuzzy set (situated well away from the zero membership space). You can see how the *positively* and *generally* hedges change the surface of a fuzzy set in Figures 5.50 and 5.51 (produced by the *LIHEDGE.CPP* program).

```
1.00                                                **         .
0.90                                              ***      ......
0.80                                            **     .......
0.70                                          ***   ......
0.60                                        ***.......
0.50                                      ***...
0.40                                  ...***
0.30                             .......***
0.20                        ......    **
0.10                 .......    ***
0.00......               ****
      0---|---|---|---|---|---|---|---|---|---|---|---|---|---|---|---0
    16.00   20.00   24.00   28.00   32.00   36.00   40.00   44.00   48.00
 .   FuzzySet:     LINE.Curve
    Description:
 *   FuzzySet:     LINE.Curve
    Description: positively LINE.Curve
    Domained:          16.00 to       48.00
```

Figure 5.50 Linear Fuzzy Set Modified with *Positively* Hedge

```
1.00                                                               .
0.90                                                          ......
0.80                                                   .......    **
0.70                                                ...***************
0.60                                     *************
0.50                                   .**...
0.40                               .******
0.30                    *************
0.20*************.....
0.10        .......
0.00......
      0---|---|---|---|---|---|---|---|---|---|---|---|---|---|---|---0
    16.00   20.00   24.00   28.00   32.00   36.00   40.00   44.00   48.00
 .   FuzzySet:     LINE.Curve
    Description:
 *   FuzzySet:     LINE.Curve
    Description: generally LINE.Curve
    Domained:          16.00 to       48.00
```

Figure 5.51 Linear Fuzzy Set Modified with *Generally* Hedge

The *FzyApplyHedge* function consolidates all the built-in system hedges into a single control package. The function also supports hedges that you define using one of the special transformation classes. Which hedge you want is determined by the contents of a hedge descriptor block (HDB). For built-in hedges, the same HDB can be used for different hedges.

```
//--hdb.hpp [EDC Rev 1.0 7/28/92] (c)1992 Metus Systems Group
//--The Hedge Descriptor Block. Each built-in and user hedge
//--is defined through one of these nodes. A hedge modifies
//--the shape of a fuzzy set.
#ifndef __hdb
#define __hdb
#include "mtypes.hpp"
struct HDB
  {
    char        HDBid[IDENLEN+1],      // Hedge identifier
                HDBdesc[DESCLEN+1];    // Hedge description
    int         HDBmode,               // n=builtin; 0=user hedge
                HDBop;                 // Type of transformer
    double      HDBvalue;              // Transformer value
    HDB        *HDBnext;               // Pointer to next hedge
  };
#endif
```

Each built-in hedge is assigned a unique hedge mode number (*HDBmode*). Hedges that are synonyms have the same hedge mode (such as *somewhat* and *rather*). The following symbolic constants and their associated hedge mode numbers are defined in the *fuzzy.hpp* file:

```
ABOUT        1
ABOVE        2
POSITIVE     3
BELOW        4
VICINITY     5
GENERALLY    6
CLOSE        7
NOT          8
SOMEWHAT     9
VERY        10
EXTREMELY   11
SLIGHTLY    12
```

On the other hand, all user-defined hedges have a hedge mode of zero, and their action is determined solely from the hedge operator (*HDBop*) and its associated transformer value (*HDBvalue*). Changing the hedge operator or the transformer value will have no effect on built-in hedges. *FzyApplyHedge* supports only simple arithmetic operations defined in the same *fuzzy.hpp* symbolics header file. The following constants with their associated values can be used:

```
ADD           1
SUBTRACT      2
MULTIPLY      3
DIVIDE        4
POWER         5
```

Care should be exercised in using any hedge mode other than `MULTIPLY` and `POWER`, since adding, subtracting, and dividing truth membership functions can result in anomalous values, which are truth memberships that are negative or greater than one. The current version of `FzyApplyHedge` does not check that the resulting truth membership value is valid and/or sensible.

The *FzyApplyHedge* hedge management code is broken into two sections for discussion: supporting user hedges and supporting built-in hedges. The built-in hedges, when necessary, are explained separately. Due to the prevalence of hedges in most information models, *FzyApplyHedge* modifies a second, incoming fuzzy set but does not create a new, hedged fuzzy set.

```
#include <math.h>
#include <stdio.h>
#include <string.h>
#include "FDB.hpp"
#include "HDB.hpp"
#include "mtypes.hpp"
#include "fuzzy.hpp"
#include "mtsptype.hpp"
void Complete_hedge(const HDB*,const FDB*,FDB*);
void FzyApplyHedge(
   const FDB* inFDBptr,
    const HDB* HDBptr,FDB* outFDBptr,int *statusPtr)
  {
   int       i,j,n;
   double    x,
             hv;
   float     local[VECMAX+1],part1,part2,part3,Rslt;
   double    EXPval;
   char     *PgmId="mtfzaph";
   char      buff[80];
   FILE     *mdllog;
```

```
*statusPtr=0;
FzyCopySet(inFDBptr,outFDBptr,statusPtr);
double  range=inFDBptr->FDBdomain[1]-inFDBptr->FDBdomain[0];
double  hedgeval=HDBptr->HDBvalue;

if(HDBptr->HDBmode==0)
  {
    switch(HDBptr->HDBop)
     {
      case ADD:
       for(i=0;i<VECMAX;i++)
        outFDBptr->FDBvector[i]=
          inFDBptr->FDBvector[i]+(float)hedgeval;
       break;
      case SUBTRACT:
       for(i=0;i<VECMAX;i++)
        outFDBptr->FDBvector[i]=
          inFDBptr->FDBvector[i]-(float)hedgeval;
       break;
      case MULTIPLY:
       for(i=0;i<VECMAX;i++)
        outFDBptr->FDBvector[i]=
          inFDBptr->FDBvector[i]*(float)hedgeval;
       break;
      case DIVIDE:
       for(i=0;i<VECMAX;i++)
        outFDBptr->FDBvector[i]=
          inFDBptr->FDBvector[i]/(float)hedgeval;
       break;
      case POWER:
       for(i=0;i<VECMAX;i++)
        outFDBptr->FDBvector[i]=
          (float)pow(inFDBptr->FDBvector[i],hedgeval);
       break;
      default:
       {
         *statusPtr=1;
         MtsSendError(18,PgmId,HDBptr->HDBid);
         return;
       }
     }/* Usertype */
    Complete_hedge(HDBptr,inFDBptr,outFDBptr);
    return;
  }
```

Listing 5.1 The User-Defined Hedge Section of `FzyApplyHedge`

If a hedge's mode is zero, we know that it is a user-defined hedge of some kind. The switch statement then selects the type of transformation. If none of the transformations matches, an error is generated. On matching the type of transformation, the truth function is modified using *HDBvalue* according to the kind of operation. This action completes the user-defined hedge, a message is written to the system log file, and the function terminates.

```
switch(HDBptr->HDBmode)
    {
     case ABOUT:
     case VICINITY:
     case CLOSE:
      if(HDBptr->HDBmode==ABOUT)    EXPval=2;
      if(HDBptr->HDBmode==VICINITY) EXPval=4;
      if(HDBptr->HDBmode==CLOSE)    EXPval=1.2;
      for(i=0;i<VECMAX;++i) local[i]=0.0;
      for(i=0;i<VECMAX;i++)
       {
        if(inFDBptr->FDBvector[i]!=0)
          {
           x=(double)i;
           for(j=0;j<VECMAX;j++)
            {
             part1=fabs(float(j)-x)*range/VECMAX;
             part2=(1.000/(1+pow(part1,EXPval)))
                    *inFDBptr->FDBvector[i];
             local[j]=max(local[j],part2);
            }
          }
       }
      for(i=0;i<VECMAX;++i) outFDBptr->FDBvector[i]=local[i];
      break;
```

The hedges that either approximate an existing fuzzy set or convert a scalar to a fuzzy set are clustered under a single switch case statement. Each semantic type requires a different exponent value to form a different type of bell-shaped set. The larger the exponent, number the broader (wider) the resulting fuzzy set. We use a local working storage area to create the new fuzzy set and then copy it into the output fuzzy region.

```
     case ABOVE:
      x=0.0;
      j=0;
      for(i=0;i<VECMAX;++i)
       if(inFDBptr->FDBvector[i] >= x)
         {
```

```
        j=i;
        x=inFDBptr->FDBvector[i];
        if(x==1.00)
          goto above_exit;
      }
    above_exit:
     for(i=0;i<VECMAX;++i)
      {
       if(i <= j)
         outFDBptr->FDBvector[i]=0.0;
        else
         outFDBptr->FDBvector[i]=1-inFDBptr->FDBvector[i];
      }
    break;
```

When we say *if x is above Y*, that means the new fuzzy set contains a curve that moves opposite the declining line of the incoming fuzzy set. The function starts with a locator variable (*x*) set to zero and continues to swap this value with higher and higher values in the fuzzy set (until it finds a [1.0] truth value). The current position of *x* in the truth array is recorded by the **cursor** *(j)*. When we complete, we set everything to the left of *j* to zero. All other values are set as their complements.

```
    case POSITIVE:
     x=.5;
     for(i=0;i<VECMAX;++i)
      if(inFDBptr->FDBvector[i] >= x)
          {
           hv=inFDBptr->FDBvector[i];
           hv=(float)2*pow(hv,TWO);
           outFDBptr->FDBvector[i]=(float)hv;
          }
        else
          {
           hv=inFDBptr->FDBvector[i];
           hv=(float)1-(2*(pow(1-hv,TWO)));
           outFDBptr->FDBvector[i]=(float)hv;
          }
     break;
```

The *positively* hedge examines each truth membership value to see if it is above or below the [.5] membership level. One of two different transformations is applied depending on the outcome of this test. (The *positively* hedge is explained in detail under The Positively Hedge on page 246 of this chapter.)

```
case BELOW:
 x=0.0;
 j=VECMAX;
 for(i=0;i<VECMAX;++i)
  if(inFDBptr->FDBvector[i] >= x)
      {
       j=i;
       x=inFDBptr->FDBvector[i];
       if(x==1.00)
         goto below_exit;
      }
 below_exit:
  for(i=0;i<VECMAX;++i)
   {
    if(i >= j)
      outFDBptr->FDBvector[i]=0.0;
     else
      outFDBptr->FDBvector[i]=1-inFDBptr->FDBvector[i];
   }
 break;
```

Except for the final transformation, this hedge works in a manner very much like the above hedge. When we say *if x is below Y*, the new fuzzy set contains a curve that moves opposite the growing line of the incoming fuzzy set. The function starts with a locator variable (x) set to zero, and it continues to swap this value with higher and higher values in the fuzzy set (until it finds a [1.0] truth value.) The current position of x in the truth array is recorded by the cursor (j). When we complete the process, we set everything to the right (and inclusive) of j to zero. All other values are set as their complements.

```
case GENERALLY:
 x=.5;
 for(i=0;i<VECMAX;++i)
  if(inFDBptr->FDBvector[i] >= x)
      {
       hv=inFDBptr->FDBvector[i];
       hv=(float)0.8*pow(hv,.5);
       outFDBptr->FDBvector[i]=(float)hv;
      }
    else
      {
       hv=inFDBptr->FDBvector[i];
       hv=(float)1-(0.8*(pow(1-hv,.5)));
       outFDBptr->FDBvector[i]=(float)hv;
      }
 break;
```

Like the *positively* hedge, the *generally* hedge examines each truth membership value to see if it is above or below the [.5] membership level. One of two different transformations is applied depending on the outcome of this test. (The *generally* hedge is explained in detail under The Generally Hedge on page 249 of this chapter.)

```
case NOT:
 for(i=0;i<VECMAX;++i)
  outFDBptr->FDBvector[i]=1-inFDBptr->FDBvector[i];
 break;
case SOMEWHAT:
 for(i=0;i<VECMAX;++i)
  outFDBptr->FDBvector[i]=
    (float)pow(inFDBptr->FDBvector[i],POINTFIVE);
 break;
case VERY:
 for(i=0;i<VECMAX;++i)
  outFDBptr->FDBvector[i]=
    (float)pow(inFDBptr->FDBvector[i],TWO);
 break;
case EXTREMELY:
 for(i=0;i<VECMAX;++i)
  outFDBptr->FDBvector[i]=
    (float)pow(inFDBptr->FDBvector[i],THREE);
 break;
case SLIGHTLY:
 for(i=0;i<VECMAX;++i)
  outFDBptr->FDBvector[i]=
    (float)pow(inFDBptr->FDBvector[i],POINTTHREE);
 break;
```

These are the standard built-in hedges. The standard Zadeh *not* operator is included as a built-in hedge for uniformity in evaluating linguistic variables such as *positively not very Long*, so that a single iterative block of code can handle the expression. The remaining hedges are power functions across the membership array and are discussed completely in Intensifying and Diluting Fuzzy Regions on page 230.

```
  default:
   {
     *statusPtr=3;
     MtsSendError(17,PgmId,HDBptr->HDBid);
     return;
   }
 }
Complete_hedge(HDBptr,inFDBptr,outFDBptr);
return;
```

```
  }
/*------------------------------------------------------------*
| The Complete_hedge routine (1) applies the current alpha-cut|
| to the hedge, (2) initiates any trace output to the log     |
*-------------------------------------------------------------*/
 void Complete_hedge(
     const HDB* HDBptr,const FDB* inFDBptr,FDB* outFDBptr)
  {
   int     status;
   char    NameBuf[DESCLEN+1];
   char    wrkBuff[80];

   sprintf(wrkBuff,"%s%s%s%s%s",
    "Hedge '",HDBptr->HDBid,"' applied to Fuzzy Set \"",
      inFDBptr->FDBid,"\"");
   MtsWritetoLog(SYSMODFILE,wrkBuff,&status);
  //--Now update the name and the description of the new
  //--hedged fuzzy set.
   strcpy(outFDBptr->FDBid,inFDBptr->FDBid);
   strcpy(NameBuf,HDBptr->HDBid);
   strcat(NameBuf," ");
   strcat(NameBuf,inFDBptr->FDBid);
   strcpy(outFDBptr->FDBdesc,NameBuf);
   return;
  }
```

Listing 5.2 The Built-In Hedges Section of `FzyApplyHedge`

The *Complete_hedge* function writes a note to the system model log indicating that a fuzzy set has been hedged. It also updates the description field *(FDBdesc)* of the outgoing fuzzy set to indicate the hedge operation. When hedge fuzzy sets are graphically displayed, this description will differentiate multiple fuzzy sets that may have the same base name (hedging a fuzzy set does not change its name, the *FDBid*).

Approximating a Scalar

There are two methods of converting scalars (numbers) into fuzzy sets. You can create a fuzzy set with the PI, beta, triangular, or other fuzzy number operators. Or you can use *FzyApplyHedge* to apply one of the approximation hedges (*about*, *near*, *around*) to a number. A hedge, however, cannot modify a raw scalar. That scalar must be passed in the form of a fuzzy singelton.

To create a fuzzy set representation of a number, you must create a *singleton*. A singleton is a fuzzy set that has all its membership values set to zero [0] except the single membership point in the domain where the scalar exists; this membership is set to one [1.0]. The following code, extracted from the *ApplyHedgeOperator* module of a (hypothetical) general expression parser, detects that the next token (in *p1*) is a number. It then fetches the variable name from the stack and forms a temporary fuzzy representation of the number (by calling *FzyFuzzyScalar*).

```
if(MtsIsNumeric(p1,&Ntype))
    {
      tempnum=atof(p1);
      p2=XRPNBlkptr->EVALstack[StkIndex-1];
      strcpy(varname,p2);
      tempFDBptr=
        FzyFuzzyScalar(varname,tempnum,FZYPOINT,statusPtr);
      FuzzyNumber=TRUE;
```

The *FZYPOINT* parameter to *FzyFuzzyScalar* indicates that the generated fuzzy set should contain a singleton representation of the numeric value passed as *tempnum*. Later in the parser code, the actual hedge is applied, and the temporary fuzzy region is released.

```
FzyApplyHedge(tempFDBptr,HDBptr,CFDBptr,statusPtr);
if(FuzzyNumber==TRUE) delete tempFDBptr;
return;
```

This code means that an expression such as *x is around 2*MfgCosts* requires that the result of the expression *(2*MfgCosts)* be converted to a singleton before the approximation hedge can be applied. In some (but not all) cases, it is easier to use the fuzzy set creation facilities to generate the required fuzzy set from one of the built-in fuzzy number operations. This, of course, is what *FzyFuzzyScalar* does if you do not indicate the singleton fuzzy set generator. You should be careful, however, not to create a fuzzy number only to have it further diffused by the hedge operator. This practice is likely to introduce an error in your model that is difficult to detect.

> The *FzyFuzzyScalar* procedure uses some of the advanced modeling capabilities to find, as an example, any previously defined domain for a scalar. It turns a raw scalar or a scalar variable into a fuzzy number. You can select a singleton representation or a PI, beta, or triangular fuzzy number. The width and inflection points are automatically calculated from the scalar's domain (or magnitude).

```
#include <stddef.h>
#include <string.h>
#include "FDB.hpp"
```

```
#include "fuzzy.hpp"
#include "mtypes.hpp"
#include "mtsptype.hpp"
FDB *FzyFuzzyScalar(
   char *Varid,double Scalar,int SetType,int *statusPtr)
 {
   FDB    *FDBptr,*nullFDB=NULL;
   char   *PgmId="mtfzfsc";
   char   *option;
   int     k;
   double  Hi,Lo,Range,Width,Center,Left,Right,Inflexpoint;

   *statusPtr=0;
//
//--Search through the variable dictionary to see if a domain
//--exists for the current number. If not, the find domain
//--function will generate a pseudo-domain for our use.
   MdlFindFzyDomain(Varid,&Hi,&Lo,statusPtr);
//
//--Allocate a new fuzzy set descriptor and complete it with
//--the information from the variable and domain/
   if(!(FDBptr=new FDB))
     {
      *statusPtr=1;
      MtsSendError(2,PgmId,Varid);
      return(nullFDB);
     }
   FzyInitFDB(FDBptr);
   strcpy(FDBptr->FDBid,Varid);
   FDBptr->FDBdomain[0]=Lo;
   FDBptr->FDBdomain[1]=Hi;
//
//--Calculate the basic curve parameters.
   Range=Hi-Lo;
   k=(int)1.00+((Scalar-Lo)/Range)*VECMAX;
   Width=(Hi-Lo)/2;
   Center=Lo+Width;
   Inflexpoint=Width;
//
//--Create the fuzzified scalar based on the  type of
//--option we have specified.
   switch(SetType)
    {
     case FZYPOINT:
       FDBptr->FDBvector[k]=1.0;
       return(FDBptr);
     case FZYPI:
```

```
            FzyPiCurve(FDBptr,Center,Width,statusPtr);
            return(FDBptr);
          case FZYBETA:
            FzyBetaCurve(FDBptr,Center,Inflexpoint,statusPtr);
            return(FDBptr);
          case FZYTRIANGLE:
            Left=Center-Width;
            Right=Center+Width;
            FzyTriangleCurve(FDBptr,Left,Center,Right,statusPtr);
            return(FDBptr);
          default:
            *statusPtr=3;
            option=MtsFormatInt(SetType);
            MtsSendError(8,PgmId,option);
            return(nullFDB);
        }
    }
```

Listing 5.3 The General Function for Creating a Fuzzy Number

When you use this routine to produce a fuzzy number, you should be using the general policy modeling environment. The *MdlFindFuzzyDomain* accesses the current policy block (*PDB*) and then searches the associated variable dictionary. If a policy is not active or the variable name does not exist, then a domain based on the general magnitude of the number is produced.

Examples of Typical Hedge Operations

In concluding this chapter on hedge operations, Figures 5.52 through 5.64 demonstrate the way various built-in hedges affect the membership function of both a standard bell-shaped fuzzy set and a linear fuzzy set. These are examples from the *PIHEDGE.CPP* program. (See the *LIHEDGE.CPP* program for examples using linear fuzzy sets and *SCHEDGE.CPP* for examples using the S-curve fuzzy sets.) To improve clarity and reduce clutter, the legends have been removed. The base fuzzy set is indicated by a dot (.), and the modified fuzzy set is represented by an asterisk (*).

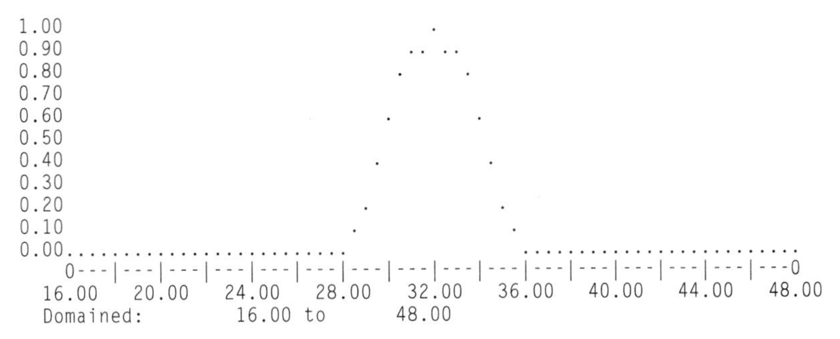

Figure 5.52 The Basic Bell-Shaped Fuzzy Set *PI.Curve*

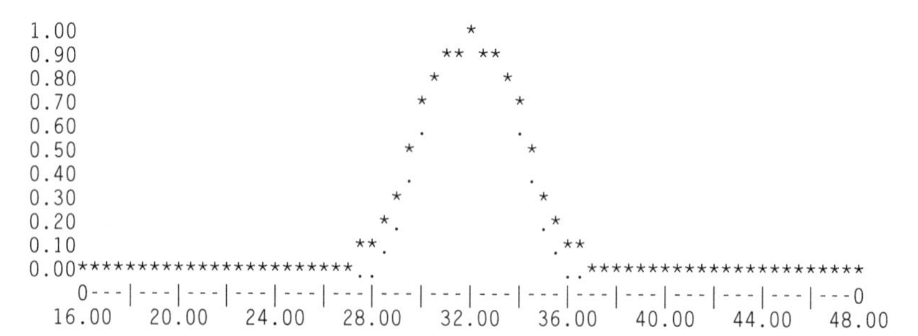

Figure 5.53 Hedge *About* Applied to Fuzzy Set *PI.Curve*

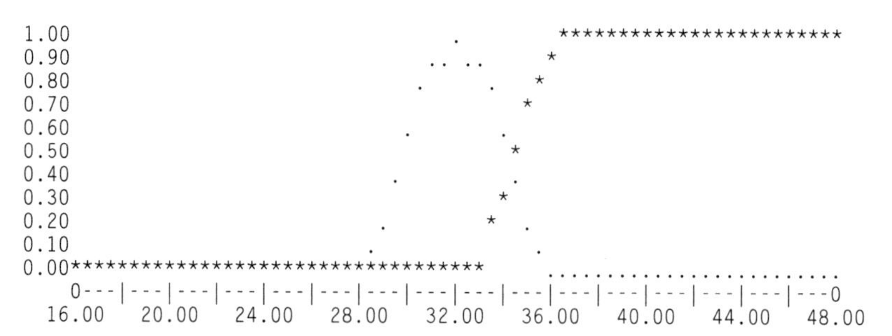

Figure 5.54 Hedge *Above* Applied to Fuzzy Set *PI.Curve*

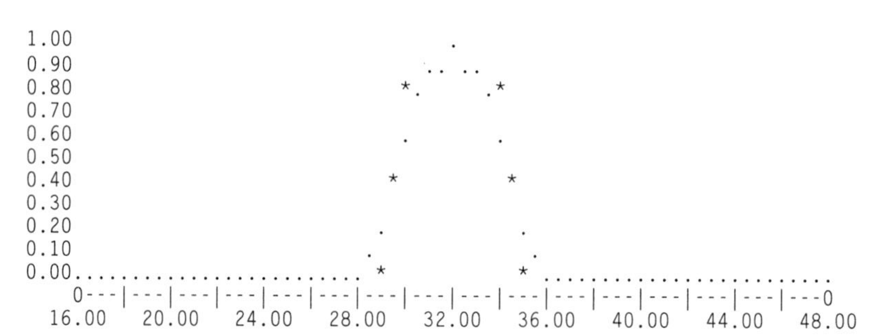

Figure 5.55 Hedge *Positively* Applied to Fuzzy Set *PI.Curve*

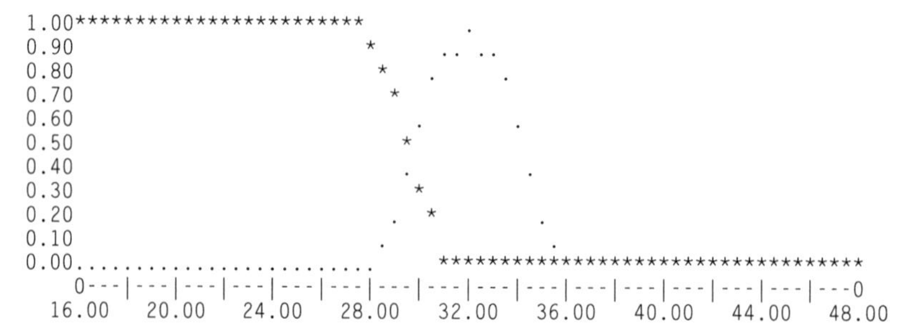

Figure 5.56 Hedge *Below* Applied to Fuzzy Set *PI.Curve*

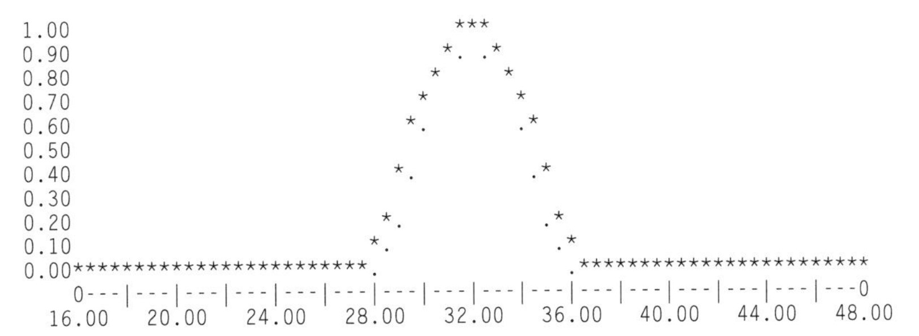

Figure 5.57 Hedge *Vicinity* Applied to Fuzzy Set *PI.Curve*

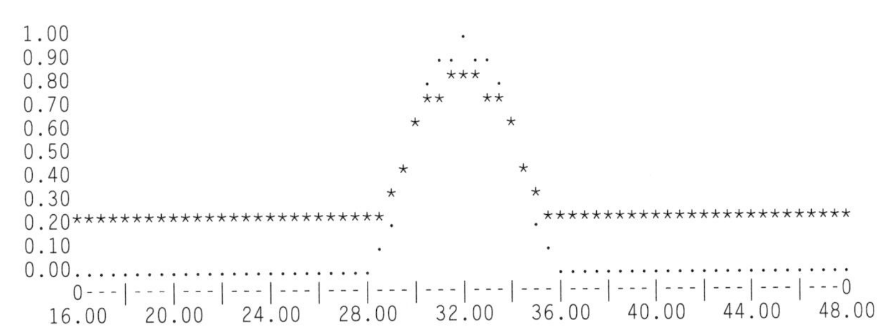

Figure 5.58 Hedge *Generally* Applied to Fuzzy Set *PI.Curve*

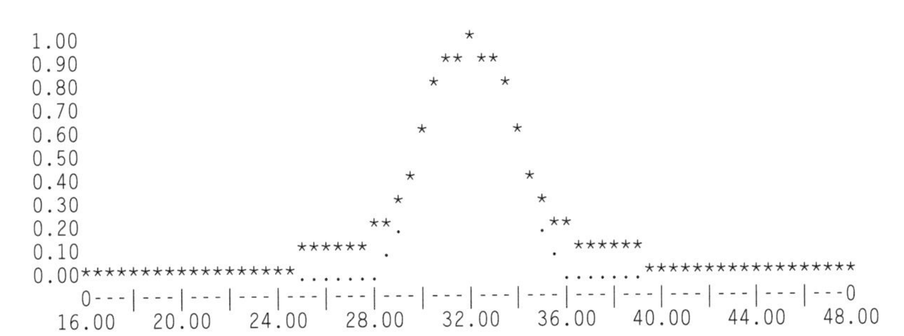

Figure 5.59 Hedge *Close* Applied to Fuzzy Set *PI.Curve*

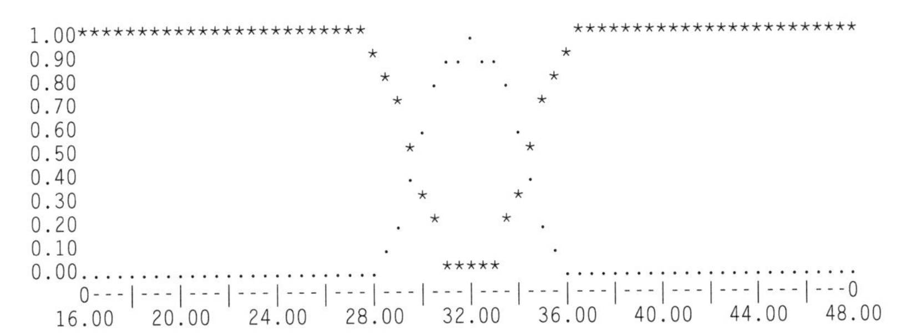

Figure 5.60 Hedge *Not* Applied to Fuzzy Set *PI.Curve*

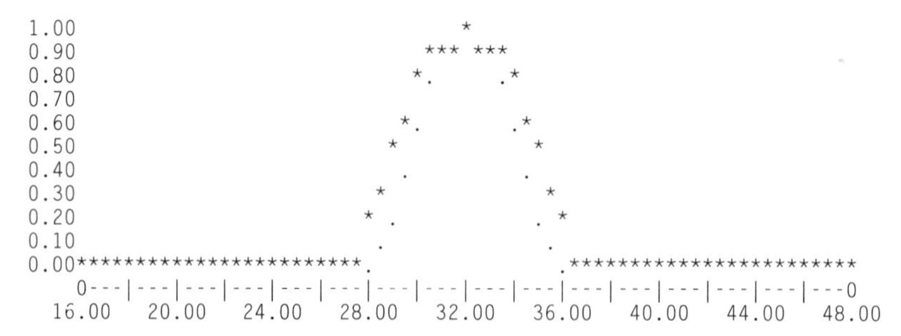

Figure 5.61 Hedge *Somewhat* Applied to Fuzzy Set *PI.Curve*

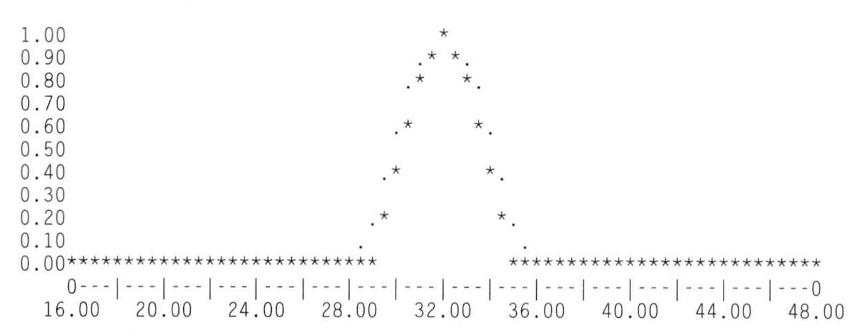

Figure 5.62 Hedge *Very* Applied to Fuzzy Set *PI.Curve*

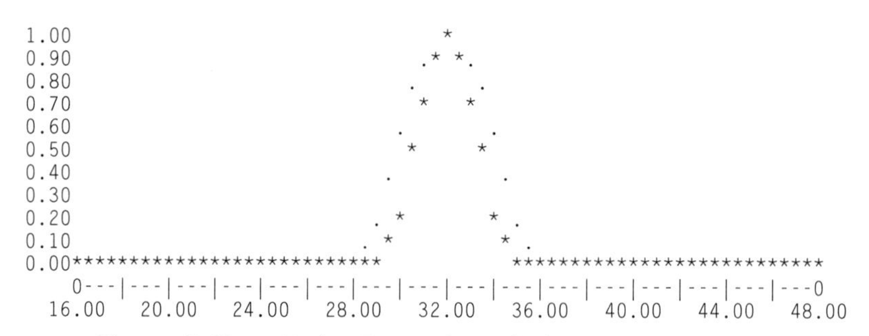

Figure 5.63 Hedge *Extremely* Applied to Fuzzy Set *PI.Curve*

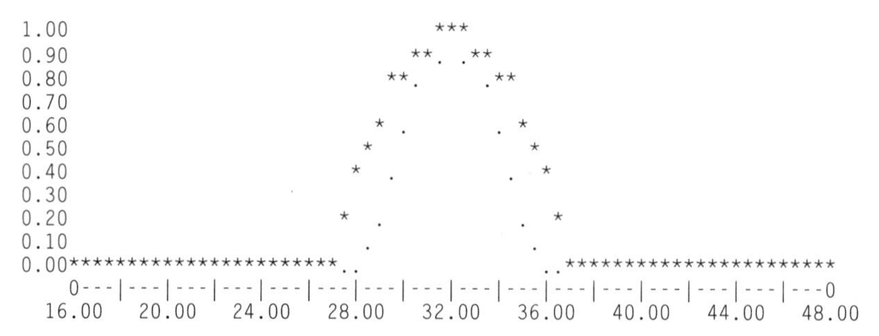

Figure 5.64 Hedge *Slightly* Applied to Fuzzy Set *PI.Curve*

As we have noted, hedges are developed from arbitrary—but nonetheless informed—choices about how the topology of a fuzzy set must be modified to correspond to the semantics of a model state. This occasionally means that the built-in hedges are insufficient to represent a particular fuzzy model transformation. You can design and implement your own arithmetic hedge by indicating a mode of transformation and a transformational value.

`ADD`	Add a value to the truth membership function. This acts as a bounded sum with the restriction that $\mu_A[x]+value \leq 1$.
`SUBTRACT`	Subtract a value from the truth membership function. This acts as a bounded difference with the restriction that $\mu_A[x]-value \geq 0$.
`MULTIPLY`	Multiply each truth membership value by a scaling factor. This is equivalent in most respects to the correlation product operation. The multiplier (k) must be in the range $k>0$ and $k\leq 1$.
`DIVIDE`	Divide each truth membership value by a proportionality factor. The divisor (k) can be any real number so long as $k>0$, and the size of k does not result in either a floating-point overflow or underflow (that is, k should not be very large or very small).
`POWER`	Raise each truth membership value by an exponent (k). Note that k can also be represented as $1/k$ so that arbitrary roots can be used. This is the technique used to support the built-in dilution and intensification hedges.

To describe and use your own hedge, you allocate a hedge descriptor block (***HDB***), initialize it to the properties of your own hedge, and then call the hedge function. The first part of this process is handled by the *FzyCreateHedge* function. As an example, the following code creates the hedge, inserts it in the current segment, and then later finds the hedge and uses it.

```
//--create a hedge for HUGE
//--as m(x) to the fifth power.
HDB *hugeHDBptr;
hugeHDBptr=FzyCreateHedge("Huge",POWER,5,statusptr);
MdlLinkHDB(hugeHDBptr,PDBptr,statusptr);

hugeHDBptr=MdlFindHDB("Huge",PDBptr,statusptr);
if(hugeHDBptr==NULL return;
FzyApplyHedge(
   BigBacklogFDBptr,hugeHDBptr,HugeBacklogFDBptr,statusptr);
```

At the low-level library level, it is necessary to locate explicitly and use the hedge. In the high-level modeling environment, the rule processor will automatically handle the application of hedges within the same context.

Notes

1. In this respect hedges belong to the heuristics of approximate reasoning rather than to the formal and mathematical domain of fuzzy set theory.

2. The process of closure, whereby a hedge forms a new fuzzy set, requires a very robust memory management scheme. In this example, the intermediate fuzzy sets generated by the hedges `very` and `not` must be released when the final fuzzy region is formed.

3. In the second example, we see an example of a hedged fuzzy comparison operator.

4. The approximation hedges use a bell-shaped fuzzy representation for reasons that we will discuss shortly. You could replace the bell curve with the triangle generator code to yield triangular-shaped fuzzy numbers. However, this procedure will not work when you apply a general approximator to an existing bell-shaped fuzzy set.

Six

Fuzzy Reasoning

δοσ μοι πoυ στω και κινω την γην. . .
Give me a place to stand,
and I will move the earth . . .

Archimedes (287–212 BC)
Pappus Alwexandrus, *Collectio*, lib. viii, prop. 10, No. xi

La raison du plus fort est toujours la meilleure.
The reason of the strongest [system] is always the best.

Jean De La Fontaine (1621–1695)
Fables, i. 10. *Le Loup et l'Agneau*

In this chapter we bring together the basic foundations of fuzzy set theory with the precepts of general approximate reasoning. We will look at the nature of fuzzy propositions and develop a method of using complex fuzzy spaces—linguistic variables—to build fuzzy reasoning systems. In examining the nature of correlating fuzzy regions, we explore two fundamental reasoning methods: monotonic and approximate (fuzzy). An approximate reasoning system combines the attributes of conditional and unconditional fuzzy propositions, correlation methods, implication (truth transfer) techniques, proposition aggregation, and defuzzification. In this chapter we also have our first look at some of the models that are further analyzed in Chapter 7, Fuzzy Models, which is about case studies.

In the previous chapters we examined the nature of fuzzy sets and the operations that can be performed on these individual sets. This discussion has been largely a review of basic fuzzy set theory. We now take up the nature of truth transfer among fuzzy regions. In determining whether or not to establish a degree of causality between different model problem states, and in making decisions about the expected value of points in these problem states, we must construct functional relationships between these states. Unlike conventional expert systems where statements are executed serially, the principal reasoning protocol behind fuzzy logic is a parallel processing paradigm. In conventional knowledge-based systems, pruning algorithms and heuristics are applied to reduce the number of rules examined. But in a fuzzy system all the rules are fired. (Some, however, have no degree of truth in their premise and so fail to contribute to the outcome.)

The root mechanism in a fuzzy model is the *proposition*—a statement of relationships between model variables and one or more fuzzy regions. A series of conditional and unconditional fuzzy associations[1] or propositions is evaluated for its degree of truth, and all those that have some truth contribute to the final output state of the solution variable set. The functional tie between the degree of truth in related fuzzy regions is called the *method of implication*. The functional tie between fuzzy regions and the expected value of a set point is called the *method of defuzzification*. Taken together, these two concepts constitute the backbone of approximate reasoning.

While developing solutions to a fuzzy model, a fuzzy reasoning system converts each solution variable into a temporary fuzzy region, thus,

$$z_i \rightarrow Z_i$$

where z is the solution (sometimes called the output) variable, and Z is the corresponding solution fuzzy set. As each proposition is evaluated, its consequent fuzzy region is used to update the solution fuzzy region (Figure 6.1). This updating process is under the control of a transfer function (g) that implements a rule of implication between the consequent fuzzy state and the output fuzzy state. Thus, we have:

$$g(w_i \otimes z_i) \rightarrow Z_i$$

There are many possible implication transfer functions, but each one attempts to correlate the semantic meaning of the antecedent (if any) with the semantics of the consequent, thus generating a solution compatible with the meaning of the fuzzy state for each output variable.

Fuzzy reasoning is not performed *in vacuo*, but within the context of a fuzzy system model. This model consists of control, solution, and working data variables; fuzzy sets; hedges; proposition (rule) statements; and the underlying control mechanisms that tie all this together into a cohesive reasoning environment. For a discussion of how variables are decomposed into their fuzzy components see the section called "Fuzzy Sets and Model Variables" on page 336 of Chapter 7. We begin our discussion of approximate reasoning with some fundamental concerns: the meaning and use of linguistic variables, and the nature of fuzzy propositions (statements of knowledge similar to rules in conventional expert systems).

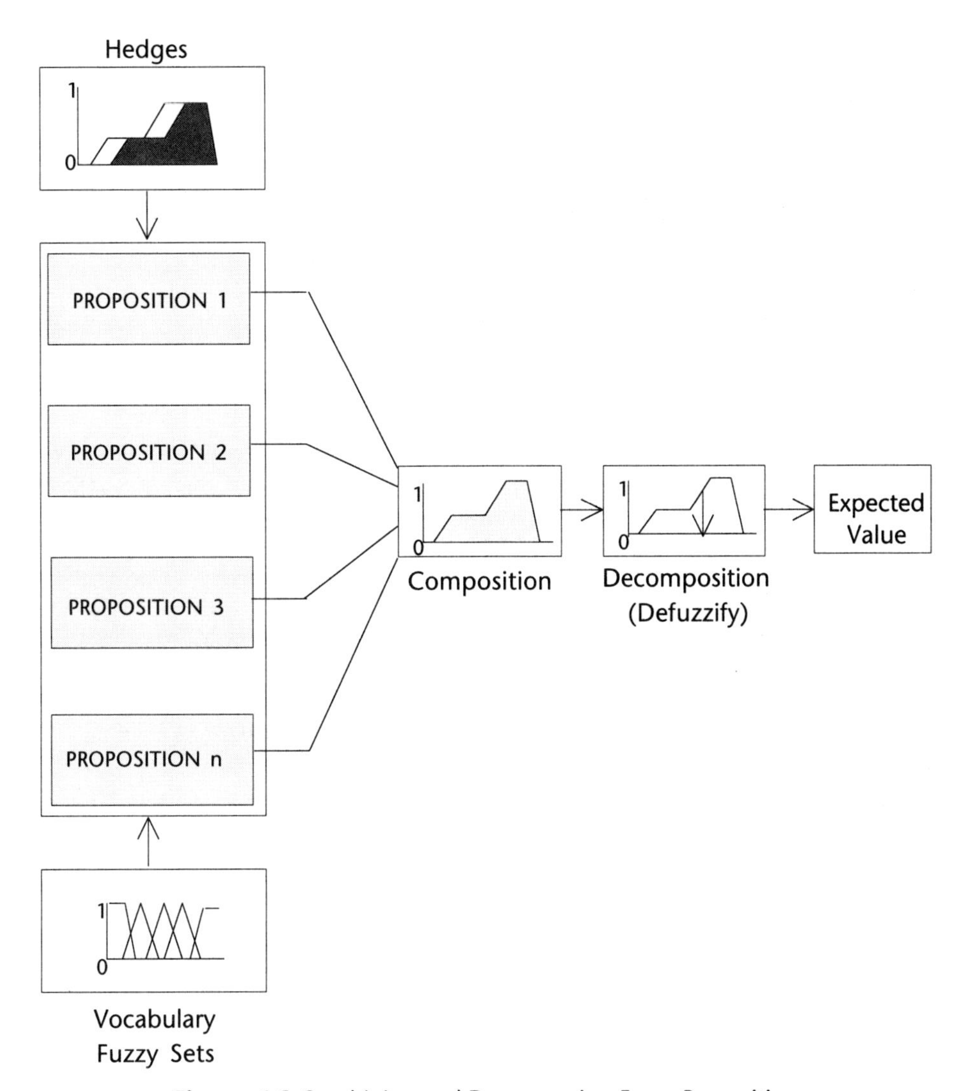

Figure 6.1 Combining and Decomposing Fuzzy Propositions

The Role of Linguistic Variables

Fuzzy models manipulate linguistic variables. A linguistic variable is the representation of a fuzzy space. This fuzzy space is, essentially, a fuzzy set derived from the evaluation of the linguistic variable. The organization of a linguistic variable is shown in Equation 6.1.

$$L_{\text{var}} \leftarrow \{q_1 \ldots q_n\}\{h_1 \ldots h_n\}F_i$$

Equation 6.1 Linguistic Variable Representation

In Equation 6.1, the predicate q represents usuality or frequency qualifiers (such as *often* or *some*), h represents a hedge (such as *very* or *somewhat*), and F is the core fuzzy set. The subscripts indicate that an element may occur any reasonable number of times. The braces indicate that this component of the linguistic variable is optional.

The simplest linguistic variable is the name of a fuzzy set directly representing a specific region in the underlying problem space. In the following rules, the **highlighted** words are the linguistic variables consisting of single fuzzy sets.

our price must be ***high***
if margins are ***thin*** *then profits will be* ***low***

A slightly more complex form of the linguistic variable includes hedges (see Chapter 5, "Fuzzy Set Hedges"). The composite *very high* or *somewhat low* are linguistic variables (consisting of a fuzzy set and a hedge) reflecting our intuitive understanding of these relationships. We could very well have extended these rules to something like

our price must be ***positively very high***
if margins are ***somewhat thin***
then profits will be ***generally low***

where *positively* and *generally* represent what Zadeh calls contrast intensification and contrast diffusion (see "Contrast Intensification and Diffusion" in Chapter 5). Hedges transform the fuzzy set to their right, forming a new fuzzy set. This process of closure is in keeping with the linguistic variable's definition as the representation of a fuzzy space.

Even more complex linguistic variables involve the use of frequency and usuality qualifiers (refer to Chapter 3, Proportional and Frequency Representations on page 113). These qualifiers reduce the derived fuzzy set by restricting the truth membership function to a range consistent with the intentional meaning of the qualifier. Fuzzy statements using these kinds of qualifiers will have a form like the following (the linguistic variables are highlighted):

our price must be ***usually high***
if margins are ***often somewhat thin***
then profits will be ***generally low***
if projects are ***mostly late***
then costs are ***frequently very high***

While useful in succinctly expressing concepts that are not allowed in the first-order predicate calculus of Boolean logic, the use of such qualifiers, in practice, seldom adds much to a model. These kinds of relationships are best modeled using functional predicates that examine the truth of a fuzzy expression. These fuzzy statements take the form:

if ***often****(margins are* ***somewhat thin****)*
then profits are ***generally low***

In either case, frequency and usuality qualifiers—unlike the class of hedge operators—are not easily used in the consequent action of a fuzzy rule. As a result, only the functional forms of the usuality and frequency qualifiers we will discuss in this handbook.

Although a linguistic variable may consist of many separate terms (such as *positively not very very high*), it should be considered a single entity in the fuzzy proposition. Any number of hedges may modify a single fuzzy set, but, in practice, the semantics (and utility) of the resulting fuzzy space is generally lost after more than three or four hedges (excluding the fuzzy complement *not*, which is treated as a hedge). Thus, a linguistic variable such as

somewhat about positively not very very high

is difficult to understand and uncertain in its actual semantics. Since we cannot reliably predict the space produced by this lengthy linguistic variable, its usefulness in the model is very limited. As a case in point, the following linguistic variable was actually used in a Reveal manufacturing model as part of the backorder analysis:

about very near generally not quite positively deep.

During a maintenance review, the original knowledge engineer spent nearly an hour attempting to remember what this expression represented!

Fuzzy Propositions

A fuzzy model consists of a series of conditional and unconditional fuzzy propositions. A proposition or statement establishes a relationship between a value in the underlying domain and a fuzzy space. Propositions are expressed in the form

$$x \quad is \quad Y$$

where x is a scalar from the domain, and Y is a linguistic variable. The effect of evaluating a fuzzy proposition is a degree or grade of membership derived from the transfer function:

$$\mu_A \leftarrow (x \in Y)$$

This is the essence of an approximate statement. The derived truth membership value establishes a compatibility between x and the generated fuzzy space Y. Such propositions answer the question "how compatible is x with Y?" or "to what degree is x a member of set Y?" This truth value is used in the correlation and implication transfer functions to create or update the output fuzzy solution space. The final solution fuzzy space is created by aggregating the collection of correlated fuzzy propositions. The correlation is based on the truth of each fuzzy proposition. (When multiple propositions are connected by *and* or *or* connectors, the truth used in the correlation process is the truth produced by applying these operators.)

Conditional Fuzzy Propositions

A conditional proposition is one that is qualified by an `if` statement. These are analogous to the rules of a conventional symbolic expert system. The proposition has the general form

$$\text{if} \quad w \quad \text{is} \quad Z \quad \text{then} \quad x \quad \text{is} \quad Y$$

where w and x are model scalar values, and Z and Y are linguistic variables. The proposition following the **if** term is the antecedent or predicate and is any arbitrary fuzzy proposition. The proposition following the **then** term is the consequent and is also any arbitrary fuzzy proposition. The statement *x is Y* is conditional on the truth of the predicate. We interpret this statement as:

x is member of Y to the degree that w is a member of Z

That is, the consequent is correlated with the truth of the antecedent. The fundamental proposition can be extended with fuzzy connectors

if (w is Z) • (y is W) • ... (u is S) then x is Y

where • is some form of the *and* or *or* operator. In the case of multiple antecedent propositions, the position of *x within Y* is determined by the composite truth of the complete antecedent.

This is the basis for fuzzy monotonic reasoning. In actual fuzzy models, x is a temporary fuzzy region (indicated by the capital X) containing the results of each proposition that specifies an elastic space for Y. Thus, we need to read the conditional fuzzy proposition as:

(X is a powerset of Y) to the degree that (w is member of Z)

This powerset is formed by first correlating the fuzzy set Y according to the truth of the antecedent proposition (*w is Z*). The solution fuzzy space is then updated by taking the union of the solution set and the newly correlated fuzzy set.

Unconditional Fuzzy Propositions

An unconditional fuzzy proposition is one that is not qualified by an `if` statement. The proposition has the general form

$$x \quad is \quad Y$$

where x is a scalar from the domain, and Y is a linguistic variable. Unconditional statements are always applied within the model and, depending on how they are applied, serve either to restrict the output space or to define a default solution space (if none of the conditional rules execute). We interpret an unconditional fuzzy proposition as:

X is the minimum subset of Y

When the output fuzzy set X is empty, then X is restricted to Y; otherwise, for the domain of Y, X becomes the *min* (X,Y). Since these propositions are unconditional, they are never correlated; that is, their truth values are never reduced before they are applied to the output space. The solution fuzzy space is updated by taking the intersection of the solution set and the target fuzzy set.

The Order of Proposition Execution

For models containing only conditional or only unconditional propositions, the order in which propositions (that is, rules) is executed is not important. However, if a model contains a mixture of these two types, then the order of execution becomes important. The effect of applying unconditional propositions changes the nature of the model solution space depending on whether the propositions are applied before or after the set of conditionals.

Unconditional propositions are generally used to establish the default support set for a model. If none of the conditional rules executes, then a value for the solution variable is determined from the space bounded by the unconditionals. For this reason, they must be executed before any of the conditionals. If none of the conditional rules that fall within the same underlying domain as the unconditionals has an antecedent strength greater than the maximum intersection of the unconditionals, they will not contribute to the model solution.

Although much less common, the unconditionals can also be used to restrict the final solution space of a model to the maximum truth of their intersection. This process is done by applying the unconditionals after all the conditional propositions have been evaluated.

Monotonic (Proportional) Reasoning

We begin the exploration of fuzzy reasoning with the simple monotonic method—a basic fuzzy implication technique. Although this form of reasoning is not often used as the

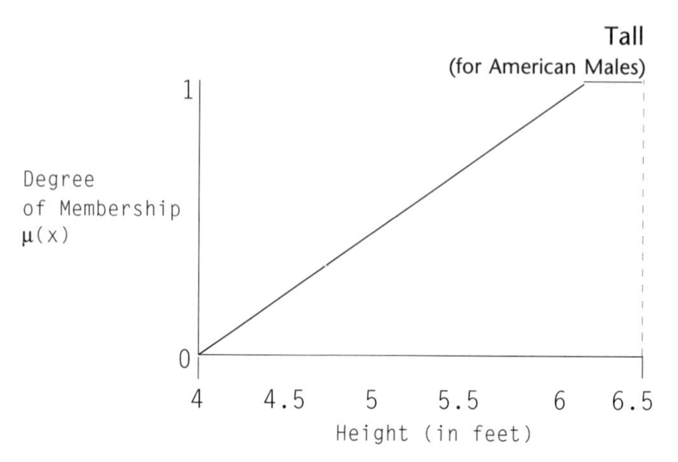

Figure 6.2 The Fuzzy Set *Tall* for American Males

primary implication method, it is often used in the chaining metric method of fuzzy scaling. (See the risk analysis case study in Chapter 8). Consider two fuzzy regions that are related through a simple proportional implication function

if x is Y then z is W

and functionally represented by the transfer function shown in Equation 6.2.

$$z = f((x, Y), W)$$

Equation 6.2 The Fuzzy Monotonic Logic Transfer Function

Bounded by these two parameters, and, under a restricted set of circumstances, a fuzzy reasoning system can develop an expected value without going through composition and decomposition. The value of the output is estimated directly from a corresponding truth membership grade in the antecedent fuzzy regions.

This form of fuzzy inference obeys a method of implication called monotonic selection. To illustrate how this method works, consider a simple weight estimating model. The underlying fuzzy vocabulary for this model consists of two fuzzy sets: *Tall* for American males (Figure 6.2) and *Heavy* for American males (Figure 6.3).

These fuzzy concepts provide the basis for a weight estimation model. The model is based on the perceived relationship between an individual's height and his weight, and it is expressed as a single production rule:

if height is Tall then weight is Heavy;

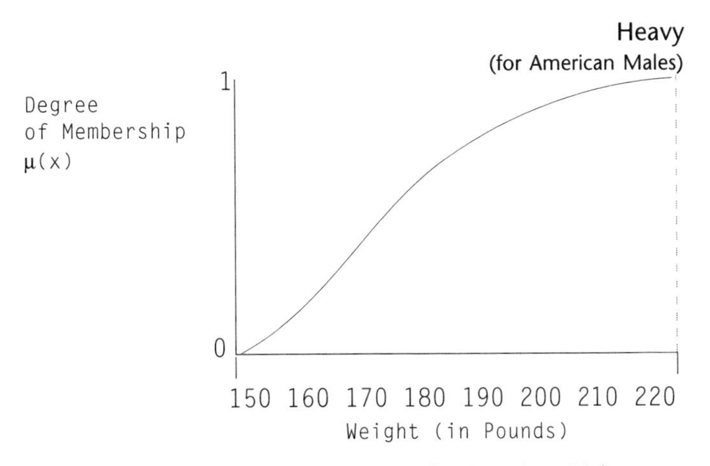

Figure 6.3 The Fuzzy Set *Heavy* for American Males

A monotonic selection implication between fuzzy regions *Y* and *W* follows the selection algorithm:

- For an element *x* in the domain of *Y*, find its membership in the fuzzy region *Y*. This is $\mu_Y[x]$.
- In fuzzy region *W* at the membership corresponding to $\mu_Y[x]$, find the surface of the fuzzy manifold. Drop a "plumb line" to the domain. The value on the domain axis, *z*, is the solution to the implication function. Functionally this is expressed as:

$$z_w = f(\mu_Y[x], D_W)$$

Figure 6.4 illustrates the implication process for this model. When height is 5'0", its membership in the fuzzy set *Tall* is [.42]. This truth function value is used with the *Heavy* fuzzy set to find a value for the solution variable weight. We "enter" the fuzzy set at the [.42] membership axis and move across until we encounter the surface of the *Heavy* fuzzy set. A value for the solution variable weight is then developed by selecting the domain value in *Heavy* for this truth membership.

A monotonic or proportional implication function is characterized by a lack of high-level orthogonality in the consequent (solution) fuzzy space. By this, we mean that while the antecedent fuzzy expression might be complex, the solution is not produced by any formal method of defuzzification, but by a direct slicing of the consequent fuzzy set at the antecedent's truth level. Figure 6.5 shows how various values of weight are derived from differing values for height.

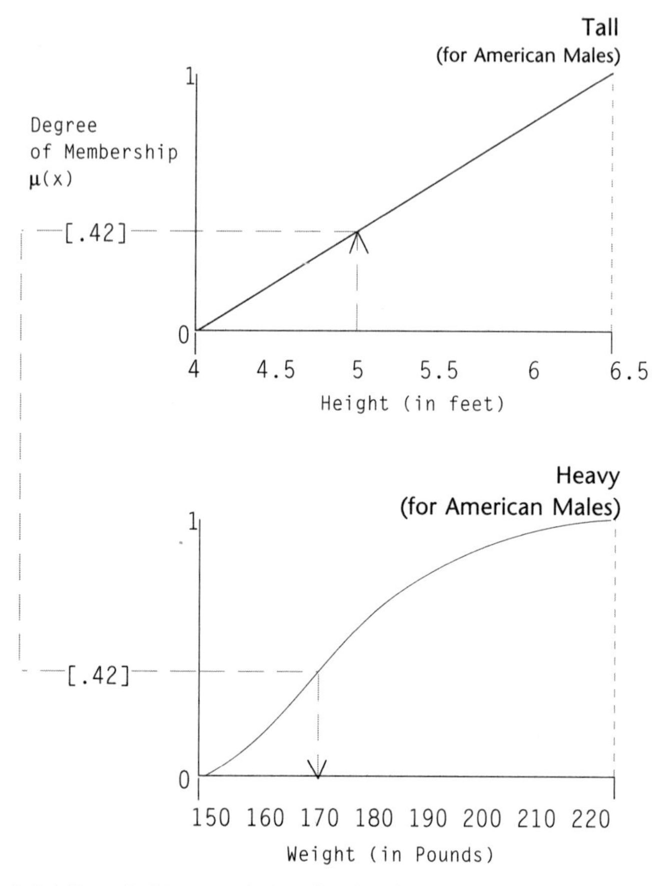

Figure 6.4 A Sample Monotonic Implication from the Weight Estimation Model

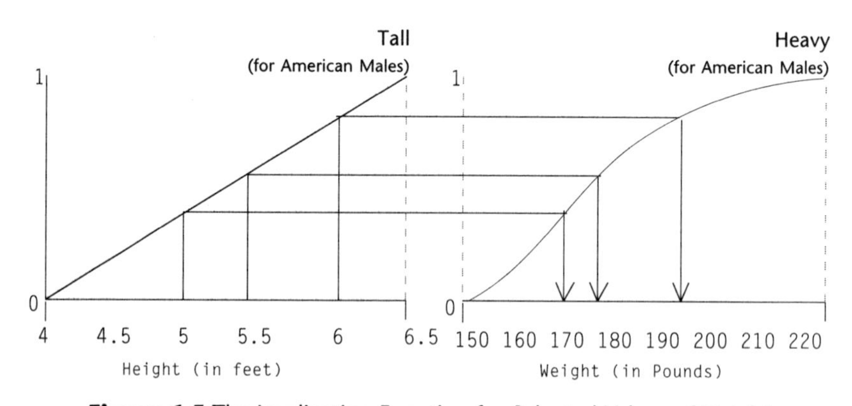

Figure 6.5 The Implication Function for Selected Values of Height

```
Fuzzy Weight Estimation Model. (c) 1998 Metus Systems Group.
Individual Height (feet):       5.50
The Rules:
R1   if height is tall then weight is heavy

Linear Increasing FuzzySet 'TALL' created.

        FuzzySet:    TALL
        Description:
      1.00                                                          .
      0.90                                                    ......
      0.80                                             .......
      0.70                                       ......
      0.60                                 .......
      0.50                           ......
      0.40                     ......
      0.30               .......
      0.20         ......
      0.10   .......
      0.00......
         0---|---|---|---|---|---|---|---|---|---|---|---|---|---|---|---|---0
        5.00    5.19    5.38    5.56    5.75    5.94    6.13    6.31    6.50
        Domained:          5.00 to        6.50
```

Figure 6.6 The Linear *Tall* Fuzzy Set

The monotonicity of the implication function is associated with the nature of fuzzy domains, not with the fuzzy set topology itself (which may or may not be monotonic). As Figure 6.5 shows, as we select values from the domain in one fuzzy set, a corresponding or proportional value is selected from the consequent fuzzy set. The following figures show how the monotonic logic process works in the actual weight model (from the `WEIGHT1.CPP` program). Figures 6.6 and 6.7 are the base fuzzy sets whose shapes are conformal.

```
Growth S-Curve FuzzySet 'HEAVY' created.

        FuzzySet:    HEAVY
        Description:
      1.00                                                          .....
      0.90                                                 ...........
      0.80                                            .....
      0.70                                       .....
      0.60                                   ...
      0.50                               ....
      0.40                            ...
      0.30                        ....
      0.20                   .....
      0.10             ......
      0.00.............
         0---|---|---|---|---|---|---|---|---|---|---|---|---|---|---|---|---0
      150.00  161.25  172.50  183.75  195.00  206.25  217.50  228.75  240.00
       Domained:         150.00 to      240.00
```

Figure 6.7 The S-Curve *Heavy* Fuzzy Set

Now the model is executed, and a new weight is produced for each height. Along with each weight estimate, the implication plateau (see Figure 6.9) is shown.This is the truth function of *Heavy* truncated at the current height's membership in *Tall*. An estimate for weight is found by taking a domain reading in *Heavy* at the left corner of the plateau formed by this truncation.This is equivalent to dropping a plumb line from the curve's surface (see Figures 6.5, 6.8, 6.9, and 6.10).

```
Model Solution:
  Height       =        5.50
  Weight       =      186.91
  Truth Grade =    0.330739

        FuzzySet:     HEAVY
        Description:
      1.00
      0.90
      0.80
      0.70
      0.60
      0.50
      0.40
      0.30                                         .......................................
      0.20                                   .....
      0.10                             ......
      0.00..............
          0---|---|---|---|---|---|---|---|---|---|---|---|---|---|---|---|---0
       150.00  161.25  172.50  183.75  195.00  206.25  217.50  228.75  240.00
        Domained:        150.00 to      240.00
```

Figure 6.8 The Implication Plateau in *Heavy* for a Height of 5.5 Feet

```
  Height       =        5.75
  Weight       =      195.00
  Truth Grade =    0.498054

        FuzzySet:     HEAVY
        Description:
      1.00
      0.90
      0.80
      0.70
      0.60
      0.50                                              ................................
      0.40                                          ...
      0.30                                     ....
      0.20                               .....
      0.10                         ......
      0.00..............
          0---|---|---|---|---|---|---|---|---|---|---|---|---|---|---|---|---0
       150.00  161.25  172.50  183.75  195.00  206.25  217.50  228.75  240.00
        Domained:        150.00 to      240.00
```

Figure 6.9 The Implication Plateau in *Heavy* for a Height of 5.75 Feet

```
Height       =       6.00
Weight       =     203.09
Truth Grade =   0.661479

      FuzzySet:     HEAVY
      Description:
   1.00
   0.90
   0.80
   0.70
   0.60                                        .............................
   0.50                                    ....
   0.40                                 ...
   0.30                             ....
   0.20                        .....
   0.10                  ......
   0.00..............
       0---|---|---|---|---|---|---|---|---|---|---|---|---|---|---|---0
   150.00  161.25  172.50  183.75  195.00  206.25  217.50  228.75  240.00
    Domained:       150.00 to      240.00
```

Figure 6.10 The Implication Plateau in *Heavy* for a Height of 6.0 Feet

The *FzyMonotonicLogic* function encapsulates the truth transfer function. We simply find the membership of the incoming scalar in one fuzzy set and then use this membership grade to select a domain value from a second fuzzy set.

```
#include "FDB.hpp"        // The Fuzzy Set descriptor
#include "mtypes.hpp"     // System constants and symbolics
#include "mtsptype.hpp"   // Metus function prototypes
double FzyMonotonicLogic(
  FDB *fromFDBptr,FDB *toFDBptr,
    double fromValue,float *PremiseTruth,int *statusPtr)
 {
  int     Idx;
  float   PTruth;
  double  Scalar;

  *statusPtr=0;
  PTruth=FzyGetMembership(fromFDBptr,fromValue,&Idx,statusPtr);
  Scalar=FzyEquivalentScalar(toFDBptr,PTruth,statusPtr);
  *PremiseTruth=PTruth;
  return(Scalar);
 }
```

Listing 6.1 The Fuzzy Monotonic Logic Function

In the monotonic reasoning function, *fromFDBptr* is the fuzzy set associated with the *fromValue* scalar. *PremiseTruth* (which is returned by the function) is the membership of *fromValue* in *fromFDBptr's* fuzzy set. The function returns the scalar out of the fuzzy set pointed to by *toFDBptr* corresponding to the membership value computed for *fromValue*.

Monotonic Reasoning with Complex Predicates

The antecedent fuzzy truth function can be generated from an arbitrarily complex approximate expression following the general form:

```
if (x is Y)•(k is U)•(s is M)... then z is W
```

The operator • represents either the conjunctive (*and*) or disjunctive (*or*) operation in any of the operator classes. Since we can view the aggregate truth of the predicate as a point in a fuzzy region bounded by the composite fuzzy sets, this reduces to the simple monotonic transfer function in Equation 6.3, which is generalized to the function in Equation 6.4.

$$z = f(\{(x,Y),(k,U),(s,M)\},W)$$

Equation 6.3 A Fuzzy Monotonic Transfer Function for Multiple Predicates

$$z = f(\sum_{i=1}^{N}(v_i, F_i)\cdot W)$$

Equation 6.4 A General Monotonic Reasoning Transfer Function

In Equations 6.3 and 6.4, the Σ operation is the general aggregation operator acting on the variable and fuzzy set tuples to produce the fuzzy predicate truth value. Note that this is distinct from the general space *Z*, the fuzzy space corresponding to solution variable *z*, which is formed by the normal min–max form of approximate reasoning.

We can see how this works by introducing another factor in the weight estimating model: the measurement of the individual's waist. The fuzzy set *Wide*, shown in Figure 6.11, couples the waist measure to the idea of *Wide*.

Our fuzzy weight estimation rule is now changed to reflect the contribution of both height and waist in determining a value for weight. Figure 6.12 shows this new multi-dimensional space bounded by height and waist size. The compensatory mean (average) *and* is used to combine the truth values from the *Tall* and *Wide* fuzzy sets.

Thus, monotonic reasoning acts as a proportional correlating function between two general fuzzy regions. We can see that the important restriction on monotonic reasoning is its requirement that the output for the model be a single fuzzy variable controlled by a single fuzzy rule (with an arbitrarily complex predicate). A two-dimensional weight estimator using fuzzy monotonic logic can be found in the `WEIGHT2.CPP` program.

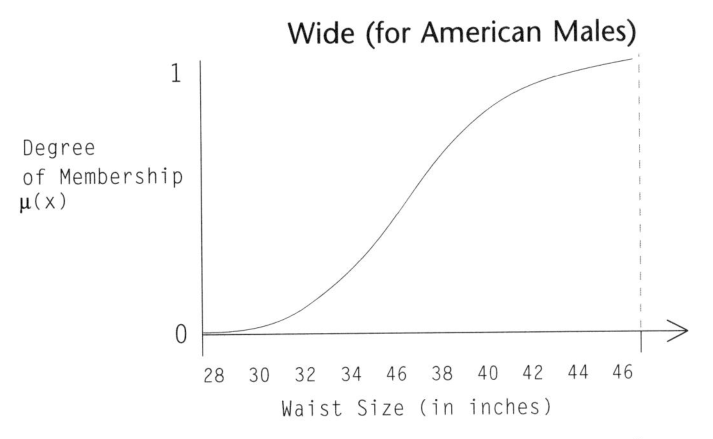

Figure 6.11 The Fuzzy Set *Wide* [Waist] for American Males

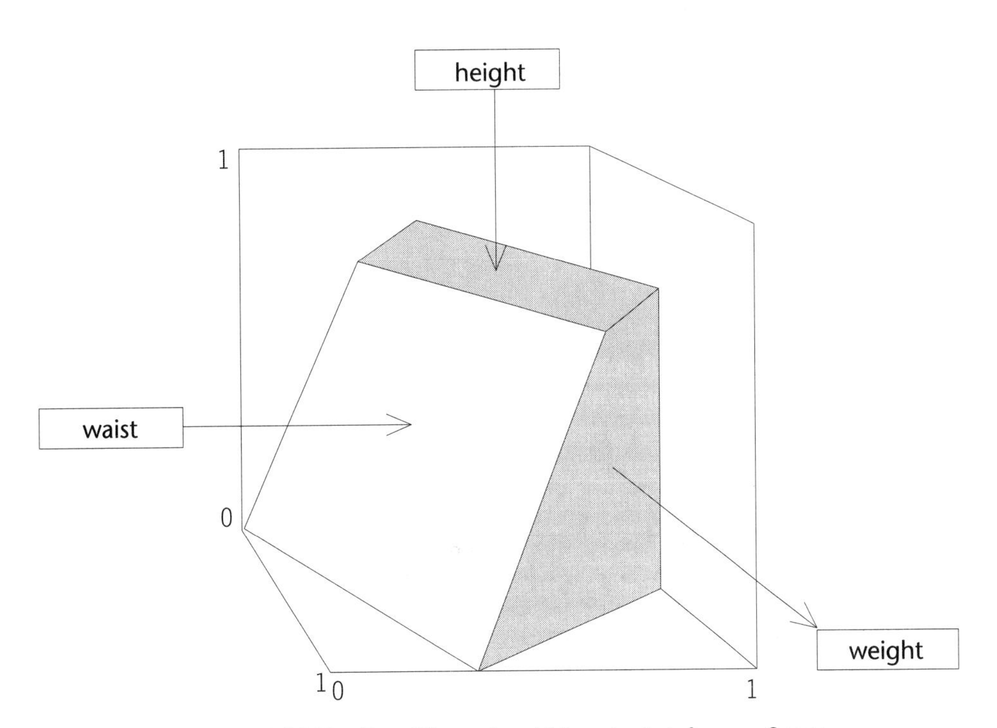

Figure 6.12 The Two-Dimensional Monotonic Inference Space

Monotonic reasoning provides a powerful method of linking the truth of two general fuzzy regions. This enables us to make estimates about the domain structure of fuzzy region *X* when we know the domain and truth value of a point in fuzzy region *Y*. However, monotonic reasoning is less general and more restrictive than the conventional min–max or min–product rules of implication. This is the result of two limitations: (1) its requirement that the output for the model be a single fuzzy variable controlled by a single fuzzy rule (with an arbitrarily complex predicate) and (2) the implication function between the two fuzzy regions is expressed as a correlated surface topology.

As the complexity of the predicate proposition increases, the degree to which monotonic reasoning will produce consistently valid results tend to decline. This is a consequence of basic complexity and information theories.

The Fuzzy Compositional Rules of Inference

Unlike monotonic reasoning, the implication space generated by the general compositional rules of inference is derived from the aggregated and correlated fuzzy spaces produced by the interaction of many statements. In effect, all the propositions are run in parallel to create an output space that contains information from all the propositions. Each conditional proposition whose evaluated predicate truth is above the current alpha-cut threshold contributes to the shape of the output solution variable's fuzzy representation. There are two principal methods of inference in fuzzy systems: the min–max method and the fuzzy additive method. These methods differ in the way they update the solution variable's output fuzzy representation.

The Min–Max Rules of Implication

The min–max compositional operation derives its name from the method for this contribution. The consequent fuzzy region is restricted to the *minimum* of the predicate truth. The output fuzzy region is updated by taking the *maximum* of these minimized fuzzy sets.[2] These steps are outlined in Equations 6.5 and 6.6.

$$\mu_{cfs}[x_i] \leftarrow \min(\mu_{pt}, \mu_{cfs}[x_i])$$

Equation 6.5 Correlation Minimum of Consequent with Predicate Truth

Equation 6.5 indicates that the consequent fuzzy set (*cfs*) is modified before it is used. This modification sets each truth function element to the minimum of either the truth function or the truth of the proposition's predicate (*pt*).

$$\mu_{sfs}[x_i] \leftarrow \max(\mu_{sfs}[x_i], \mu_{cfs}[x_i])$$

Equation 6.6 Maximum of Current Solution with Correlated Consequent

Equation 6.6 indicates that the solution fuzzy set (*sfs*) is updated by taking, for each truth function value, the maximum of either the truth value of the solution fuzzy set or the fuzzy set that was correlated in Equation 6.5. These steps result in reducing the strength (that is, the effective height) of the fuzzy set output to equal the maximum truth of the predicate. Then, using this modified fuzzy region, these steps apply it to the output by using the `or` (union) operator. When all the propositions have been evaluated, the output contains a fuzzy set that reflects the contribution from each proposition.

The Fuzzy Additive Rules of Implication

The fuzzy additive compositional operation takes a slightly different approach to updating the solution variable's fuzzy region. The consequent fuzzy region is still reduced by the maximum truth value of the predicate (see Equation 6.5). But the output fuzzy region is updated by a different rule, shown in Equation 6.7.

$$\mu_{sfs}[x_i] = (\mu_{sfs}[x_i] + \mu_{cfs}[x_i])$$

Equation 6.7 Maximum of the Correlated Solution Using the Fuzzy Additive Technique

We can see that the sum of the membership values will easily exceed [1.0]. You might be tempted to implement this operation as the bounded-sum operation applied to the output fuzzy region, as shown in Equation 6.8.

$$\mu_{sfs}[x_i] = (\mu_{sfs}[x_i] + \mu_{cfs}[x_i])$$

Equation 6.8 The Bounded Sum Operation

This would be a mistake because important information about the shape of the outcome fuzzy set is lost. There are two approaches to solving this problem, depending on how you defuzzify the output and how you want to display the results. If you use the centroid method of defuzzification (see "Composite Moments (Centroid)" on page 307), and you are not graphing the solution fuzzy set, then the output fuzzy region can be processed without any changes. However, if you are using some other method of defuzzification (see

"Composite Maximum (Maximum Height)" on page 309) or you are graphing the output fuzzy set, then the maximum membership height must be brought back to [1.0]. You can do this simply by normalizing the fuzzy set (see "Fuzzy Set Height and Normalization" on page 84).

Accumulating Evidence with the Fuzzy Additive Method

The fuzzy additive method solves many problems in decision models where we want all the rules to contribute something to the final model solution. With the min–max implication method, only rules that have a truth greater than the current "high water mark" in the output fuzzy set will make any contribution to the solution. The additive method overcomes this problem in a large number of cases. To see this, consider the following propositions:

[P1] *if project.duration is Long then risk is Increased;*
[P2] *if project.staffing is Large then risk is Increased;*
[P3] *if project.funding is Low then risk is Increased;*
[P4] *if project.complexity is High then risk is Increased;*
[P5] *if project.priority is High then risk is Increased;*
[P6] *if project.visibility is High then risk is Increased;*

We want to determine the degree of risk in assuming a project from a series of fuzzy propositions that update the degree of risk based on such factors as project length, complexity, staffing size, project funding levels, etc. As each proposition is evaluated, the associated degree of risk is tied to the truth of the proposition (the truer the proposition, the higher the project risk). The *Increased* fuzzy set (see Figure 6.13) is simply a metric on an arbitrary scale of [0,1000.] For more information about this kind of measurement, see "Use of Psychometric Domains" on page 90.

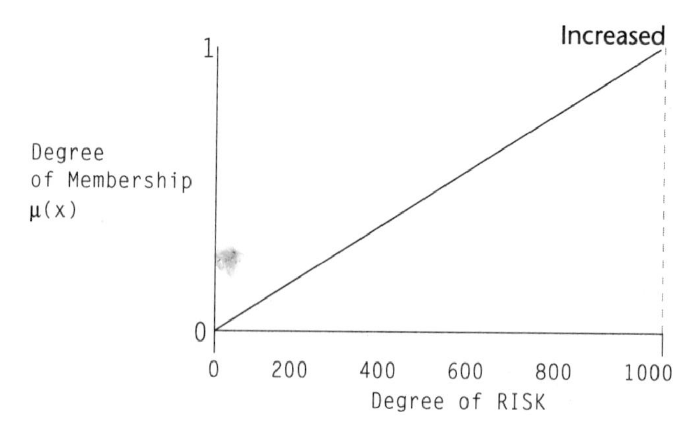

Figure 6.13 The *Increased Risk* Fuzzy Set

When the first proposition is evaluated, the premise

project.duration is Long

has a truth of [.48]. The *Risk* solution fuzzy region is updated with *Increased* correlated at this level. Figure 6.14 shows the *Risk* solution fuzzy set at this point. (Since *Risk* was an empty output fuzzy region, the *Increased* fuzzy set is copied into *Risk*.) When the second proposition is executed, its predicate

project.staffing is Large

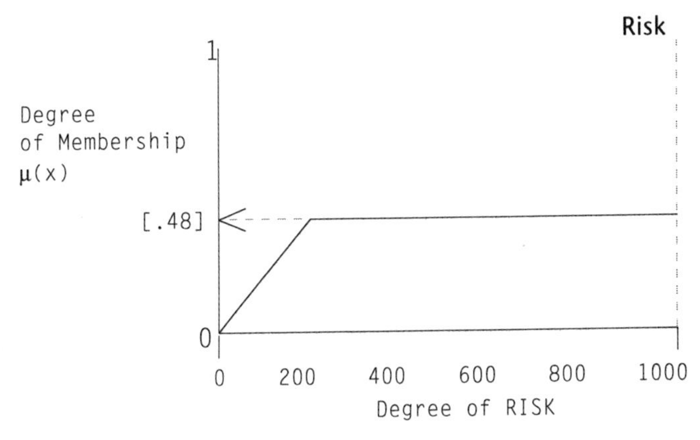

Figure 6.14 Risk After Executing the First Proposition

has a truth of [.33]. Figure 6.15 shows how this correlation reduces the height of *Increased* to the maximum truth of the proposition's predicate.

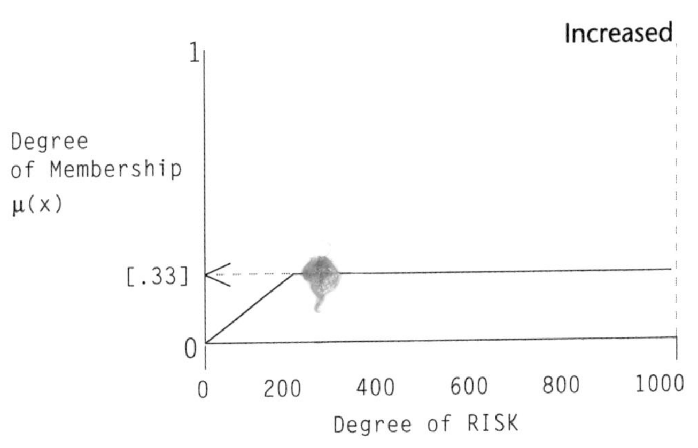

Figure 6.15 *Increased* Correlated at the [.33] Level

Applying this region to *Risk* using the min–max implication method has no effect on the solution (since *max([.48],[.33]) = [.48]*). Only propositions that have a truth greater than [.48] will contribute to the solution value. Accumulating the support space for *Risk* in this manner seems inappropriate since *Risk* is intensified by additional pieces of evidence. The fuzzy additive method combines the truth membership functions of the current *Risk* fuzzy set with the correlated *Increased* set. The *Risk* is now:

```
[.48]+[.33] = [.81]
```

Figure 6.16 shows the state of the *Risk* solution variable after the application of the first two propositions. The *Risk* continues to grow as each proposition introduces new evidence about the degree of risk in accepting the project.

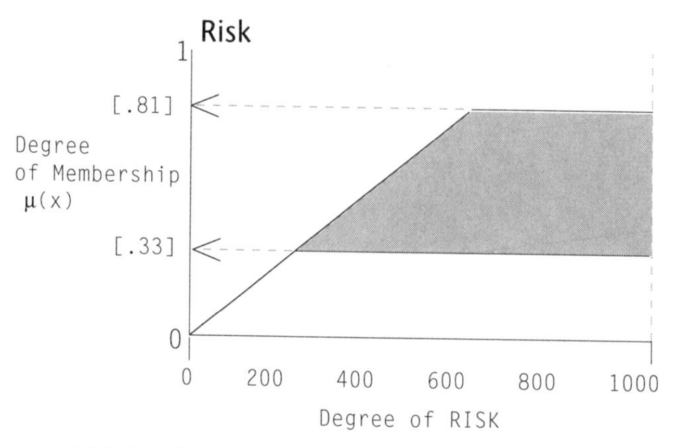

Figure 6.16 Solution Variable *Risk* After Additive Implication Update

Fuzzy systems whose actions select different fuzzy regions can use the min–max inference process (as is the case in most control engineering applications, but is certainly not the case in nearly all business and related decision support problems). This problem with accumulating evidence in a fuzzy system is particularly important when, like the project risk example, the proposition draws on a collection of fuzzy sets that cover the same domain (or have such wide overlaps that they effectively coincide). Figure 6.17 illustrates a number of difficult-to-use fuzzy regions that have significant overlaps.

The use of the fuzzy additive implication method in these cases can provide a better representation of the problem state than systems that rely solely on the min–max inference scheme.

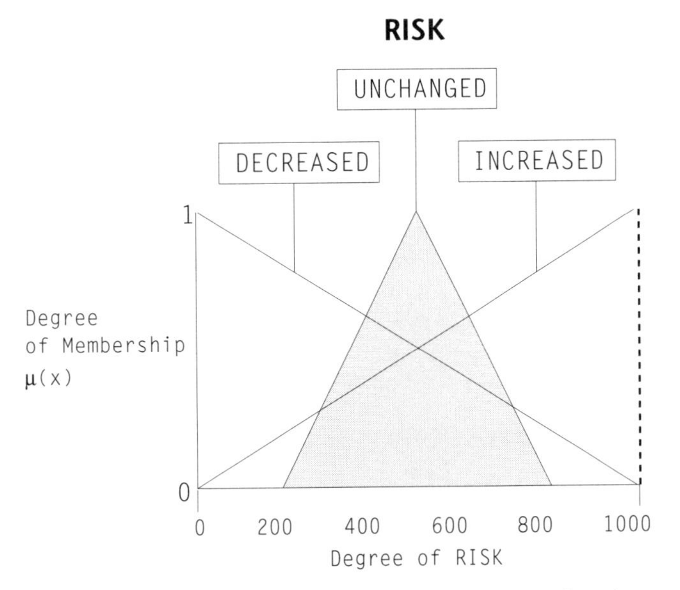

Figure 6.17 Significantly Overlapping Fuzzy Control Regions

FUZZY IMPLICATION EXAMPLE

Figure 6.18 illustrates the method used in the min–max inference technique to handle conditional fuzzy propositions. This figure shows the effect on the consequent *THROTTLE.ACTION* of applying two rules:

```
if pressure(t) is LOW and temperature(t) is COOL
   then throttle.action is POSITIVEMODERATE;

if pressure(t) is OK and temperature(t) is COOL
   then throttle.action is ZERO;
```

Each rule updates the fuzzy solution variable's output region by applying the conditional fuzzy proposition function. A complete discussion of the correlation process can be found in the section "Correlation Methods" on page 293.

In the first rule, the pressure reading at time *P(t)* has a value of [.57] in *Low*, and the temperature reading at time *T(t)* has a value of [.48] in *Cool*. The minimum ([.48]) is used to truncate the *PM* (positive moderate) fuzzy set at this level. The truncated fuzzy region is then copied to the *THROTTLE.ACTION* working fuzzy set.

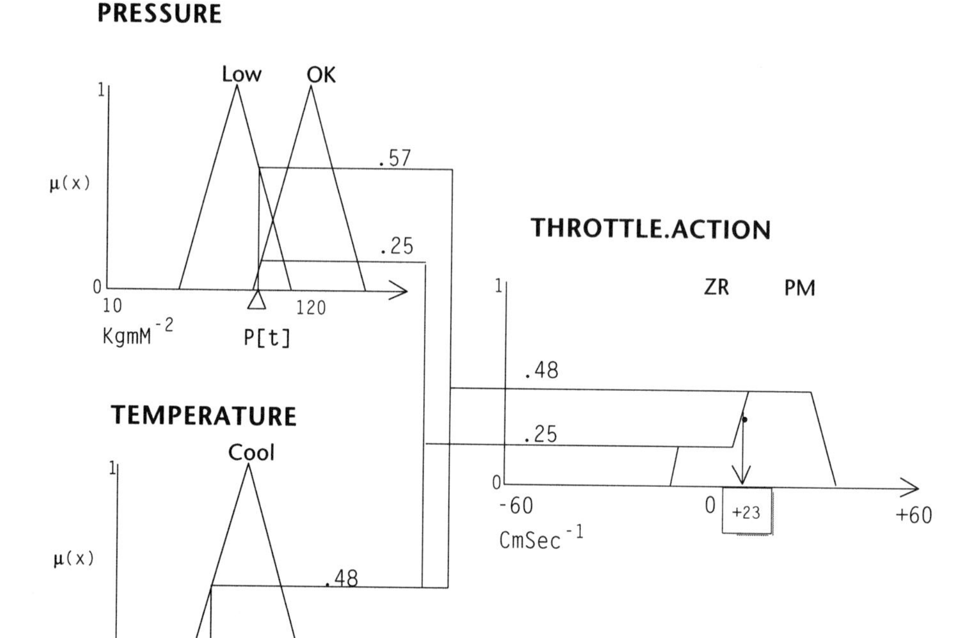

Figure 6.18 Apply Rules Using the Min–Max Inference Technique

In the second rule, the pressure reading at time *P(t)* has a value of [.25] in *OK*, and the temperature reading at time *T(t)* has a value of [.48] in *Cool*. The minimum ([.25]) is used to truncate the *ZR* (zero) fuzzy set at this level. The truncated fuzzy region is then applied to the *THROTTLE.ACTION* by *or*ing it with the current fuzzy region. This is done by taking the maximum of the truncated *ZR* and the truncated *PM* fuzzy sets at each point in the working area.

The *FzyCondProposition* function encapsulates the various implication methods. This routine does not actually evaluate a fuzzy proposition—that must be done prior to calling this function. The conditional fuzzy proposition handler selects the correct correlation technique, generates a working fuzzy set for the consequent, and updates the solution variable in the fuzzy control block. Since this function is tightly coupled to the modeling system, comments on the code have been inserted in-line.

```
#include "FDB.hpp"
#include "XFZYctl.hpp"
#include "XSYSctl.hpp"
#include "fuzzy.hpp"
#include "mtypes.hpp"
#include "mtsptype.hpp"
void FzyCondProposition(
  FDB *inFDBptr,FSV *FSVptr,
      int CorrMethod,float PredTruth,int *statusPtr)
 {
   FDB   *outFDBptr;
   char   wrkBuff[80];
   int    i,thisCorrMethod,thisImplMethod;
   float  memvector[VECMAX];

   *statusPtr=0;
   if(PredTruth<XSYSctl.XSYSalfacut)
     {
      sprintf(wrkBuff,"%s%6.4f",
       "Rule fails. Premise truth is below alpha threshold: ",
         XSYSctl.XSYSalfacut);
      MtsWritetoLog(SYSMODFILE,wrkBuff,statusPtr);
      *statusPtr=RULEBELOWALFA;
      return;
     }
```

The current system alpha threshold is stored in the *XSYSalfacut* field of the system control block (set, by default, to [.20] when the *MdlConnecttoFMS* function is executed). If the predicate truth is less than the alpha-cut threshold, a message is entered in the log, and the function exits with the proper return code.

```
//
//--Apply any correlation restriction to the incoming fuzzy set.
//--We make a copy of the set and then change it if necessary.
//--If CORRNONE is specified we simply drop through without any
//--restriction on the incoming fuzzy set.
   FzyCopyVector(memvector,inFDBptr->FDBvector,VECMAX);
   thisCorrMethod=CorrMethod;
   if(CorrMethod==CORRDEFAULT)
       thisCorrMethod=FSVptr->FzySVimplMethod;
   if(thisCorrMethod==CORRMINIMUM)
      FzyCorrMinimum(memvector,PredTruth,statusPtr);
   if(thisCorrMethod==CORRPRODUCT)
      FzyCorrProduct(memvector,PredTruth,statusPtr);
   //
```

This code segment handles the correlation process (discussed fully in the next section, Correlation Methods on page 293). A copy of the incoming fuzzy set's truth function is made (we don't want to actually modify the source fuzzy set). If a specific correlation type is specified, it is used. If the *CORRDEFAULT* value was entered, then the correlation method associated with the current solution variable is selected from the fuzzy set working area. This default is established after the *FzyInitFZYctl* function is executed to set up the fuzzy modeling environment, and *FzyADDFZYctl* is invoked to insert a solution variable. When the correct correlation method has been found, the copy of the truth function in *memvector* is modified. As noted in the code comments, if *CORRNONE* is indicated, then no correlation is performed.

```
//--Now find the solution fuzzy set and apply the conditional
//--proposition method. This is equivalent to ORing the two
//--fuzzy regions.
   thisImplMethod=FSVptr->FzySVimplMethod;
   outFDBptr=FSVptr->FzySVfdbptr;
   if(outFDBptr->FDBempty)
     {
      FzyCopyVector(outFDBptr->FDBvector,memvector,VECMAX);
      outFDBptr->FDBempty=FALSE;
      return;
     }
    else
     {
      switch(thisImplMethod)
       {
        case MINMAX:
          for(i=0;i<VECMAX;i++)
           outFDBptr->FDBvector[i]=
             max(outFDBptr->FDBvector[i],memvector[i]);
          break;
        case BOUNDEDADD:
          for(i=0;i<VECMAX;i++)
           outFDBptr->FDBvector[i]=
             min(1,outFDBptr->FDBvector[i]+memvector[i]);
          break;
        default:
          sprintf(wrkBuff,"%s%4d",
           "Unknown Implication Function: ",thisImplMethod);
          MtsWritetoLog(SYSMODFILE,wrkBuff,statusPtr);
          *statusPtr=5;
          return;
       }
     }
```

This code section actually performs the implication function. The type of implication method is found in the fuzzy control block, and a pointer to the undergeneration solution fuzzy set is initialized (*outFDBptr*). Note that the pointer *FSVptr*, which you pass to this function, contains the address of the fuzzy solution variable descriptor block (*FSV*) for the current solution variable. If the working fuzzy set is empty, then the correlated fuzzy truth vector is simply copied across; otherwise, the appropriate implication action is performed.

Refer to Chapter 10, *Using the Fuzzy Code Libraries*, for complete details on setting up both the system and the fuzzy logic control blocks. Both of these external structures must be initialized before the fuzzy reasoning functions will work.

```
//
//--Now update the possibility density arrays so that if we are
//--using the preponderance of evidence technique the statistics
//--will be available.
   FSVptr->FzySVupdcnt++;
   for(i=0;i<VECMAX;i++)
    if(outFDBptr->FDBvector[i]>0) FSVptr->FzySVcntarray[i]++;
   return;
 }
```

Listing 6.2 The Conditional Fuzzy Proposition Handler (`FzyCondProposition`)

The function terminates by counting the number of times the output fuzzy region was modified and by updating an array that counts the number of overlapping fuzzy sets in each truth membership cell. This information will be used later as part of the action that determines the scalar value of the solution variable.

Correlation Methods

Both the min–max and the fuzzy additive implication mechanisms for conditional fuzzy propositions involve reducing the truth of a consequent fuzzy region by the truth of the rule's premise before the solution variable's working fuzzy region is updated. This process of correlating the consequent with the truth of the predicate stems from the observation that the truth of the fuzzy action cannot be any greater than the truth of the proposition's premise. There are two principal methods of restricting the height of the consequent fuzzy set: correlation minimum and correlation product.

Correlation Minimum

The most common method of correlating the consequent with the premise truth truncates the consequent fuzzy region at the truth of the premise. This method is called *corre-*

lation minimum, since the fuzzy set is minimized by truncating it at the maximum of the predicate's truth.

In Figure 6.19, the controller for a steam turbine adjusts the throttle depending on the current pressure and temperature setting. This figure shows the results of applying the following rule for a particular reading of temperature and pressure:

```
if pressure[t] is LOW and temperature[t] is COOL
   then throttle.action is POSITIVEMODERATE;
```

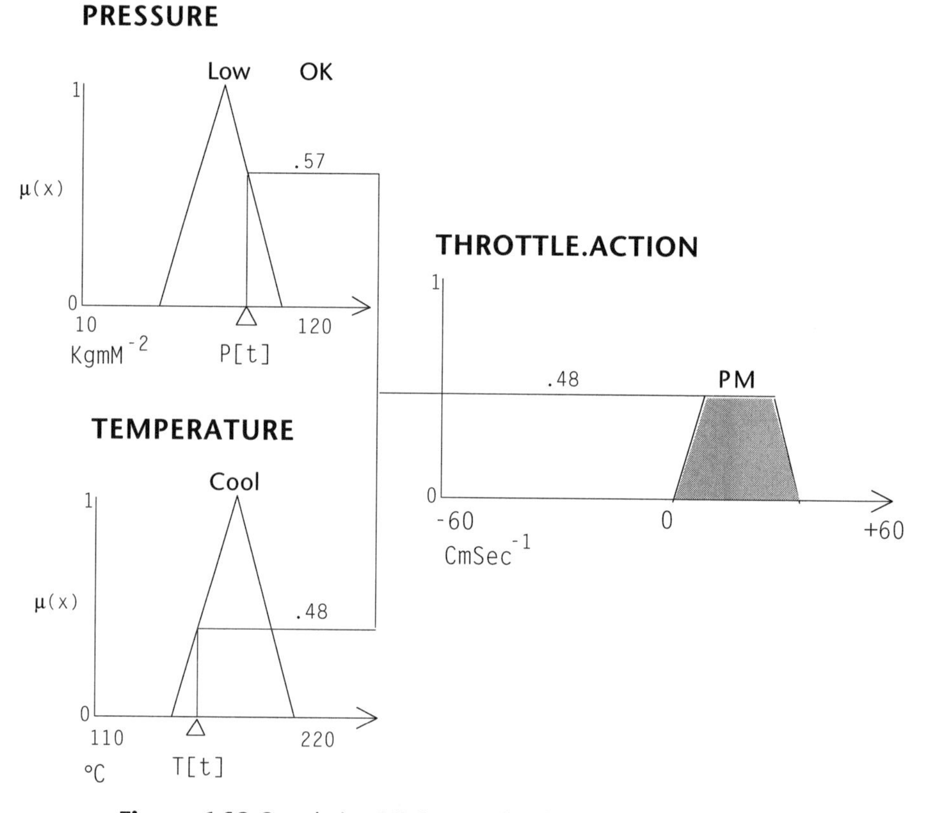

Figure 6.19 Correlation Minimum of Turbine THROTTLE.ACTION

The model is solving for a current throttle setting action (itself decomposed into a set of overlapping fuzzy sets).

A pressure reading at time *P(t)* has a [.57] degree of membership in the fuzzy set *Low*. A temperature reading at time *T(t)* has a [.48] degree of membership in the fuzzy set *Cool*. The fuzzy implication function requires the minimum of these two values ([.48]). We then use the correlation minimum to truncate the *TROTTLE.ACTION's PM* (positive moderate) fuzzy set at this [.48] level. This becomes the current output fuzzy set for the throttle action solution variable.

The correlation minimum mechanism usually creates a plateau since the top of the fuzzy region is sliced by the predicate truth value. This action introduces a certain amount of information loss. If the truncated fuzzy set is multimodal or otherwise irregular, the surface topology above the predicate truth level is discarded. The correlation minimum method, however, is often preferred over the correlation product (which does preserve the shape of the fuzzy region) for three main reasons: it intuitively reduces the truth of the consequent by the maximum truth of the predicate; it involves less complex and faster arithmetic (an important consideration for microprocessors and microcontrollers); and it generally generates an aggregated output surface that is easier to defuzzify using the conventional techniques of composite moments (centroid) or composite maximum (center of maximum height).

Correlation Product

While correlation minimum is the most frequently used technique, *correlation product* offers an alternative and, in many ways, better method of achieving this goal. With correlation product, the intermediate fuzzy region is scaled instead of truncated. The truth membership function is adjusted by applying the following transformation:

```
for(i=0;i<VECMAX;i++)FDBvector[i]=FDBvector[i]*PredTruth;
```

That is, the membership is scaled using the truth of the predicate. This has the effect of shrinking the fuzzy region (when *PredTruth*<1) while still retaining the original shape of the fuzzy set.

In Figure 6.20 the controller for a steam turbine adjusts the throttle depending on the current pressure and temperature setting (also see Figure 6.19 for the corresponding correlation minimum representation). Figure 6.30 shows the results of applying the following rule for a particular reading of temperature and pressure:

```
if pressure(t) is LOW and temperature(t) is COOL
   then throttle.action is POSITIVEMODERATE;
```

The model is solving for a current throttle setting action (itself decomposed into a set of overlapping fuzzy sets).

A pressure reading at time *P(t)* has a [.57] degree of membership in the fuzzy set *Low*. A temperature reading at time *T(t)* has a [.48] degree of membership in the fuzzy set *Cool*. The min–max implication function requires that we take the minimum of these two values ([.48]). We then use the correlation product to scale the *THROTTLE.ACTION PM* (positive moderate) fuzzy set at this [.48] level. This becomes the current output fuzzy set for the throttle action solution variable.

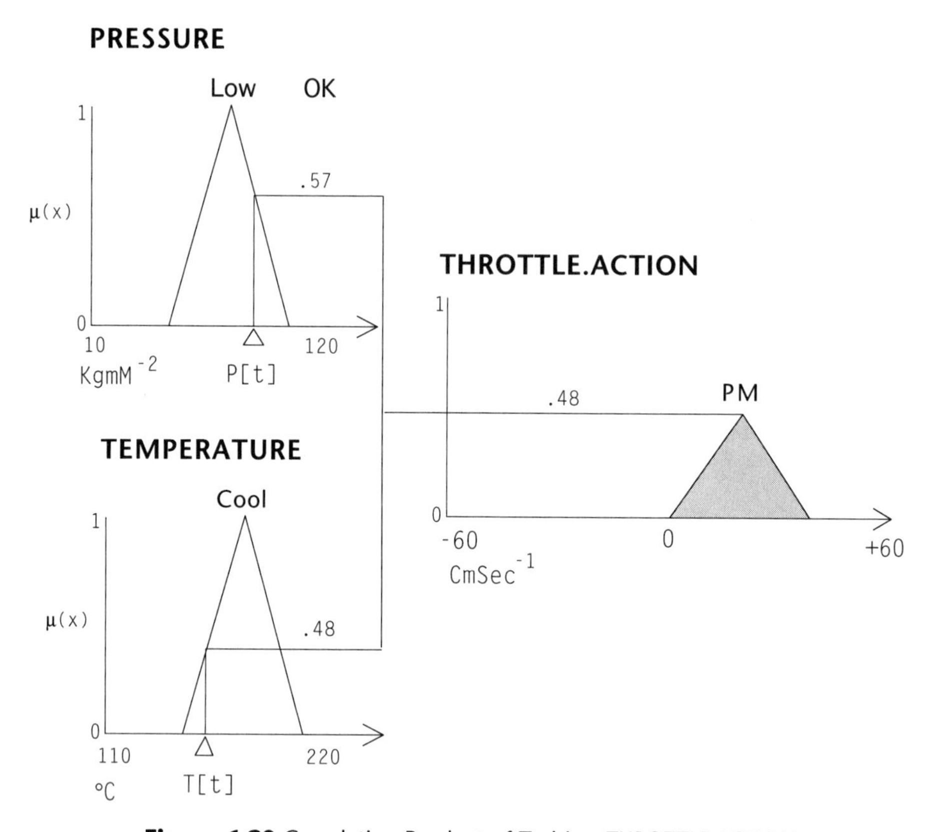

Figure 6.20 Correlation Product of Turbine *THROTTLE.ACTION*

The correlation product mechanism does not introduce plateaus into the output fuzzy region, although its does increase the irregularity of the fuzzy region and could affect the results obtained from composite moments or composite maximum defuzzification. This lack of explicit truncation has the consequence of generally reducing information loss. If the intermediate fuzzy set is multimodal, irregular, or bifurcated in other ways, this surface topology will be retained when the final fuzzy region is aggregated with the output variable's undergeneration fuzzy set.

The *FzyCorrMinimum* and *FzyCorrProduct* functions reduce the height of the specified fuzzy set either by truncating the truth membership or by scaling the truth membership.

```
#include <stdlib.h>
#include "fuzzy.hpp"
#include "FDB.hpp"
#include "mtypes.hpp"
#include "mtsptype.hpp"
void FzyCorrMinimum(
    FDB* FDBptr,const truthval Plateau,int *statusPtr)
   {
    int i;
    char *PgmId="mtfzcor";
    *statusPtr=0;
    if(Plateau<0||Plateau>1)
      {
       *statusPtr=1;
       MtsSendError(26,PgmId,FDBptr->FDBid);
       return;
      }
    for(i=0;i<VECMAX;i++)
      FDBptr->FDBvector[i]=min(FDBptr->FDBvector[i],Plateau);
    return;
   }
```

Listing 6.3 The Correlation Minimum Function

```
void FzyCorrProduct(
    FDB* FDBptr,const truthval Scale,int *statusPtr)
   {
    int i;
    char *PgmId="mtfzcor";
    *statusPtr=0;
    if(Scale<0||Scale>1)
      {
       *statusPtr=1;
       MtsSendError(26,PgmId,FDBptr->FDBid);
       return;
      }
    for(i=0;i<VECMAX;i++)
      FDBptr->FDBvector[i]=FDBptr->FDBvector[i]*Scale;
    return;
   }
```

Listing 6.4 The Correlation Product Function

The Minimum Law of Fuzzy Assertions

An unconditional fuzzy proposition is a fuzzy assertion in the form:

$$x \quad is \quad Y$$

In an unconditional fuzzy proposition, the truth of the consequent fuzzy region is not restricted by any predicate. The consequent fuzzy region is used to restrict the current solution variable's output fuzzy set using the minimum functional relationship (the maximum output fuzzy set truth is limited by the minimum truth of the consequent). The method of applying a fuzzy assertion is specified in Equation 6.9.

$$\mu_{sfs}[x] \leftarrow \min(\mu_{sfs}[x_i], \mu_{cfs}[x_i])$$

Equation 6.9 Minimum of Current Solution with Consequent

Unconditionals provide broad support for what might otherwise be an undefined model. When none of the conditional propositions executes, the unconditionals define the model solution. As an example, consider a simple two-rule inventory policy for determining the amount of `WIDGETS` that should be backordered (from the `INVMOD1.CPP` program):

[P1] backorderAmt must be near ERP
[P2] if orders are High then backorderAmt must be Large;

Proposition one [*P1*] indicates that the backorder amount should be near the economic reorder point (*ERP*). This is unconditional. Proposition two [*P2*] says that if the number of orders (for widgets) is *High*, then the backorder amount should be *Large*. The basic fuzzy set vocabulary for this model appears in Figures 6.21 and 6.22.

The economic reorder point (*ERP*) is part of the organization's formal replenishment policy for inventory and indicates the quantity of goods that must be kept in stock both to satisfy customer demand and to minimize carrying and stockage costs. This often includes a provision for buffer or safety stock. The inventory model specifies that the backorder amount should be kept close to this *ERP*, which is a sensible provision if we have no other contra-indicating evidence (see Figure 6.23). In this model, the economic reorder point is set to a constant of one hundred and fifty (`150`).

```
  FuzzySet:    HIGH.ORDERS
  Description:
1.00                                                                .
0.90                                                           ......
0.80                                                     .......
0.70                                                ......
0.60                                          .......
0.50                                     ......
0.40                                ......
0.30                          .......
0.20                     ......
0.10           .......
0.00......
     0---|---|---|---|---|---|---|---|---|---|---|---|---|---|---|---0
  90.00  122.50  155.00  187.50  220.00  252.50  285.00  317.50  350.00
  Domained:         90.00 to      350.00
```

Figure 6.21 The Inventory Model *HIGH.ORDERS* Fuzzy Set

```
  FuzzySet:     LARGE.BOAmt
  Description:
1.00                                                            .....
0.90                                                 ...........
0.80                                            .....
0.70                                       .....
0.60                                    ...
0.50                                ....
0.40                             ...
0.30                         ....
0.20                    .....
0.10              ......
0.00.............
    0---|---|---|---|---|---|---|---|---|---|---|---|---|---|---|---|---0
   0.00   37.50   75.00  112.50  150.00  187.50  225.00  262.50  300.00
  Domained:          0.00 to     300.00
```

Figure 6.22 The Inventory Model *LARGE.BOAmt* Fuzzy Set

```
  FuzzySet:     NEAR.ERP
  Description:
1.00                                .
0.90                              .. ..
0.80                             .     .
0.70
0.60                            .       .
0.50
0.40                           .         .
0.30
0.20                          .           .
0.10                         .             .
0.00.........................               .........................
    0---|---|---|---|---|---|---|---|---|---|---|---|---|---|---|---|---0
   0.00   37.50   75.00  112.50  150.00  187.50  225.00  262.50  300.00
  Domained:          0.00 to     300.00
```

Figure 6.23 The Inventory Model *NEAR.ERP* Fuzzy Set

A few examples of the inventory model illustrate the relationship between conditional and unconditional propositions in a fuzzy model. Figure 6.25 shows the outcome of the inventory model when the previous quarter's orders were sixty (*60*). The fuzzy set shown is the working fuzzy set for the solution variable *BOAmt* (backorder amount).

```
Fuzzy WIDGET Inventory Model. (c) 1998 Metus Systems Group.
[Mean] Quarterly Orders :      60.00
Economic Reorder Point  :     150.00
The Rules:
P1   BackorderAmt must be near ERP
P2   if orders are high then backorderAmt must be large
P1   BackorderAmt must be near ERP


       FuzzySet:     BOAmt
       Description:
     1.00                                  .
     0.90                                .. ..
     0.80                               .     .
     0.70
     0.60                              .       .
     0.50
     0.40                             .         .
     0.30
     0.20                            .           .
     0.10                           .             .
     0.00........................               ........................
         0---|---|---|---|---|---|---|---|---|---|---|---|---|---|---|---0
        0.00   37.50   75.00  112.50  150.00  187.50  225.00  262.50  300.00
       Domained:          0.00 to     300.00


P2   if orders are high then backorderAmt must be large
PremiseTruth=       0.00


       FuzzySet:     BOAmt
       Description:
     1.00                                  .
     0.90                                .. ..
     0.80                               .     .
     0.70
     0.60                              .       .
     0.50
     0.40                             .         .
     0.30
     0.20                            .           .
     0.10                           .             .
     0.00........................               ........................
         0---|---|---|---|---|---|---|---|---|---|---|---|---|---|---|---0
        0.00   37.50   75.00  112.50  150.00  187.50  225.00  262.50  300.00
       Domained:          0.00 to     300.00
Model Solution:
  BackOrderAmt=      148.83
  CompIdx     =        1.00
  surface hght=        1.00
```

Figure 6.24 An Inventory Model Execution with Orders=60

When proposition [*P1*] executes unconditionally, the working fuzzy set for *BOAmt* is updated to the value of *near ERP.* Proposition two [*P2*] does not execute since the value of orders (*60*) is below the minimum membership threshold for the *High.Orders* fuzzy set (which starts at *90*). The value for *BOAmt*, [148.83], is taken only from the *near ERP* fuzzy space. The model exhibits the same kind of behavior until the magnitude of orders enters the *High.Orders* fuzzy set. In fact, Table 6.1 shows the backorder amounts calculated by the model for representative order quantities.

Orders	BackOrder	CIX
60	148.83	0.9986
80	148.83	0.9986
100	155.86	0.9661
130	172.27	0.5104
150	179.30	0.2435
190	189.84	0.2785
200	191.02	0.3093

Table 6.1 Inventory Backorder Policy for Various Order Values

The backorder quantity only changes when the value for orders causes the conditional proposition [*P2*] to execute by evaluating with some degree of truth the predicate proposition *orders are High.* The underlying mechanism behind this process can be observed in the following model executions (Figures 6.25 through 6.28). Only the *BOAmt* solution space for the second proposition is displayed.

```
[Mean] Quarterly Orders :      100.00
R2   if orders are high then backorderAmt must be large
PremiseTruth=         0.04
FuzzySet:     BOAmt
        Description:
      1.00                                   .
      0.90                                 .. ..
      0.80                                .     .
      0.70
      0.60                               .       .
      0.50
      0.40                              .         .
      0.30
      0.20                             .           .
      0.10                            .             .
      0.00.........................               .........................
         0---|---|---|---|---|---|---|---|---|---|---|---|---|---|---|---0
        0.00   37.50   75.00  112.50  150.00  187.50  225.00  262.50  300.00
        Domained:          0.00 to      300.00
Model Solution:
  BackOrderAmt=       155.86
  CompIdx      =        0.97
  surface hght=         1.00
```

Figure 6.25 An Inventory Model Execution with Orders=100

```
[Mean] Quarterly Orders :     130.00
R2   if orders are high then backorderAmt must be large
PremiseTruth=       0.15

         FuzzySet:    BOAmt
         Description:
      1.00                                       .
      0.90                                     .. ..
      0.80                                    .     .
      0.70
      0.60                                   .       .
      0.50
      0.40                                  .         .
      0.30
      0.20                                 .           .
      0.10                                .             ..........................
      0.00.........................
          0---|---|---|---|---|---|---|---|---|---|---|---|---|---|---|---0
         0.00   37.50   75.00  112.50  150.00  187.50  225.00  262.50  300.00
         Domained:          0.00 to      300.00

Model Solution:
  BackOrderAmt=      172.27
  CompIdx     =        0.51
  surface hght=        1.00
```

Figure 6.26 An Inventory Model Execution with Orders=130

```
[Mean] Quarterly Orders :     190.00
R2   if orders are high then backorderAmt must be large
PremiseTruth=       0.38

         FuzzySet:    BOAmt
         Description:
      1.00                                       .
      0.90                                     .. ..
      0.80                                    .     .
      0.70
      0.60                                   .       .
      0.50
      0.40                                  .         .
      0.30                                                 ........................
      0.20                                 .          ....
      0.10                             ....
      0.00......................
          0---|---|---|---|---|---|---|---|---|---|---|---|---|---|---|---0
         0.00   37.50   75.00  112.50  150.00  187.50  225.00  262.50  300.00
         Domained:          0.00 to      300.00

Model Solution:
  BackOrderAmt=      189.84
  CompIdx     =        0.28
  surface hght=        1.00
```

Figure 6.27 An Inventory Model Execution with Orders=190

```
[Mean] Quarterly Orders :     200.00

R2   if orders are high then backorderAmt must be large
PremiseTruth=       0.42

        FuzzySet:     BOAmt
        Description:
     1.00                                    .
     0.90                                  .. ..
     0.80                                 .     .
     0.70
     0.60                                .       .
     0.50
     0.40                               .         .             .............
     0.30                                          .............
     0.20                              .           .
     0.10                          .....
     0.00.....................
         0---|---|---|---|---|---|---|---|---|---|---|---|---|---|---|---0
        0.00   37.50   75.00  112.50  150.00  187.50  225.00  262.50  300.00
        Domained:          0.00 to      300.00

Model Solution:
  BackOrderAmt=      191.02
  CompIdx     =        0.31
  surface hght=        1.00
```

Figure 6.28 An Inventory Model Execution with Orders=200

Methods of Decomposition and Defuzzification

Using the general rules of fuzzy inference, the evaluation of a proposition produces one fuzzy set associated with each model solution variable. As an example, the following propositions, when evaluated, will correlate the consequent fuzzy sets *A*, *B*, and *C* to produce a fuzzy set representing the solution variable D:

```
if w is Y then D is A
if x is X then D is B
if y is Z then D is C
```

To find the actual value for the corresponding scalar *d*, we must find a value that best represents the information contained in the fuzzy set *D*. This process, illustrated in Figure 6.29, is called *defuzzification*. Such a process yields the expected value of the variable for a particular execution of a fuzzy model.

Defuzzification is the final phase of fuzzy reasoning. As Figure 6.30 illustrates, the evaluation of the model propositions is handled through an aggregation process that pro-

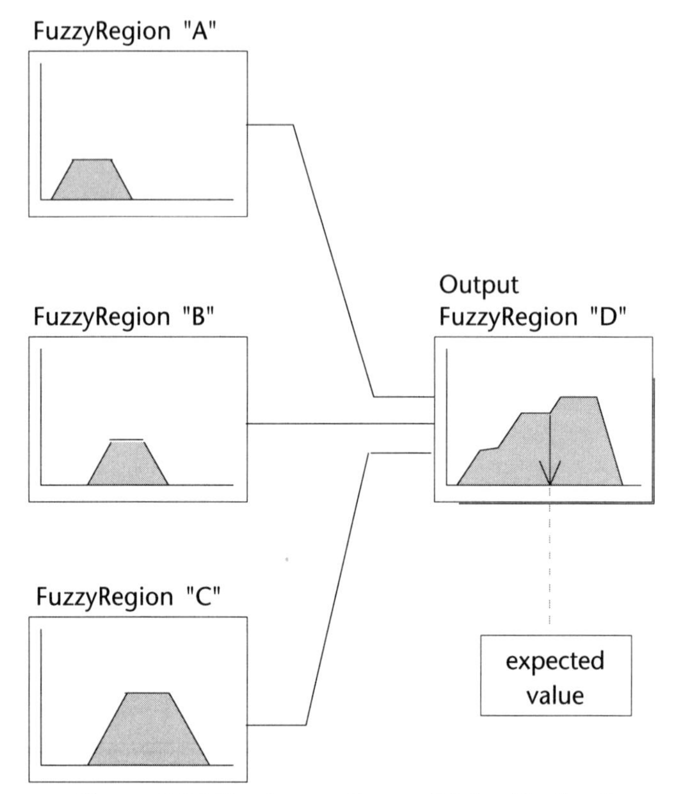

Figure 6.29 The Aggregation and Defuzzification Process

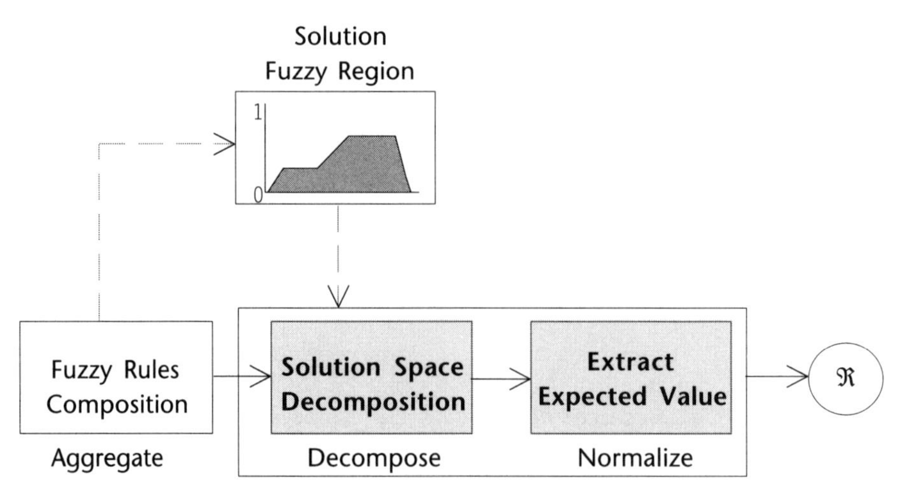

Figure 6.30 The Decomposition Process

duces the final fuzzy regions for each solution variable. This region is then decomposed using one of the defuzzification methods.

In fuzzy models, there are several methods of determining the expected value of the solution fuzzy region. These are the *methods of decomposition*, also called *methods of defuzzification,* and they describe the ways we can derive an expected value for the final fuzzy state space. Our current understanding of decomposition relies more on heuristics than on algorithms derived from "first principles." This is probably the best we can achieve since it does not seem possible to represent a complex, multidimensional space with a single number.

As Figure 6.31 illustrates, defuzzification means dropping a "plumb line" to some point on the underlying domain. At the point where this line crosses the domain axis, we read the expected value of the fuzzy set.

For a given solution space (S_i) in three possible degrees of freedom—domain values (d), truth membership gradient (t), and space possibility density or weight (w)—we could reduce the expected representation of the space by one dimension. We can see that the solution space is represented by a matrix of domain, truth, and density properties, and the reduced space is the dot-product of the mean domain value and the Cartesian product of the truth/density space:

$$\Re_i\left[\frac{d}{t \otimes w}\right] \leftarrow decomp(S_i \cdot \begin{bmatrix} d_1 & t_1 & w_1 \\ d_n & t_n & w_n \end{bmatrix})$$

A reduction by another dimension means that we lose information about the shape of the fuzzy region (width, height, and point density). This loss of information increases the amount of entropy in the defuzzified space.

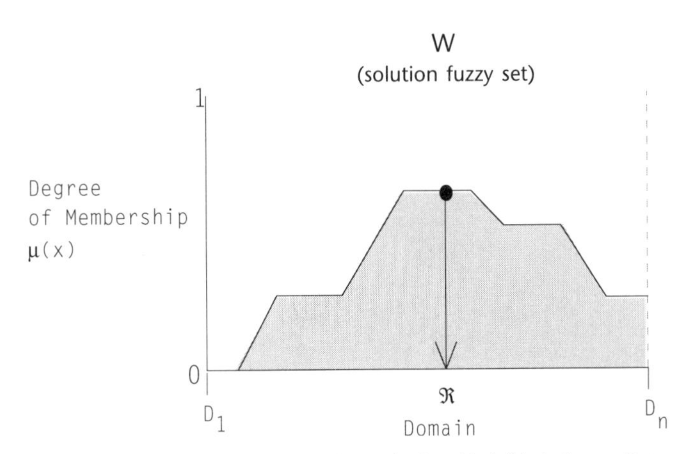

Figure 6.31 Defuzzifying the Solution Variable's Fuzzy Set

Underlying all the defuzzification functions is the process of finding the best place along the surface of the fuzzy set to drop this plumb line. This generally means that defuzzification algorithms are a compromise with or a trade-off between the need to find a single point result and the loss of information such that a process entails. This kind of information loss or increase in system entropy is a natural consequence of a reduction in the representational dimensionality of the fuzzy region.

To illustrate how the decomposition process is integrally tied to the composition of the fuzzy space, we can decompose the weight estimation model into a series of interlocking fuzzy spaces associated with the weight and height control variables. This can be viewed as a [3x3] Cartesian space (Figure 6.32).

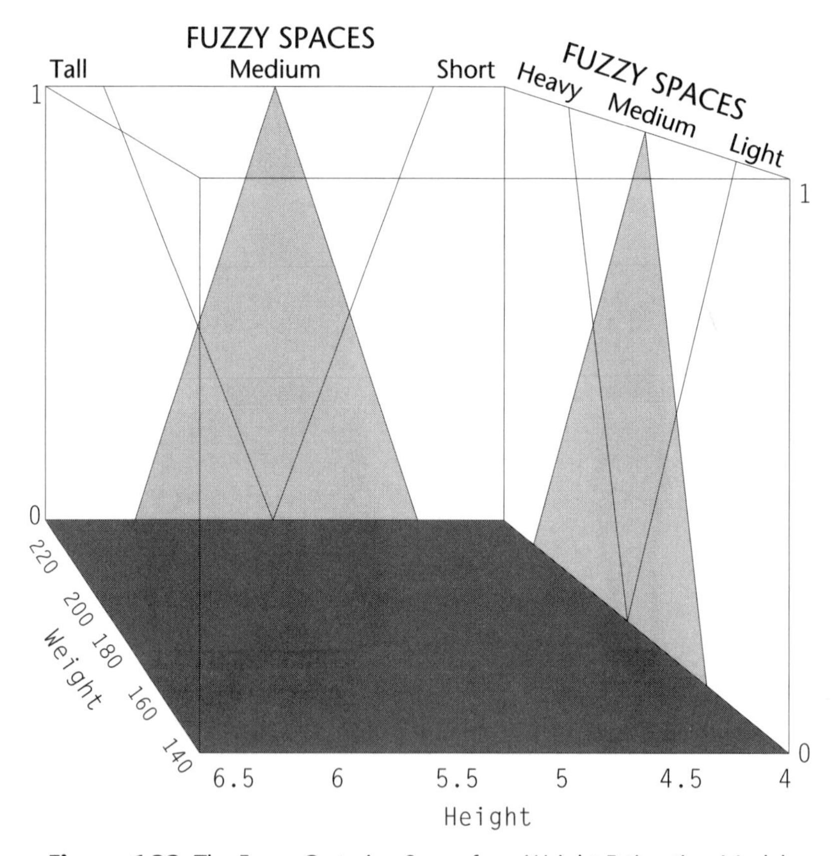

Figure 6.32 The Fuzzy Cartesian Space for a Weight Estimating Model

In this model organization, *Tall* and *Heavy* have been replaced with fuzzy spaces describing distinct (but still fuzzy) regions related to the underlying control variables of height and weight. The central region of the cube provides a visual representation of the fuzzy product from rules interacting along these control domains.

A weight estimation policy uses these fuzzy spaces to compute the expected value for weight (the consequent fuzzy space) from the product of the height values and their associated spaces. The rules or propositions that determine a weight from a given height appear as:

[P1] if height is Tall then weight is Heavy;
[P2] if height is Medium then weight is Medium;
[P3] if height is around Medium then weight is near Medium;
[P4] if height is short then weight is light

Each proposition establishes a fuzzy region bounded by the Cartesian product of the linguistic variables and the maximum truth of the antecedent condition (the degree to which height is compatible with the associated fuzzy space). These rule or proposition spaces construct a multidimensional hyperspace, if you will, that connects the various edges of each fuzzy antecedent and consequent space. The region appears similar to the one shown in Figure 6.33.

The idea behind decomposition involves finding the singleton or scalar value that best represents the composite structure of the final fuzzy space. This process, in essence, derives the expected value of the fuzzy set. Let's consider the two most frequently used decomposition methods—composite moments (centroid) and composite maximum.

Composite Moments (Centroid)

The *centroid* or center of gravity technique finds the "balance" point of the solution fuzzy region by calculating the weighted mean of the fuzzy region. Arithmetically, for fuzzy solution region A, this procedure is formulated as

$$\Re = \frac{\sum_{i=0}^{n} d_i \cdot \mu_A(d_i)}{\sum_{i=0}^{n} \mu_A(d_i)}$$

where d is the i'th domain value, and $\mu(d)$ is the truth membership value for that domain point. As Figure 6.34 shows, a centroid or composite moments defuzzification finds a point representing the fuzzy set's center of gravity.

Centroid defuzzification is the most widely used technique because it has several desirable properties: (1) the defuzzified values tend to move smoothly around the output fuzzy region; that is, changes in the fuzzy set topology from one model frame to the next

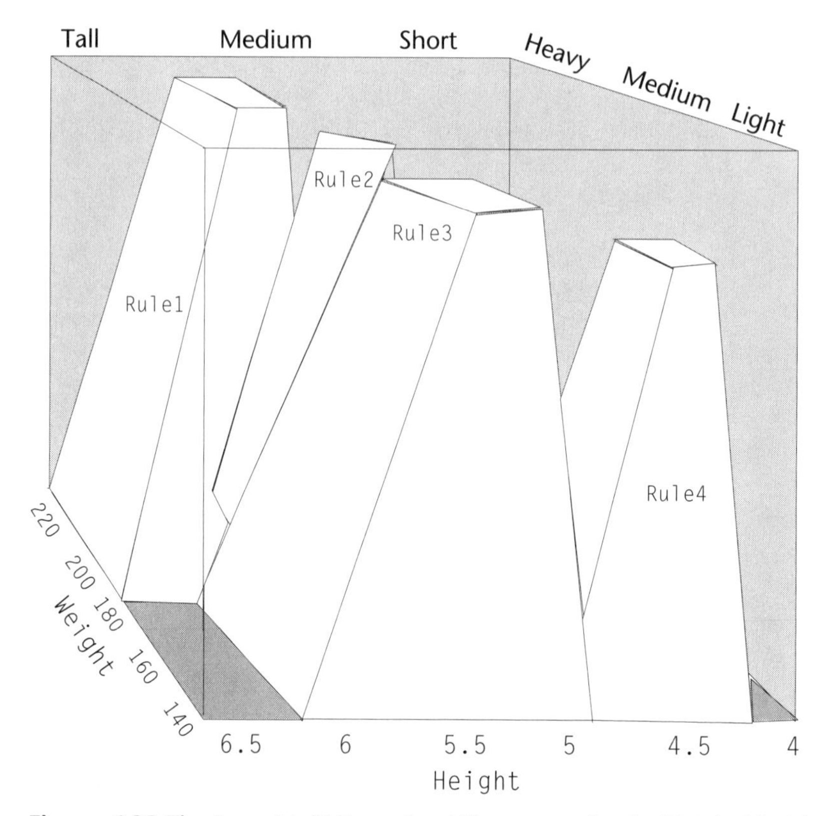

Figure 6.33 The Fuzzy Multidimensional Hyperspace for the Weight Model

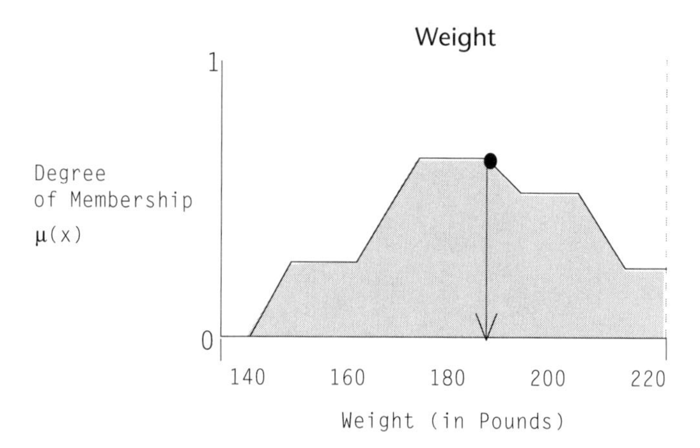

Figure 6.34 The Centroid Method of Defuzzification

usually result in smooth changes in the expected value; (2) it is relatively easy to calculate; and (3) it can be applied to both fuzzy and singleton output set geometries.

Composite Maximum (Maximum Height)

There are three closely related kinds of composite maximum techniques: the *average maximum*, the *center of maximums*, and the simple *composite maximum*. A maximum decomposition, as illustrated in Figure 6.35, finds the domain point with the maximum truth. If this point is ambiguous (that is, it lies along a plateau), then these methods employ a conflict resolution approach such as averaging the values or finding the center of the plateau.

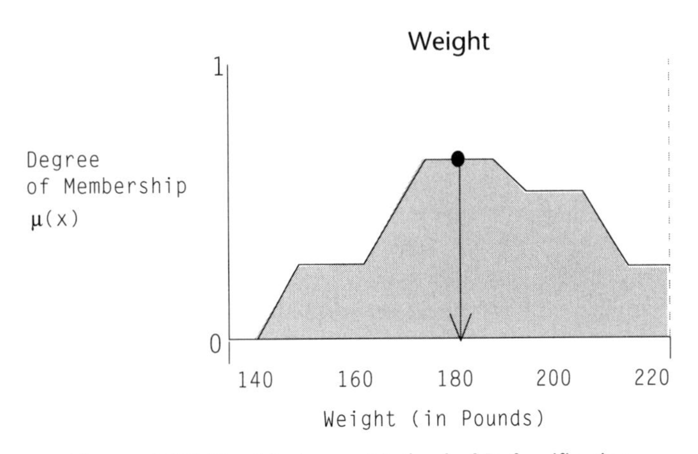

Figure 6.35 The Maximum Method of Defuzzification

Unlike the centroid technique, however, the maximum decomposition has some attributes that are generally applicable to a narrower class of problems because: (1) the expected value is sensitive to a single rule that dominates the fuzzy rule set and (2) the expected value tends to jump from one frame to the next as the shape of the fuzzy region changes. The maximum defuzzification techniques, however, are important for a broad class of models that assess the maximum extent of some fuzzy property. Risk assessment is an example of such models. When the following risk policy rules are evaluated

[P1] if age is young then risk is High
[P2] if distance.to.work is far then risk is Moderate;
[P3] if accidents are above acceptable
then risk is Excessive;
[P4] if DWI.convictions are above near zero
then risk is Unacceptable;

the model should be sensitive to the proposition that has the greatest degree of truth. Such decisions about risk are usually discontinuous; in other words, risk is a class of values. The composite maximum tends to settle on the plateau created by the correlation process and thus gives a consistent class of risk assessments for a closely clustered set of proposition truths.

Hyperspace Decomposition Comparisons

In the hyperspace representation, we can view the process of defuzzification as finding the hyperplane that bisects the space geometry along the axis of highest possible density. The hyperspace represents the area created by a proposition under a particular degree of evaluation. Thus, given a particular input height [y], various defuzzification methods produce different measures for the output value of weight [x].

Figure 6.35 reflects the possibility space bounded by each proposition within the weight estimation model and shows the results of applying different defuzzification techniques. The composite maximum selects the value of the fuzzy surface that is represented by the maximum truth value. For example, if the maximum lies on a plateau, then the center of the plateau is selected. The composite moments, calculating the centroid or the "first moment of inertia," select a value corresponding to the center of gravity of the fuzzy space. In constructing a fuzzy model, these two principal decomposition methods achieve different results. The maximum method is sensitive to the area circumscribed by the single rule with the maximum truth function. Fuzzy model solution variables tend to move in jumps from one plateau to the next. The moments method, on the other hand, finds the average density of the fuzzy surface and produces a smooth movement from one solution value to the next.

Preponderance of Evidence Technique

A fuzzy model can also be responsive to the rules of evidence for a particular fuzzy consequent state. We interpret the rules of evidence as the number of votes received by a particular region in the solution fuzzy space. This area is analogous to the center of mass function for a solid region bounded by the fuzzy space that has a membership truth function exceeding [.5]. In a knowledge-based system, a production rule

if p then x is Y

represents a piece of evidence p for the existence of x. As multiple rules execute, they contribute to the description of x in the model. This process can be pictured as shown in Figure 6.37.

By exploiting the center of mass concept associated with fuzzy solution spaces, we can elicit expected values that reflect the region with the most votes. As the visualization space (see Figure 6.37) shows, the expected value is taken from a region inside the rule

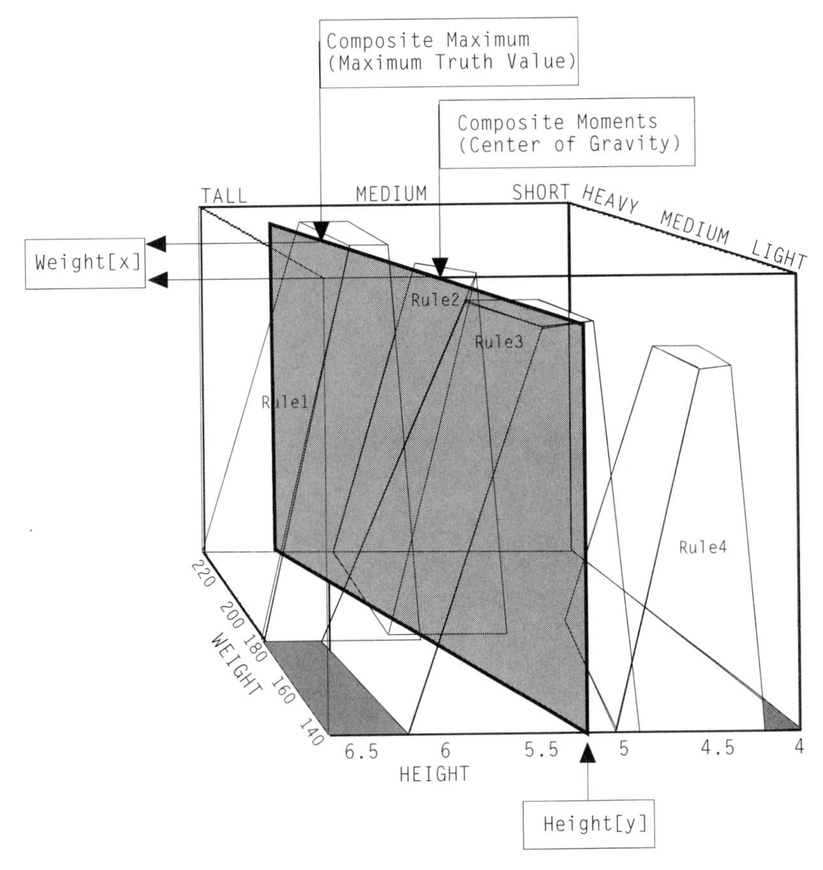

Figure 6.36 The Maximum Method of Defuzzification

execution space where the number of support fuzzy sets has the highest density. This means that the expected value is selected from a region that is supported by the largest number of rules.

The preponderance of evidence method employs a change in the geometry of the solution fuzzy set to achieve better results with the standard composite moments defuzzification. The technique attempts to minimize the effect of "spuriously" skewed fuzzy sets in the solution space by giving increased emphasis to the region that is supported by the most overlapping sets. The preponderance of evidence method works best in models where many propositions are executed to derive the space for the solution variable.

We can see this clearly by examining a projection along the weight estimation model's fuzzy region for a given value of height (Figure 6.38).

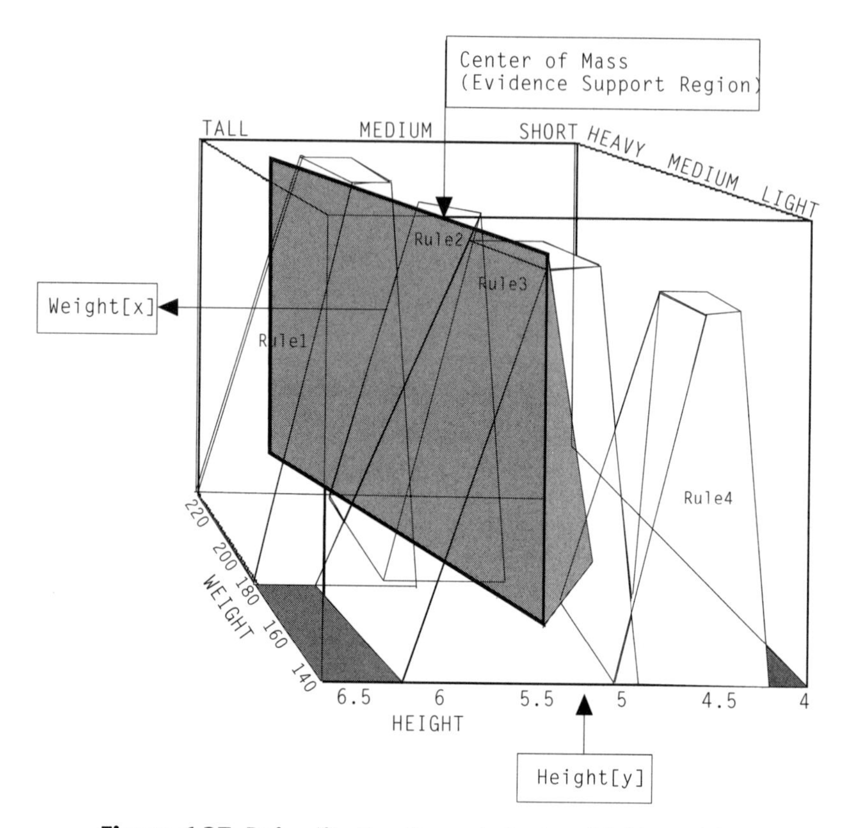

Figure 6.37 Defuzzification Through Center of Evidence Mass

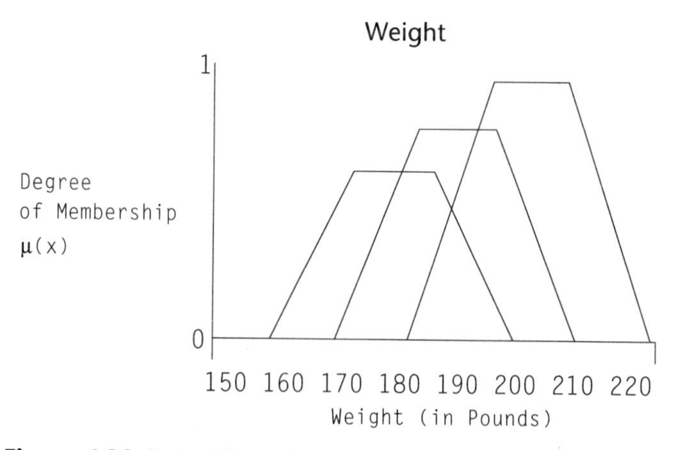

Figure 6.38 Output Fuzzy Support Regions in the Weight Model

These intersecting regions are established by the activation of rules that have some degree of truth in the antecedent height assertion. The topology of the final fuzzy surface is the edge of the space defined by the maximum points of intersection along each individual fuzzy space (Figure 6.39).

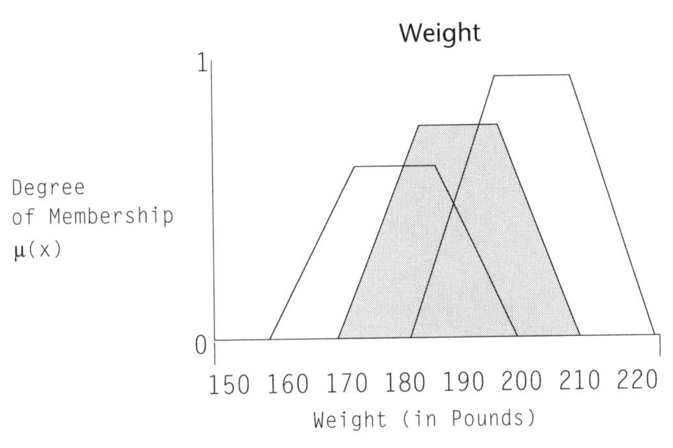

Figure 6.39 The Maximum Support Region in Center of Evidence

The rules of fuzzy implication use this consequent space in finding the expected value of the solution variable. We have already seen that fuzzy models can be responsive to the maximum truth of the rule set, as well as to the center of gravity of the surface itself. This generates quite different expected values, as Figure 6.40 shows.

On the other hand, a defuzzification process using the center of mass finds the area that has the highest density of intersecting fuzzy spaces. The first moment of this area is then used to derive an expected value, as illustrated in the membership transition graph of Figure 6.41 for the consequent space.

A fuzzy model, then, provides a direct mechanism for modeling the type of profound judgments that are typically associated with both multiple experts in a common domain and samples from a decision space, where the combined rules of evidence for a particular state lead to a correlation between the model process and the final solution recommendation.

Using correlation product instead of correlation maximum as an implication transfer function can have a significant impact on some of the defuzzification techniques. In particular, since the consequent fuzzy regions are scaled rather than truncated, the output (solution) fuzzy region will lack definite plateaus. The output region can be highly modal (filled with many "peaks and valleys") so that those defuzzifications that rely on the truncation effect will not work as predicted. When you use correlation product, you should restrict the defuzzification to composite moments.

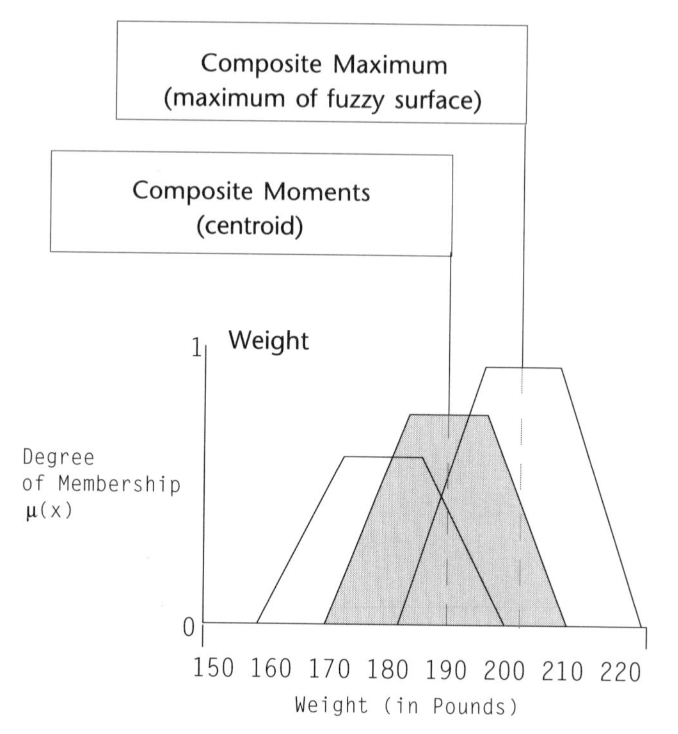

Figure 6.40 Different Defuzzifications on an Evidence Geometry

Other Defuzzification Techniques

The composite maximum and moments represent the most widely used forms of defuzzification. In particular, the moments or centroid method is preferred in a large majority of fuzzy models since it appears to assimilate all the information in the output fuzzy set. The centroid behaves in a manner similar to Bayesian estimates in that it selects a value supported by the knowledge accumulated from each executed proposition. In nearly all cases, the composite moments or the composite maximum will provide the best overall expected value.

Unless you have reason to believe that your model requires a more advanced or a specialized method of defuzzification, you should limit your model to either the composite moments (centroid) or the composite maximum defuzzification technique. At the very least, one of these choices should be the starting point for your method of model validation and verification.

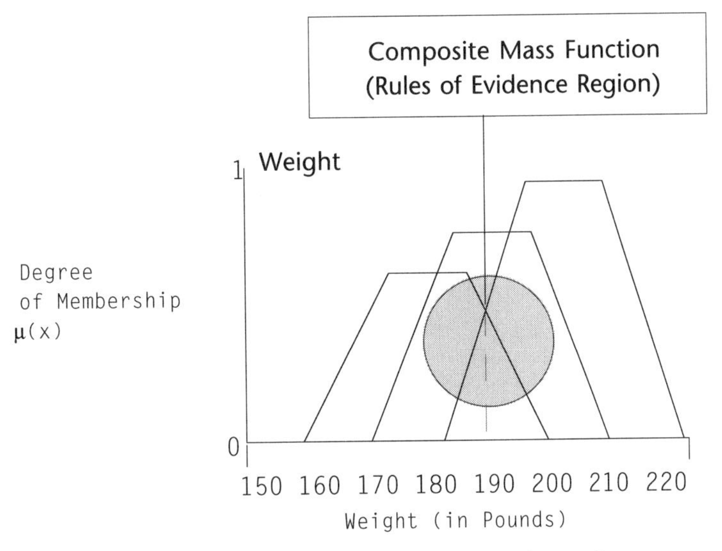

Figure 6.41 First Moment for the Rules of Evidence Geometry

There are, however, other techniques for decomposing a fuzzy set into an expected value. The brief survey of alternative defuzzification techniques presented here is not inclusive, and more exotic methods exist, such as those that attempt to measure and use the output set's degree of fuzziness (its information entropy) and those that deal with high-order fuzzy sets. These methods are not as general as the maximum or centroid, and often they require special knowledge about the model infrastructure. The average of the maximum defuzzification, presented first, has been used successfully in place of the strict composite maximum.

The Average of Maximum Values

The average of the maximum value defuzzification finds the mean maximum value of the fuzzy region, as shown in Figure 6.42.

If this value is a single point, then it is returned; otherwise, the average of the plateau is calculated and returned. An average value often provides a better approximation for a composite maximum approach when the plateaus tend to be clipped at the left or right edges. See Composite Maximum (Maximum Height) on page 309 for additional details.

The Average of the Support Set

The average of the nonzero region is the same as taking the average of the support set for the output fuzzy region (see Figure 6.43).

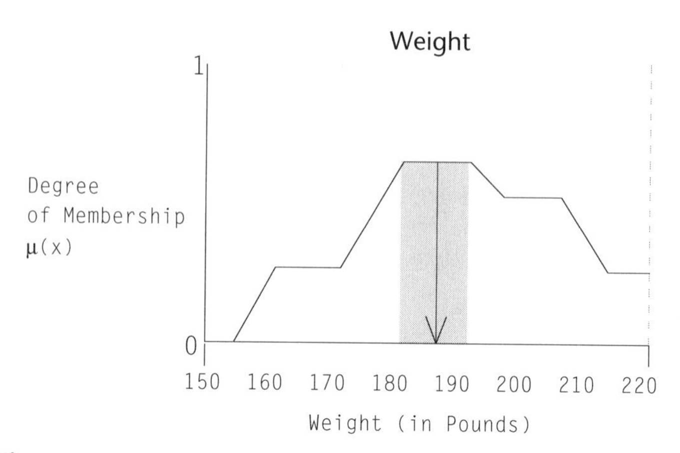

Figure 6.42 The Average of the Maximum Plateau Defuzzification

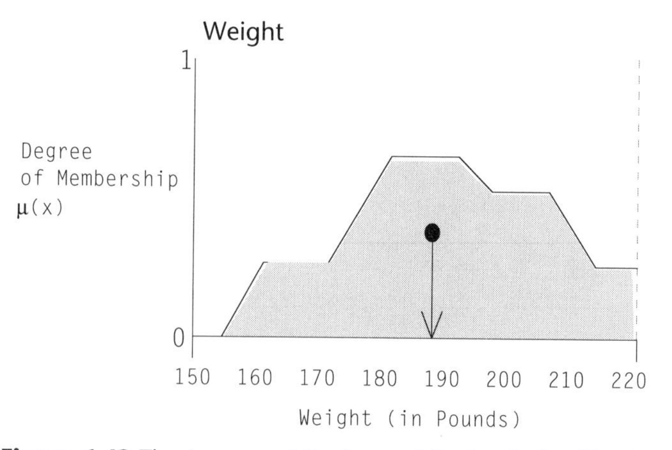

Figure 6.43 The Average of the Support Region Defuzzification

This method is similar to the composite moments defuzzification, except that it is not weighted by the individual truth membership gradients of each nonzero domain value. Since this method is based on the support set, the defuzzification is sensitive to the current alpha-cut threshold, which is applied prior to defuzzification so that all values below the alpha-cut are set to zero.

The Far and Near Edges of the Support Set

These techniques select the value at the right or left fuzzy set edge and are of most use when the output fuzzy region is structured as a single-edged plateau, as shown in Figures 6.44 and 6.45.

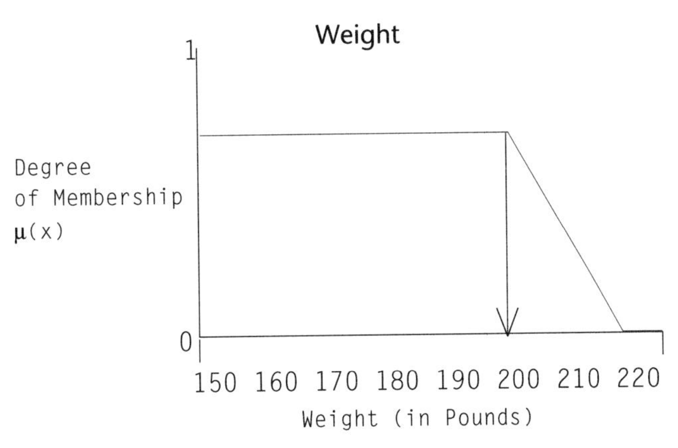

Figure 6.44 The Far Edge of the Support Set Defuzzification

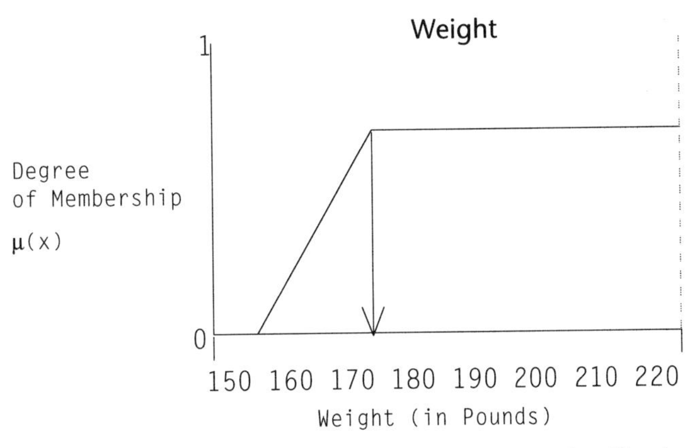

Figure 6.45 The Near Edge of the Support Set Defuzzification

This kind of defuzzification is occasionally useful when you have only one or two highly overlapping fuzzy sets that are truncated at either end of the domains. However, these techniques do not work well when you use correlation product with bell-shaped fuzzy sets. These edge techniques look for a plateau by moving toward the left from the right edge (or to the right from the left edge) of the fuzzy region. The current truth level is not counted as a plateau. Figure 6.46 illustrates the consequence of this discovery process.

The Center of Maximums

In a multimodal or multiplateau fuzzy region, the center of maximums technique finds the highest plateau and then the next highest plateau. The midpoint between the centers of these plateaus is selected, as we see in Figure 6.47.

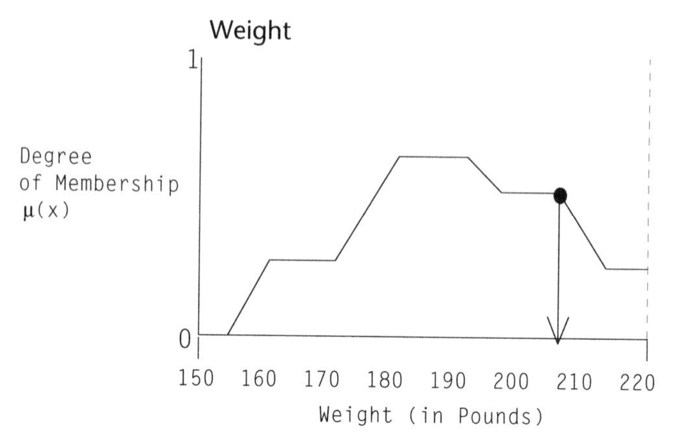

Figure 6.46 The Discovery of a New Plateau Region

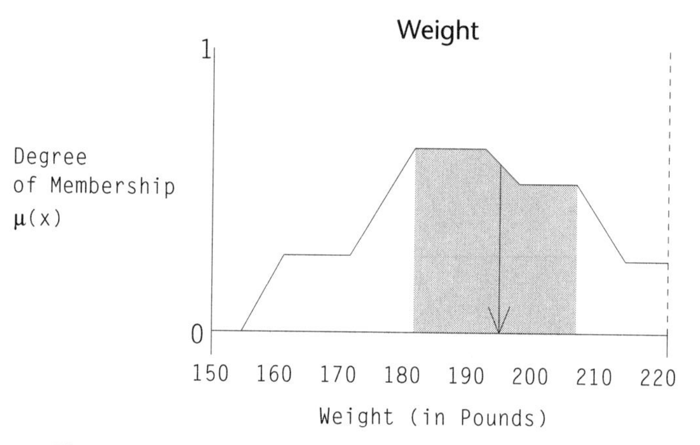

Figure 6.47 The Center of Maximums Defuzzification

If there is no plateau, then the maximum point in the fuzzy region is used. If only a single plateau is found, the process switches to the standard maximum defuzzification method. In general, this technique provides a slightly better and more robust defuzzification method than the composite maximum when you suspect that the output region may contain many peaks or plateaus.

The *FzyDefuzzify* function encapsulates all the defuzzification techniques into a single package. The function returns a double-precision, floating-point value as the expected value of the indicated solution fuzzy set. Since the defuzzification process expects a fuzzy solution variable control block pointer (*FSV*), you will need to understand the architecture of the modeling software to use this facility properly. See Chapter 10, *Using the Fuzzy Code Libraries,* for complete details.

```
#include <stdio.h>
#include <string.h>
#include "FDB.hpp"
#include "fuzzy.hpp"
#include "mtypes.hpp"
#include "mtsptype.hpp"
static char  wrkBuff[128];
static char *DefuzzNames[]=
 {
  "Centroid",
  "Maximum Plateau",
  "Maximum Height",
  "Average Support Set",
  "Best Evidence",
  "Least Entropy",
  "Near Edge",
  "Far Edge"
 };
void CompleteDefuzz(char*,double,float);
double FzyDefuzzify(
     FDB* FDBptr,
      const int DefuzzMethod,float *Grade,int *statusPtr)
   {
    int        i,j,k,n,tvpos,EdgeCnt;
    int        Edges[2];
    float      x,y,tval;
    double     Scalar,SumofScalars;
    const int  Left = 0;
    const int  Right = 1;
    char       *PgmId="mtfzdfz";
*statusPtr=0;
    EdgeCnt=0;

  /* Apply alpha-cut pruning to the solution fuzzy set */
    FzyApplyAlfa(FDBptr,FDBptr->FDBalfacut,STRONG);
    if(FzyGetHeight(FDBptr)==0)
      {
       *Grade=0.0;
       Scalar=0.0;
       sprintf(wrkBuff,"%s%s%s%s%s%s",
        "Cannot Apply \"",DefuzzNames[DefuzzMethod-1],
          "\" Defuzzification--",
        "'",FDBptr->FDBid,"' is empty. [height==0].");
       MtsWritetoLog(SYSMODFILE,wrkBuff,statusPtr);
       *statusPtr=99;
       return(Scalar);
      }
```

Before applying any defuzzification technique, you must first apply the alpha-threshold cut associated with the solution variable's fuzzy set. Pruning the fuzzy set by setting all truth membership values equal to or less than the alpha threshold to zero could result in an empty set—or the set could have been empty when *FzyDefuzzyify* was called. In either case, if the fuzzy set is empty, a message is issued, and the function terminates.

```
switch(DefuzzMethod)
 {
    case BESTEVIDENCE:
     FzyIsolatePDR(FDBptr);
    case CENTROID:
     x=0.0;
     y=0.0;
     for(i=0;i<VECMAX;i++)
      {
        x=x+FDBptr->FDBvector[i]*(float)i;
        y=y+FDBptr->FDBvector[i];
      }
     if(y==0)
       {
        *Grade=0.0;
        Scalar=0.0;
        sprintf(wrkBuff,"%s%s%s", "CENTROID Failed. '",
         FDBptr->FDBid,"' is empty. [height==0].");
        MtsWritetoLog(SYSMODFILE,wrkBuff,statusPtr);
        *statusPtr=99;
        return(Scalar);
       }
    tvpos=(int)(x/y);
    Scalar=FzyGetScalar(FDBptr,tvpos,statusPtr);
    *Grade=FDBptr->FDBvector[tvpos];
    CompleteDefuzz("CENTROID",Scalar,*Grade);
    return(Scalar);
```

The composite moments, or centroid defuzzification, finds the scalar position (*tvpos*, the truth vector position) in the truth membership array by computing the weighted average of the membership. This value is essentially the center of gravity of the membership surface. The scalar domain value and its membership value are returned.

If the preponderance of evidence defuzzification (*BESTEVIDENCE*) is specified (a special case of the centroid), then *FzyIsolatePDR* is called to reduce the nonzero part of the membership array to the region with the maximum rule overlaps. The counts for this type of defuzzification are automatically maintained by the *FzyUncondProposition* function. The standard centroid is then performed. Note that we also protect against division by zero, which can occur if the fuzzy set is empty. The test for a height of zero immediately after applying the alpha-cut should protect against this division, but a little defensive paranoia never hurts.

```
case MAXPLATEAU:
 FzyFindPlateau(FDBptr,Edges,&EdgeCnt,statusPtr);
 if(EdgeCnt==2)
     tvpos=(Edges[Left]+Edges[Right])/2;
    else
     {
      tvpos=Edges[Left];
      if(Edges[Left]==0)        tvpos=Edges[Right];
      if(Edges[Right]>=VECMAX) tvpos=Edges[Left];
     }
 Scalar=FzyGetScalar(FDBptr,tvpos,statusPtr);
 *Grade=FDBptr->FDBvector[tvpos];
 CompleteDefuzz("MAXPLATEAU",Scalar,*Grade);
 sprintf(wrkBuff,"%s%s%2d%s",
   FDBptr->FDBid," is a ",EdgeCnt," edged surface.");
 MtsWritetoLog(SYSMODFILE,wrkBuff,statusPtr);
 return(Scalar);
```

This code represents the standard composite maximum defuzzification technique. Relying to a large degree on the clipping action of the correlation minimum technique, the method begins by finding the plateau with the highest truth membership. The *FzyFindPlateau* also returns the number of edges on the plateau, as well as the *FDBvector* cell position of the right and left edges. If the plateau has two edges (see Figure 6.48), then the center of the plateau is returned.

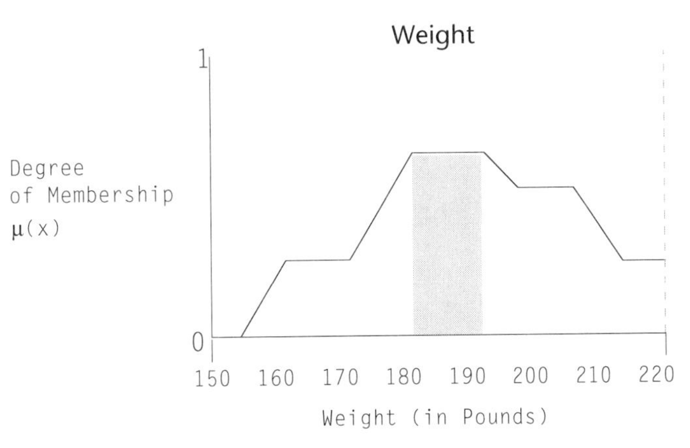

Figure 6.48 A Double-Edged Plateau at Highest Truth Level

If the plateau is single-edged (see Figure 6.49), indicating that the top of the plateau is truncated at the edge of the fuzzy set's explicit domain, the value at the corresponding edge is returned.

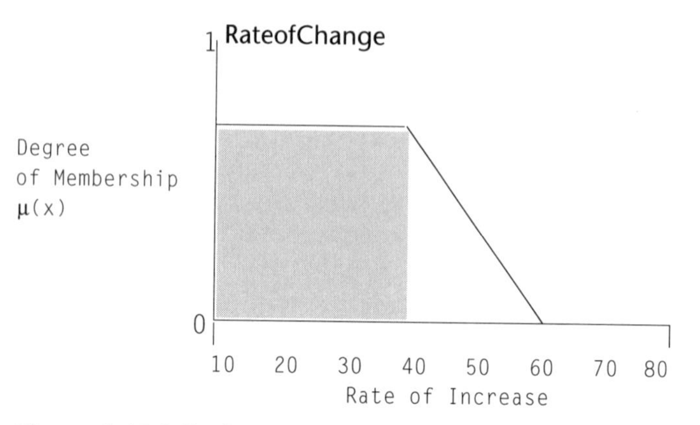

Figure 6.49 A Single-Edged Plateau with the Edge on the Right

```
case MAXIMUM:
 x=FDBptr->FDBvector[0];
 k=0;
 for(i=0;i<VECMAX;i++)
  if(FDBptr->FDBvector[i]>x)
    {
     x=FDBptr->FDBvector[i];
     k=i;
    }
 for(j=k+1;j<VECMAX;j++)
   if(FDBptr->FDBvector[j]!=x) break;
 tvpos=(int)(((float)j-(float)k)/2);
 if(tvpos==0) tvpos=k;
 Scalar=FzyGetScalar(FDBptr,tvpos,statusPtr);
 *Grade=FDBptr->FDBvector[tvpos];
 CompleteDefuzz("MAXIMUM",Scalar,*Grade);
 return(Scalar);
```

The simple maximum defuzzification method finds the point in the membership array with the maximum truth value. If this is a plateau, like the composite maximum, the center of the plateau is used. This is the preferred method of maximum defuzzification when correlation product is used to scale the consequent fuzzy sets. The solution fuzzy sets tend to have a "sawtooth" appearance without the broad plateaus that accompany the correlation minimum method. The search for plateaus in the simple maximum also assures that solution fuzzy sets with scaled trapezoidal or platykurtic bell-curves will be resolved properly.

```
case AVGSUPPORT:
 tval=0,SumofScalars=0;
 n=0;
```

```
for(i=0;i<VECMAX;i++)
 {
   if(FDBptr->FDBvector[i]>0)
     {
      n++;
      tval=tval+FDBptr->FDBvector[i];
      Scalar=FzyGetScalar(FDBptr,i,statusPtr);
      SumofScalars=SumofScalars+Scalar;
     }
 }
if(n==0)
  {
   *statusPtr=3;
   return((double)0);
  }
*Grade=(tval/n);
Scalar=SumofScalars/n;
CompleteDefuzz("AVGSUPPORT",Scalar,*Grade);
return(Scalar);
```

The support set average defuzzification method sums the domain values and the truth membership for each membership value that is greater than zero. The result of this defuzzification is the average value of the support set and the average membership value. This is not a weighted average (as in the composite moments) but a simple mean of the domain values across the support set. If the solution fuzzy set tends to span a significant portion of the set's domain, this function will not be generally useful.

```
case NEAREDGE:
 FzyFindPlateau(FDBptr,Edges,&EdgeCnt,statusPtr);
 tvpos=Edges[Left];
 Scalar=FzyGetScalar(FDBptr,tvpos,statusPtr);
 *Grade=FDBptr->FDBvector[tvpos];
 CompleteDefuzz("NEAREDGE",Scalar,*Grade);
 return(Scalar);
case  FAREDGE:
 FzyFindPlateau(FDBptr,Edges,&EdgeCnt,statusPtr);
 tvpos=Edges[Right];
 *Grade=FDBptr->FDBvector[tvpos];
 Scalar=FzyGetScalar(FDBptr,tvpos,statusPtr);
 CompleteDefuzz("FAREDGE",Scalar,*Grade);
 return(Scalar);
```

Defuzzification at the edges of the solution fuzzy set is straightforward: find the edge of the plateau, and return its scalar value and membership. Edge defuzzifications are not particularly common in information models, process engineering, and signal systems, but they are often important in robotics and other intelligent autonomous vehicle systems.

```
          default:
           Scalar=0.0;
           *statusPtr=99;
           MtsSendError(28,PgmId,MtsFormatInt(DefuzzMethod));
           return(Scalar);
       }

 }
 void CompleteDefuzz(char *DefuzzType,double Scalar,float Grade)
  {
   int   status;
   char  tmpBuff[28];

   memset(tmpBuff,'\0',28);
   strcpy(tmpBuff,"'");
   strcat(tmpBuff,DefuzzType);
   strcat(tmpBuff,"' ");
   sprintf(wrkBuff,"%-14.14s%s%10.3f%s%6.4f%s",
     tmpBuff," defuzzification. Value: ",Scalar,", [",Grade,"]");
   MtsWritetoLog(SYSMODFILE,wrkBuff,&status);
   return;
  }
```

Listing 6.5 The Defuzzification Function

A `CompleteDefuzz` procedure sends a message to the model activity log file indicating the type of defuzzification performed, the expected value, and its grade of membership This grade of membership is the unit compatibility index.

Singleton Geometry Representations

In a singleton geometry output space, the terms associated with the fuzzy regions are represented as single vertical points instead of fuzzy set membership functions. In essence, we are designating a unique and crisp "support value" for the output. The compositional inference method for singleton geometry models is slightly different than those for fuzzy region based models. As Figure 6.50 illustrates, the aggregation step is removed since the singletons are proportionally modified in place rather than combined.

The method of implication using singletons remains generally the same as that using fuzzy sets except that the transfer function is direct. Each singleton represents a point in the output space and is connected through what amounts to linear interpolation with its neighbors for purposes of decomposition.

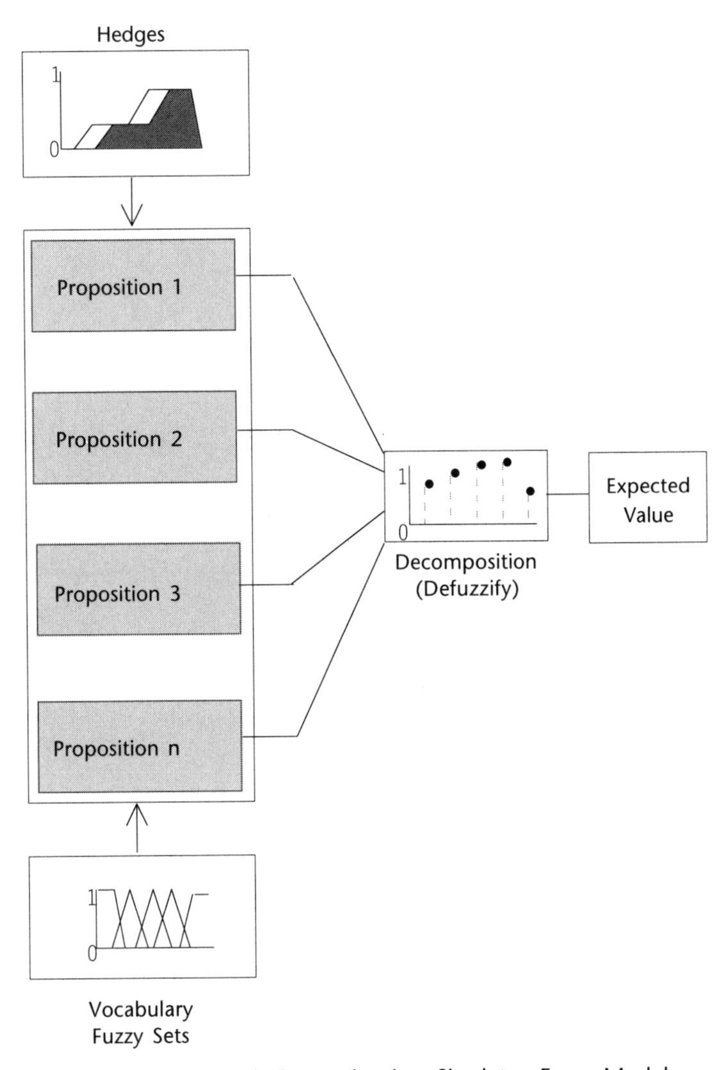

Figure 6.50 Logic Processing in a Singleton Fuzzy Model

Caution should be exercised in selecting an alternative representational geometry for a fuzzy model. The singleton model has several important restrictions on the use of operators in a fuzzy system:

An output singleton space cannot be hedged. This means that a rule proposition such as

```
if x is Y then z is very W
```

has no specific meaning in a singleton system since a singleton's topology can only be changed by increasing or decreasing its height. Hedges can, however, be used in the predicate of a rule.

An output singleton space cannot be produced from a fuzzy production. This means that a rule proposition such as

if x is Y then z is W but not W

does not contain a valid singleton consequent, even though the expression *W but not W* is a valid fuzzy expression. Singleton output representations only have a valid surface when they are evaluated for decomposition.

Unconditional fuzzy propositions cannot be used with a singleton representation.

Although we can show that the singleton representation is a special case of the more general fuzzy (or Mamdani) representation, its ability to defuzzify into a set of distinct values is dependent on the number and distribution of singletons.

Singletons are specified as support points along the output variable space. Each singleton, like a fuzzy set term, is identified by a label. This label is interpreted as a unit locus in the output. Figure 6.51 shows how the output variable *Price* can be composed on individual single points.

In singleton geometry, we make a distinction between the representational scheme for the model support vocabulary and the solution output vocabulary. We still need to handle the predicate propositions as fuzzy set–based expressions in order to produce the space-specific truth function. However, once we have a truth value for the predicate, this can be applied directly to the output singletons as a scaling factor. In a singleton geometry, the propositions or rules are expressed in a manner identical to the fuzzy set–based model

$$\text{if } u_i \text{ is } V_i \bullet x_i \text{ is } Y_i \text{ then } z_i \text{ is } S_i$$

where S_i is any of the singleton support terms. When the model is initialized, each singleton point has a value [1.0], and its future value is determined by the proportional scaling due to the predicate truth function:

$$S_i \leftarrow p_{tv} \cdot S_i$$

In effect, the topology of the output fuzzy set is determined by a set of scaled vertical singletons. Each singleton is adjusted by the truth of the proposition predicate for each proposition that specifies a particular point in the output space. Figure 6.52 shows how the *Price* output variable space could be adjusted in this fashion.

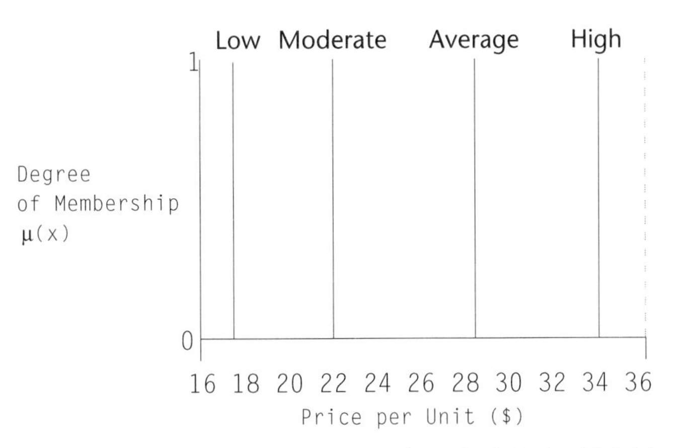

Figure 6.51 Singleton Representation of a Solution Variable

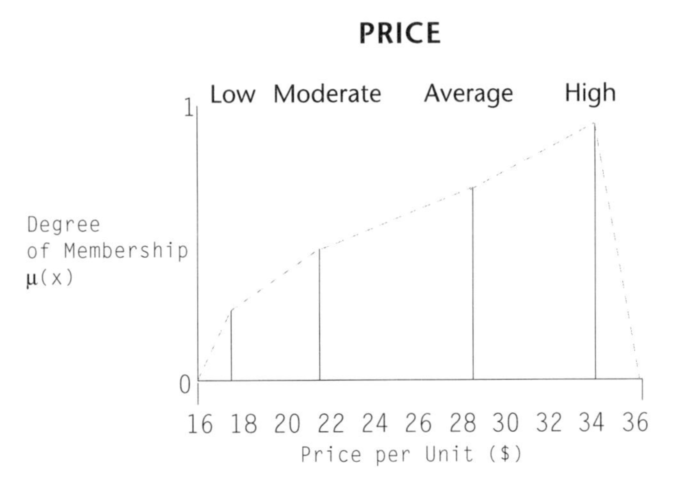

Figure 6.52 Singleton Proportioned by Proposition Truth

Since piecewise linear interpolation between the singleton points is used to construct an output fuzzy set representation, defuzzification in singleton and fuzzy set–based systems is equivalent. In fact, the centroid, or center of gravity, defuzzification method is slightly simplified. The centroid is calculated as follows:

$$\Re = \frac{\sum_{i=0}^{n} d_i \cdot \mu_A S_i}{\sum_{i=0}^{n} \mu_A S_i}$$

Equation 6.10 Calculating the Centroid in a Singleton Model

In this case, we need not perform a numerical integration of the entire fuzzy set surface. Instead, the domain value at each singleton is multiplied by the height (truth membership value) of the singleton. This general technique produces a centroid defuzzification that is generally equivalent to the centroid found by taking the area across an entire fuzzy set.

NOTES

1. The idea of fuzzy rules or propositions as *associations* is common in control literature and forms the nucleus of the fuzzy associative memory (FAM) developed by Kosko et al. The term *proposition* is used here as a more descriptive label for the kinds of extended fuzzy statements employing unconditional assertions, hedges, and n-order fuzzy regions.

2. A brief forward reference is in order here. Although the min(A,B) operation is used to correlate the consequent fuzzy set with the predicate truth, this is not the only way of performing a correlation. An alternative correlation technique is discussed in the section "Correlation Methods" on page 293.

Seven

Fuzzy Models

Nam et ipsa scientia potestas est.
Knowledge itself is power.

> Francis Bacon (1561–1626)
> Religious Meditations.
> Of Heresies

Let us consider the reason of the case.
For nothing is law that is not reason.

> Sir John Powell (1645–1713)
> Coggs v. Bernard, 2 Lord Raymond, 911

This chapter begins the coupling of theory with practice. We start with a discussion of the fuzzy modeling software and its functional organization, such as the concept of a policy and the use of the various control blocks. This is the infrastructure for building and using fuzzy models and will introduce many of the underlying features and facilities used in the case studies. The fuzzy model architecture is integrated with modeling techniques for representing control and solution variables, defining and using hedges, managing the execution of fuzzy propositions (rules), installing and using the explanatory facility, understanding alpha threshold cuts, and interpreting the outcome of models.

The Basic Fuzzy System

The fuzzy modeling software included with this handbook is more than just a set of routines for creating and managing fuzzy sets. The code represents a cohesive and integrated set ıs, which, taken together, comprise a complete fuzzy modeling environment.[1] ıg system is an integral part of how the handbook addresses the issues associ- gning, organizing, and implementing fuzzy systems. As a complete modeling and to provide the highest degree of flexibility, the modeling system is organized in two ways: as a relatively simple approach for embedding fuzzy reasoning into application programs and as a more complex modeling environment supporting the decomposition of problems into logical units (called *policies*). A complete discussion of the actual software and the control blocks is found in Chapter 10, *Using the Fuzzy Code Libraries*.

Fuzzy Models and the Code Libraries

The CD that accompanies this book includes five code libraries: *met16dos* (C/C++ code for DOS-based models), *met16win* (C/C++ code for Windows 3.1 DLL calls), *met32win* (C/C++ code for Windows 95 DLL calls), *met16vb3* (API definitions for access to the *met16win* dynamic link libraries), and *met32vb5* (API definitions for access to the *met32win* dynamic link libraries). The dynamic link library modules are designed to work through application program interface (API) calls within such rapid application development environments as Microsoft's Visual Basic (for which the API definitions have been provided).

In this chapter, we are fundamentally concerned with the basic architecture of fuzzy models and what happens inside them. We do not deal with such matters as graphical user interfaces, the ergonomics of using a model, and the display of model properties. As a result, I have coded all examples using the basic *met16dos* library version, thus providing a (generally) platform-independent version of the models. With the exception of the *printf* command and services that prompt for user values, the code can be easily ported to Windows 3.1 or Windows 95 with only minor changes (such a replacing a DOS prompt with a Windows dialog box).

THE FUZZY MODEL OVERVIEW

A fuzzy model, like traditional expert and decision support systems, is based on the input, process, output flow concept. A fuzzy model differs, however, in two important properties: what flows into and out of the process, and the fundamental transformation activity embodied in the process itself. Figure 7.1 is an overview of the fuzzy modeling environment.

The model accepts a set of inputs (x_1, x_2, ..., x_n) as its knowledge about the outside world. These inputs can be ordinary arithmetic scalars, arrays, and matrices; strings; or fuzzy regions. The fuzzy input might be fuzzy output regions from previous models, fuzzy sets that are globally available throughout the model, or fuzzy numbers (such as information that has a certain degree of uncertainty due to noise or probable sampling error).

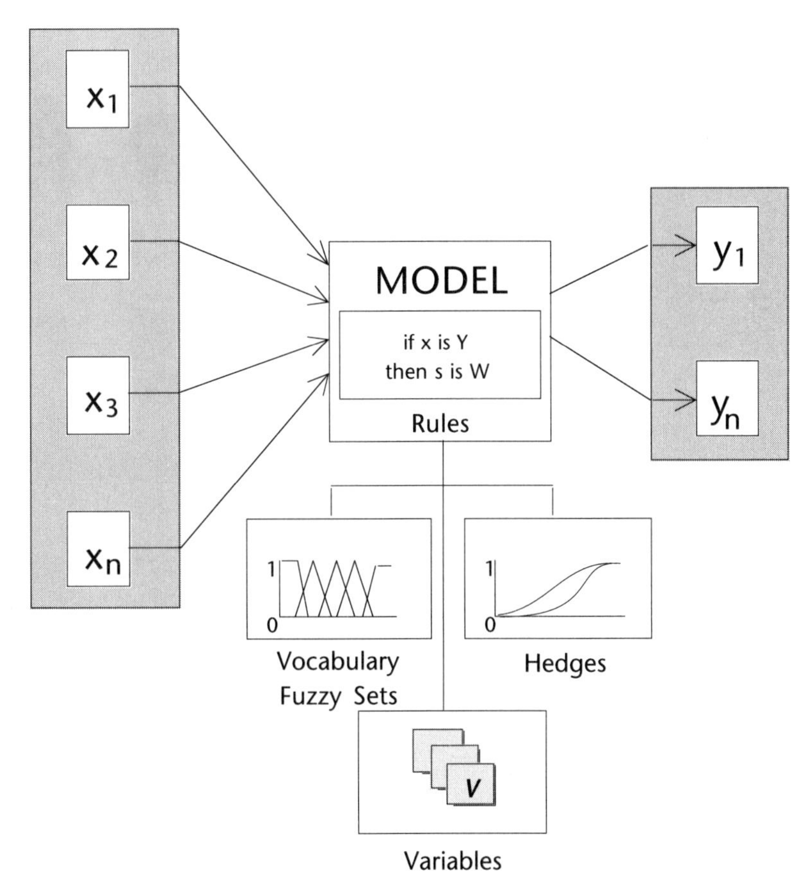

Figure 7.1 Overview of the Fuzzy Modeling Environment

The model itself executes a series of fuzzy propositions or rules[2] using control structures stored in the application code—the vocabulary of fuzzy sets, hedges, and formally defined variables. (As we will shortly see, solution variables must be formally encoded in a special structure so the modeling software can access them in a consistent way.) The model executes these propositions in a way that simulates parallel processing (all the rules contribute to the final value of the solution variable). This means that the values of the solution variables are not available for use until all the rules have been executed. These propositions are logically interpreted as

$$if\ L_{p1} \bullet L_{p2} \bullet \ldots L_{pn}\ then\ L_c$$

where L_{pn} is a predicate linguistic variable and L_c is the consequent linguistic variable specifying the solution variable (y_n). When all the rules have been executed, a temporary fuzzy region (Y_n) exists for each solution variable. When we want a discrete value for any solution variable, we use the process of defuzzification to find its expected value. These temporary fuzzy sets, unless explicitly saved by the model builder, are deleted when the model is closed.

THE MODEL CODE VIEW

At a somewhat simple level, a fuzzy model consists of two principal components: the application code containing the actual fuzzy reasoning logic and a group of control blocks that manage the actual modeling environment. Figure 7.2 schematically illustrates how the model is organized.

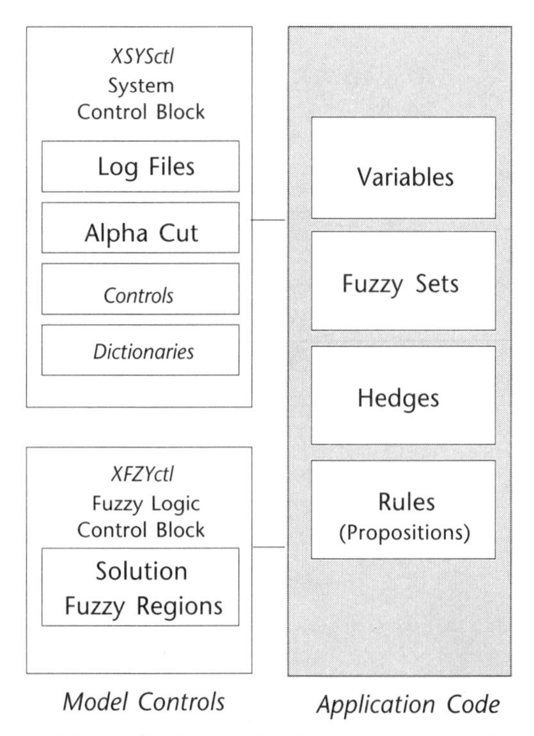

Figure 7.2 The Code View of the Simple Fuzzy Model Environment

The control block structures require some explanation. The *XSYSctl* control block is used throughout the modeling environment and must be attached to the application modeling code in all cases. (The control block is initialized by the *MdlConnecttoFMS* function.) This external control block holds information about the location of the model, the active log and trace files, the value of global modeling statistics such as the alpha-cut threshold, and, for more complex models, a series of dictionary hash tables containing various model components.

The *XFZYctl* control block plays a fundamental role in approximate reasoning. Information about each of the solution variables in the model is stored in this control block. When a formal variable (represented by a variable definition block, or *VDB*) is added to the *XFZYctl* block, the model automatically allocates a working fuzzy set and initializes the associated methods for correlation, implication, and defuzzification. The conditional and unconditional fuzzy proposition handlers, as well as the defuzzification functions, use the working fuzzy set stored in this control block.

Code Representation of Fuzzy Variables

Each rule in a fuzzy model specifies a relationship between a predicate fuzzy region and a fuzzy region associated with one of the solution variables. Each time a rule is evaluated and has a nonzero truth, the fuzzy space associated with the solution variable is updated. In the modeling software, this relationship is maintained through a formal declaration process using the *XFZYctl* control block (see Figure 7.3).

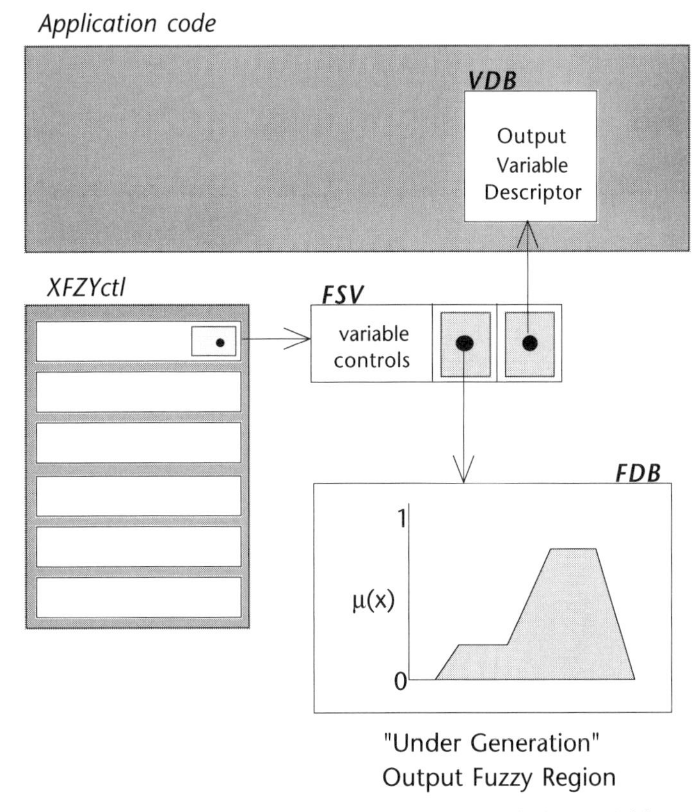

Figure 7.3 The Control Mechanism for Fuzzy Solution Variables

The *XFZYctl* control block has room for thirty-two solution variables in one model. Each slot in the control block is a pointer to a fuzzy solution variable (*FSV*) structure. This structure contains pointers to the variable definition block (*VDB*) for the variable and to the fuzzy set descriptor block (*FDB*) for the undergeneration fuzzy region associated with the solution variable. When a fuzzy model is initialized (after the global *XSYSctl* block has been properly connected), a *VDB* must be allocated for each solution variable and inserted into the *XFZYctl* block through the *FzyAddFZYctl* function. Listing 7.1 shows the basic application code declarations for the important model variables.

```
FDB    *HiFDBptr,
       *LgFDBptr,
       *aeFDBptr,
       *BOAFDBptr;
VDB    *VDBptr;
FSV    *FSVptr;
```

Listing 7.1 Defining Program Variables for a Basic Fuzzy Model

```
//--Initialize the fuzzy modeling environment. We set up the
//--model external control space, create the output fuzzy
variable,
//--and add it to the fuzzy control region.
  Domain[0]=0;
  Domain[1]=300;
  VDBptr=VarCreateScalar("BOAmt",REAL,Domain,"0",statusPtr);
  FzyInitFZYctl(statusPtr);
  if(!(FzyAddFZYctl(VDBptr,&BOAFDBptr,&FSVptr,statusPtr)))
    {
     *statusPtr=1;
     MtsSendError(12,PgmId,"BOAmt");
     return(NoBOAmt);
    }
  FzyDisplayFSV(FSVptr,"The FSV pointer");
  thisCorrMethod  =FSVptr->FzySVimplMethod;
  thisDefuzzMethod=FSVptr->FzySVdefuzzMethod;
//
```

Listing 7.2 Allocating and Inserting a Solution Variable into the Fuzzy Model

In Listing 7.2 we see how a formal variable for *BOAmt* (backorder amount) is created and inserted into the *XFZYctl* control block. *VarCreateScalar* created the variable of type REAL (single precision floating point) with a domain of values between 0 and 300. *VarCreateScalar* also initializes the default fuzzy logic controls for correlation (composite minimum), implication (min/max), and defuzzification (composite moments). *FzyInitFZYctl* initializes the fuzzy logic working area. This function is called once at the beginning of each model. *FzyAddFZYctl* inserts the *VDB* for *BOAmt* into the fuzzy work area. A pointer to the created undergeneration fuzzy set is returned (*BOAFDBptr*) as well as a pointer to the FSV block for this variable (*FSVptr*). The pointer to the *FSV* is an important model parameter—it is used by the proposition handler to actually update the actions of a rule.

Note that data variables in the modeling system are not stored in the control block; they are handled, as illustrated in Figure 7.4, through your application code when each fuzzy rule is executed.

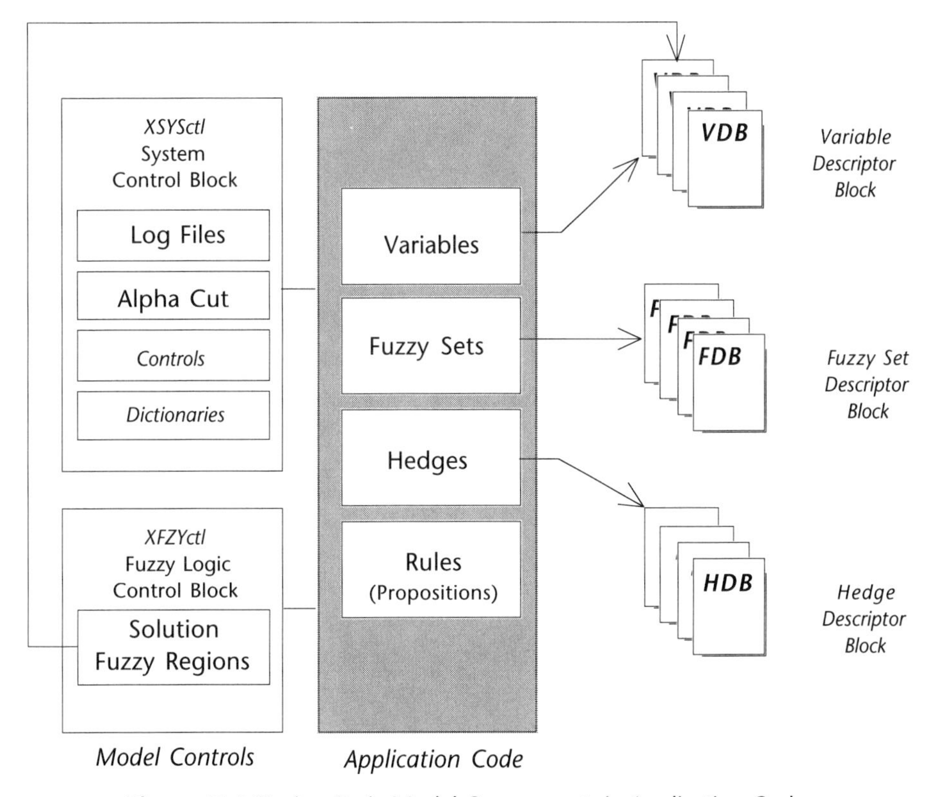

Figure 7.4 Storing Basic Model Components in Application Code

In this kind of simple fuzzy model, conventional application code variables are used to hold and manage the model components. This means, as illustrated in Listing 7.2 and Listing 7.3, that fuzzy sets, variables, and hedges are allocated and deleted through the standard C++ program language constructs. (These examples are from the `INVPOL1.CPP` program.)

```
//--Hi Orders Fuzzy set
   HiFDBptr=FzyCreateSet(
     "HIGH.ORDERS",INCREASE,Domain,Parms,0,statusPtr);
//--Large BOAmt fuzzy set
   LgFDBptr=FzyCreateSet(
     "LARGE.BOAmt", GROWTH,Domain,Parms,0,statusPtr);
//--Around the ERP
  Parms[0]=ERP;Parms[1]=ERP*.30;
  aeFDBptr=FzyCreateSet(
     "AROUND.ERP", PI     ,Domain,Parms,2,statusPtr);
```

Listing 7.3 Creating Application Fuzzy Sets

The program variables *HiFDBptr* and *LgFDBptr* are pointers of type *FDB*. As with variable and hedge pointer variables, you must issue the C++ delete to release this storage when the model has completed.

Incorporating Hedges in the Fuzzy Model

A hedge changes the shape of a fuzzy set (see "Hedges and Fuzzy Surface Transformers" on page 217). In the basic fuzzy modeling system, each hedge is described by a hedge descriptor block (*HDB*) that is allocated and initialized according to the hedge type. The result of applying a hedge to a fuzzy set is another fuzzy set. Figure 7.5 illustrates the fundamental hedging process.

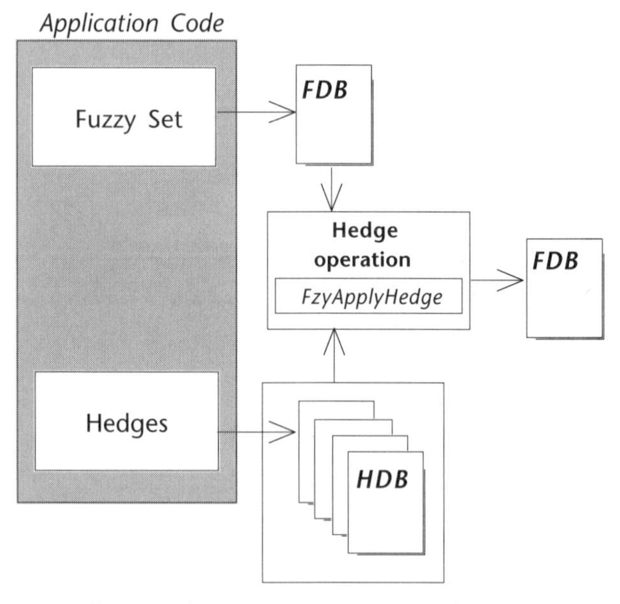

Figure 7.5 Applying Hedges to a Fuzzy Set to Produce a New Fuzzy Set

Listing 7.4 shows the process (from the new product pricing model) of allocating and initializing a hedge structure. This hedge is then used to modify the fuzzy set *High* (pointed to by the *HiFDBptr* program variable).

```
//--Rule 4. If the Competition.price is not very high,
//            then our price should be near the competition.price
//
  WrkFDBptr=new FDB;
  FzyInitSet(WrkFDBptr);
  HDBptr=new HDB;
  FzyInitHedge(HDBptr);
  strcpy(HDBptr->HDBid,"VERY");
  HDBptr->HDBmode =VERY;
//--Apply the hedge VERY to the High fuzzy set
  FzyApplyHedge(HiFDBptr,HDBptr,WrkFDBptr,statusPtr);
  FzyDrawSet(WrkFDBptr,SYSMODFILE,statusPtr);
```

Listing 7.4 Allocating a Hedge, and Hedging the Fuzzy Set `High`

If the hedge uses one of the built-in transformation types, then *HDBmode* is set to the hedge symbolic identifier. No other information is required, although storing a literal string with the hedge's name will prove helpful when the hedged fuzzy set is displayed. (The hedge operation changes the description of the new fuzzy set to include the name of the hedge.)

Representing and Executing Rules in Code

A rule in the fuzzy model is not maintained as a structure but consists of application code that performs many tasks: it applies any hedges to the vocabulary fuzzy sets, determines the membership of a scalar in a fuzzy set, performs any fuzzy *and* or *or* operations, and uses the predicate truth in a call to either a fuzzy conditional or unconditional proposition handler. Figure 7.6 shows the flow of processing control in rule execution.

Unconditional rules simply match the consequent fuzzy set with the current undergeneration fuzzy region. Listing 7.5 shows how this process is done. As Listing 7.6 shows, the evaluation process for conditional fuzzy propositions involves finding the degree of membership for a predicate scalar in the control fuzzy set. Once we have this truth, the proposition manager is invoked to actually update the contents of *XFZYctl* for the particular solution variable. This variable's undergeneration fuzzy set is referenced by the *FSVptr* variable, which is automatically set (see Code Representation of Fuzzy Variables on page 333) when the solution variable's VDB is added to the work area.

```
// Rule 1--BackorderAmt must be near ERP
  fprintf(mdlout,"%s\n",Rules[Rulecnt]);
  FzyUnCondProposition(aeFDBptr,FSVptr);
```

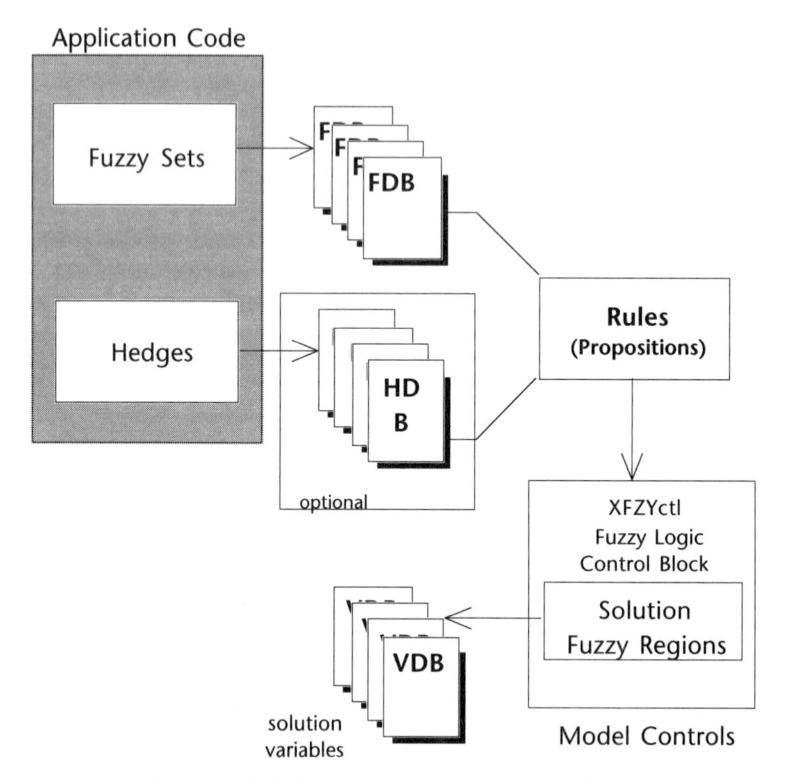

Figure 7.6 The Rule Execution Process Flow

```
   FzyDrawSet(BOAFDBptr,SYSMODFILE,statusPtr);
   Rulecnt++;
```

Listing 7.5 Evaluating an Unconditonal Fuzzy Rule in the Inventory Model

For conditional rules, the process is only slightly more involved. A conditional fuzzy rule needs the truth of the antecedent so that the correct correlation can be applied. Listing 7.6 shows the two steps in evaluating a conditional rule.

```
// Rule 2--if orders are high then backorderAmt must be large
   fprintf(mdlout,"%s\n",Rules[Rulecnt]);
   PremiseTruth=FzyGetMembership(
      HiFDBptr,Orders,&Idxpos,statusPtr);
   fprintf(mdlout,"%s%10.2f\n","PremiseTruth= ",PremiseTruth);
   FzyCondProposition(
        LgFDBptr,FSVptr,thisCorrMethod,PremiseTruth,statusPtr);
```

```
    FzyDrawSet(BOAFDBptr,SYSMODFILE,statusPtr);
    Rulecnt++;
  //
```

Listing 7.6 Evaluating a Conditional Fuzzy Rule in the Inventory Model

The *FzyGetMembership* function finds the degree of membership for orders in the *High.Orders* fuzzy set (whose address is stored in *HiFDBptr*). The result of this is the resolution of the predicate proposition, *orders are High.* Using this truth membership, *FzyUncondProposition* is invoked to complete the fuzzy rule and update the working fuzzy set for *BOAmt*.

Physical rules are not the same as logical (or "visual") rules. Each model maintains an array of text strings containing the logical rules in standard if–then syntax. The corresponding logical rule is displayed on the model activity log file before the conditional or unconditional proposition function is applied. This practice improves model maintenance and aids in model verification.

Setting Alpha-Cut Thresholds

Alpha-cut thresholds control the firing of rules and the outcome of operations on fuzzy sets. An alpha cut essentially makes all truth membership functions below a specific value equal to zero.When an alpha-cut threshold is applied to the truth of a rule's predicate, it determines whether or not the truth is sufficient to fire the rule. When an alpha-cut threshold is applied to a fuzzy set, it establishes a line through the truth membership function equal to the alpha cut. Truth membership values below this line are considered equal to zero. For detailed conceptual information, see Chapter 3, *Fuzzy Sets,* "Fuzzy Alpha-Cut Thresholds," page 94.

There are several alpha-cut thresholds maintained in the fuzzy modeling system. Figure 7.7 schematically shows their locations. Each of the alpha-cut levels controls some function in the modeling protocol (although not all of the alpha-cut thresholds are used automatically by the code).

The alpha threshold stored in the *XSYSctl* block (*XSYSalfacut*) is the primary threshold value in the fuzzy system and is initialized to zero [0.0] when the *MdlConnecttoFMS* function is issued. This value is automatically used by the *FzyCondProposition* and *FzyUnCondProposition* to decide whether or not a predicate is true. When fuzzy sets (*FDBs*) and variables (*VDBs*) are initialized (through *FzyInitFDB* and *VarInitVDB*, respectively), their alpha-cut thresholds are inherited from the current value of *XSYSalfacut.*

The alpha cut stored with the fuzzy set descriptor block (*FDB*) is used to control the truth membership array resulting from the intersection, union, and negation of sets. This value, inherited from the threshold associated with the variable (*VDBalfacut*), is used to adjust the solution fuzzy set prior to defuzzification. When you use the *VarAttachFDB* function to associate fuzzy sets with a variable, the current *VDB* alpha-cut threshold replaces the *FDB* alpha-cut threshold.

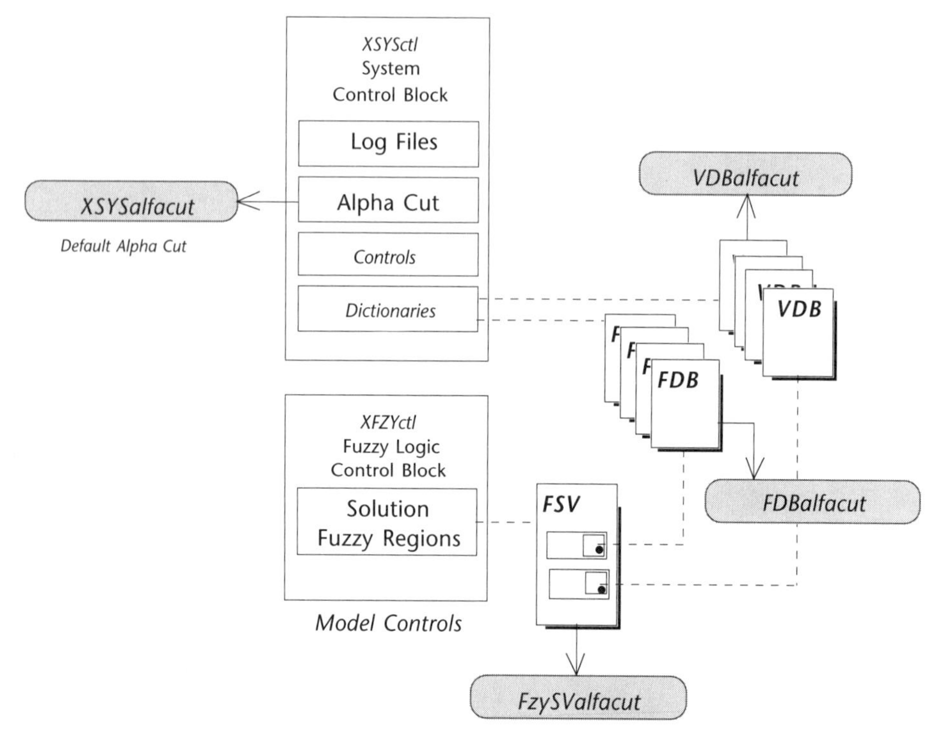

Figure 7.7 Location of Alpha-Cut Thresholds in the Fuzzy Modeling System

The alpha cut stored in the fuzzy solution variable (*FSV*) structure is inherited from the variable's *VDB* when it is added to the fuzzy work area (using *FzyAddXFZYctl*). This value is assigned to the output fuzzy set (as *FDBalfacut*) when the defuzzification process is called.

Including a Model Explanatory Facility

An explanatory facility explains the reasoning that underlies a particular model recommendation by answering important questions such as:

Why did the model make that recommendation?
How much confidence can I have in that recommendation?
Was any significant evidence overlooked?
Is this model working properly?

Fuzzy models are extremely difficult to explain since the results represent the combined action of many rules that are, in effect, executed in parallel. Because of the parallel nature of fuzzy systems, the explanatory process cannot simply roll back the rule execution tree. Rather, a system that attempts to explain its reasoning must reflect the larger semantics of the model itself. The facilities included in the fuzzy system code provide the platform for a comprehensive explanatory system. Figure 7.8 illustrates how the explanatory services are integrated with the fuzzy model.

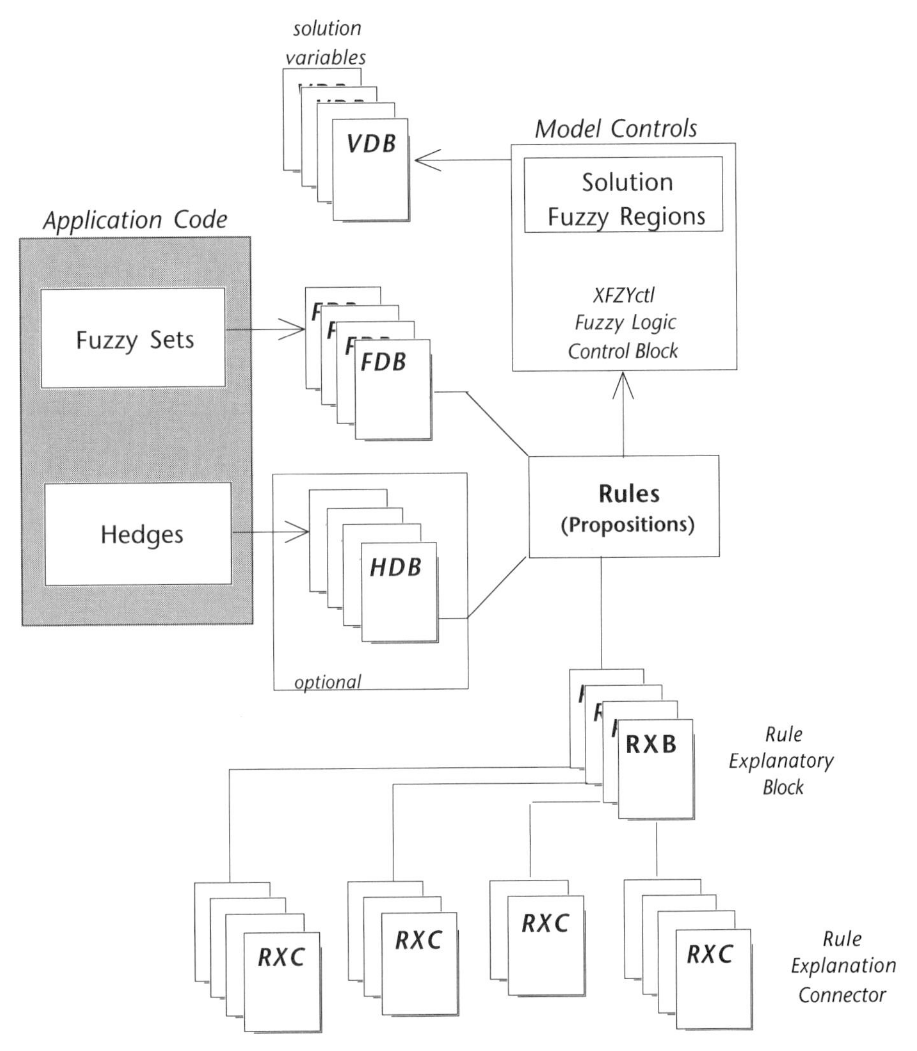

Figure 7.8 The Fuzzy Model Explanatory Services Schematic

Incorporating an explanatory facility involves a good deal of complicated code, in fact, this is the most complicated code logic we have encountered yet in this book. The facility is broken down into two cooperating segments. This first is a function (*ConcludeRule*) that does some rule housekeeping and allocates the root rule explanatory block (*RXB*) for the current rule. This new *RXB* is linked into a chain of *RXB* nodes, one for each rule. The head of this chain is linked into the *XSYSexplainptr* field of the *XSYSctl* control block. The second segment is a set of explanatory function calls that occur immediately after the proposition is evaluated. These calls—one for each predicate and consequent proposition—actually build the explanation as a linked list of rule explanation connectors (*RXC*). This chain of proposition explanations is linked into the current *RXB* node. Figure 7.9 shows how these chains are created and maintained.

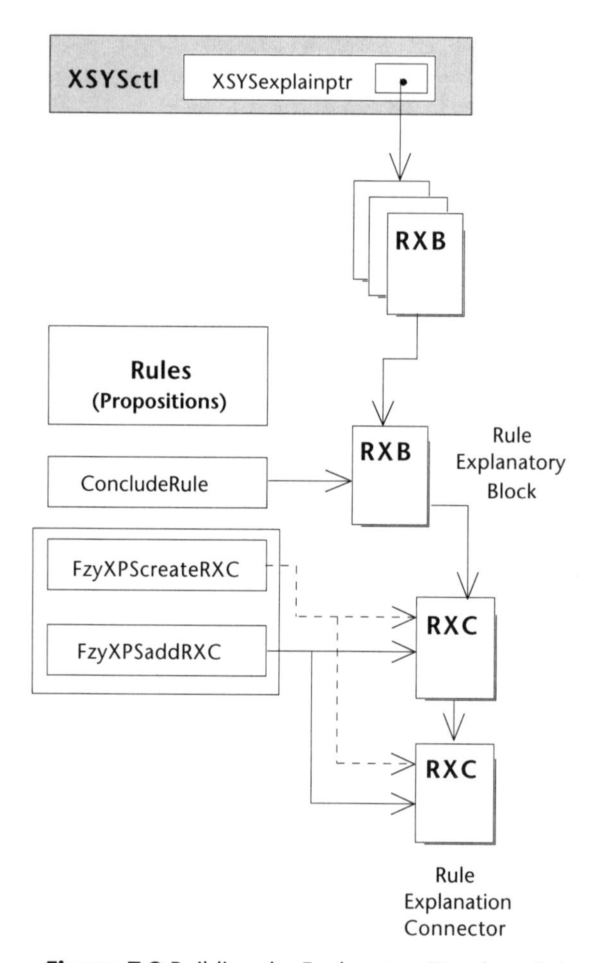

Figure 7.9 Building the Explanatory Tree for a Rule

Before the explanatory services are used, the structure information must be included and the head and tail pointers to the *RXB* chain initialized. Listing 7.7 shows how this process should be done. Note that *RXBhead* and *RXBlastptr* must be declared as static since they are used by, but not passed to, the *ConcludeRule* function. (See the last few lines of the procedure in Listing 7.9.)

```
#include "RXB.hpp"
•
•
RXBhead    =NULL;
RXBplastptr=NULL;
```

Listing 7.7 The `include` File and Initializations for the Explanatory Facility

Listing 7.8 shows a rule from a project assessment model with the explanatory functions. The explanatory system requires the addition of five extra statements to this model for this rule. These are shown following the *FzyCondProposition* function.

```
//R1--if project.length is long then ProjRisk is High
  PremiseTruth=FzyGetMembership(
   pjLenFDBptr[1],ProjectDuration,&Idxpos,statusPtr);
  FzyCondProposition(
    riskFDBptr[1],FSVptr,thisCorrMethod,PremiseTruth,statusPtr);
  ConcludeRule(riskFDBptr,statusPtr);
  RXCptr=FzyXPScreateRXC(
    "the Project Length","Long","",PremiseTruth);
  FzyXPSaddRXC(RXBptr,RXCptr,RXCpredtype,statusPtr);
  RXCptr=FzyXPScreateRXC("ProjRisk1","High","",PremiseTruth);
  FzyXPSaddRXC(RXBptr,RXCptr,RXCconstype,statusPtr);
  Rulecnt++;
```

Listing 7.8 A Rule with Its Associated Explanatory Functions

The explanatory facility relies on the association of a textual representation of the rule contents with the truth memberships found during the evaluation of each individual proposition in the overall rule. See Listing 7.9.

```
static void ConcludeRule(FDB *RiskFDBptr,int *CondPropstatus)
  {
   char   wrkBuff[128];
   int    status;

   ruletruth[Rulecnt]=PremiseTruth;
   sprintf(wrkBuff,"%s%6.4f","Premise Truth= ",PremiseTruth);
   MtsWritetoLog(SYSMODFILE,wrkBuff,&status);
   if(*CondPropstatus!=RULEBELOWALFA)
     {
```

```
        rulefired[Rulecnt]=TRUE;
        FzyDrawSet(RiskFDBptr,SYSMODFILE,&status);
        ++Firedcnt;
      }
    if((RXBptr=FzyXPScreateRXB(Rulecnt+1,0,PremiseTruth))==NULL)
        {
         sprintf(wrkBuff,"%s%4d","RXB for Rule ",Rulecnt);
         MtsSendError(2,PgmId,wrkBuff);
         return;
        }
    if(RXBhead==NULL) RXBhead=RXBptr;
      else            RXBplastptr->RXBnext=RXBptr;
    RXBplastptr=RXBptr;
  }
```

Listing 7.9 The *ConcludeRule* Function that Allocates an Explanatory Node

The code in Listing 7.9 is only one example (albeit from an actual risk assessment model for a financial services company). A brief explanation of its workings may be necessary. The first part of the procedure uses a static *Rulecnt* variable to store the current rule's predicate truth in an array for later use. The truth for this rule is then written to the current model log. (The rule identifier and other statistics have previously been sent to the model log, so this premise truth is associated with the current rule.) The status pointer variable (*statusPtr*) from *FzyCondProposition* is passed to and used by this procedure. If the truth of the rule's predicate is above the alpha-cut threshold (that is, the rule was executed), we indicate that the rule was fired, draw the current image of the working fuzzy region for the solution variable, and increment the count of actually fired rules.

The last part of *ConcludeRule* handles the explanatory services by allocating a new top-level rule node (*RXB*). In this version of the explanatory system, we assume that the model rules are numbered numerically: 1, 2, 3, etc. If the head of the explanatory rule nodes is *NULL* (no other nodes exist), we link this node into the head; otherwise, we link it into the *RXBnext* pointer of the node at the tail of the current chain. The pointer to this tail (*RXBplastptr*) is then set to the new node.

When the model execution is complete, you can display the explanations. As Listing 7.10 shows, this is done in two parts.

```
fprintf(mdlout,"\f\n\n%s\n","EXPLANATIONS:");
FzyXPSshowrules(RXBhead);
FzyXPSconclusion(
    RXBhead->RXBconsptr,
    Risklvl[dfuzzIX],RiskFDBptr->FDBdomain,RiskCIX[dfuzzIX]);
```

Listing 7.10 Displaying the Model Explanations

The first function, *FzyXPSshowrules*, displays all the rules that contributed to the solution, ranked by their strength of contribution. This information is currently placed on the model log. The second function, *FzyXPSconclusion*, displays the value of the solution variable, its degree of certainty based on the overall rule support, and its place (*high*, *medium*, *low*, etc.) in the domain of the underlying solution variable.

The Advanced Fuzzy Modeling Environment

The advanced fuzzy modeling facilities change the structural organization and memory utilization properties of a model by introducing the concept of a *policy*. In the basic fuzzy model architecture, the fuzzy sets, variables, and hedges are stored as ordinary variables in the application code. This structure has the advantage of compactness and relative simplicity. Such simple models, however, make the logical decomposition of a problem very difficult, and they significantly limit the amount of knowledge sharing between the various parts of a complex model. The advanced modeling facilities, through the policy concept, solve many of these problems. As Figure 7.10 shows, policies are maintained in a special dictionary (*XSYSpolicies*) in the *XSYSctl* control block.

The policy dictionary contains all the policy descriptor blocks (*PDB*) currently active for the running model. Pointers in the *XSYSctl* (*XSYScurrMDBptr* and *XSYScurrPDBptr*) act as cursors, pointing to the current model descriptor block (*MDB*) and to the policy that's currently in focus. You may notice that the policy dictionary is one of several dictionaries maintained in the *XSYSctl* area. The remaining dictionaries are used by both the underlying repository system and the integrated blackboard communication system. We do not discuss these facilities in this book.

The Policy Concept

A policy is a logical segment of a model. This segmentation can be functional, operational, or organizational (in the sense that a policy can be used to share model elements across program code boundaries or among other policy blocks). The policy concept removes from your application code the native representation of variables, fuzzy sets, and hedges. Fuzzy rules remain in the application code as functional evaluations of conditional and unconditional propositions.)[3] Figure 7.11 shows the schematic organization of a policy.

Fuzzy sets, hedges, and variables are stored, located, and removed through a set of standard dictionary management functions. This process allows a consistent method of referencing model information. Since policies are accessed through the global *XSYSctl* control block, your model can be decomposed into separate C/C++ programs with the ability to easily move around the modeling environment regardless of program code boundaries.

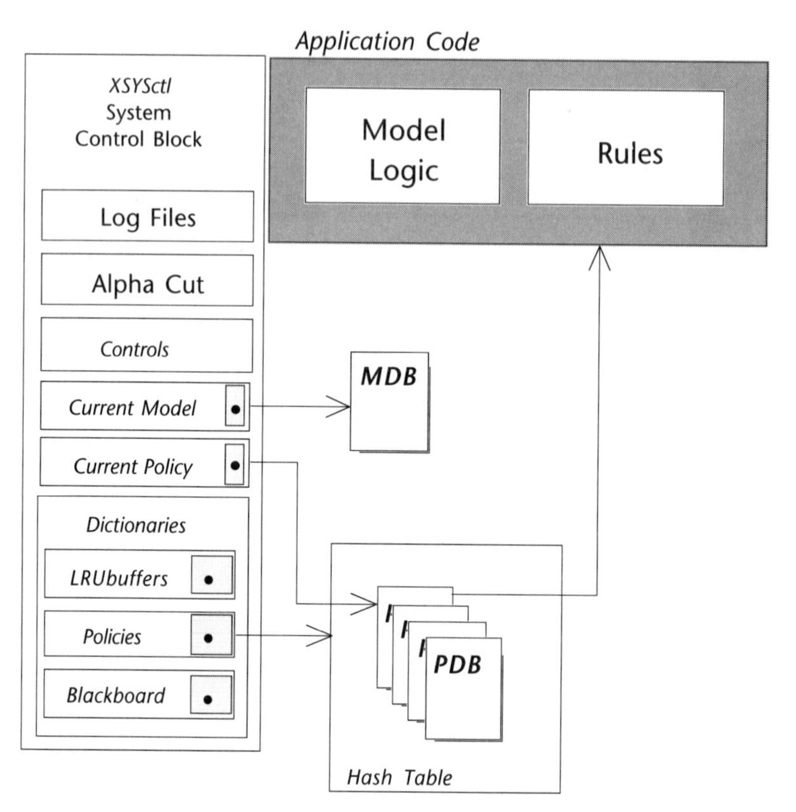

Figure 7.10 Overview of the Advanced Fuzzy Modeling Environment

UNDERSTANDING HASH TABLES AND DICTIONARIES

The dictionaries in the *XSYSctl* control block and the *PDB* use a hashing technique to store and retrieve members. Identifier hashing evenly distributes identifier keys across a dictionary space and provides one of the fastest known methods of direct addressing. The idea behind a hashing algorithm is simple:

$$h(k, S^*) \rightarrow S_k$$

That is, the hashing function (h)—using the key (k) and the total address space (S^*)—produces the address slot (S) in the dictionary. While there are many approaches to hashing, the easiest and, in fact, one of the better approaches simply takes the residue of the modulus for the key value and the number of available dictionary slots. In our discussion, we treat keys as character strings (since this is the common data type for the storage of knowledge-based components). Listing 7.11 shows the standard hashing function used in the fuzzy model library.

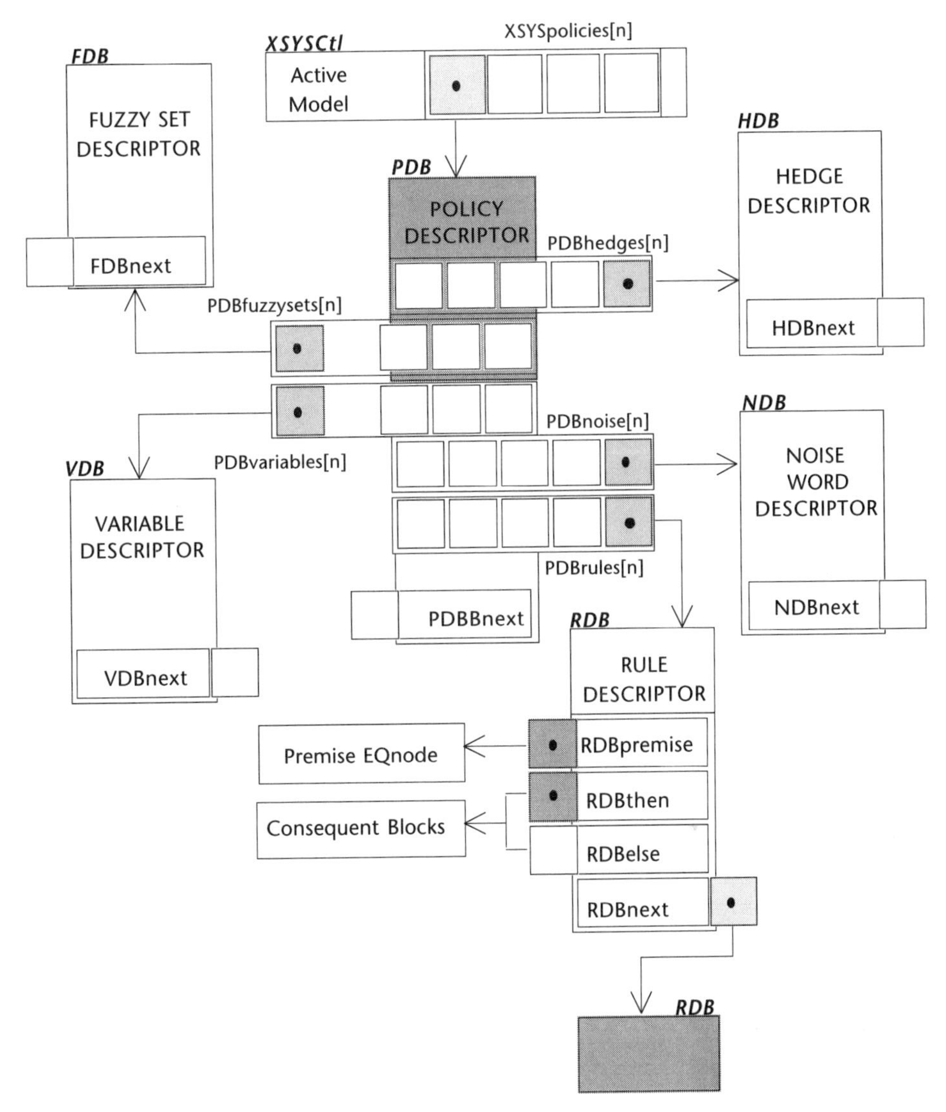

Figure 7.11 Schematic Organization of a Fuzzy Model Policy

The *MtsHashString* function is an implementation of this hashing algorithm and forms the basis for all dictionary addressing in the fuzzy modeling system. The dictionary sizes (*MaxSlots*) are stored in the *mtypes.hpp* header file.

```
#include <stddef.h>
long MtsHashString(char *StrItem,long MaxSlots)
  {
   if(StrItem==NULL) return((long)-1);
   if(MaxSlots<1)    return((long)-2);
   MtsNormalizeKey(StrItem);
   unsigned long Hval=0;                   //initialize hash
   char *Byteptr=StrItem;                 //get address of identifier
   for(;*Byteptr;++Byteptr)
    {
      Hval<<=1;                            //shift one bit to left
      Hval+=*Byteptr;                      //add in the byte value
    }
   return((long)(Hval%MaxSlots));          //take the modulo
  }
```

Listing 7.11 The Standard Dictionary Hashing Function

This *MtsHashString* takes the character string containing the key and the total number of slots (which, for reasons we will not discuss, should be a prime number) and returns a slot number (an unsigned long integer). We first check to make sure that valid key and slot arguments have been passed. We then call a routine, *MtsNormalizeKey*, that removes duplicate blanks, converts all key characters to uppercase (making the hash process case-insensitive), and left-justifies the key. The algorithm treats each byte as an integer, shifting its bits one place to the left before accumulating the current value, so that closely related character groups are diffused in the address space. Finally, the remainder from the division of the hash value and the slot space is used as the returned slot address.

In general, a good hashing algorithm distributes keys evenly across the address space, which minimizes the chance that two identifiers have the same hash address. In practice, of course, hashing almost always results in these address conflicts, called *collisions*. There are two ways of resolving a collision (assuming we can't simply throw away the colliding key): either place the key in another, vacant dictionary slot or store the key in the slot along with its collision keys.

In the first solution, called *alternate probing*, if the address is already occupied we simply look at slots `S+1, S+2, ..., S+n` until we find a vacant slot. Retrieval follows the same process—if the hashed address does not contain the required key, we probe the dictionary in the same manner until the key is found (or until we hit a vacant slot, in which case we know the key is not in the dictionary). Note that this probing can be implemented either linearly or through an alternative hashing algorithm.

In the second solution, collisions are simply stored at the same address. This is called *hash-chaining* or *bucket-chaining* (from the convention of calling a multikey slot space a "bucket"). The collision is chained linearly into the existing entries. Each hash entry node (such as the *FDB* for fuzzy sets, the *VDB* for variables, or the *PDB* for policies) must have

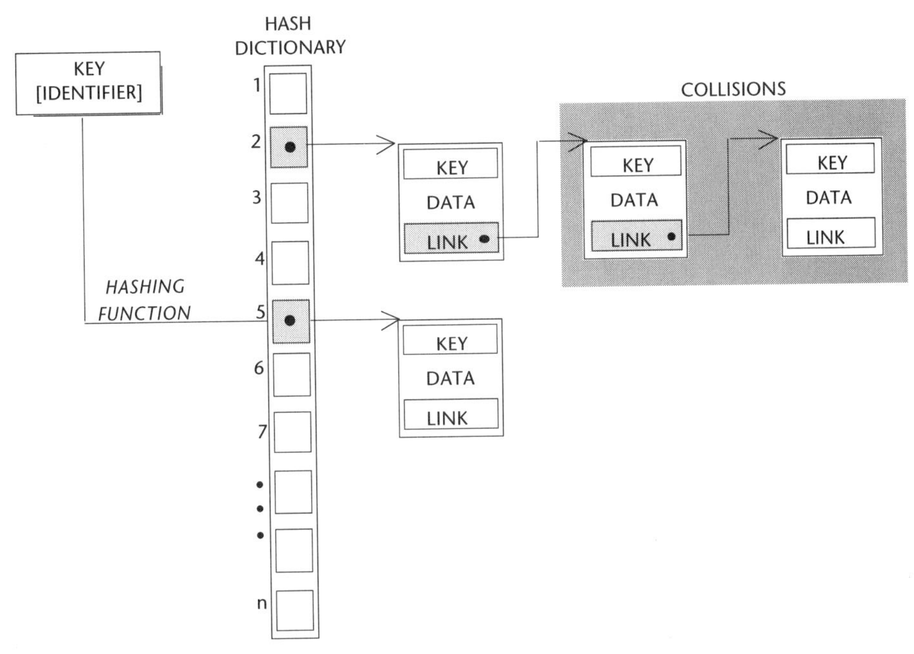

Figure 7.12 A Hash Dictionary Using Collision Chaining

a pointer to the next node (such as the **PDBnext* in the *PDB* structure). Figure 7.12 illustrates how the hash dictionaries are organized through collision chaining.

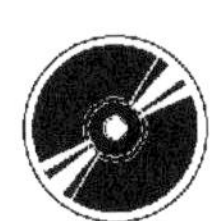

The *MdlLinkFDB* function inserts a fuzzy set descriptor into the specified policy's dictionary. *MdlFindFDB* searches the policy's fuzzy set dictionary for a fuzzy set by the specified name. *MdlRemoveFDB* deletes a fuzzy set from the dictionary. These functions illustrate how the advanced modeling facility uses the dictionaries associated with a policy to capture, maintain, and use model components.

```
void MdlLinkFDB(FDB *FDBptr,PDB *PDBptr,int *statusPtr)
 {
  FDB    *FDBwrkptr;
  char   *PgmId="mtmdlkf";
  char    Fsetid[IDENLEN+1];
  int     i;
  long    slot;

  *statusPtr=0;
```

```
//
//--Copy the variable identifier into a working area. We use
//--this with the hashing function to generate an address in
//--the PDB variable dictionary.
  strcpy(Fsetid,FDBptr->FDBid);
  slot=MtsHashString(Fsetid,FDBmax);
  if(slot<0||slot>FDBmax)
    {
     *statusPtr=1;
     MtsSendError(97,PgmId,Fsetid);
     return;
    }
//
//--Look at the FDB dictionary slot. If it's empty, then
//--simply link in the variable and return. Otherwise, find
//--the last node and link this one into the end of the chain.
  FDBwrkptr=PDBptr->PDBfuzzysets[slot];
  if(FDBwrkptr==NULL)
    {
     PDBptr->PDBfuzzysets[slot]=FDBptr;
     return;
    }
  for(;;)
    {
     if(strcmp(FDBwrkptr->FDBid,FDBptr->FDBid)==0)
       {
        *statusPtr=3;
        MtsSendError(65,PgmId,FDBptr->FDBid);
        return;
       }
     if(FDBwrkptr->FDBnext==NULL)    // Is there a next FDB node?
       {                             // No...
        FDBwrkptr->FDBnext=FDBptr;   // Link this FDB into chain
        return;                      // Return to caller
       }
     FDBwrkptr=FDBwrkptr->FDBnext; // Move to next FDB node
    }
}
```

Listing 7.12 Inserting a Fuzzy Set into a Policy's Fuzzy Set Dictionary

The *MdlLinkFDB* function takes the fuzzy set whose address is in *FDBptr* and inserts it into the fuzzy set dictionary for the policy whose address is in *PDBptr*. This shows the simplicity of working with hash tables. If the dictionary cell (indicated by the slot index) is *NULL*, then we just link in the *FDB*; otherwise, we traverse the chain of *FDB* nodes until we find the end of the chain. This *FDB's FDBnext* field is set to the address of the incoming fuzzy set. Along the way we also check for a duplicate fuzzy set name.

```
FDB *MdlFindFDB(char *Fzyid,PDB *PDBptr,int *statusPtr)
 {
  FDB    *FDBptr,*FDBnull=NULL;
  char   *PgmId="mtmdfdf";
  int     i;
  long    slot;

  *statusPtr=0;
  if(PDBptr==NULL)
    {
     *statusPtr=1;
     MtsSendError(102,PgmId,Fzyid);
     return(FDBnull);
    }
//
//--Now hash the variable name to get its slot location in
//--the PDB's variable dictionary.
  slot=MtsHashString(Fzyid,FDBmax);
  if(slot<0||slot>FDBmax)
    {
     *statusPtr=3;
     MtsSendError(97,PgmId,Fzyid);
     return(FDBnull);
    }
//
//--Look at the FDB dictionary slot. If it's empty, then
//--return a null pointer saying that the fuzzy set doesn't
//--exist. Otherwise look through the remainder of the chain
//--for the variable name until we come to the chain end.
  *statusPtr=5;
  FDBptr=PDBptr->PDBfuzzysets[slot];
  if(FDBptr==NULL) return(FDBnull);
  for(;;)
    {
     if(strcmp(FDBptr->FDBid,Fzyid)==0)
       {
        *statusPtr=0;
        return(FDBptr);
       }
     if(FDBptr->FDBnext==NULL) return(FDBnull);
     FDBptr=FDBptr->FDBnext;
    }
 }
```

Listing 7.13 Finding a Fuzzy Set in a Policy's Fuzzy Set Dictionary

Locating an entry in a hash dictionary is easy and fast. In this example of finding a specified fuzzy set in a policy's fuzzy set dictionary (Listing 7.13), the proper index in the dictionary is generated by the hash function. If the cell is empty (*NULL*), we return *NULL* indicating that the fuzzy set does not exist. Otherwise, we traverse the chain looking for the fuzzy set. If found, its address is returned. If we reach the end of the chain, the fuzzy set does not exist, so we return a *NULL* address.

```
void MdlRemoveFDB(char *Fzyid,PDB *PDBptr,int *statusPtr)
 {
  FDB     *FDBptr,*FDBprevptr=NULL;
  char    *PgmId="mtmdrmf";
  int      i;
  long     slot;

  *statusPtr=0;
//
//--Now hash the variable name to get its slot location in
//--the PDB's variable dictionary.
  slot=MtsHashString(Fzyid,FDBmax);
  if(slot<0||slot>FDBmax)
    {
     *statusPtr=1;
     MtsSendError(97,PgmId,Fzyid);
     return;
    }
//
//--Look at the FDB dictionary slot. If it's empty, then
//--return a null pointer saying that the fuzzy set doesn't
//--exist. Otherwise look through the remainder of the chain
//--for the variable name until we come to the chain end.
  *statusPtr=3;
  FDBptr=PDBptr->PDBfuzzysets[slot];
  if(FDBptr==NULL) return;
  for(;;)
    {
     if(strcmp(FDBptr->FDBid,Fzyid)==0)
       {
        *statusPtr=0;
        if(FDBprevptr==NULL)
          {
           PDBptr->PDBfuzzysets[slot]=FDBptr->FDBnext;
           delete FDBptr;
           return;
          }
        FDBprevptr->FDBnext=FDBptr->FDBnext;
        delete FDBptr;
        return;
```

```
      }
     FDBprevptr=FDBptr;
     if(FDBptr->FDBnext==NULL) return;
     FDBptr=FDBptr->FDBnext;
    }
}
```

Listing 7.14 Removing a Fuzzy Set from a Policy's Fuzzy Set Dictionary

Deleting an entry from a hash-based dictionary is only slightly more complicated than finding the entry. If the slot address generated by the hash function is *NULL*, we return. Otherwise, we traverse the hash chain keeping a pointer (*FDBprevptr*) to the previous *FDB* node in the chain. When we find the desired fuzzy set, we examine the previous pointer. If *FDBprevptr* is *NULL*, then the node is first on the chain. We make the next *FDB* node (whose address is in *FDBnext*) the head of the chain and delete the memory held by the deleted node. If *FDBprevptr* is not *NULL*, then we skip over this node by coupling the previous *FDB* with the one immediately after the located *FDB*. Memory is then released for the found *FDB*. Although this sounds complicated, you can see from the code in Listing 7.14 that the process is actually quite straightforward.

While hashing is easy to implement and provides very fast retrieval and insertion times, it does have several drawbacks. First, the size of the dictionary cannot be dynamically increased very easily. Since the current number of slots in the dictionary is used in the hashing algorithm, if we change this size, then previously stored keys cannot be located (there are, however, extensible hash algorithms). Another drawback to hashing is its random nature. While we can quickly find a specific key or look at all the keys by following the collision chain for each non-*NULL* slot, we cannot retrieve an ordered set of keys (without sorting them or maintaining a sequential order pointer in each structure). This means we cannot use hash dictionaries as a means of primary or secondary indexing where key order is important or significant. Although threaded hash algorithms exist, when this kind of access control is needed, tree-structured representations such as AVL binary trees and B*Tree organizations should be used.

Creating a Model and Associated Policies

A model descriptor block (`MDB`) is an optional component in the advanced fuzzy modeling code framework. It does, however, provide a repository for global model properties such as the primary policy associated with the model, an access password, the type of model, and other attributes that control how the model is used.

```
if(((MDBptr=MdlCreateModel(
   "RiskAssessment",MODELADD,&status))==NULL) return;
strcpy(MDBptr->MDBpsw,"SOCRATES");
MDBptr->MDBprivate=TRUE;
```

Listing 7.15 Creating and Adding Attributes to a Model Description

A policy descriptor block (*PDB*) is the essential segmentation mechanism. You need to create and add the policy description to the *XSYSctl* control block before moving on to organize the remainder of your fuzzy model. In fact, the organization of the model is centered around the population and deployment of policies. Like the model description, a policy is created and linked into the modeling system with a single function.

```
if(((PDBptr=MdlCreatePolicy(
   "ProjectRisk",MODELADD,&status))==NULL) return;
strcpy(PDBptr->PDBid,"Assessment of Project Risk Factors");
```

Listing 7.16 Creating and Adding Attributes to a Model Description

This function returns a pointer to the newly allocated (and initialized) *PDB*. The *MODELADD* parameter indicates that the newly created block should be added to the policy dictionary in the *XSYSctl* control block. The pointer returned by the function must be retained if you want to populate the associated dictionaries. (All the dictionary functions that add, delete, and locate elements require the address of the containing policy; see Listing 7.12 for an example.)

The *XSYSctl* control block is organized to help you keep track of the currently active model and policy. After creating the model descriptor, its address should be assigned to the *XSYScurrMDBptr*. When you create a policy or activate an existing policy for work, its address should be assigned to the *XSYScurrPDBptr*. Listing 7.17 illustrates this concept.

```
if(((PDBptr=MdlCreatePolicy(
   "ProjectRisk",MODELADD,&status))==NULL) return;
strcpy(PDBptr->PDBid,"Assessment of Project Risk Factors");
XSYSctl.XSYScurrPDBptr=PDBptr;
```

Listing 7.17 Storing a Policy's Address in the `XSYSctl` Control Block

Since the *XSYSctl* block is defined as an external structure, it is available to all program modules linked with your main model. By setting *XSYScurrPDBptr* to a particular policy, you can synchronize the behavior of individual code units with the overall model strategy. The values of the current model and policy pointers in *XSYSctl* can be determined with the *MdlGetcurrentEnvironment* function shown in Listing 7.18:

```
MdlGetcurrentEnvironment(&curMDBptr,&currPDBptr,&status);
if(currPDBptr==NULL)
  {
    printf("%s\n","Error. No Policy is curently active!");
    return;
  }
```

Listing 7.18 Finding the Current Model and Policy Environments

Notice that the address of the pointers (***curMDBptr*) must be passed to this function since the pointer values themselves are updated by the function.The corresponding *MdlSetcurrentEnvironment* can be used to update the *XSYSctl* with values for a particular model and policy.

The *MdlCreateModel* and *MdlCreatePolicy* functions allocate storage for an *MDB* and a *PDB*, respectively. They can also store the blocks in the *XSYSctl* control block.

```
void Complete_Model(const MDB*);
MDB *MdlCreateModel(char *Mdlid,int StoreAction,int *statusPtr)
 {
  MDB*    MDBptr;
  char   *PgmId="mtmdpcr";

  *statusPtr=0;
  MDBptr=NULL;
//--Create a new model and initialize it to the default state
  if(!(MDBptr=new MDB))
    {
     *statusPtr=1;
     MtsSendError(2,PgmId,Mdlid);
     return(MDBptr);
    }
   MdlInitMDB(MDBptr);
//
//--Now copy in the parameters to complete the general properties
//--of the Model
    strcpy(MDBptr->MDBid,Mdlid);
    if(StoreAction==MODELADD)  MdlLinkMDB(MDBptr,TRUE,statusPtr);
    Complete_Model(MDBptr);
    return(MDBptr);
  }
void Complete_Model(const MDB* MDBptr)
 {
  int      status;
  char     wrkBuff[180];

  sprintf(wrkBuff,"%s%s%s","Model '",MDBptr->MDBid,"' created.");
  MtsWritetoLog(SYSMODFILE,wrkBuff,&status);
  return;
 }
```

Listing 7.19 Creating and Initializing an MDB

The *MdlCreateModel* function encapsulates several functions. The *MDB* is allocated and initialized to its default values. If the *StoreAction* parameter indicates that the *MDB* should be stored in the *XSYSctl* block, then *MdlLinkMDB* is invoked. If another model is already active, it is replaced by the current model (a warning notice is issued, however).

```
void Complete_Policy(const PDB*);
PDB *MdlCreatePolicy(char *Polid,int StoreAction,int *statusPtr)
 {
  PDB*    PDBptr;
  char   *PgmId="mtmdpcr";

  *statusPtr=0;
  PDBptr=NULL;
//--Create a new policy and initialize it to the default state
  if(!(PDBptr=new PDB))
    {
     *statusPtr=1;
     MtsSendError(2,PgmId,Polid);
     return(PDBptr);
    }
   MdlInitPDB(PDBptr);
//
//--Now copy in the parameters to complete the general properties
//--of the Policy
    strcpy(PDBptr->PDBid,Polid);
    if(StoreAction==MODELADD)   MdlLinkPDB(PDBptr,statusPtr);
    Complete_Policy(PDBptr);
    return(PDBptr);
  }
void Complete_Policy(const PDB* PDBptr)
 {
  int      status;
  char     wrkBuff[180];

  sprintf(wrkBuff,"%s%s%s",
   "Policy '",PDBptr->PDBid,"' created.");
  MtsWritetoLog(SYSMODFILE,wrkBuff,&status);
  return;
 }
```

Listing 7.20 Creating and Initializing a PDB

In a fashion similar to creating a model, the *MdlCreatePolicy* function allocates memory for the *PDB* and initializes it to a set of default values. When the *StoreAction* parameter is set to MODELADD, the policy node is stored in the *XSYSctl* policy dictionary.

Managing Policy Dictionaries

In our previous discussion of hashing techniques (see Understanding Hash Tables and Dictionaries on page 346). the code for inserting, locating, and removing fuzzy sets from the policy dictionary was presented. These functions are part of a family of C++ programs for managing the contents of policy dictionaries. As Figure 7.13 illustrates, each type of storable object—variables, fuzzy sets, and hedges—has an associated set of dictionary management functions.

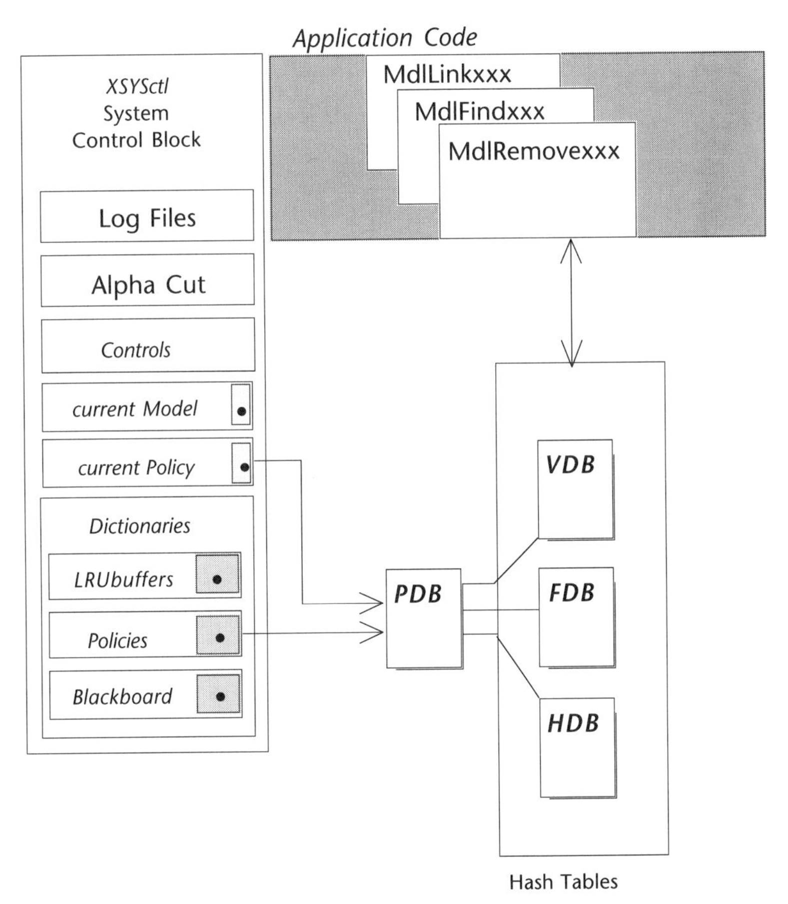

Figure 7.13 Inserting, Finding, and Removing Objects from the Policy Dictionaries

Now, as you build your model, its operational logic will revolve around the contents of the various policies comprising its functional logic. When you create a new element, such as a fuzzy set or a hedge, these are stored in the corresponding policy. Listing 7.21 illustrates this approach.

```
if(((pjriskPDBptr=MdlCreatePolicy(
   "ProjectRisk",MODELADD,&status))==NULL) return;
strcpy(pjriskPDBptr->PDBid,"Assessment of Project Risk Factors");
XSYSctl.XSYScurrPDBptr=prjriskPDBptr;
•
•
•
LgFDBptr=FzyCreateSet(
  "LARGE.DEBT",GROWTH,domain,parms,0,&status);
MdlLinkFDB(LgFDBptr,pjriskPDBptr,&status);
exHDBptr=FzyCreateHedge("EXTREMELY",POWER,3,&status);
MdlLinkHDB(exHDBptr,pjriskPDBptr,&status);
```

Listing 7.21 Storing a Fuzzy Set and a Hedge Definition in a Policy

Loading Default Hedges

You can install all the default hedges for the fuzzy modeling system (there are twelve of them) into a specified policy with a single function. As Figure 7.14 schematically illustrates, the *MdlInsertHedges* function populates the policy hedge dictionary with the *HDB* for each hedge.

You can install the default hedges in each policy (advisable if you plan to modify the operation of any built-in hedge) or in a single policy that is shared across your model. For a complete discussion of hedges and related surface transformers see Chapter 5, and in particular, the section called "Hedges and Fuzzy Surface Transformers" on page 217. Also see "Incorporating Hedges in the Fuzzy Model" on page 336.

The *MdlInsertHedges* function (Listing 7.22) works by maintaining a serial correspondence between the symbolic values of the hedges and a counter in the program code. You should exercise extreme caution in modifying this program so that this relationship is maintained.

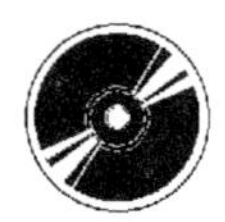

MdlInsertHedges populates a policy with all the default hedges. These hedges represent a class of types (that is, *somewhat* is the dilution hedge and could be invoked for instances of *rather*, *sort-of*, or *quite*).

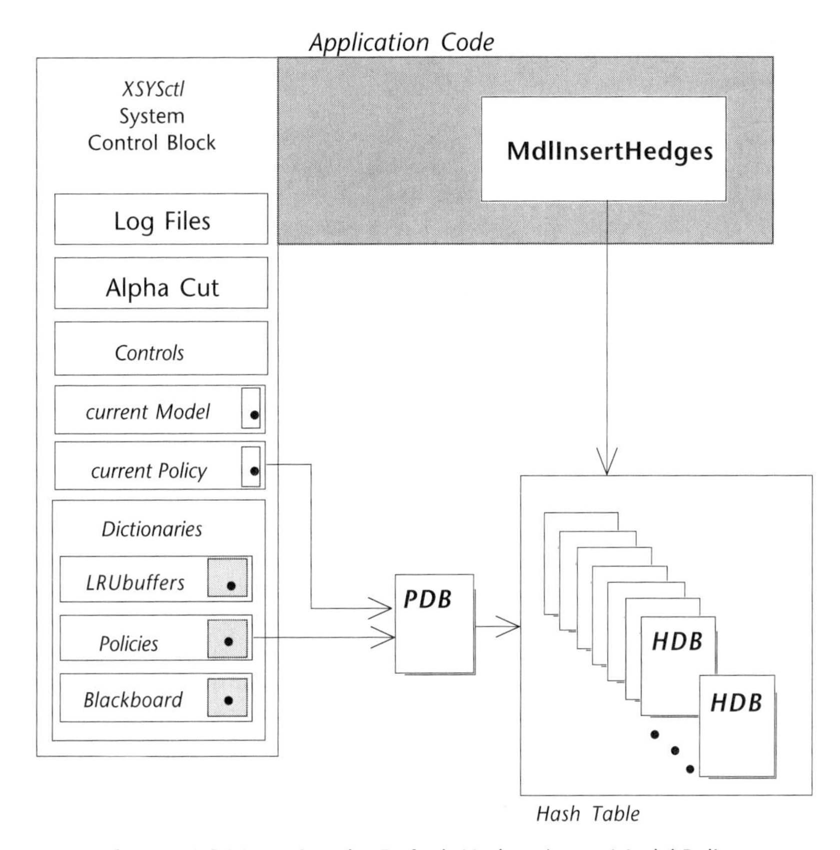

Figure 7.14 Inserting the Default Hedges into a Model Policy

```
void MdlInsertHedges(PDB *PDBptr,int *HdgCnt,int *statusPtr)
 {
  HDB  *HDBptr;
  int   i,Hedgemax=12;
  char *PgmId="mtmdihd";
  char  wrkBuff[80];
  char *HedgeNames[]=
   {
    "about",
    "above",
    "positively",
    "below",
    "vicinity",
    "generally",
```

```
    "close",
    "not",
    "somewhat",
    "very",
    "extremely",
    "slightly"
   };

  *statusPtr=0;
  for(i=0;i<Hedgemax;i++)
    {
     if(!(HDBptr=new HDB))
       {
        *statusPtr=1;
        MtsSendError(2,PgmId,HedgeNames[i]);
        return;
       }
     FzyInitHDB(HDBptr);
     strcpy(HDBptr->HDBid,HedgeNames[i]);
     HDBptr->HDBmode   =i+1;
     HDBptr->HDBop     =0;
     HDBptr->HDBvalue  =0.0;
     MdlLinkHDB(HDBptr,PDBptr,statusPtr);
     if(*statusPtr!=0) return;
     (*HdgCnt)++;
    }
  sprintf(wrkBuff,"%s%s%s",
   "Default Hedges installed in Policy '",PDBptr->PDBid,"'");
  MtsWritetoLog(SYSMODFILE,wrkBuff,statusPtr);
  return;
 }
```

Listing 7.22 Installing the Default Hedges in a Policy

Fundamental Model Design Issues

Having examined the architectures of the basic and advanced fuzzy modeling environments, we now turn to some of the fundamental issues, techniques, and problems associated with building and using fuzzy systems. A fuzzy model represents the synthesis of expert knowledge in the form of rules and fuzzy sets with their corresponding machine representations. This synthesis involves an understanding of how a model is actually represented in procedural code; how the various components are abstracted and encoded; how the relationships between model parameters, rules, and fuzzy sets are constructed; and how to approach the issues involved in representing Boolean and semi-fuzzy information.

In this section, we will address many of these issues by examining the system structure and practical examples from actual code models.

INTEGRATING APPLICATION CODE WITH THE MODELING SYSTEM

An embedded fuzzy model couples C/C++ program logic with the data structures and the fuzzy system library functions to produce a finished, production model. We now turn to how this coupling is actually done. The schematic in Figure 7.15 summarizes, in a general way, the activities performed in your application code.

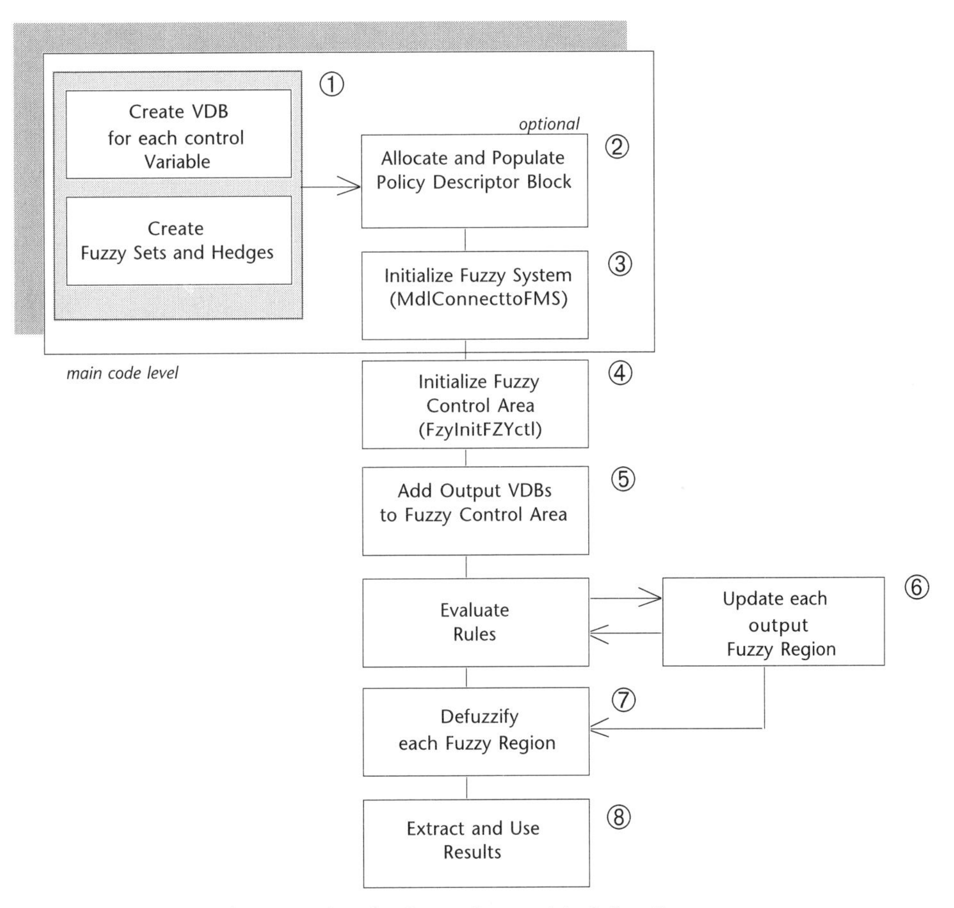

Figure 7.15 The Fuzzy System Modeling Process

The boxed region at the top of the schematic represents the functions that are normally performed within the main code block. Each logical model segment process is repeated for every policy in the model and is often represented as an external function. This is the general architecture of a fuzzy model, but many variations are possible and often necessary. In particular, the use of formal policy constructs for small and relatively simple models is often not needed and unnecessarily complicates the code.

TASKS AT THE MODULE MAIN PROGRAM LEVEL

Each C/C++ program has a `main` program. This program receives arguments (such as the name of the model, passwords, the location of data files, etc.), performs memory allocations for elements needed during the entire model execution, and establishes the high-level control logic for the model itself (such as invoking the proper series of policies, examining return status codes, and making decisions about the next processing step). There are several import tasks performed in the `main` program, as we'll see in the next six subsections of this chapter.

> The term **policy** has two distinct meanings in the fuzzy modeling system. First, it is a physical structure defined by a policy descriptor block (PDB) and stored in the *XSYSctl* external block. Second, it is a logical and self-contained unit of knowledge within the overall model. Models can be decomposed into many policies. A project risk assessment model might have, as an example, a policy that assesses risk based on the physical nature of the project (length, complexity, organization, funding, similarity to previous projects, time of year, etc.). Another policy might evaluate risk based on project management factors (experience and age of the project manager, skill sets of critical project members, visibility of project within organization, actual and perceived project priority, etc.). A third policy could take the results of the first two and formulate an over risk assessment.
>
> In the discussions that follow (unless otherwise indicated), the term *policy* is used to mean a functional or logical unit of the model represented as a complete C/C++ function, independent of its association with a *PDB*. In model policies the variables, fuzzy sets, and hedges might be stored in local or static program variables or in formal *PDB* dictionaries.

Connecting the Model to the System Control Blocks

Before the fuzzy modeling system can be used, the *XSYSctl* control block must be properly initialized. This is one of the first actions your application code should perform. The connection to the system control block is handled through a single function,

```
MdlConnecttoFMS(&status);
```

and must be included at the very top of the main program code. Within a single code space for your model, *MdlConnecttoFMS* should only be called once. This function sets up the entire fuzzy modeling infrastructure by initializing the following tasks:

- all the performance and operating statistics
- the debugging and execution trace switches
- the global alpha-cut threshold [0.0]
- the model policies dictionary
- the current model and policy pointers

This function also establishes the active file pointers for the system and model tracing logs as well as the system and user error diagnostic files. Unless these file pointers are properly initialized, your fuzzy modeling system will terminate (or behave in unpredictable ways) when the first error or log message is written. Also, any attempts to store policy structures will cause errors since the dictionary hash array has not been properly set to *NULL* values.

Allocating and Installing the Policy Structure

In normal practice, there is a functional equivalence between a logical PDB and a C/C++ function so that, when we talk about a model policy, we are discussing not only the physical policy structure but also the code that performs some functional component of the model. At the main program level, all the model policy structure *PDBs* are allocated and stored in the *XSYSctl* dictionary. Allocating all the *PDB* blocks at the main level ensures that we have a consistent view of the model and provides the framework for associating code blocks (i.e., functions) with each policy.

Defining Solution (Output) Variables

A solution variable appears on the left-hand side of a fuzzy rule's consequent and is determined through a defuzzification process after the model policy has executed. In the following conditional proposition, the variable z is a solution variable:[4]

$$\textit{if } x \textit{ is } Y \textit{ then } z \textit{ is } W$$

After all the rules that specify z have been evaluated, the defuzzification process is used to find the solution variable's expected value. A single policy can contain up to thirty-two solution variables (based on the current array size in the *XFZYctl* block). This step essentially determines what the policy does. For each solution variable, you must allocate and complete a variable descriptor block (*VDB*). This formal variable is inserted into the *XFZYctl* work area and is managed by the conditional and unconditional proposition functions. For details on the formal variable concept, see Code Representation of Fuzzy Variables on page 333.

Creating and Storing Fuzzy Sets in Application Code

Fuzzy sets can be stored in one of two ways: in program code or in a *PDB* dictionary. Perhaps the easiest way of using fuzzy sets, at least for models of small to moderate complexity, is through arrays of FDBs in the policy function. As an example, in the risk assessment model, the fuzzy sets are declared and initialized in pointer arrays (see Listing 7.23).

```
FDB  *prlFDBptr[FzySetMax],     // project risk level (total)
     *pmrFDBptr[FzySetMax],     // project manager's risk
     *porFDBptr[FzySetMax];     // project organization risk
    //
 for(i=0;i<FzySetMax;i++)
    {
     prlFDBptr[i] =NULL;
     pmrFDBptr[i] =NULL;
     porFDBptr[i] =NULL;
     FDBarray[i]  =NULL;
    }
```

Listing 7.23 Defining and Initializing of Some Project Risk Fuzzy Sets

The symbolic constant *FzySetMax* is defined to dimension the FDB arrays to the appropriate number of fuzzy sets. Each fuzzy set associated with a model variable is stored in an array associated with that variable. Listing 7.24 illustrates how the fuzzy sets for the solution variable *PRL* (project risk level) are stored in the corresponding *FDB* pointer array.

```
//---(PRL) PROJECT RISK LEVEL
  Domain[0]=0;Domain[1]=1000;
  prlFDBptr[0]=FzyCreateSet(
    "hiPRL",GROWTH,Domain,Parms,0,statusPtr);
  prlFDBptr[1]=FzyCreateSet(
    "loPRL",DECLINE,Domain,Parms,0,statusPtr);
  strcpy(prlFDBptr[0]->FDBdesc,"High for Project Risk");
  strcpy(prlFDBptr[1]->FDBdesc,"Low for Project Risk");
  FDBcnt=2;
  FzyPlotVar(
     "PrjRiskk--Project-based Risk level Assessment",
     prlFDBptr,FDBcnt,SYSMODFILE,statusPtr);
```

Listing 7.24 Creating and Storing Fuzzy Sets in Application Code

From a model verification, maintenance, and debugging viewpoint, this storage system has the added advantage of code compatibility with the *FzyPlotVar* routine. The entire set of overlapping fuzzy sets associated with a variable can be displayed by passing the *FDB* array to this function.

When the policy is complete, memory for the fuzzy sets should be released. The array-based approach also makes memory management easier since, as Listing 7.25 illustrates, we can loop through the arrays and delete any existing sets.

```
for(i=0;i<FzySetMax;i++)
 {
  if(prlFDBptr[i]!=NULL) delete prlFDBptr[i];
  if(pmrFDBptr[i]!=NULL) delete pmrFDBptr[i];
  if(porFDBptr[i]!=NULL) delete porFDBptr[i];
 }
```

Listing 7.25 Releasing Memory Held by Application Code Fuzzy Sets

You must ensure that each of the cells in the fuzzy set array has been initialized to *NULL* (see Listing 7.23). Of course, if this is a dynamically allocated array, the C++ statement *delete []prlFDBptr* could be used to release memory for the array after all the sets have been released. Because fuzzy sets represent a significant resource commitment in the model, their management is important for models that use large numbers of fuzzy sets.

Creating and Storing Fuzzy Sets in a Policy's Dictionary

Policies provide a mechanism for storing, sharing, and using model components across the code boundaries of large, complex systems. Working with policies requires less code within the application space since much of the structural architecture and processing functions have been moved to the *PDB* and the functions that manage the block's contents. Listing 7.24 shows how the project risk assessment fuzzy sets are stored in the containing policy.

```
//---(PRL) PROJECT RISK LEVEL
  Domain[0]=0;Domain[1]=1000;
  FDBptr[0]=FzyCreateSet(
    "hiPRL",GROWTH,Domain,Parms,0,statusPtr);
  strcpy(prlFDBptr[0]->FDBdesc,"High for Project Risk");
  MdlLinkFDB(FDBptr[0],PDBptr,statusPtr);
  FDBptr[1]=FzyCreateSet(
    "loPRL",DECLINE,Domain,Parms,0,statusPtr);
  strcpy(prlFDBptr[1]->FDBdesc,"Low for Project Risk");
  MdlLinkFDB(FDBptr[1],PDBptr,statusPtr);
  FDBcnt=2;
  FzyPlotVar(
     "PrjRiskk--Project-based Risk level Assessment",
     FDBptr,FDBcnt,SYSMODFILE,statusPtr);
```

Listing 7.26 Creating and Storing Fuzzy Sets in Application Code

In this example, the generated fuzzy sets are still stored initially in a program *FDB* array so that it can be used easily by the *FzyPlotVar* function. In this case, however, only a single working array, *FDBptr[]*, is required for all the fuzzy sets created in this model (or policy). Memory storage is also simplified. The *MdlClosePolicy* function will release all the memory held by a specified policy.

At a higher level, *MdlRemovePDB* deletes a *PDB* from the *XSYSctl* dictionary. This function does not, however, release the memory associated with the dictionaries. Failure to issue the *MdlClosePolicy* function prior to deleting a policy's descriptor will result in significant memory leaks.

Storing fuzzy sets in the policy dictionary distributes the sets randomly across the dictionary so that they are not implicitly associated with a particular model variable. You can construct an explicit relationship between a set of fuzzy sets and a model variable by using a special array (*VDBfuzzysets[]*) built into the *VDB*. Listing 7.24 illustrates how this association is made.

```
//---(PRL) PROJECT RISK LEVEL
  Domain[0]=0;Domain[1]=1000;
  FDBptr=FzyCreateSet("hiPRL",GROWTH,Domain,Parms,0,statusPtr);
  strcpy(prlFDBptr[0]->FDBdesc,"High for Project Risk");
  MdlLinkFDB(FDBptr[0],PDBptr,statusPtr);
  prlVDBptr->VDBfuzzysets[FDBcnt]=FDBptr;
  FDBcnt++;
  FzyPlotVar(
     "PrjRiskk--Project-based Risk level Assessment",
      prlVDBptr->VDBfuzzysets,FDBcnt,SYSMODFILE,statusPtr);
```

Listing 7.27 Associating Policy Fuzzy Sets with a Variable

This method has the added advantage of eliminating the need to create fuzzy sets with an array of *FDB* pointers. The variable *VDBfuzzysets[]* array can be passed directly to *FzyPlotVar*. If the variable descriptor is stored in the *XSYSctl* dictionary, then the fuzzy sets associated with a variable are available throughout the model environment.

Loading and Creating Hedges

Hedges, like fuzzy sets, can be stored in application code (as arrays or specific hedge variables) or in the *PDB* hedge dictionary. The same kinds of trade-offs exist.[5] In the case of hedges, a set of default or built-in hedges exists—those that are recognized implicitly by the *FzyApplyHedge* function. The fuzzy modeling system contains two functions for loading the default hedges into your model. *MdlInsertHedges* installs the default hedges in the specified policy's hedge dictionary. (see Listing 7.28).

```
pjriskPDBptr=MdlCreatePolicy("ProjectRisk",MODELADD,statusPtr);
if(pjriskPDBptr==NULL) return;
•
•
//--install the default hedges into the policy
MdlInsertHedges(PDBptr,&HdgCnt,statusPtr);
```

Listing 7.28 Loading Default Hedges into a Policy Structure

For a discussion of the policy hedge dictionary, see Loading Default Hedges on page 358. *FzyInsertHedges*, on the other hand, installs the hedges in a specified array of *HDB* pointers. As Listing 7.29 illustrates, you can create an array of default hedges in your model code. One note of caution: the array must be sufficiently large to hold all the default hedges.

```
HDB *defaultHedges[16];
int       defHdgCnt;

FzyInsertHedges(defaultHedges,&defHdgCnt,statusPtr);
if(*statusPtr!=0) return;
```

Listing 7.29 Loading Default Hedges into Application Code

To find a particular hedge, you must search through the array looking for a match with the hedge identifier (or the hedge's *HDBmode* number). Listing 7.30 illustrates a small function that searches the hedge arrays and returns a pointer to the correct *HDB* for the required hedge.

```
HDB *FindDefHedge(char *Hedgeid,HDB **Hedges,int HdgCnt)
  {
    int  i;
    for(i=0;i<HdgCnt;i++)
       if(strcmp(Hedgeid,Hedges[i]->HDBid)==0) return(Hedges[i]);
    return(NULL);
  }
```

Listing 7.30 Finding a Default Hedge in Application Code

The overhead of this direct look-up approach is minimal since there are only twelve default hedge types. If you increase the number of built-in hedges by adding new hedge classes, incorporate aliases for the hedge class name, or include your own hedges in the array of *HDBs*, then more sophisticated forms of look-up might become appropriate.

In addition to the built-in fuzzy modeling system hedges, you can define your own hedges that will be recognized by *FzyApplyHedge*. These can be new hedges or aliases for built-in hedges (such as *rather* and *sort-of* as synonyms for *somewhat*). An alias hedge has the same *HDBmode* as its corresponding built-in hedge. When you create a new hedge, it can be stored in a policy or in an application code variable just like the built-in hedges. Listing 7.31 illustrates how a new hedge is inserted into either a policy or an existing application code array.

```
HDBptr=FzyCreateHedge("TREMENDOUSLY",POWER,4,statusPtr);
MdlLinkHDB(HDBptr,PDBptr,statusPtr);

HdgCnt++;
defaultHedges[HdgCnt]=HDBptr;
```

Listing 7.31 Inserting a User Hedge into a Policy and into Application Code

A model that uses one or two specific hedges does not need to load the default hedges. Instead, a single *HDB* is allocated, initialized, and reused when required. As an example, Listing 7.32 illustrates how the built-in hedge *very* is created and used.

```
if(!(HDBptr=new HDB))
    {
     *statusPtr=1;
     MtsSendError(2,PgmId,"VERY");
     return(NoPrice);
    }
  FzyInitHedge(HDBptr);
   strcpy(HDBptr->HDBid,"VERY");
   HDBptr->HDBmode=VERY;
   //--Predicate (Premise) of the rule)
//--Apply the hedge VERY to the High fuzzy set
  FzyApplyHedge(HiFDBptr,HDBptr,WrkFDBptr,statusPtr);
  FzyDrawSet(WrkFDBptr,SYSMODFILE,statusPtr);
```

Listing 7.32 Creating and Using a Hedge on an As-Required Basis

Since this is a built-in hedge, only the symbolic value *very* is needed to identify the hedge function, but the same kind of in-place hedge use can work for hedges that you define as part of a model or policy. Listing 7.32 shows how a new hedge, *tremendously*, is defined and then used.

```
 strcpy(HDBptr->HDBid,"TREMENDOUSLY");
   HDBptr->HDBop   =POWER;
   HDBptr->HDBvalue=4.0;
//--Predicate (Premise) of the rule)
//--Apply the hedge TREMENDOUSLY to the High fuzzy set
  FzyApplyHedge(HiFDBptr,HDBptr,WrkFDBptr,statusPtr);
```

Listing 7.33 Creating and Using a Hedge on an As-Required Basis

Although you should allocate and store all the *PDBs* for each model segment in the main program block, you might want to defer actually loading the variables, fuzzy sets, and hedges into less frequently used or large policies since these consume a considerable amount of memory (a factor that might be especially important in DOS-based applications). When invoked, the policy function can allocate and populate its own structures.

Segmenting Application Code into Modules

A physical model can be and (in most cases) should be decomposed into separated code modules that are linked to produce the complete fuzzy system. Figure 7.16 illustrates how a model is organized as a series of external functions, with each code segment performing some functional part of the model.

Maintaining Addressability to the Model

Addressability to the model knowledge representation components (such as fuzzy sets and hedges) is maintained through the *XSYSctl* external control block. The storage version of this block (*SSYSCTL.hpp*) is included in the main program. Each module includes the *XSYSCTL*.hpp header to incorporate the shared, external version of the block. The overall communication is established at the main program through the *MdlConnecttoFMS* function. See Tasks at the Module Main Program Level on page 362 for details on this connection process.

Each linked code module is usually (but not necessarily) associated with a policy definition in the model—that is, a linked program executes the logic associated with the variables, fuzzy sets, and hedges stored in the policy. Within the separate policies there are several preliminary tasks necessary to maintain information linkage with the main model/ program.

Establishing the Policy Environment

There are two general methods of properly establishing a policy environment. If the main program has set the *XSYScurrPDBptr* field in the *XSYSctl* block to the address of the

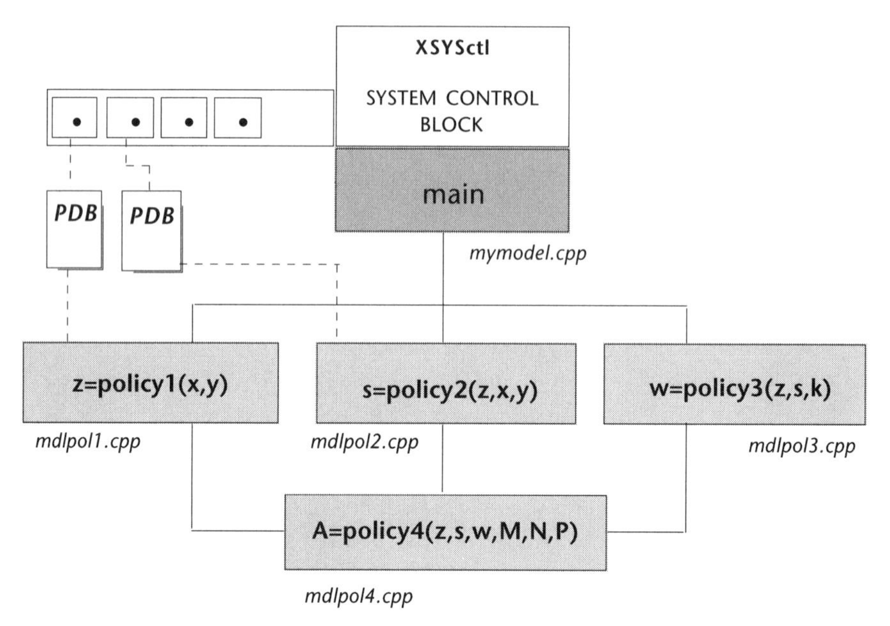

Figure 7.16 Model Organization as Several C++ Modules

policy before calling the function, then this environment can be established by a single statement such as

```
prcPDBptr=XSYSctl.XSYScurrPDBptr;
```

where *prcPDBptr* is the local pointer variable containing the address of a *PDB* structure. In some cases, however, this method may not be possible, in which case the individual modules find their policy information through the *MdlFindPDB* function, also invoked with a single statement

```
prcPDBptr=MdlFindPDB("PRICING",statusPtr);
```

where the string *PRICING* is the identification of the policy. To reduce errors in this kind of linking, a good naming convention mandates that the policy name and the external function name be the same. This minimizes the chance of typographic errors and also establishes a disciplined relationship between structure and logic.

Initializing the Fuzzy Logic Work Area for the Policy

Each policy shares the external control block *XFZYctl*. This control block holds the working ("undergeneration") fuzzy set and other parameters for each solution variable in the policy. Since this block is external, it can be shared among several modules. However, any module that initiates a policy must initialize the *XFZYctl* block. Listing 7.34 illustrates how this task is accomplished.

```
VDBptr=MdlFindVDB("PRICE",prcPDBptr,statusPtr);
FzyInitFZYctl(statusPtr);
if(!(FzyAddFZYctl(VDBptr,&PriceFDBptr,&FSVptr,statusPtr)))

   {
    *statusPtr=1;
    MtsSendError(12,PgmId,"PRICE");
    return(NoPrice);
   }
thisCorrMethod  =FSVptr->FzySVimplMethod;
thisDefuzzMethod=FSVptr->FzySVdefuzzMethod;
```

Listing 7.34 Initializing the `XFZYctl` Block at the Start of a Policy

In this example, the output or solution variable *PRICE* is located in the policy's variable dictionary, and its address is stored in a local *VDBptr* variable. The fuzzy work area is initialized (through *FzyInitFZYctl*) removing all active fuzzy set references and initializing all the counters. The *PRICE* variable is added to the working fuzzy logic area (through the *FzyAddFZYctl* function) creating a new fuzzy solution variable structure inside the control block. The *FSV* structure contains information about the fuzzy logic process as well as a new and empty working fuzzy set with the same name as the variable. The address of the new fuzzy set is returned in the local *PriceFDBptr* variable and the address of the new *FSV* structure is returned in *FSVptr.*

Locating the Necessary Fuzzy Sets and Hedges

Once the associated policy structure has been located and stored, you can find and store the addresses for each fuzzy set and hedge used in the policy. Listing 7.35 shows how a set of fuzzy sets and the `very` hedge are retrieved from the policy dictionaries.

```
//
//--Find the fuzzy sets in the policy dictionary
//
  HiFDBptr=MdlFindFDB("HIGH",           prcPDBptr,statusPtr);
  LoFDBptr=MdlFindFDB("LOW",            prcPDBptr,statusPtr);
  amFDBptr=MdlFindFDB("Near.2*MfgCosts",prcPDBptr,statusPtr);
  ncFDBptr=MdlFindFDB("Near.CompPrice" ,prcPDBptr,statusPtr);
  wkFDBptr=MdlFindFDB("NULL",           prcPDBptr,statusPtr);
//
//--Find the hedge very in the policy dictionary
```

```
//
  veryHDBptr=MdlFindHDB("very",prcPDBptr,statusPtr);
//
```

Listing 7.35 Finding and Storing the Policy's Fuzzy Sets and Hedges

If you are not using policy mechanisms, you have several alternative choices: define the fuzzy sets and hedges inside the policy code itself, store the fuzzy sets in a static external array for use by each linked external function, or pass the fuzzy sets and hedges into the policy through an array parameter in a manner similar to the following:

```
price=pricingpolicy(
    mfgcosts,compcosts,
      fzysetArray,fzycnt,hedgeArray,hdgcnt,statusPtr)
```

Here, the parameters *fzysetArray* and *hedgeArray* contain an array of fuzzy sets and hedges, respectively. The parameters are followed by count variables indicating the extent of each array (or the array could be allocated to *n+1* with the last position containing a *NULL* value as an end-of-list indicator).

Exploring a Simple Fuzzy System Model

We conclude this section on fundamental design issues with a detailed examination of an actual fuzzy model's code. The operation and use of this pricing model is explored in more depth in Chapter 8, *Fuzzy Systems: Case Studies*, but it will serve here to illustrate many of the concepts we have discussed in building a physical model. Figure 7.17 shows how the pricing model is organized.

The main program model defines all the knowledge structures used in the model except the rules themselves. This information is stored in a single policy. The actual pricing policy is then invoked by the main program as an external C++ function. The policy function returns an estimated product price based on the rules.

The `fshprcl.cpp` code block is the main program for the pricing model. In this top-level function, we establish the organization of the model and prepare the environment so the actual pricing policy can be called. Each logical segment of the program code is described as we proceed rather than after the entire listing.

```
#include <stdio.h>
#include <string.h>
```

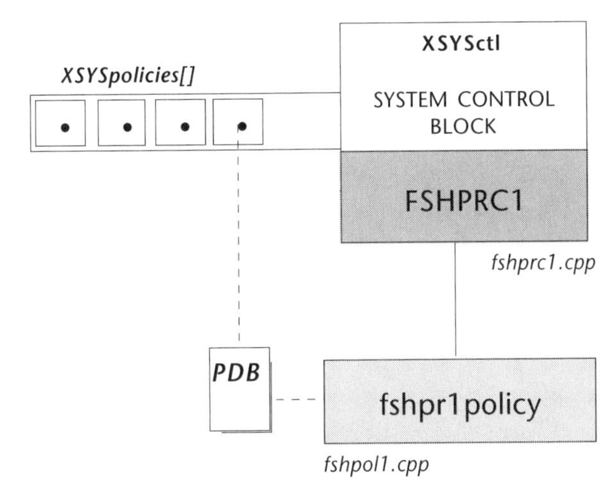

Figure 7.17 Organization of the Pricing Model

```
#include "PDB.hpp"
#include "FDB.hpp"
#include "SFZYctl.hpp"
#include "SSYSctl.hpp"
#include "fuzzy.hpp"
#include "mtypes.hpp"
#include "mtsptype.hpp"
//
//--These are the "logical" rules.
//
static const char *Rules[]=
 {
  "R1   our price must be high",
  "R2   our price must be low",
  "R3   our price must be around 2*mfgcosts",
  "R4   if the competition.price is not very high",
  "        then our price must be around the competition.price"
 };
static const Rulemax=5;
double fshpr1policy(const char**,double,double,float*,int*);
```

The program prologue includes all the standard libraries, the definitions for various structures, the storage version of the *XSYSctl* block, and the header files containing important symbolic constants for fuzzy logic (fuzzy.hpp) and the general system (*mtypes.hpp*). The mtsptype.hpp header contains prototype definitions for all the library functions. This is followed by a static array containing the logical representation of the rules. The model does not actually process rules in this form, but uses this array to display the rules. As each rule is evaluated, its logical form is written to the system model log file.

The pricing policy *fshpr1policy* has its prototype declaration inserted before the main program begins. All internal prototypes would also appear at this juncture in the code:

```
void main(void)
 {
  PDB      *prcPDBptr;
  FDB      *FDBptr[FDBvecmax];
  VDB      *VDBptr;
  int       i,status,FDBcnt,Hdgcnt,TryCtl[2];
  float     compidx;
  double    TwiceMfgCosts,mfgcosts=0,compprice=0,Price;
  double    Domain[2],Parms[4];
  FILE     *mdlout;

  MdlConnecttoFMS(&status);
  prcPDBptr=MdlCreatePolicy("PRICING",MODELADD,&status);
  XSYSctl.XSYScurrPDBptr=prcPDBptr;
  MdlInsertHedges(prcPDBptr,&Hdgcnt,&status);
  mdlout=MtsGetSystemFile(SYSMODFILE);
```

The main program begins the actual model. We start by defining all the local variables needed by the main program. This includes **prcPDBptr* to hold the address of the policy structure when it is allocated, an array of *FDB* structures (**FDBptr[]*) to hold fuzzy sets when they are created, and **VDBptr* to hold the address of the formal solution variable's *VDB* when it is allocated. The jobs of other local variables will be evident as they are used.

Following the declaration and initialization of variables, the actual model code begins. *MdlConnecttoFMS* is called first. This initializes all the fields in the *XSYSctl* control block. You must do this at the top of the model in order to use any of the standard files or the policy dictionaries. (Even if you do not want to use the policy concept, the modeling system will fail if the system log files are not properly initialized.) We then create a policy named *PRICING*, store its address in **prcPDBptr*, and automatically add it to the system policy dictionary. *XSYScurrPDBptr* is set to the address of this policy so it can be used by any of the subsequent policy modules. The set of default hedges is then inserted into the policy dictionary (through the function *MdlInsertHedges*). Finally, the *mdlout* file is set to the current system model log file:

```
//
//--create and insert the control variable
//
  Domain[0]=16;
  Domain[1]=36;
  VDBptr=VarCreateScalar("PRICE",REAL,Domain,"0",&status);
  MdlLinkVDB(VDBptr,prcPDBptr,&status);
```

A *VDB* is created for the *PRICE* solution variable. This is a *REAL* variable (that is, double precision floating point) with a permissible range of values between 16 and 36. The variable is initialized to zero (0). The *PRICE* variable is linked into the policy's variable dictionary:

```
//--Create the basic fuzzy sets (High and Low for Price)
//
  FDBptr[0]=FzyCreateSet("HIGH",INCREASE,Domain,Parms,0,&status);
  strcpy(FDBptr[0]->FDBdesc,"High for Price");
  MdlLinkFDB(FDBptr[0],prcPDBptr,&status);
  FDBptr[1]=FzyCreateSet("LOW", DECREASE,Domain,Parms,0,&status);
  strcpy(FDBptr[1]->FDBdesc,"Low for Price");
  MdlLinkFDB(FDBptr[1],prcPDBptr,&status);
  FDBcnt=2;
  FzyPlotVar("PRICE",FDBptr,FDBcnt,SYSMODFILE,&status);
//--Create an empty fuzzy set as a working area
  FDBptr[0]=FzyCreateSet("NULL",EMPTYSET,Domain,Parms,0,&status);
  MdlLinkFDB(FDBptr[0],prcPDBptr,&status);
```

We now allocate, initialize, and store into the policy's fuzzy set dictionary the two basic vocabulary sets: *High* and *Low* (as a measure of price). Both of these are linear fuzzy sets (one increasing from left to right, the other decreasing from left to right) that stretch across the entire domain. The *Domain* array has already been set for the *PRICE* variable and is used here without change. (If the fuzzy sets are mapped across a different domain, the Domain array values would be changed.)

When the fuzzy sets have been created, the *FzyPlotVar* function is called to draw them on the model log file. Figure 7.18 shows how this graphical representation appears on the model log.

```
PRICE
Domain [UofD]:      16.00 to      36.00

      1.00*                                                               .
      0.90 *******                                                  ......
      0.80        ******                                     .......
      0.70              ******                         ......
      0.60                    *******           .......
      0.50                           ******.....
      0.40                          ...... *******
      0.30                   .......              ******
      0.20             ......                           *******
      0.10      .......                                        ******
      0.00......                                                     ******
          0---|---|---|---|---|---|---|---|---|---|---|---|---|---|---|---0
        16.00   18.50   21.00   23.50   26.00   28.50   31.00   33.50   36.00
     .   FuzzySet:    HIGH
        Description: High for Price
        Domained:          16.00 to      36.00
     *   FuzzySet:    LOW
        Description: Low for Price
        Domained:          16.00 to      36.00
```

Figure 7.18 The *High* and *Low* Fuzzy Sets in the Pricing Model

Storing the fuzzy sets in a small array facilitates this plotting function since it expects the fuzzy sets as an array. In any case, displaying each fuzzy set (or a family of fuzzy sets such as those associated with *PRICE*) is always a good idea when you build a fuzzy model. The shape and properties of the underlying fuzzy sets, as well as those created during the modeling process, give us confidence that the modeling logic is correct and the results are representative of and consistent with the fuzzy set apparatus. Now we turn to the actual fuzzy model code.

```
//
//--Prompt for the model Data
//
  TryCtl[0]=8;
  TryCtl[1]=0;
  Domain[0]=8;Domain[1]=16;
  mfgcosts=MtsAskforDBL(
    "mfgcosts",
    "Enter the Manufacturing Costs:",
    Domain,TryCtl,&status);
  if(status!=0) return;
  Domain[0]=16;Domain[1]=36;
  compprice=MtsAskforDBL(
    "compprice",
    "Enter the Competition's Price:",
    Domain,TryCtl,&status);
  if(status!=0) return;
```

In this simple fuzzy pricing model, the values used by the rules (the manufacturing costs of the product and the average value of the competition's products in the same geographic or demographic region) are requested from the terminal. The *MtsAskforDBL* utility function accepts a variable identifier, a prompt string, the allowable domain for the answer, and the maximum number of attempts allowed before prompting stops. This maximum limit is stored in *TryCtl[0]*. The actual number of prompts before a correct answer was received is returned in *TryCtl[1]*. If an error in the prompting is detected, the model terminates.

```
//
//--Now create the fuzzy number for around 2*mfgcosts
//
  Domain[0]=16;Domain[1]=36;
  TwiceMfgCosts=2*mfgcosts;
  Parms[0]=TwiceMfgCosts;
  Parms[1]=TwiceMfgCosts*.25;
  FDBptr[0]=FzyCreateSet(
    "Near.2*MfgCosts",PI,Domain,Parms,2,&status);
```

```
   strcpy(FDBptr[0]->FDBdesc,"Near (Around) Twice MfgCosts");
   MdlLinkFDB(FDBptr[0],prcPDBptr,&status);
   FzyDrawSet(FDBptr[0],SYSMODFILE,&status);
   FDBcnt++;
 //
 //--Now create the fuzzy number for around compprice
   Parms[0]=compprice;
   Parms[1]=compprice*.15;
   FDBptr[0]=FzyCreateSet(
     "Near.CompPrice",PI,Domain,Parms,2,&status);
   strcpy(FDBptr[0]->FDBdesc,"Near (Around) Competition Price");
   MdlLinkFDB(FDBptr[0],prcPDBptr,&status);
   FzyDrawSet(FDBptr[0],SYSMODFILE,&status);
```

Once the values for the manufacturing costs and the competition price are available, the fuzzy sets based on these values are allocated, initialized, and stored in the policy fuzzy set dictionary. The domains for both sets are the same so it is set only once. We make the around *2*MfgCosts* a moderately wide fuzzy set (with a 25% diffusion) while the around compprice fuzzy set is somewhat narrower (having only a 15% diffusion).

When you run the model, the values for the manufacturing costs and the competition's price are requested. As an example:

```
Enter the Manufacturing Costs: 12
Enter the Competition's Price: 32
```

Given these particular values, Figures 7.19 and 7.20 show the kind of fuzzy graph written to the current model activity log.

```
  FuzzySet:    Near.2*MfgCosts
  Description: Near (Around) Twice MfgCosts
1.00                          ..
0.90                       ...  ....
0.80                     ..       .
0.70                    .          ..
0.60                  ..             .
0.50                 .                .
0.40                .                  .
0.30               .                    .
0.20              .                      .
0.10            ..                        ..
0.00...........                            ........................
    0---|---|---|---|---|---|---|---|---|---|---|---|---|---|---|---0
  16.00   18.50   21.00   23.50   26.00   28.50   31.00   33.50   36.00
  Domained:        16.00 to      36.00
```

Figure 7.19 around *2*MfgCosts* in the Pricing Model

```
 FuzzySet:     Near.CompPrice
 Description: Near (Around) Competition Price
1.00                                                     ..
0.90                                                  ...  ..
0.80                                                 .       ..
0.70                                                .          .
0.60                                               .            .
0.50                                              .
0.40                                             .               .
0.30                                            .                 .
0.20                                           .                   .
0.10                                          .                     .
0.00.........................................                        ..
     0---|---|---|---|---|---|---|---|---|---|---|---|---|---|---|---0
 16.00   18.50   21.00   23.50   26.00   28.50   31.00   33.50   36.00
 Domained:          16.00 to        36.00
```

Figure 7.20 *around compprice* in the Pricing Model

In this model, we generate and store the *around 2*MfgCosts* and the *around compprice* fuzzy set at the main program level. Unlike the base *High* and *Low* fuzzy sets, the values for these fuzzy sets will change from model run to model run. In instances where the values are read from database systems or system files, you may place this data management function in the governing policy. In this case, you will need to move the generation of these fuzzy sets into the policy module itself.

```
//
//--Display the rules on the model log and invoke the policy
//
fprintf(mdlout,"%s\n",
  "Fuzzy Pricing Model. (c) 1992 Metus Systems Group.");
fprintf(mdlout,"%s%10.2f\n",
  "[Mean] Competition Price: ",compprice);
fprintf(mdlout,"%s%10.2f\n",
  "Base Manufacturing Costs: ",mfgcosts );
//
  fprintf(mdlout,"%s\n","The Rules:");
  for(i=0;i<Rulemax;i++) fprintf(mdlout,"%s\n",Rules[i]);
  fputc('\n',mdlout);
  Price=fshpr1policy(Rules,mfgcosts,compprice,&compidx,&status);
  fprintf(stdout,"\n\n\n%s%8.2f%s%8.3f\n",
   "The Recommended Price is: ",
        Price," with a CIX of ",compidx);
  return;
 }
```

Listing 7.36 The Main Program Code for the FSHPRC1 Pricing Model

We have now set up all the variables, fuzzy sets, and hedges. The data necessary to interpret the model (manufacturing costs and competition's price) are available to the model. We now, as shown in Figure 7.21, display the name of the model, show the data values, and list the rules.

```
Fuzzy Pricing Model. (c) 1998 Metus Systems Group.
[Mean] Competition Price:      32.00
Base Manufacturing Costs:      12.00
The Rules:
R1   our price must be high
R2   our price must be low
R3   our price must be around 2*mfgcosts
R4   if the competition.price is not very high
        then our price must be around the competition.price
```

Figure 7.21 Model Banner, Data, and Rules Display

When the model has finished executing, its results are sent to the communications terminal. This message

```
The Recommended Price is: 27.02 with a CIX of 0.45
```

indicates the recommended price found by defuzzifying the solution fuzzy set associated with the *PRICE* variable. Along with the price is the unit compatibility index of the solution variable. This value indicates the strength of the final solution.

The main program has established the fuzzy modeling environment by initializing the system control block, allocating a formal variable structure for each solution variable, and allocating and storing the vocabulary and run-dependent fuzzy sets. At this point, the actual fuzzy model, *fshpr1policy*, is called to find an estimated price. In this next section we look at how this policy is organized (in slightly less detail since much of this code has been covered in previous sections of this chapter).

The *fshpr1policy* code (in *fshpol1.cpp*) is the actual price estimating policy. In this function, we execute the model's rules, perform the approximate reasoning activities, defuzzify the solution variable's fuzzy region, and return an estimated price for the new product based on known data and the model rules.

```
#include <stdio.h>
#include <string.h>
#include "PDB.hpp"        // The Policy descriptor
#include "FDB.hpp"        // The Fuzzy Set descriptor
#include "HDB.hpp"        // The Hedge descriptor
```

```
#include "VDB.hpp"       // A Metus Variable descriptor
#include "XFZYctl.hpp"   // The fuzzy parallel processor work area
#include "XSYSctl.hpp"   // The Metus System control region
#include "mtypes.hpp"    // System constants and symbolics
#include "fuzzy.hpp"     // Fuzzy Logic constants and symbolics
#include "mtsptype.hpp"  // Metus function prototypes
//
double fshprlpolicy(
   const char **Rules,
      double  MfgCosts,double  CompPrice,float  *CIXptr,int  *sta-
tusPtr)
 {
  PDB      *prcPDBptr;
  FDB      *HiFDBptr,
           *LoFDBptr,
           *amFDBptr,
           *ncFDBptr,
           *PriceFDBptr,
           *wkFDBptr;
  HDB      *veryHDBptr;
  VDB      *VDBptr;
  FSV      *FSVptr;
```

The standard libraries, as well as the fuzzy modeling header files, are included. The policy accepts a constant character array containing the rules, the policy external data (manufacturing costs and competitor's price), a float pointer to the calculated unit compatibility index, and the general status code parameter. We then need to define pointer variables for the enclosing policy (**prcPDBptr*), the fuzzy sets needed in the rules (such as **HiFDBptr* and **amFDBptr*), the *very* hedge (**veryHDBptr*), and the *PRICE* solution variable (**VDBptr*). The **wrkFDBptr* will point to an empty working fuzzy set used in the system hedge operations.

Although the *FzyAddFZYctl* function creates the solution variable's fuzzy set and allocates the new *FSV* structure, you must declare appropriate pointer variables in the policy to hold their addresses. These are set and returned by the function. In this case, **PriceFDBptr* is a pointer to the *PRICE* working fuzzy region, and **FSVptr* is a pointer to the new *FSV* for the *PRICE* variable.

```
  char     *PgmId="fshpol1";
  int       i,
            Rulecnt,
            thisCorrMethod,
            thisDefuzzMethod;
  float     fsetheight;
  double    Price,
            NoPrice=0,
```

```
          PremiseTruth;
FILE     *mdlout;

*statusPtr=0;
prcPDBptr=XSYSctl.XSYScurrPDBptr;
mdlout=MtsGetSystemFile(SYSMODFILE);
fprintf(mdlout,"%s\n","Price Estimation Policy Begins....");
VDBptr=MdlFindVDB("PRICE",prcPDBptr,statusPtr);
FzyInitFZYctl(statusPtr);
if(!(FzyAddFZYctl(VDBptr,&PriceFDBptr,&FSVptr,statusPtr)))
  {
   *statusPtr=1;
   MtsSendError(12,PgmId,"PRICE");
   return(NoPrice);
  }
thisCorrMethod  =FSVptr->FzySVimplMethod;
thisDefuzzMethod=FSVptr->FzySVdefuzzMethod;
```

The policy begins by establishing addressability to its *PDB*. Since this model has only a single policy, the main program placed the address of this structure in the *XSYScurrPDBptr* field of the *XSYSctl* block. We need only store this address in the **prcPDBptr*. The system model activity log file is then retrieved. The next lines of code initialize the approximate reasoning facilities. The *VDB* associated with the *PRICE* variable is located in the policy's variable dictionary. The fuzzy working control region (*XFZYctl*) is initialized, and this solution variable is inserted into the control region. The modeling system creates a new *FSV* block and a fuzzy set for *PRICE*. Their addresses are returned in **PriceFDBptr* and **FSVptr*. The model then extracts the stated correlation and defuzzification methods associated with this variable.

The *VarCreateScalar* function (called in the main program) establishes a set of default values for each control variable. These values establish the default correlation method (correlation maximum), implication function (min/max), and defuzzification technique (composite moments). You can alter these defaults at this point in the model by assigning different values to the control fields in the *FSV* block.

```
//
//--Find the fuzzy sets in the policy dictionary
//
  HiFDBptr=MdlFindFDB("HIGH",           prcPDBptr,statusPtr);
  LoFDBptr=MdlFindFDB("LOW",            prcPDBptr,statusPtr);
  amFDBptr=MdlFindFDB("Near.2*MfgCosts",prcPDBptr,statusPtr);
  ncFDBptr=MdlFindFDB("Near.CompPrice" ,prcPDBptr,statusPtr);
  wkFDBptr=MdlFindFDB("NULL",           prcPDBptr,statusPtr);
```

```
//
//--Find the hedge very in the policy dictionary
//
  veryHDBptr=MdlFindHDB("very",prcPDBptr,statusPtr);
```

The fuzzy sets and hedges created and stored by the main program are retrieved from the policy and assigned to the corresponding local variables. Note that the dictionary functions are case sensitive ("*HIGH*" and "*high*" are two different fuzzy sets). The fuzzy identifier *NULL* has no intrinsic meaning to the fuzzy modeling system. The name was chosen simply to indicate that the set is empty.

```
//
//-------B E G I N   M O D E L   P R O C E S S I N G---------
//
  Rulecnt=0;
  fprintf(mdlout,"%s\n","Rule Execution....");
//--Rule 1. Our Price must be High
  fprintf(mdlout,"%s\n",Rules[Rulecnt]);
  FzyUnCondProposition(HiFDBptr,FSVptr);
  FzyDrawSet(PriceFDBptr,SYSMODFILE,statusPtr);
  Rulecnt++;
//
//--Rule 2. Our Price must be Low
  fprintf(mdlout,"%s\n",Rules[Rulecnt]);
  FzyUnCondProposition(LoFDBptr,FSVptr);
  FzyDrawSet(PriceFDBptr,SYSMODFILE,statusPtr);
  Rulecnt++;
//
//--Rule 3. Our Price must be around 2*MfgCosts
  fprintf(mdlout,"%s\n",Rules[Rulecnt]);
  FzyUnCondProposition(amFDBptr,FSVptr);
  FzyDrawSet(PriceFDBptr,SYSMODFILE,statusPtr);
  Rulecnt++;
```

The heart of the fuzzy model is the evaluation of its basic propositions. These are equivalent to and work in the same way as rules in conventional expert systems except that fuzzy system rules are effectively run in parallel. This means that the variable specified in the consequent of a rule is not available until all the rules have been executed and the final fuzzy set defuzzified. In this section of the model, the first three unconditional fuzzy rules are evaluated. The model prints the logical rule to the model activity log, evaluates the proposition, and draws the resulting solution fuzzy set for *PRICE*. The *FzyUnCondProposition* function forms the intersection of the specified target fuzzy set (such as *HIGH*) with the solution variable's undergeneration fuzzy set. The fuzzy model draws the new solution fuzzy region for *PRICE* after each rule is evaluated. This process, like the convention of drawing all the vocabulary fuzzy sets, visually confirms the model process flow. Figure

7.22 illustrates how the *PRICE* solution fuzzy set appears after the first two policy rules are executed.

```
  FuzzySet:    PRICE
  Description:
1.00
0.90
0.80
0.70
0.60
0.50                                         .
0.40                                   ...... .......
0.30                            .......             ......
0.20                      ......                          .......
0.10              .......                                        ......
0.00......                                                             ......
     0---|---|---|---|---|---|---|---|---|---|---|---|---|---|---|---|---0
  16.00    18.50    21.00    23.50    26.00    28.50    31.00    33.50    36.00
  Domained:          16.00 to        36.00
```

Figure 7.22 The *PRICE* Solution Region After Executing the First Two Pricing Rules

```
//
//--Rule 4. If the Competition.price is not very high,
//              then our price should be near the competition.price
  fprintf(mdlout,"%s\n",Rules[Rulecnt]);
  fprintf(mdlout,"%s\n",Rules[Rulecnt+1]);
  FzyInitFDB(wkFDBptr);
//
//--------------Evaluate Predicate of rule---------------------
//
//--Apply the hedge VERY to the High fuzzy set
  FzyApplyHedge(HiFDBptr,veryHDBptr,wkFDBptr,statusPtr);
  FzyDrawSet(wkFDBptr,SYSMODFILE,statusPtr);
//--Apply the operator NOT to the hedged fuzzy set
  FzyApplyNOT(ZADEHNOT,0,wkFDBptr,statusPtr);
  FzyDrawSet(wkFDBptr,SYSMODFILE,statusPtr);
//--computed membership of competiton price in this fuzzy region
  PremiseTruth=FzyGetMembership(
     wkFDBptr,CompPrice,&Idxpos,statusPtr);
  fprintf(mdlout,"%s%10.2f\n","PremiseTruth= ",PremiseTruth);
//
//--------------Perform consequent proposition-----------------
//
  FzyCondProposition(
     ncFDBptr,FSVptr,thisCorrMethod,PremiseTruth,statusPtr);
  FzyDrawSet(PriceFDBptr,SYSMODFILE,statusPtr);
```

The code above shows a conditional fuzzy rule involving a hedged predicate. The working fuzzy set is initialized so that it can be used in the hedge operation. (Since this is the only hedged rule in this policy, we need not initialize the working fuzzy set though initializing working sets is a good habit. The modeling system updates the description of the hedged set by appending a string describing the hedge—if the set is not initialized then erroneous information could appear in the displays.) This rule hedges the *HIGH* fuzzy set, producing a new fuzzy region stored in **wkFDBptr*. The standard *ZADEH not* operation is used to take the complement of this set.[6] When we have created this new fuzzy region, the proposition is evaluated, and its truth (*PremiseTruth*) is used by *FzyCondProposition* to complete the rule. Notice that we draw the new fuzzy region after each hedge or complement operation is applied. For hedge operations, this is very important since we need to assure ourselves that the generated fuzzy region correctly maps to the semantic meaning of the rule.

```
//
//--Defuzzify to find expected value for price
//
  Price=FzyDefuzzify(
     PriceFDBptr,thisDefuzzMethod,CIXptr,statusPtr);
  fsetheight=FzyGetHeight(PriceFDBptr);
  fprintf(mdlout,"%s\n","Model Solution:");
  fprintf(mdlout,"%s%10.2f\n","  Price       = ",Price       );
  fprintf(mdlout,"%s%10.2f\n","  CompIdx     = ",*CIXptr     );
  fprintf(mdlout,"%s%10.2f\n","  SurfaceHght = ",fsetheight  );
//
  FzyCloseFZYctl(statusPtr);
  return(Price);
 }
```

Listing 7.37 The Program Code for the FSHPOL1 Pricing Policy

In this final code segment (Listing 7.37), the *PRICE* fuzzy region is defuzzified. Using the composite moments (also called the *centroid*), *FzyDefuzzify* finds the value that is best represented by the combined action of all the rules. We also retrieve the height of the fuzzy set. Along with the unit compatibility index, the height of the solution fuzzy region is an important metric in deciding how confident we are in the final model recommendation. The policy results are written to the model activity log, the fuzzy work area is closed (this releases the storage for the *PRICE* fuzzy set), and we return the value for the estimated price.

Exploring a More Extensive Pricing Policy

The relatively simple fuzzy model we examined in the previous section lacked some of the support features that we might want to include in production systems. In this section we briefly dissect a slightly more extensive model that incorporates the explanatory facility, investigates alternate methods of defuzzification, and provides a slightly more extensive rule tracking facility. Figure 7.23 is a schematic of the extended model architecture.

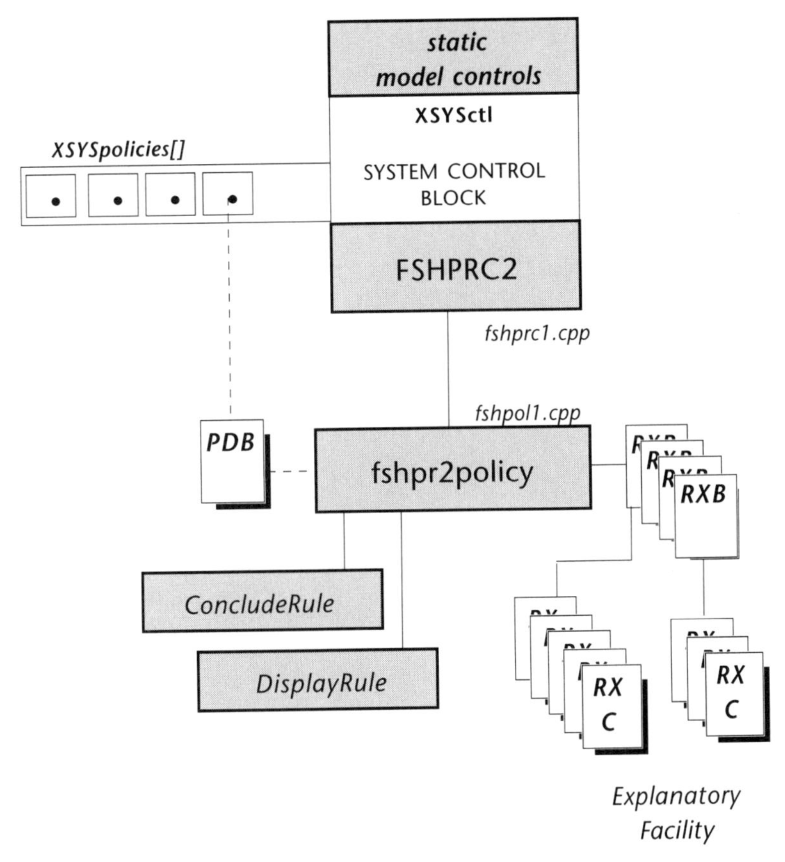

Figure 7.23 Organization of a More Extensive Pricing Policy

In addition to these new features, the extended pricing model also includes the *mdltypes.hpp* header file. This header contains a large number of static variables used to support more advanced model and policy construction. Here we are only examining the fuzzy policy (*fshpr2policy*) since the main program is essentially the same as *fshprc1.cpp* except that main program for *fshprc2.cpp* does not pass the array of rules to this policy. Rule display and management are done within the policy. For models with a number of policies called from a main model, this is the preferred method of handling rules (each policy has its own set of rules).

The *fshpol21.cpp* code block is the actual fuzzy policy for computing an estimated new product price. In many respects it has the same logical organization as *fshprlpolicy*. Generally, we will discuss only the major differences. Large blocks of code that are substantially the same in the two policies are shown in small print.

```
#include <stdio.h>
#include <string.h>
#include "PDB.hpp"        // The Policy descriptor
#include "FDB.hpp"        // The Fuzzy Set descriptor
#include "HDB.hpp"        // The Hedge descriptor
#include "VDB.hpp"        // A Metus Variable descriptor
#include "XFZYctl.hpp"   // The fuzzy parallel processor work area
#include "XSYSctl.hpp"   // The Metus System control region
#include "mtypes.hpp"    // System constants and symbolics
#include "fuzzy.hpp"     // Fuzzy Logic constants and symbolics
#include "mdltypes.hpp"  // Definitons for the modeling system
#include "mtsptype.hpp"  // Metus function prototypes
//
//--------static variables used in the policy procedures-------
//
//--These are the "logical" rules.
//
static const char *Rules[]=
 {
  "R1   our price must be high",
  "R2   our price must be low",
  "R3   our price must be around 2*mfgcosts",
  "R4   if the competition.price is not very high",
  "        then our price must be around the competition.price"
 };
static int Rulecnt=0;
static int Rulemax=4;
//                          R1    R2   R3   R4
static int far RuleIDX[4]={ 0,    1,   2,   3  };
static int far RuleSLN[4]={ 1,    1,   1,   2  };
static char  *PgmId="fshpol2";
```

The rules for this policy have been moved from the main program into the policy itself—a desirable feature for models with more than one or two policies. The static scalar and array variables following the rules array are used to support the *DisplayRule* function. This is simply an internal version of the *FzyDisplayRule* function. These variables (including the program identifier *PgmId*) are static since they are shared by *ConcludeRule* and *DisplayRule*. The Rulemax variable indicates the maximum number of rules in this policy.

RuleIDX[i] indicates where in the *Rules[]* array the *i*'th rule begins. The *RuleSLN[i]* (rule source lines) indicates how many lines in the *Rules[]* array the *i*'th rule occupies.

```
//
double fshpr2policy(
   double MfgCosts,double CompPrice,float *CIXptr,int *statusPtr)
 {
  PDB      *prcPDBptr;
  FDB      *HiFDBptr,
           *LoFDBptr,
           *amFDBptr,
           *ncFDBptr,
           *PriceFDBptr,
           *wkFDBptr;
  HDB      *veryHDBptr;
  VDB      *VDBptr;
  FSV      *FSVptr;
  int       i,
            status;
  long      Memory;
  float     PriceCIX[8];
  double    Price,
            PriceEXV[8],
            NoPrice=0;

  *statusPtr=0;
  prcPDBptr=XSYSctl.XSYScurrPDBptr;
  mdlout=MtsGetSystemFile(SYSMODFILE);
  fprintf(mdlout,"%s\n","Price Estimation Policy Begins....");
  VDBptr=MdlFindVDB("PRICE",prcPDBptr,statusPtr);
  FzyInitFZYctl(statusPtr);
  if(!(FzyAddFZYctl(VDBptr,&PriceFDBptr,&FSVptr,statusPtr)))
    {
     *statusPtr=1;
     MtsSendError(12,PgmId,"PRICE");
     return(NoPrice);
    }
  thisCorrMethod  =FSVptr->FzySVimplMethod;
  thisDefuzzMethod=FSVptr->FzySVdefuzzMethod;
  fprintf(mdlout,"%s\n","The Rules:");
  for(i=0;i<Rulemax;i++) fprintf(mdlout,"%s\n",Rules[i]);
  fputc('\n',mdlout);
//
//--Find the fuzzy sets in the policy dictionary
//
  HiFDBptr=MdlFindFDB("HIGH",          prcPDBptr,statusPtr);
```

```
    LoFDBptr=MdlFindFDB("LOW",              prcPDBptr,statusPtr);
    amFDBptr=MdlFindFDB("Near.2*MfgCosts",prcPDBptr,statusPtr);
    ncFDBptr=MdlFindFDB("Near.CompPrice" ,prcPDBptr,statusPtr);
    wkFDBptr=MdlFindFDB("NULL",             prcPDBptr,statusPtr);
  //
  //--Find the hedge very in the policy dictionary
  //
    veryHDBptr=MdlFindHDB("very",prcPDBptr,statusPtr);
  //
  //--Set up explanatory controls
  //
    RXBhead    =NULL;
    RXBplastptr=NULL;
```

These static variables are defined in the *mdltypes.hpp* header file. They are used to chain together the rule explanatory nodes for each rule and must be initialized to *NULL* before any rule is actually evaluated.

```
  //
  //--------B E G I N    M O D E L    P R O C E S S I N G---------
  //
    Rulecnt     =0;
    Firedcnt    =0;
    PremiseTruth=1.0;
    fprintf(mdlout,"%s\n","Rule Execution....");
  //--Rule 1. Our Price must be High
    DisplayRule(Rulecnt);
    FzyUnCondProposition(HiFDBptr,FSVptr);
    ConcludeRule(PriceFDBptr,statusPtr);
    RXCptr=FzyXPScreateRXC("Price","High","",1.0);
    FzyXPSaddRXC(RXBptr,RXCptr,RXCconstype,statusPtr);
    FzyDrawSet(PriceFDBptr,SYSMODFILE,statusPtr);
    Rulecnt++;
  //
  //--Rule 2. Our Price must be Low
    DisplayRule(Rulecnt);
    FzyUnCondProposition(LoFDBptr,FSVptr);
    ConcludeRule(PriceFDBptr,statusPtr);
    RXCptr=FzyXPScreateRXC("Price","Low","",1.0);
    FzyXPSaddRXC(RXBptr,RXCptr,RXCconstype,statusPtr);
    FzyDrawSet(PriceFDBptr,SYSMODFILE,statusPtr);
    Rulecnt++;
  //
  //--Rule 3. Our Price must be around 2*MfgCosts
    DisplayRule(Rulecnt);
    FzyUnCondProposition(amFDBptr,FSVptr);
```

```
    ConcludeRule(PriceFDBptr,statusPtr);
    RXCptr=FzyXPScreateRXC("Price","Around 2*MfgCosts","",1.0);
    FzyXPSaddRXC(RXBptr,RXCptr,RXCconstype,statusPtr);
    FzyDrawSet(PriceFDBptr,SYSMODFILE,statusPtr);
    Rulecnt++;
```

The core logic for the unconditional fuzzy rules has not changed in this policy, but we have added some additional supporting features. Before the policy rule execution begins, we initialize the *Rulecnt* and *Firedcnt* variables to zero. *Rulecnt* is used by the *DisplayRule* function, and *Firedcnt* is updated by the *ConcludeRule* function and is used after all the rules have been evaluated. The static *PremiseTruth*, not used by the unconditionals in the previous policy, is used in the *ConcludeRule* function, so it is set to [1.0] for the duration of the unconditionals.

After the rule has been evaluated (through *FzyUnCondProposition*) the explanatory facility is invoked. *ConcludeRule*, as we will shortly see, creates the top explanatory node for the rule. We then create and add a rule explanatory clause node. Since there is no predicate for an unconditional rule, this clause is added as a consequent node. (The *RXCconstype* parameter indicates whether this is a predicate or consequent node.)

```
//
//--Rule 4. If the Competition.price is not very high,
//             then our price should be near the competition.price
  DisplayRule(Rulecnt);
  FzyInitFDB(wkFDBptr);
//
//--------------Evaluate Predicate of rule-----------------------
--
//
//--Apply the hedge VERY to the High fuzzy set
  FzyApplyHedge(HiFDBptr,veryHDBptr,wkFDBptr,statusPtr);
  FzyDrawSet(wkFDBptr,SYSMODFILE,statusPtr);
//--Apply the operator NOT to the hedged fuzzy set
  FzyApplyNOT(ZADEHNOT,0,wkFDBptr,statusPtr);
  FzyDrawSet(wkFDBptr,SYSMODFILE,statusPtr);
//--computed membership of competiton price in this fuzzy region
  PremiseTruth=FzyGetMembership(
    wkFDBptr,CompPrice,&Idxpos,statusPtr);
  fprintf(mdlout,"%s%10.2f\n","PremiseTruth= ",PremiseTruth);
//
//--------------Perform consequent proposition-------------------
--
//
  FzyCondProposition(
     ncFDBptr,FSVptr,thisCorrMethod,PremiseTruth,statusPtr);
```

```
ConcludeRule(PriceFDBptr,statusPtr);
RXCptr=FzyXPScreateRXC(
   "CompetitionPrice","High","not very",PremiseTruth);
FzyXPSaddRXC(RXBptr,RXCptr,RXCpredtype,statusPtr);
RXCptr=FzyXPScreateRXC(
   "Price","CompetitionPrice","near",PremiseTruth);
FzyXPSaddRXC(RXBptr,RXCptr,RXCconstype,statusPtr);
FzyDrawSet(PriceFDBptr,SYSMODFILE,statusPtr);
```

For an unconditional rule, the explanatory facility requires an explanation for each predicate and one for the consequent. We thus have two calls to *FzyXPSCreateRXC* and *FzyXPSaddRXC*—one for the predicate and one for the consequent.

```
//
//------S O L V E   F O R   M O D E L   S O L U T I O N------
//
  if(Firedcnt==0)
   {
    *statusPtr=88;
    fprintf(stdout,"%s\n",
     "CFRT Context is UNDECIDED. No Rules Fired!");
    *CIXptr=0.0;
    FzyCloseFZYctl(statusPtr);
    return(0);
   }
```

Since this policy contains unconditional rules, *Firedcnt* will never be zero; however, this type of code used to detect an undecided model should be part of all fuzzy system models.

```
//
//--Defuzzify to find expected value for price
//
  printf("..Defuzzifying Result:\n");
  strcpy(PriceFDBptr->FDBdesc,"FINAL Price Estimate");
  FzyDrawSet(PriceFDBptr,SYSMODFILE,statusPtr);
  printf("....Applying Defuzzification Strategies\n");
  dfuzzIX=0;
  for(i=0;i<MaxDefuzz;i++)
   {
    PriceEXV[i]=FzyDefuzzify(
       PriceFDBptr,DefuzzMethods[i],CIXptr,statusPtr);
    PriceCIX[i]=*CIXptr;
    if(thisDefuzzMethod==DefuzzMethods[i]) dfuzzIX=i;
   }
```

Instead of a single defuzzification process, this model explores the results of applying several defuzzification strategies and then storing the results in an array of price estimates (with their associated unit compatibility indexes). The list of defuzzification methods and other control variables (such as *MaxDefuzz*), shown below, are maintained in the *mdl-types.hpp* header file.

```
static int   far DefuzzMethods[]=
{CENTROID,MAXPLATEAU,MAXIMUM,AVGMAXIMUM,NEAREDGE};
static char far * far DefuzzNames[] =
   {
     "Centroid",
     "Maximum Plateau",
     "Maximum Height",
     "Average of Maximums",
     "Near Edge"
   };
static int MaxDefuzz=5;
```

During these strategy explorations, the stated defuzzification method (stored in *this-DefuzzMethod*) selected for the solution variable is identified, and its location in the array is indicated by the *dfuzzIX* variable. Having a policy evaluate several defuzzification schemes is not a normal component of production models,[7] but it provides an excellent method of observing the effects of different methods on the expected value.

```
Rulecnt--;
fprintf(mdlout,"%s\n","The Rules:");
sumTruth=0;
for(i=0;i<Rulemax;i++)
  if(rulefired[i]==TRUE)
    {
     fprintf(mdlout,"%4d%s%7.4f%s%s\n",
        i+1,". ",ruletruth[i]," ",Rules[RuleIDX[i]]);
     sumTruth+=ruletruth[i];
    }
avgTruth=sumTruth/Firedcnt;
fprintf(mdlout,"%s\n",     "        ------");
fprintf(mdlout,"%s%7.4f\n","       ",avgTruth);
fputc('\n',mdlout);
```

What is the execution profile of our policy? This next code section lists each rule that actually fired and its predicate truth value. The average truth of all the executed rules (`0.9380`) is also displayed. Figure 7.24 shows how this profile appears.

```
The Rules:
   1.  1.0000 R1   our price must be high
   2.  1.0000 R2   our price must be low
   3.  1.0000 R3   our price must be around 2*mfgcosts
   4.  0.7519 R4   if the competition.price is not very high
       ------
       0.9380
```

Figure 7.24 The Summary of Rule Execution Strengths in the Policy

The rules appear in the order of execution. Unconditional rules will always have a truth value of [1.0] and tend to bias the average truth of the model. For this reason, you might want to exclude unconditional rules when evaluating how well the model performed against data (since unconditionals are insensitive to actual data!).

```
fsetheight=FzyGetHeight(PriceFDBptr);
fprintf(mdlout,"%s\n","------------------------------------");
fprintf(mdlout,"%s\n","Model Solution:");
for(i=0;i<MaxDefuzz;i++)
 {
  fprintf(mdlout,"%s\n",DefuzzNames[i]);
  fprintf(mdlout,"%s%10.2f\n","  Price       : ",PriceEXV[i]  );
  fprintf(mdlout,"%s%10.2f\n","  CIX         : ",PriceCIX[i]  );
 }
fprintf(mdlout,"%s\n","------------------------------------");
fprintf(mdlout,"%s%10.2f\n",  "SurfaceHght  : ",fsetheight  );
```

The defuzzified values, along with their unit compatibility values, are displayed with this piece of code. The name of the defuzzification technique is printed, followed by the expected value and its strength. The actual height of the solution fuzzy set is also displayed. Figure 7.25 shows how this appears in the model activity log.

```
------------------------------------
Model Solution:
Centroid
  Price      :      24.75
  CIX        :       0.60
Maximum Plateau
  Price      :      26.00
  CIX        :       0.75
Maximum Height
  Price      :      26.00
  CIX        :       0.75
```

Figure 7.25 The Defuzzifications Display in the Pricing Policy

```
Average of Maximums
  Price       :      24.01
  CIX         :       0.34
Near Edge
  Price       :      23.62
  CIX         :       0.75
------------------------------------
SurfaceHght  :       0.75
```

Figure 7.25 The Defuzzifications Display in the Pricing Policy *(continued)*

The two most common defuzzification techniques are shown first: the *centroid* or composite moments and the *maximum plateau* or the composite maximum. These are followed by the maximum height (if the output fuzzy set has a plateau, then this figure and the composite maximum will normally be the same), the average of maximums, and the near edge (the edge at the left-hand side of the plateau).

```
//
  fprintf(mdlout,"\f\n\n%s\n","EXPLANATIONS:");
  FzyXPSshowrules(RXBhead);
  FzyXPSconclusion(
    RXBhead->RXBconsptr,
    PriceEXV[dfuzzIX],PriceFDBptr->FDBdomain,PriceCIX[dfuzzIX]);
  FzyCloseFZYctl(statusPtr);
  MdlClosePolicy(prcPDBptr,&Memory,statusPtr);
  printf("....Selecting Stated Defuzzification Strategy\n");
  *CIXptr=PriceCIX[dfuzzIX];
  return(PriceEXV[dfuzzIX]);
 }
```

The policy is ended by displaying the explanations, releasing all memory held by the fuzzy work area and the policy itself (*MdlClosePolicy* releases the memory held in each of the policy dictionaries), and returning the defuzzified value corresponding to the required defuzzification method. The explanatory facility provides two levels of detail. You can request an explanation on a rule-by-rule basis (*FzyXPSshowrules*) or simply as an explanation of the policy's conclusion (*FzyXPSconclusion*). The rules are sorted in descending order by their associated truth values and printed on an indented scale. Figure 7.26 shows the explanations generated for a typical pricing policy execution.

```
EXPLANATIONS:
1   .75  .50  .25   0
|----+----+----+----+------------------------R U L E S----|
 The contribution of rule 1 means that Price is certainly High
```

Figure 7.26 Explanations from the Pricing Policy

```
 According to rule 2, Price is definitely Low
 According to rule 3, Price is certainly Around 2*MfgCosts
    According to rule 4, we have positively suggested that
    because CompetitionPrice is, for the most part, not very High
     the Price is, to a large degree, near CompetitionPrice

--------------------------------------S O L U T I O N S------

In Conclusion....From these rules we find that "Price"
has a value of '24.750' which is, in part, supported.
This is an extremely high value.
```

Figure 7.26 Explanations from the Pricing Policy *(continued)*

The explanatory facility's conclusion is based on the strength of the solution variable's fuzzy set and the magnitude of the number compared to the underlying domain. The annotation that '24.750' is an extremely high value indicates that it lies somewhere in the extreme right-hand side of the solution variable's domain.

```
//
static void ConcludeRule(FDB *PriceFDBptr,int *CondPropstatus)
  {
   char   wrkBuff[128];
   int    status;

   ruletruth[Rulecnt]=PremiseTruth;
   sprintf(wrkBuff,"%s%6.4f","Premise Truth= ",PremiseTruth);
   MtsWritetoLog(SYSMODFILE,wrkBuff,&status);
   if(*CondPropstatus!=RULEBELOWALFA)
     {
      rulefired[Rulecnt]=TRUE;
      FzyDrawSet(PriceFDBptr,SYSMODFILE,&status);
      ++Firedcnt;
     }
   if((RXBptr=FzyXPScreateRXB(Rulecnt+1,0,PremiseTruth))==NULL)
      {
       sprintf(wrkBuff,"%s%4d","RXB for Rule ",Rulecnt);
       MtsSendError(2,PgmId,wrkBuff);
       return;
      }
   if(RXBhead==NULL) RXBhead=RXBptr;
     else              RXBplastptr->RXBnext=RXBptr;
   RXBplastptr=RXBptr;
  }
```

We make a call to *ConcludeRule* after each rule has been evaluated and after the solution variable's fuzzy region has been updated. This function records and displays the predicate truth and, if the truth is nonzero (it lies above the current system alpha-cut threshold), marks the rule as fired and displays the current solution fuzzy set. Finally, a new rule explanatory block (*RXB*) is allocated, initialized, and linked into the tail of the current rule explanation list.

```
//
static void DisplayRule(int WhichRule)
 {
  int i,j,k;

  j=RuleIDX[WhichRule];
  k=RuleSLN[WhichRule];
  i=0;
  fprintf(mdlout,"%s\n",
      "  -----------------------------------------------");
  for(i=0;i<k;i++,j++)  fprintf(mdlout,"%s\n",Rules[j]);
  return;
 }
```

Listing 7.38 The Policy Program Code for `fshpr2policy`

This final function, as shown in Listing 7.38, displays the current rule. It uses the *RuleSLN[]* array to determine how many lines in the *Rules[]* array must be printed for the current rule.

The Interpretation of Model Results

Fuzzy models, like any conventional system, must be validated and verified through their predictive behavior against known cases or against the reasonable judgments of experts (or, failing that, against common sense). Knowledge engineers and systems analysts, however, must still determine whether or not the output of a fuzzy model is consistent with the relationships between rules and fuzzy sets, and whether or not the expected value of a solution variable is, within the context of the model's internal logic, valid. These kinds of decisions are necessary in order to decide when we can begin a thorough stress test of the model itself. In this section, we look at two factors associated with the model's internal logic: the idea of undecidable models, and the general notion of compatibility between the fuzzy spaces in the model, the rule productions, and the model's actual data. This latter is a measurement is based on the principle of the unit and statistical compatibility index.

UNDECIDABLE MODELS

The result of executing a fuzzy model can be undecidable. This occurs when the output fuzzy region for one or more solution variables is either all zeros or all ones. Figure 7.27 shows the results of executing the INVMOD3 program. This model contains a single rule:

```
if orders are high then backorderAmt must be large
```

Orders	BackOrder	CIX
60	0.00	0.0000
80	0.00	0.0000
100	212.11	0.0290
130	212.11	0.1257
150	212.11	0.1902
190	212.11	0.3159
200	212.11	0.3481

Figure 7.27 Undecidable Backorder Values for the *INVMOD3* Mode

When the value for *orders* is below the minimum truth for HIGH, the rule does not execute. Since there are no unconditional rules, the fuzzy output space associated with *backorderAmt* is left unchanged. Notice that in Figure 7.27 the first two entries are undecidable.

The result of an undecidable model is a zero expected value and a zero unit compatibility index. In Figure 7.27, no special action was taken when an undecidable policy result was returned. In production models (especially where zero might be a valid expected value[8]) some exception process should be initiated.

> The *FzyIsUndecidable* function examines the contents of a fuzzy set. If the truth function is all zero or all one, then it reports *True;* otherwise, *False* is returned. Note that *FzyDefuzzify* makes a check for an empty fuzzy set so the greater preponderance of undecidables will be caught at that point.

```
bool FzyIsUndecidable(FDB *FDBptr)
 {
  int i;
  if(FzyGetHeight(FDBptr)==0)      return(TRUE);
  for(i=0;i<VECMAX;i++)
   if(FDBptr->FDBvector[i]!=1.0) return(FALSE);
  return(TRUE);
 }
```

Listing 7.39 The Undecidable Policy Function

Figure 7.29 is an excerpt from the inventory policy execution. Note that the empty fuzzy set is detected during the defuzzification process. Since most undecidable models are output fuzzy sets with a height of zero, no other tests are made.

Here we can see how the undecidable solution region is produced. When the rule fails to execute, the output fuzzy region is left in its empty condition. The defuzzification function detects this condition and indicates that the output cannot be processed. Although the centroid (composite moments) method is shown, none of the defuzzification methods will work on an empty fuzzy set.

```
Fuzzy WIDGET Inventory Model. (c) 1998 Metus Systems Group.
[Mean] Quarterly Orders :      60.00
R1   if orders are high then backorderAmt must be large
PremiseTruth=       0.00

FuzzySet:     BOAmt
        Description:
      1.00
      0.90
      0.80
      0.70
      0.60
      0.50
      0.40
      0.30
      0.20
      0.10
      0.00.................................................................
          0---|---|---|---|---|---|---|---|---|---|---|---|---|---|---|---0
         0.00   37.50   75.00  112.50  150.00  187.50  225.00  262.50  300.00
        Domained:          0.00 to      300.00

---Cannot Apply "Centroid" Defuzzification--'BOAmt' is empty. [height==0].
Model Solution:
  BackOrderAmt=       0.00
  CompIdx      =      0.00
  surface hght=       0.00

[Mean] Quarterly Orders :      80.00
R1   if orders are high then backorderAmt must be large
PremiseTruth=       0.00

        FuzzySet:     BOAmt
        Description:
      1.00
      0.90
      0.80
      0.70
      0.60
      0.50
      0.40
      0.30
      0.20
      0.10
      0.00.................................................................
          0---|---|---|---|---|---|---|---|---|---|---|---|---|---|---|---0
         0.00   37.50   75.00  112.50  150.00  187.50  225.00  262.50  300.00
        Domained:          0.00 to      300.00
```

Figure 7.28 Execution of the *INVMOD3* Inventory Program for Selected Orders

```
---Cannot Apply "Centroid" Defuzzification--'BOAmt' is empty. [height==0].
Model Solution:
  BackOrderAmt=      0.00
  CompIdx     =      0.00
  surface hght=      0.00


[Mean] Quarterly Orders :     130.00
R1   if orders are high then backorderAmt must be large
PremiseTruth=      0.15

        FuzzySet:    BOAmt
        Description:
      1.00
      0.90
      0.80
      0.70
      0.60
      0.50
      0.40
      0.30
      0.20
      0.10                                     .............................
      0.00.....................................
          0---|---|---|---|---|---|---|---|---|---|---|---|---|---|---|---0
         0.00   37.50   75.00  112.50  150.00  187.50  225.00  262.50  300.00
        Domained:          0.00 to      300.00

09/28/1993  1.36pm---'CENTROID'     defuzzification. Value:    212.109, [0.1257]
Model Solution:
  BackOrderAmt=    212.11
  CompIdx     =      0.13
  surface hght=      0.15

[Mean] Quarterly Orders :     150.00
R1   if orders are high then backorderAmt must be large
PremiseTruth=      0.23

        FuzzySet:    BOAmt
        Description:
      1.00
      0.90
      0.80
      0.70
      0.60
      0.50
      0.40
      0.30
      0.20                                               ...................
      0.10                              .................
      0.00.............................
          0---|---|---|---|---|---|---|---|---|---|---|---|---|---|---|---0
         0.00   37.50   75.00  112.50  150.00  187.50  225.00  262.50  300.00
        Domained:          0.00 to      300.00

09/28/1993  1.36pm---'CENTROID'     defuzzification. Value:    212.109, [0.1902]
Model Solution:
  BackOrderAmt=    212.11
  CompIdx     =      0.19
  surface hght=      0.23
```

Figure 7.28 Execution of the *INVMOD3* Inventory Program for Selected Orders *(continued)*

```
[Mean] Quarterly Orders :      190.00
R1   if orders are high then backorderAmt must be large
PremiseTruth=        0.38

        FuzzySet:     BOAmt
        Description:
      1.00
      0.90
      0.80
      0.70
      0.60
      0.50
      0.40
      0.30                                              .......................
      0.20                                    ..........
      0.10                          ..........
      0.00......................
          0---|---|---|---|---|---|---|---|---|---|---|---|---|---|---|---0
         0.00   37.50   75.00  112.50  150.00  187.50  225.00  262.50  300.00
        Domained:          0.00 to      300.00

09/28/1993  1.36pm---'CENTROID'     defuzzification. Value:    212.109, [0.3159]
Model Solution:
  BackOrderAmt=      212.11
  CompIdx     =        0.32
  surface hght=        0.38
```

Figure 7.29 Execution of the *INVMOD3* Inventory Program for Selected Orders

Compatibility Index Metrics

How can we measure the robustness[9] of a fuzzy model? One interesting way is through the strength of its recommendations. How well does the model's logic and its data fit together, and how consistent is the strength of its recommendations across time? Factors such as these contribute to the overall responsiveness of the model and are tied directly to our belief that the model is working correctly. A critical, observable aspect of the model is the structure of the fuzzy sets associated with each solution variable. Unlike the base vocabulary sets, the solution fuzzy sets are created through the rule aggregation process. They indicate, consequently, the degree of truth in the underlying rule productions, which indicate how well the rules respond to the model data. The relationship between these factors is expressed in the *compatibility index.*

The Idea of a Compatibility Index

Fuzzy models, using the idea of a compatibility index (CIX), have an intrinsic way of measuring their compatibility with model data. There are two kinds of model compatibility: *statistical* and *unit.* The statistical compatibility measures how well the model performs over a wide range of data (and is a true measure of actual system compatibility), while the unit compatibility measures the recommendation strength of a single model execution. The idea behind the compatibility index is simple: if the height of an output fuzzy region is close to [0] or [1], then the model assumes the properties of a Boolean space. This very high or low height indicates that the data lies at the extremes of the fuzzy sets causing the predicate truth to be consistently close to zero or one. As Figure 7.30 illustrates, there are regions of incompatibility in solution fuzzy sets.

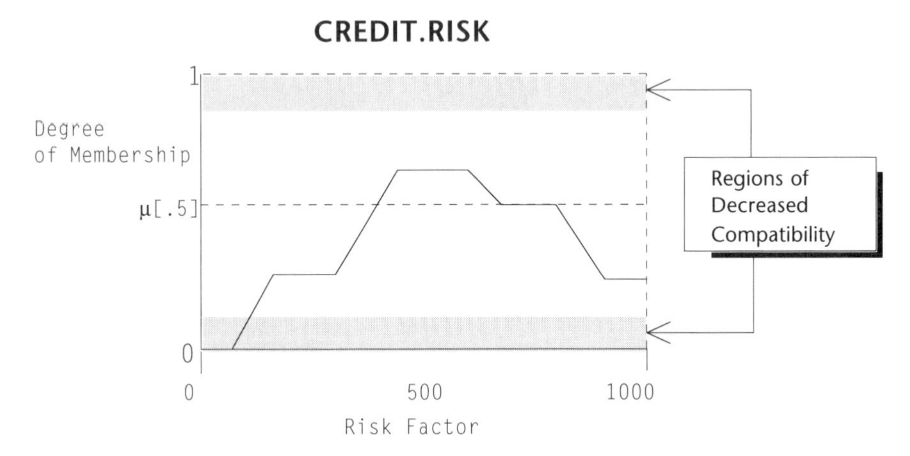

Figure 7.30 Regions of Incompatibility in the Solution Fuzzy Set

The Unit Compatibility Index

There is one unit compatibility index for each solution variable in a model or policy. The unit compatibility measures the height of the solution variable's fuzzy region. This index reading is often important because the composite moments and composite maximum techniques are generally insensitive to the strength of the output fuzzy region. Figure 7.31 shows the centroid defuzzification of *Risk* for a fuzzy set whose height is $\mu[.72]$. The center of gravity elicits an expected value of "492" in the risk factor domain.

Figure 7.32 shows the same defuzzification when the height of the output fuzzy set is only $\mu[.42]$. The center of gravity is essentially the same; consequently, the expected value is the same.

When the fuzzy region is very weak (see Figure 7.33), at a height of $\mu[.15]$ the centroid defuzzification process still locates the same balance point and produces the same expected value for the policy.

The composite maximum, while producing a slightly different answer, would also generate the same expected value for each of these truth levels. Defuzzification methods are essentially sensitive to the width of the fuzzy region but not to its height. Yet the height of a solution variable's output region indicates the strength of the model's recommendation.

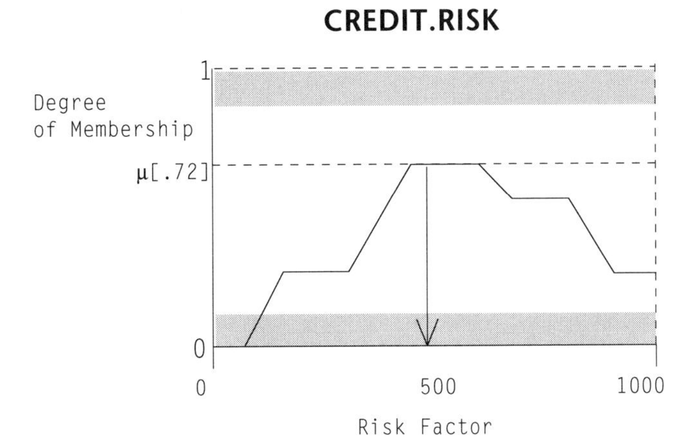

Figure 7.31 Defuzzification of *Risk* at μ[.72]

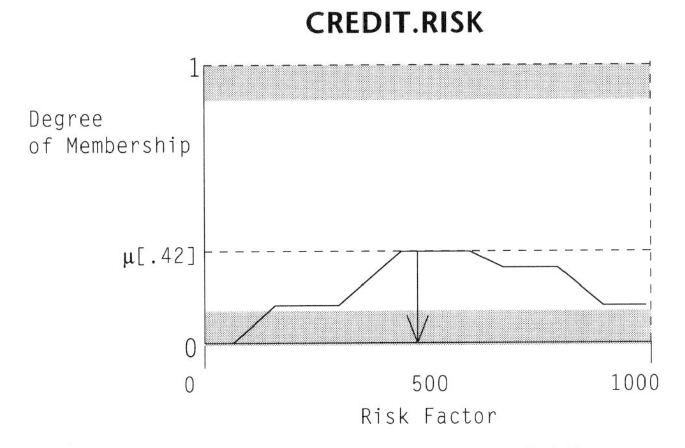

Figure 7.32 Defuzzification of *Risk* at μ[.42]

The implication of the compatibility index on the strength of a model's recommendation can be seen clearly from the output of the single rule inventory model:

if orders are high then backorderAmt must be large

Figure 7.34 shows the results of executing the INVMOD4 program, which shows the compatibility index for selected values of orders.

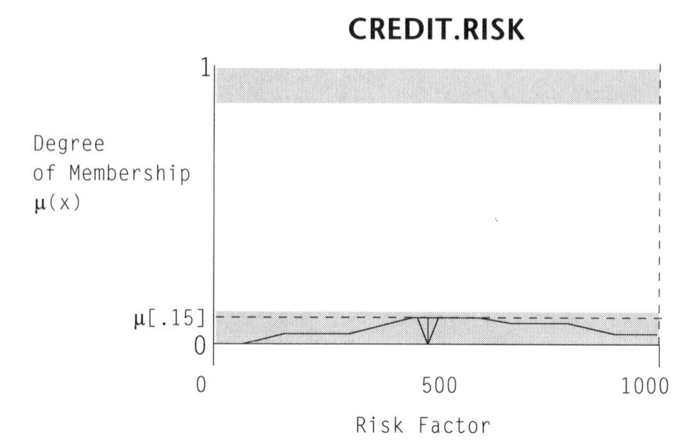

Figure 7.33 Defuzzification of *Risk* at μ[.15]

Orders	BackOrder	CIX
60	0.00	0.0000
80	0.00	0.0000
92	212.11	0.0032
95	212.11	0.0129
97	212.11	0.0193
100	212.11	0.0290
130	212.11	0.1257
150	212.11	0.1902
190	212.11	0.3159
200	212.11	0.3481
240	212.11	0.4738
280	212.11	0.6027
300	212.11	0.6640

Figure 7.34 Compatibility for *Backorder Amounts* from the *INVMOD4* Program

The single rule produces approximately the same shaped output fuzzy region at different heights, depending on the truth of the rule's predicate: *orders are high*. The centroid defuzzification method finds the same expected value for each backorder amount regardless of the solution fuzzy set's height.[10] To see this clearly, we can follow the execution of the single rule inventory system for various order levels. Figure 7.35 shows the fuzzy set that the model uses to determine when orders are considered high. Figure 7.36 shows the fuzzy set for a large backorder amount. Figure 7.37 contains the execution trace for the model.

```
      FuzzySet:     HIGH.ORDERS
      Description:
   1.00                                                                    .
   0.90                                                               ......
   0.80                                                         .......
   0.70                                                   ......
   0.60                                            .......
   0.50                                      ......
   0.40                                ......
   0.30                         .......
   0.20                  ......
   0.10           .......
   0.00......
        0---|---|---|---|---|---|---|---|---|---|---|---|---|---|---|---|---0
      90.00  122.50  155.00  187.50  220.00  252.50  285.00  317.50  350.00
      Domained:          90.00 to      350.00
```

Figure 7.35 The Base Fuzzy Set *HIGH.ORDERS* in the *INVMOD4* Inventory Model

```
      FuzzySet:     LARGE.BOAmt
      Description:
   1.00                                                                   .....
   0.90                                                         ...........
   0.80                                                    .....
   0.70                                               .....
   0.60                                           ...
   0.50                                       ....
   0.40                                   ...
   0.30                              ....
   0.20                         .....
   0.10                   ......
   0.00..............
        0---|---|---|---|---|---|---|---|---|---|---|---|---|---|---|---|---0
        0.00    37.50   75.00  112.50  150.00  187.50  225.00  262.50  300.00
      Domained:          0.00 to      300.00
```

Figure 7.36 The Base Fuzzy Set *LARGE.BOAmt* in the *INVMOD4* Inventory Model

```
Fuzzy WIDGET Inventory Model. (c) 1998 Metus Systems Group.
[Mean] Quarterly Orders :       92.00
R1   if orders are high then backorderAmt must be large
PremiseTruth=        0.00

      FuzzySet:     BOAmt
      Description:
   1.00
   0.90
   0.80
   0.70
   0.60
   0.50
   0.40
   0.30
   0.20
   0.10
   0.00.................................................................
        0---|---|---|---|---|---|---|---|---|---|---|---|---|---|---|---|---0
        0.00    37.50   75.00  112.50  150.00  187.50  225.00  262.50  300.00
      Domained:          0.00 to      300.00
```

Figure 7.37 Execution of *INVMOD4* with Backorder Amounts for Various Order Levels

```
---'CENTROID'     defuzzification. Value:     212.109, [0.0032]
Model Solution:
  BackOrderAmt=     212.11
  CompIdx     =       0.00
  surface hght=       0.00


[Mean] Quarterly Orders :       95.00
R1   if orders are high then backorderAmt must be large
PremiseTruth=       0.02

        FuzzySet:    BOAmt
        Description:
      1.00
      0.90
      0.80
      0.70
      0.60
      0.50
      0.40
      0.30
      0.20
      0.10
      0.00.................................................................
          0---|---|---|---|---|---|---|---|---|---|---|---|---|---|---|---0
         0.00   37.50   75.00  112.50  150.00  187.50  225.00  262.50  300.00
        Domained:          0.00 to      300.00

---'CENTROID'     defuzzification. Value:     212.109, [0.0129]
Model Solution:
  BackOrderAmt=     212.11
  CompIdx     =       0.01
  surface hght=       0.02

[Mean] Quarterly Orders :       97.00
R1   if orders are high then backorderAmt must be large
PremiseTruth=       0.02

        FuzzySet:    BOAmt
        Description:
      1.00
      0.90
      0.80
      0.70
      0.60
      0.50
      0.40
      0.30
      0.20
      0.10
      0.00.................................................................
          0---|---|---|---|---|---|---|---|---|---|---|---|---|---|---|---0
         0.00   37.50   75.00  112.50  150.00  187.50  225.00  262.50  300.00
        Domained:          0.00 to      300.00
```

Figure 7.37 Execution of *INVMOD4* with Backorder Amounts for Various Order Levels *(continued)*

```
---'CENTROID'     defuzzification. Value:     212.109, [0.0193]
Model Solution:
  BackOrderAmt=     212.11
  CompIdx     =       0.02
  surface hght=       0.02


[Mean] Quarterly Orders :     100.00
R1   if orders are high then backorderAmt must be large
PremiseTruth=       0.04

        FuzzySet:    BOAmt
        Description:
      1.00
      0.90
      0.80
      0.70
      0.60
      0.50
      0.40
      0.30
      0.20
      0.10
      0.00..............................................................
          0---|---|---|---|---|---|---|---|---|---|---|---|---|---|---|---0
         0.00   37.50   75.00  112.50  150.00  187.50  225.00  262.50  300.00
        Domained:          0.00 to      300.00

09/28/1993  4.55pm---'CENTROID'     defuzzification. Value:     212.109, [0.0290]
Model Solution:
  BackOrderAmt=     212.11
  CompIdx     =       0.03
  surface hght=       0.04

[Mean] Quarterly Orders :     130.00
PremiseTruth=       0.15

        FuzzySet:    BOAmt
        Description:
      1.00
      0.90
      0.80
      0.70
      0.60
      0.50
      0.40
      0.30
      0.20
      0.10                                   ...........................
      0.00...................................
          0---|---|---|---|---|---|---|---|---|---|---|---|---|---|---|---0
         0.00   37.50   75.00  112.50  150.00  187.50  225.00  262.50  300.00
        Domained:          0.00 to      300.00
```

Figure 7.37 Execution of *INVMOD4* with Backorder Amounts for Various Order Levels *(continued)*

```
---'CENTROID'     defuzzification. Value:    212.109, [0.1257]
Model Solution:
  BackOrderAmt=     212.11
  CompIdx     =       0.13
  surface hght=       0.15


[Mean] Quarterly Orders :     150.00
R1   if orders are high then backorderAmt must be large
PremiseTruth=       0.23

        FuzzySet:    BOAmt
        Description:
      1.00
      0.90
      0.80
      0.70
      0.60
      0.50
      0.40
      0.30
      0.20                                             ...................
      0.10                             .................
      0.00............................
          0---|---|---|---|---|---|---|---|---|---|---|---|---|---|---|---0
         0.00   37.50   75.00  112.50  150.00  187.50  225.00  262.50  300.00
        Domained:         0.00 to     300.00

---'CENTROID'     defuzzification. Value:    212.109, [0.1902]
Model Solution:
  BackOrderAmt=     212.11
  CompIdx     =       0.19
  surface hght=       0.23


[Mean] Quarterly Orders :     190.00
R1   if orders are high then backorderAmt must be large
PremiseTruth=       0.38

        FuzzySet:    BOAmt
        Description:
      1.00
      0.90
      0.80
      0.70
      0.60
      0.50
      0.40
      0.30                                         .......................
      0.20                                ..........
      0.10                      ..........
      0.00......................
          0---|---|---|---|---|---|---|---|---|---|---|---|---|---|---|---0
         0.00   37.50   75.00  112.50  150.00  187.50  225.00  262.50  300.00
        Domained:         0.00 to     300.00
```

Figure 7.37 Execution of *INVMOD4* with Backorder Amounts for Various Order Levels *(continued)*

```
---'CENTROID'     defuzzification. Value:     212.109, [0.3159]
Model Solution:
  BackOrderAmt=      212.11
  CompIdx     =        0.32
  surface hght=        0.38

[Mean] Quarterly Orders :      200.00
R1   if orders are high then backorderAmt must be large
PremiseTruth=        0.42

        FuzzySet:    BOAmt
        Description:
      1.00
      0.90
      0.80
      0.70
      0.60
      0.50
      0.40                                                            .............
      0.30                                                .............
      0.20                                       ........
      0.10                              ..........
      0.00.....................
          0---|---|---|---|---|---|---|---|---|---|---|---|---|---|---|---0
         0.00   37.50   75.00  112.50  150.00  187.50  225.00  262.50  300.00
        Domained:         0.00 to     300.00

---'CENTROID'     defuzzification. Value:     212.109, [0.3481]
Model Solution:
  BackOrderAmt=      212.11
  CompIdx     =        0.35
  surface hght=        0.42

[Mean] Quarterly Orders :      240.00
R1   if orders are high then backorderAmt must be large
PremiseTruth=        0.57

        FuzzySet:    BOAmt
        Description:
      1.00
      0.90
      0.80
      0.70
      0.60
      0.50                                                       ..................
      0.40                                               ........
      0.30                                         ......
      0.20                                   ......
      0.10                          .........
      0.00..................
          0---|---|---|---|---|---|---|---|---|---|---|---|---|---|---|---0
         0.00   37.50   75.00  112.50  150.00  187.50  225.00  262.50  300.00
        Domained:         0.00 to     300.00
```

Figure 7.37 Execution of *INVMOD4* with Backorder Amounts for Various Order Levels *(continued)*

```
---'CENTROID'     defuzzification. Value:    212.109, [0.4738]
Model Solution:
  BackOrderAmt=     212.11
  CompIdx     =       0.47
  surface hght=       0.57

[Mean] Quarterly Orders :     280.00
R1   if orders are high then backorderAmt must be large
PremiseTruth=       0.73

        FuzzySet:    BOAmt
        Description:
      1.00
      0.90
      0.80
      0.70                                                      ...........
      0.60                                              .........
      0.50                                        ......
      0.40                                   .....
      0.30                              .....
      0.20                         .....
      0.10                 ........
      0.00................
          0---|---|---|---|---|---|---|---|---|---|---|---|---|---|---|---0
         0.00   37.50   75.00  112.50  150.00  187.50  225.00  262.50  300.00
        Domained:         0.00 to     300.00

---'CENTROID'     defuzzification. Value:    212.109, [0.6027]
Model Solution:
  BackOrderAmt=     212.11
  CompIdx     =       0.60
  surface hght=       0.73


[Mean] Quarterly Orders :     300.00
R1   if orders are high then backorderAmt must be large
PremiseTruth=       0.80

        FuzzySet:    BOAmt
        Description:
      1.00
      0.90
      0.80
      0.70                                                           ......
      0.60                                                ...........
      0.50                                         .......
      0.40                                    .....
      0.30                                ....
      0.20                            ....
      0.10                       .....
      0.00                ........
      0.00................
          0---|---|---|---|---|---|---|---|---|---|---|---|---|---|---|---0
         0.00   37.50   75.00  112.50  150.00  187.50  225.00  262.50  300.00
        Domained:         0.00 to     300.00
```

Figure 7.37 Execution of *INVMOD4* with Backorder Amounts for Various Order Levels *(continued)*

```
---'CENTROID'    defuzzification. Value:    212.109, [0.6640]
Model Solution:
  BackOrderAmt=     212.11
  CompIdx     =       0.66
  surface hght=       0.80
```

Figure 7.37 Execution of *INVMOD4* with Backorder Amounts for Various Order Levels *(continued)*

The meaning of unit compatibility is clear: the lower the height of the solution fuzzy set, the less compatibility there is between the decision and the underlying production rules. In the first nonzero model execution (refer back to Figure 7.34), when orders are 92, the recommended backorder amount is 212.11 with a unit compatibility of .0032.[11] Intuitively we can see that this recommendation is not supported. We can interpret the compatibility as a measure of how far into the Cartesian product of the rule matrix the fuzzy solution space sits.

You should not assume that a low unit compatibility value on any particular execution of a model is cause for concern. Fuzzy models, like any decision system, respond to data points across the entire spectrum of their operating range. You must decide within the model context whether a low unit compatibility affects your confidence in the model results. Also, as a general rule, the more rules that contribute to a solution, the less pronounced are the compatibility problems (perhaps a common sense view—the more evidence you bring to bear on a problem, the less likely that all the evidence will lie at the edges of your rule set).

Scaling Expected Values by the Compatibility Index

One approach that has been successful in risk assessment, resource allocation, inventory, and anomaly detection models uses the compatibility index as a scaling factor on the expected value. This idea stems from the view that the index falls within the range [0,1], and the smaller the compatibility index, the less strength we have in the output fuzzy sets. Figure 7.38, produced by the *INVMOD5* inventory program, shows the results of scaling the expected backorder amounts by the compatibility index (the far right column).

Orders	BackOrder	CIX	CIX*BOAmt
60	0.00	0.0000	0.00
80	0.00	0.0000	0.00
92	212.11	0.0032	0.68

Figure 7.38 Expected Backorders from *INVMOD5* Scaled by the Compatibility Index

95	212.11	0.0129	2.73
97	212.11	0.0193	4.10
100	212.11	0.0290	6.15
130	212.11	0.1257	26.66
150	212.11	0.1902	40.34
190	212.11	0.3159	67.00
200	212.11	0.3481	73.83
240	212.11	0.4738	100.50
280	212.11	0.6027	127.84
300	212.11	0.6640	140.83

Figure 7.38 Expected Backorders from *INVMOD5* Scaled by the Compatibility Index *(continued)*

Intuitively this approach appears to work well; the low index values scale the backorders to a reasonable level for orders that are just slightly within the *High.Orders* fuzzy set. Quarterly orders of 95 (just five orders into the fuzzy set) yield an adjusted backorder of 3 Widgets. (2.73 rounded to the next highest integer—we can't order fractional Widgets!)

The Statistical Compatibility Index

The unit compatibility index is important when deciding whether or not to accept a particular model recommendation or when we need to adjust the recommendation based on the strength of the production rule set. But unit compatibility index values say nothing about the overall robustness and fitness of the fuzzy model. To assess how the model performs over a wide spectrum of data points, we must examine the statistical compatibility index. This index is calculated as the average of the unit compatibility index values over a significantly large execution space. As a rule of thumb, the statistical compatibility index, over time, should tend to a mean value between $\mu[.40]$ and $\mu[.80]$. Average values above or below these limits—especially if they are well below or well above the low and high points, respectively—indicate that there is a long-term, endemic weakness in the production rules. Too many data points in the Cartesian space are triggering unit compatibilities of close to zero or close to one. As Figure 7.39 indicates, a model that has a preponderance of its compatibility indices either high or low begins to adopt a Boolean behavior pattern. The outcome of the rule set centers around the Boolean truth states of one or zero.

A trend line through the compatibility index generated over a large number of model executions provides an insight into the model performance. As previously noted, a model that is working well should have a long-term average compability index in the interval [.40,.80]. This also indicates a trend line that is roughly perpendicular to a region around the [.5] membership in the outcome fuzzy set. Figure 7.40 illustrates this concept.

On the other hand, a trend that proceeds upward or downward in the long term indicates a model that is becomming less and less well-behaved. In these instances, the data is falling more and more away from the central measures of the underlying fuzzy sets and

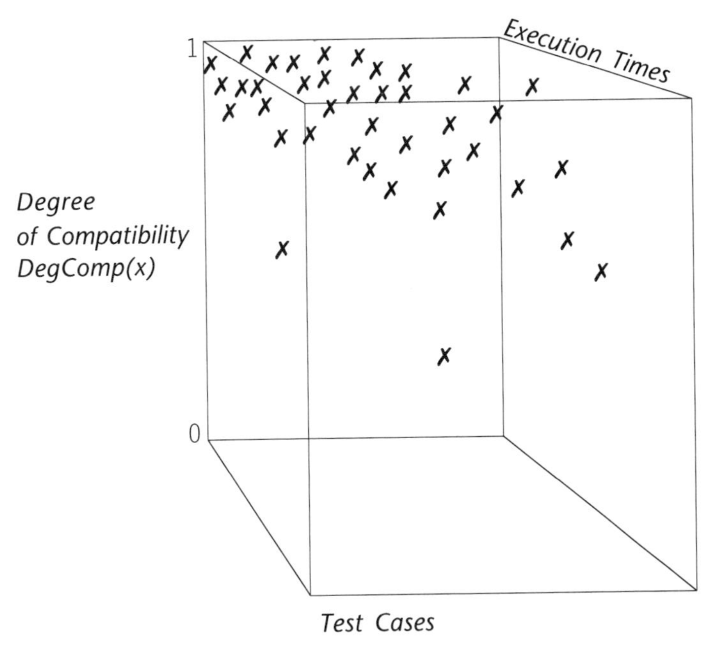

Figure 7.39 A Fuzzy Model with Moderately Boolean Behavior Properties

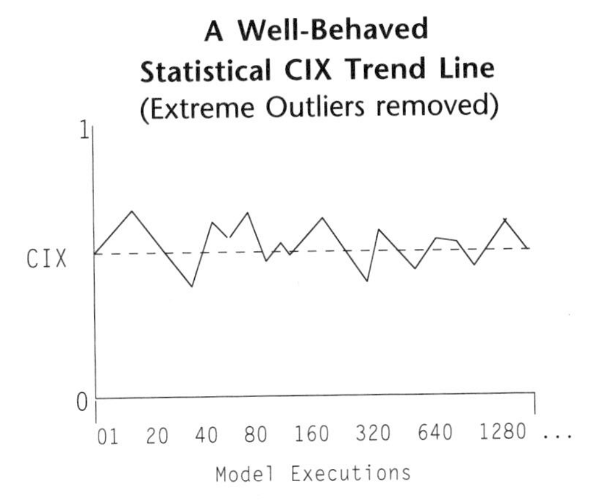

Figure 7.40 The CIX Trend Line (Well-Behaved Model)

into the set extremes. As an example, Figure 7.41 illustrates two poorly behaved compatibility trend lines.

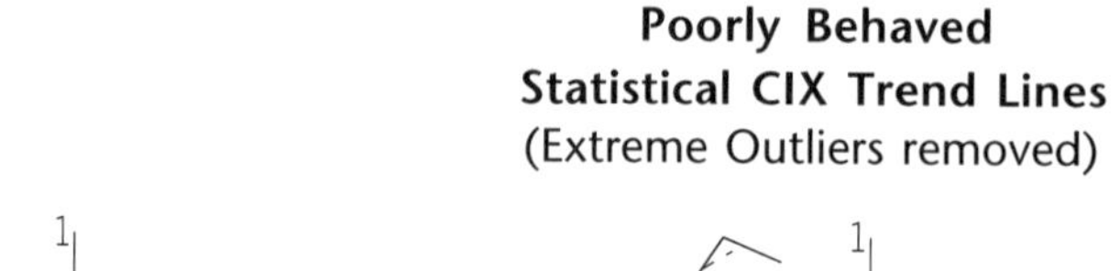

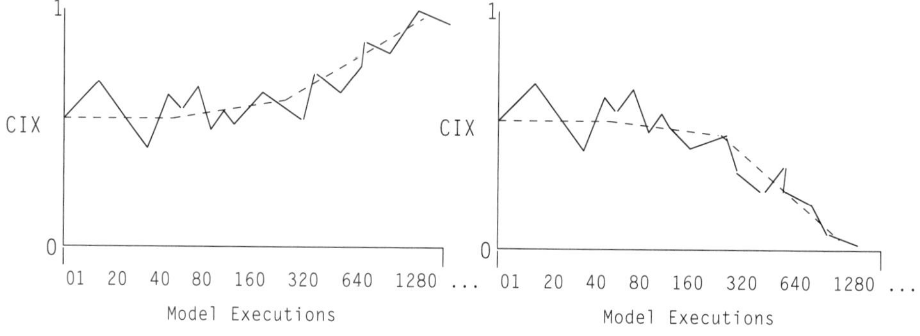

Figure 7.41 CIX Trend Lines (Poorly Behaved)

What exactly is the statistical compatibility index telling us? As the name implies, it measures the compatibility between the model's outcome prediction and the data used to make that prediction. Consider a very simple credit risk assessment model based solely on the candidate's pre-tax income. Figure 7.42 shows some of the fuzzy sets for this model.

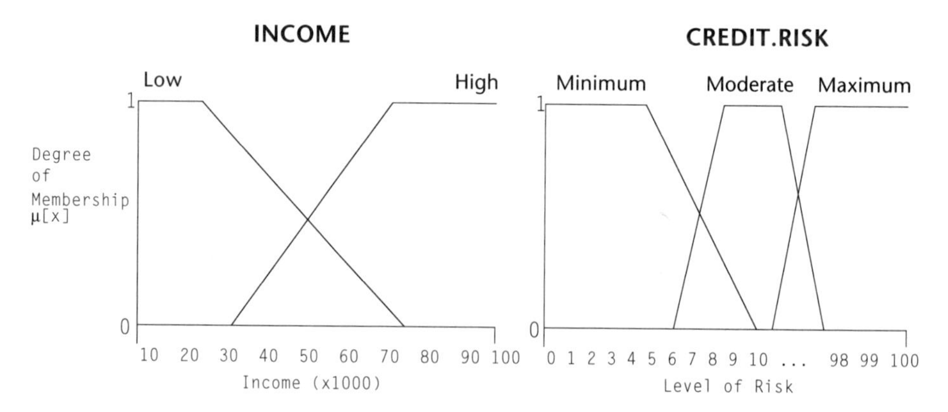

Figure 7.42 The Credit Risk Assessment Fuzzy Sets

Associated with these fuzzy sets are the actual production rules that describe the knowledge base for the model. To keep matters simple and focus on the use of the compatibility index, we will examine just a handful of the underlying rules:

```
if Income is Low then Credit.Risk is Moderate
if Income is High then Credit.Risk is Minimum
```

We will assume that the income variable's range represents the distribution of incomes in the underlying demographics for which the model was designed, with a mean of roughly $45K to $50K. Thus, we would expect most incomes to fall in the range [25,75], so that the majority of the membership values will fall in the range [.2,.8], with a corresponding frequency in the compatibility index. However, if the demographics of the population shift so that the mean income moves toward the $60K to $70K range, then the model execution centers around a small range of values. Figure 7.43 illustrates this shift.

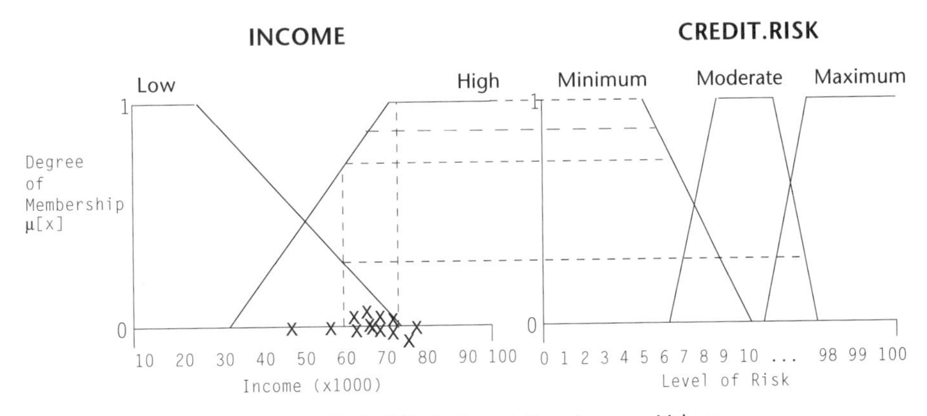

Figure 7.43 A Shift in Population Income Values

From Figure 7.41 we can see that, as the model executes, most of the data elements are drawn from the new distribution of data causing a preponderance of truth values in a localized region. Thus, an average value (the middle line in the high to medium transfer region) is produced. Over a period of time, or over a large number of executions, this average value begins to dominate the compatibility index for the model.

When we select a typical value from the clustered region, say $68K, and run the rules, both will fire with differing degrees of predicate truth values, as illustrated in the following rule:

```
if Income is Low [.30] then Credit.Risk is Moderate
if Income is High [.92] then Credit.Risk is Minimum
```

As Figure 7.44 shows, the combined outcome of the rules, using composite maximum, generates a fuzzy region that is biased toward the minimum risk region. The height of the outcome fuzzy region is [.92], and the height at the point of defuzzification is [.88]. For more information, see "Selecting Height Measurements" on page 414.

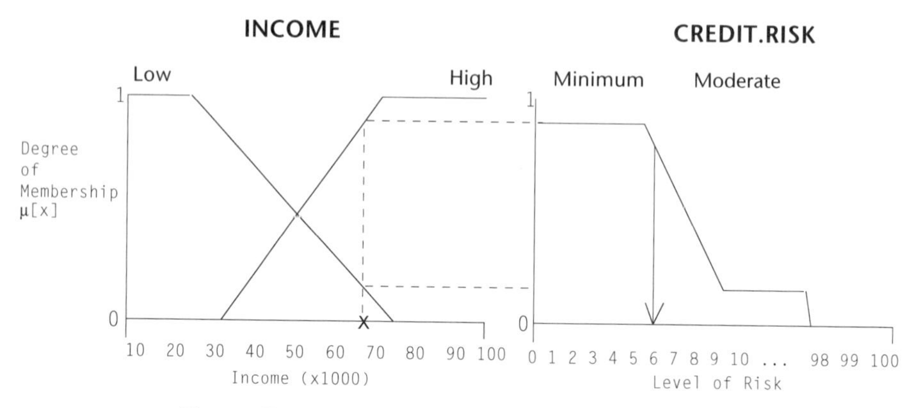

Figure 7.44 Credit Risk Defuzzification at CIX of [.92]

What the compability index is telling us, then, is whether or not the model is still robust given the current input parameters. If the range of the input parameters begins to shift, we should expect to see a shift in the outcome variable's compatibility measurement—either toward zero or toward one. In either case, this shift tells us that we need to re-evaluate the mechanics of our model (or, for fuzzy systems built through a data mining process, we need to re-train the system, thereby generating new rules).

Selecting Height Measurements

There are two height measurements in the solution fuzzy set: the actual height of the fuzzy set and the height at which the defuzzification process occurs. From methods such as the composite maximum, these two measurements will generally coincide. But for the composite moments or centroid technique, the height of the fuzzy set is not the height of defuzzification. The compatibility index is based on the height at the point of defuzzification since this gives you the minimum working altitude in the fuzzy region. (This point can be equal to, but never greater than the fuzzy set's height.) It is the height at defuzzification that indicates the strength of the decision process that generates the solution variable's expected value. All model examples in this book show the actual height of both the fuzzy region and the point of defuzzification.

Measuring Variability in the Model

The statistical compatibility index function (*FzyStatCompIndex*) collects a wide spectrum of statistics about how a model performs. One important measurement is the degree of variation between the minimum and maximum truth values in the unit compatibility domain. This measures the variability of the model across time. Little variability (the mean variance is less than one or two truth values) indicates a model that is insensitive to data. This can be due to the granularity of the fuzzy sets (they are too large or excessively overlapped) or a lack of rules that discriminate between different model states.

NOTES

1. This code is derived from the Metus Fuzzy Modeling System, a commercial fuzzy expert and decision support system designed and developed by the author. In fact, most of the code was copied without modification from the Metus embedded modeling libraries. The code is therefore more tightly linked to an overall control structure than what might be found in a collection of arbitrary fuzzy logic routines.

2. In discussing fuzzy models, we will relax the formal definition of a fuzzy statement as a logical proposition and frequently use the more common term "rule." If there is a possibility of confusing crisp, symbolic rules with fuzzy rules, then the term "proposition" will be used to indicate a fuzzy statement.

3. Figure 7.11 shows a rule descriptor block (`RDB`) chained into the policy. Compiled rules are not discussed in this book.

4. The variable `X` is a control variable. However, `X` might have been determined through a previous policy (`X` was a solution variable in the previous policy). Whether a variable is used in a control or solution sense is policy-dependent.

5. Hedges are optional components of a fuzzy model, used to transform linguistically the shape of a fuzzy set. If your model does not use hedges, then they can be omitted.

6. The *FzyApplyNOT* function does not create a new fuzzy set. If you want to create a third fuzzy set to hold the complement of *very HIGH*, then you can use the *FzyApplyHedge* function once more with the `Not` hedge.

7. Naturally there are many exceptions to this observation. A policy that seeks the largest or smallest expected value or that applies some other heuristic to selecting a final solution value might very well examine the results of several defuzzification techniques.

8. If you allow the defuzzification function to detect an undecidable model, the difference between a valid "0" return and an undecidable fuzzy set can be detected by examining the unit compatibility index (CIX). The unit CIX will also be zero for undecidables but should be nonzero for valid model executions.

9. In the ordinary and dictionary sense of the word: *stout, strong and sturdy; constitutionally healthy; vigorous.*

10. Defuzzification techniques, such as centroid and composite maximum, which take their values from the topology of the solution fuzzy region, consistently fail on single rule models such as the inventory example. This inventory model (`INVMOD4`) is, however, correctly handled by the monotonic logic inference technique.

11. We are ignoring here the obvious effects of an alpha threshold cut. In fact, alpha cuts are designed to address explicitly this kind of extremely weak inference state.

Eight

Fuzzy Systems

Case Studies

EYPHKA!
Eureka!
(I have found it!)

Archimedes (287–212 BC)
fr. Vitruvius Pollio, *De Architectura*, ix, 215

Nil posse creari
De nilo.
Nothing can be created out of nothing.

Lucretius (99–55 BC)
De Rerum Natura, i, 155

This chapter surveys a number of fuzzy models and explores how they are designed and how they work. To see some of the mechanics of classical fuzzy logic, we start with a control engineering example—a simple steam turbine controller. Following this example, we plunge into a detailed analysis and dissection of a new product pricing model, look at how the system works, explore different variations on the model, and construct a small fuzzy profit and loss model based on price and demand. The pricing system is followed by a project risk assessment model that uses a form of monotonic fuzzy logic to chain and scale a collection of independent risk factors.

A Fuzzy Steam Turbine Controller

A steam turbine controller adjusts an injector nozzle so that the amount of fuel reaching the turbine maintains a constant speed. The fuel rate is increased or decreased based on the temperature and pressure in the steam containment vessel. The fuzzy logic controller senses the current temperature and pressure and, using a set of heuristics encoded in the knowledge-base rules, makes positive (increasing the injector throttle's rate) or negative (decreasing the injector throttle's rate) adjustments. Before examining the turbine controller, we need to review how a fuzzy logic controller works.

The Fuzzy Control Model

Fuzzy logic controllers (FLCs) serve the same function as the more conventional proportional–integral–derivative controllers (PIDs)[*]. PIDs manage a complex control surface by reading sensor information, executing a mathematical model, and making changes to the device actuators. FLCs, however, manage this complex control surface through heuristics rather than a mathematical model. Further, a fuzzy system is able to approximate, to any level of precision, any continuous linear or nonlinear function (see Kosko, Bart, "Fuzzy Systems as Universal Approximators." *IEEE Transactions on Computers*, 1993). A fuzzy controller is a fuzzy system model. It employs fuzzy sets to represent the semantic properties of each control and solution variable, and processes its input and output by using a set of production if–then rules that associate an input value, through a collection of fuzzy sets, into a new output representation.

The Fuzzy Logic Controller

In Figure 8.1 we see the process flow logic for a typical fuzzy contoller. The process starts at the bottom of this figure: the input is read from the sensors as an electrical signal. The signal is converted to a meaningful representation and then "fuzzified," that is, the values are converted to their fuzzy representations. (See "Uncertain and Noisy Data" on page 532.) These sensor values execute all the rules in the knowledge repository that have the fuzzified input in their premise, resulting in a new fuzzy set representation for each output variable. Centroid defuzzification is used, in the majority of cases, to develop the expected value for each of these output variables. The output value adjusts the setting of an actuator, which adjusts the state of the physical system. The new state is picked up by the sensors, and the entire process begins once more.

[*] a controller whose output is the sum of three factors: a proportional term, an integrating term, and a derivative or differentiating term—each modified by its own adjustable gain.

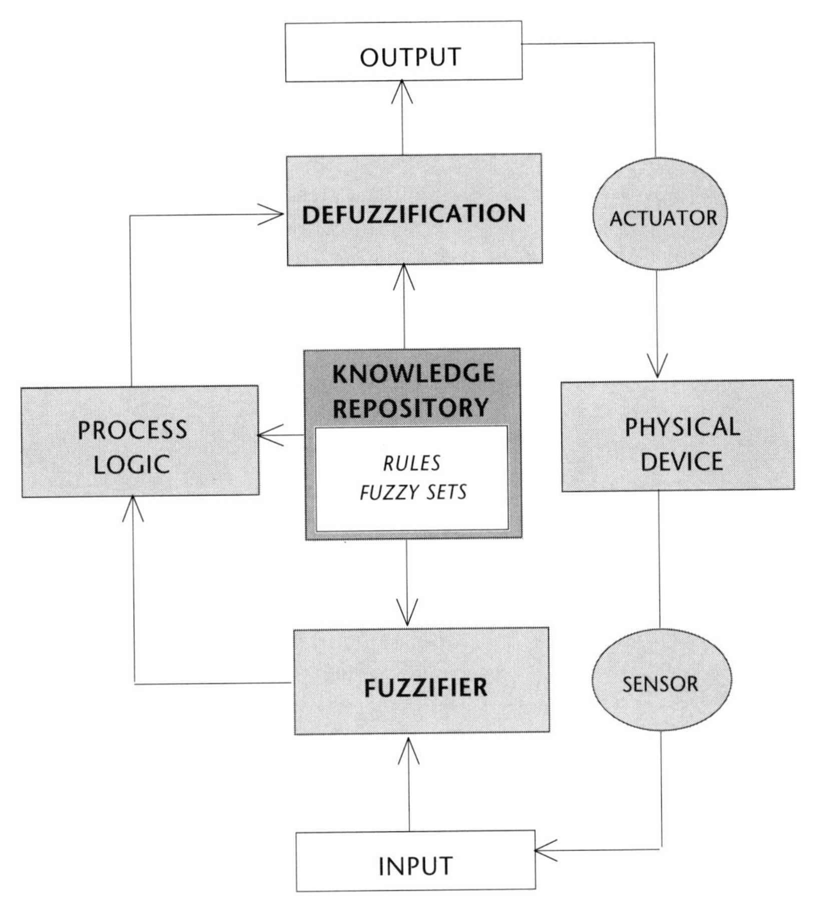

Figure 8.1 The FLC Process Flow

The Conventional PID Controller

In contrast to an FLC, a conventional PID controller is based on a rigorous mathematical model of some linear process. These models use a set of equations that describe the stable equilibrium state of the control surface through coefficients assigned to the proportional, integral, and derivative aspects of the system. As Figure 8.2 illustrates, a conventional controller reads a sensor value, applies the mathematical model, and produces an output from the mathematical algorithm.

The PID model may appear simpler and thus, perhaps, more economical, but we should not easily make this assumption. In fact, fuzzy controllers are often more easily prototyped and implemented, generally have an equal performance profile with PID systems, are simpler to describe and verify, can be maintained and extended with a higher

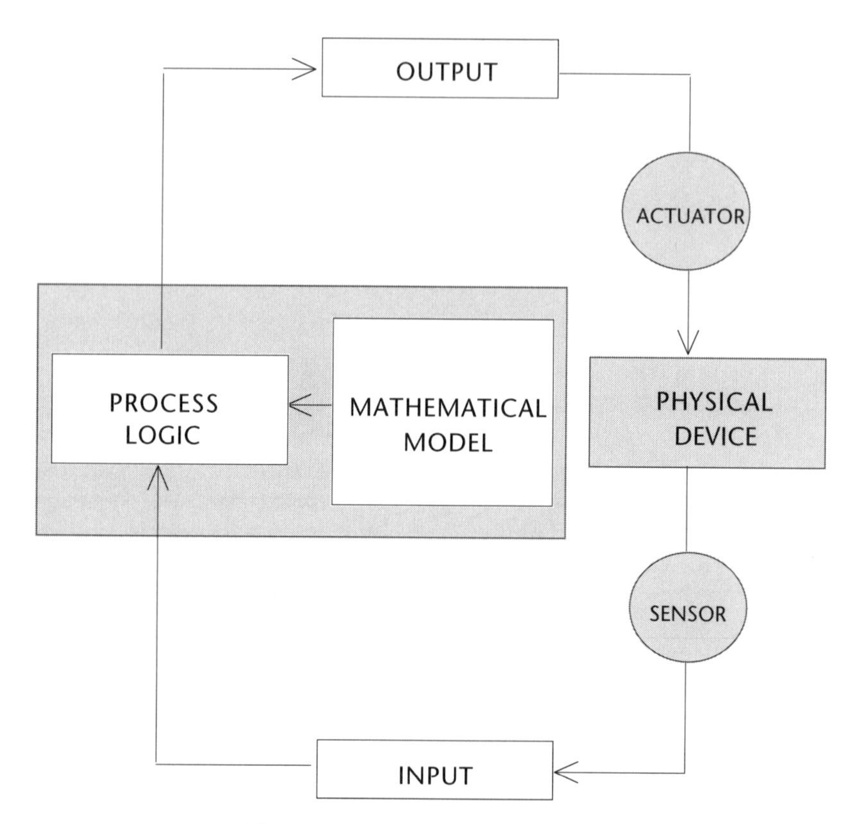

Figure 8.2 The PID Model Process

degree of accuracy in less time, and, due to their reliance on rules and knowledge, give their environments what Lotfi Zadeh calls a higher *machine intelligence quotient.*

The Steam Turbine Plant Process

To illustrate how a fuzzy logic controller works, we will look at a simple steam turbine plant. The operation of the turbine's injector throttle action (TA) is based on two control variables: the temperature (T) and pressure (P) at sensor time t, indicated by T[t] and P[t], respectively. Figure 8.3 shows the process flow for the simplified turbine plant.

While we are exploring the fuzzy logic controller for the injector, you should bear in mind that an actual steam turbine is much more complicated and has many different sensors and feedback systems. Figure 8.4 shows a schematic representation of the steam turbine with a few more feedbacks.

In the extended model shown in Figure 8.4, the controller for the injector aperture not only processes the temperature and pressure but considers the change in the temperature

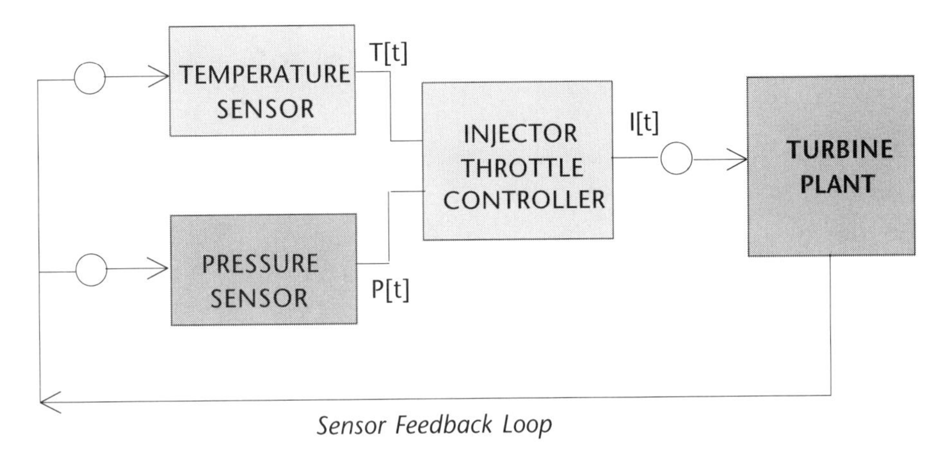

Figure 8.3 A Simplified Steam Turbine Control Process

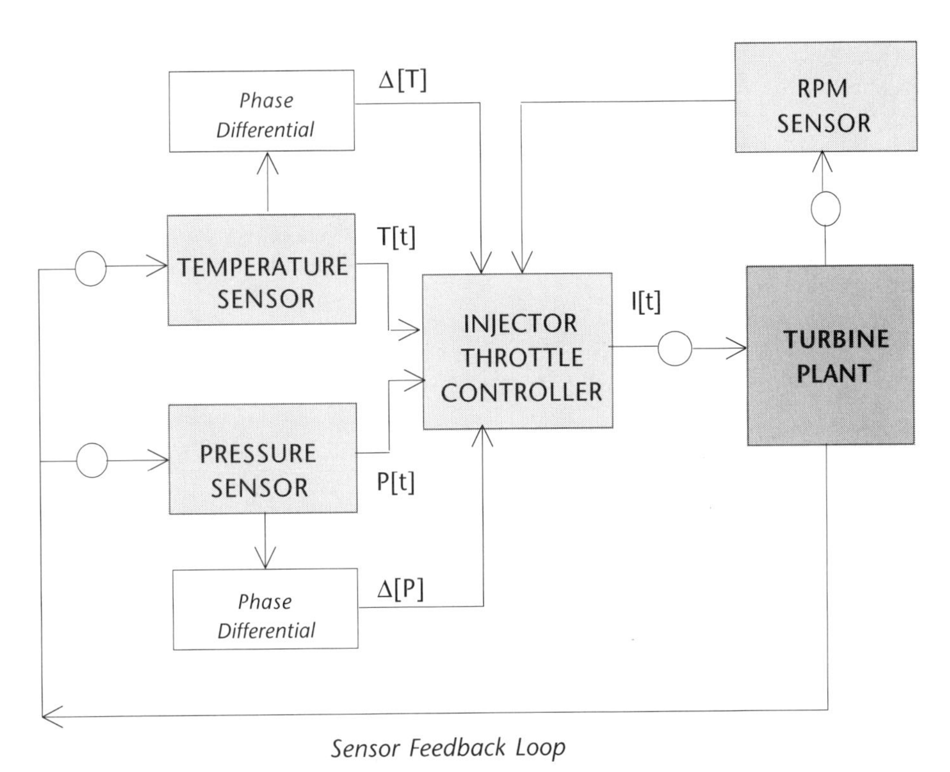

Figure 8.4 A More Realistic (and Complex) Steam Turbine Control Process

and pressure since the last sensor reading and also the load on the turbine in the form of the rotor RPM rate. In fact, in most control systems, it is the change (Δt, Δp, and Δrpm) that provides input into the fuzzy logic controller.

Designing the Fuzzy Logic Controller

The human throttle controller knows that an operating relation exists, perhaps expressed functionally as:

$$I_{ta} = g(T, P)$$

This equation is simply a mathematical way of saying that throttle injector action is roughly a function [g(x)] of temperature and pressure. Given a pair of temperature and pressure readings, the operator can specify a throttle action. A fuzzy logic controller encodes the actions, in a set of rules, that the operater takes as temperature and pressure change.

To build a controller representing this relationship, we decompose each model variable into a set of fuzzy control regions called the *term set.* Figures 8.5 and 8.6 show how the variables *Temperature* and *Pressure* have been redefined into sets of component fuzzy regions.

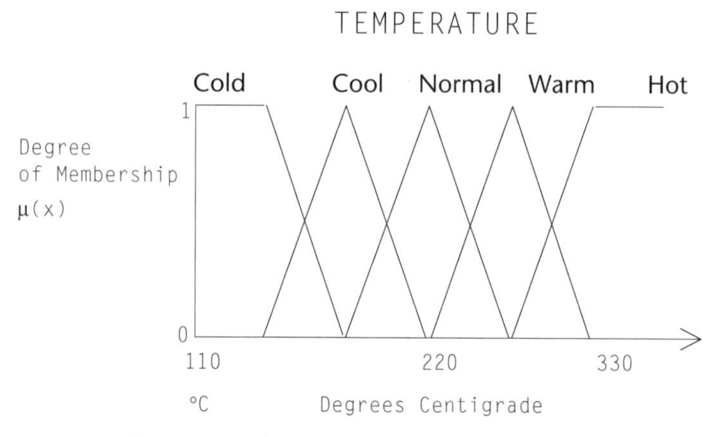

Figure 8.5 The *Temperature* Control Variable

Finally, we must also redefine the output or solution variable into a set of fuzzy regions. Figure 8.7 illustrates how the throttle action is represented as a series of movements in centimeters per second from *Negative.Large* (*NL*) through *Zero* (*ZR*) to *Positive.Large* (*PL*).

After isolating the control features of the turbine and decomposing each variable into fuzzy sets, the conceptual model is completed by writing the production rules describing the action taken on each combination of control variables. Figure 8.8 shows how some of these rules might appear.

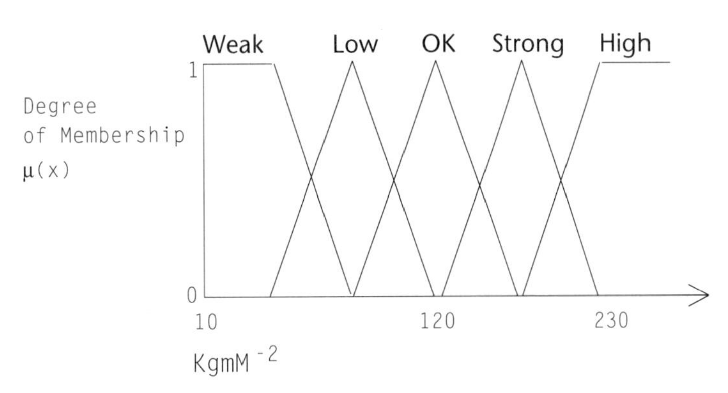

Figure 8.6 The *Pressure* Control Variable

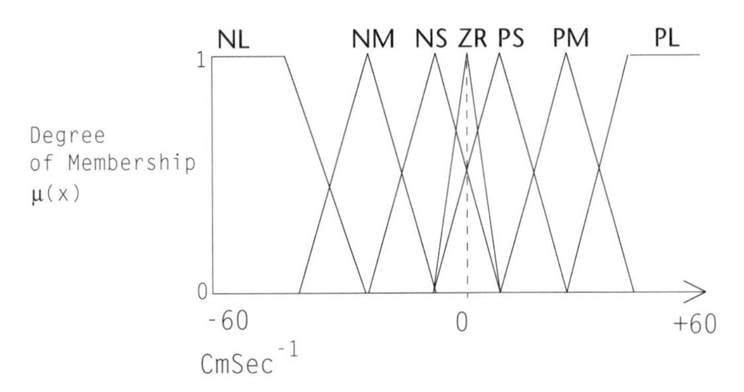

Figure 8.7 The *Throttle.Action* Solution Variable

```
[R1] if temperature is COOL and pressure is WEAK
   then throttle action is PL

[R2] if temperate is COOL and pressure is LOW
   then throttle action is PM

[R3] if temperature is COOL and pressure is OK
   then throttle action is ZR

[R4] if temperature is COOL and pressure is STRONG
   then throttle action is NM
```

Figure 8.8 Sample Rules from the Steam Turbine Controller

The steam turbine controller is a *2x1* (a "two-by-one") control system. That means there are two control variables and one solution variable. It is often convenient to think of such control systems as a matrix of actions in an *MxN* array. The fuzzy states of one control variable form the horizontal axis, and the fuzzy states of the other control variable form the vertical axis. At the intersection of a row and column is the fuzzy state of the solution variable. For for a three-by-one system, the representation assumes the shape of an *MxNxK* cube. Systems of higher order are more difficult to conceptualize but, in principle, the same kind of correspondence between control and solution variables exists.

This form of representation, very common in the control engineering field, is called a *fuzzy associative memory*, or FAM. Figure 8.9 shows a fuzzy associative memory for the steam turbine controller.

PRESSURE / TEMP.	Weal	Low	OK	Strong	High
Cold	*PL*	*PM*	*PS*	*NS*	*NM*
Cool	*PL*	*PM*	*ZR*	*NM*	*NM*
Normal	*PM*	*PS*	*ZR*	*NS*	*NM*
Warm	*PM*	*PS*	*NS*	*NM*	*NL*
Hot	*PS*	*PS*	*NM*	*NL*	*NL*

Figure 8.9 FAM for the Steam Turbine Controller

Running the Steam Turbine FLC Logic

A fuzzy logic controller, like all fuzzy systems, is essentially a parallel processor. All rules with any truth in their predicates fire and contribute to the fuzzy set region being built to represent the throttle action solution variable. In the steam turbine, we make a reading of pressure (*P[t]*) and temperature (*T[t]*). Each rule finds a membership grade for pressure and temperature in the corresponding fuzzy regions. Fuzzy controller rules specify a general region of control—the actual degree of control is not determined until a sensor reading value is available. In fact, a fuzzy rule such as

```
[R2] if temperature is COOL and pressure is LOW
   then throttle action is PM
```

is interpreted as "to the degree that temperature is [*somewhere*] in the region of *cool* and pressure is [*somewhere*] in the region of *low*, the throttle action is [*somewhere*] in the region of *PM*."

The intersection of the actual reading (along the domain or horizontal axis of the fuzzy set) and the surface of any particular fuzzy set tells us where in the region a reading lies. It also establishes the location of throttle action in the consequent fuzzy region. Fuzzy controllers use standard fuzzy inferencing techniques—the correlation minimum and the min/max inferencing methods (although some researchers, such as Kosko,[2] are proponents of the correlation product and the fuzzy additive inferencing method, with some experiments indicating that these methods yield better results). The minimum degree of membership for all the predicates in a rule is used to reduce the *Throttle.Action* fuzzy set. This reduced set is then combined with the *Throttle.Action* using the maximum of the combined degrees of membership. Here are the steps:

1. For all the predicate expressions connected by an *and*, we take the minimum of their collective membership truths. This final truth is the truth of the rule premise:

$$P_{truth} = \min(E_1, E_2, E_3, \ldots, E_n)$$

   In the previous rule [*R2*], this means evaluating the fuzzy propositions (temperature is *Cool*) and pressure is *Low*) to find their degree of truth. The minimum of the two truth values forms the truth of the predicate.

2. The fuzzy set on the right-hand side of the consequent action (the "control" fuzzy set) is then reduced in height by this amount:

$$\mu_{Control}[x] = \min(\mu_{Control}[x], P_{truth})$$

   In [*R2*] the control fuzzy set is *PM*. We would then reduce the height of *PM* to the height of the fuzzy predicate. This is the correlation minimum process.

3. This newly modified fuzzy set (*PM* correlated to the value of the predicate) is then copied into the output variable's fuzzy set. If that region is not empty, then it is *OR*'d with the current contents by taking the maximum of the new fuzzy region and the currently existing fuzzy region at each point along the domain (the horizontal axis):

$$\mu_{Solution}[x] = \max(\mu_{Solution}[x], \mu_{Control}[x])$$

Let's see how this process works. At some time *t,* the sensors read a pressure and temperature. The pressure reading falls within the domain of two pressure fuzzy sets, and the temperature reading falls within the *Cool* fuzzy set region.

```
[R2] if temperature is COOL and pressure is LOW
   then throttle action is PM

[R3] if temperature is COOL and pressure is OK
   then throttle action is ZR
```

The first rule causes the *PM* throttle action to be copied into the (currently empty) output throttle action fuzzy set. But before doing this step, its height is truncated at the truth of the rule's premise (the minimum of [*.57*] and [*.48*]). Figure 8.10 shows how this step works and how the throttle action output fuzzy region looks after the first rule is fired

The second rule is selected because the pressure reading has some significant degree of membership in *Low* as well a small degree of membership in *OK*:

```
[R3] if temperature is COOL and pressure is OK
   then throttle action is ZR
[R3] if temperature is COOL and pressure is OK
   then throttle action is ZR
```

When [*R3*] is fired, the *ZR* fuzzy set is also truncated at the truth of its premise (the minimum of [*.25*] and [*.48*]) and then copied into the output throttle action region. But since the region is not empty, this modified fuzzy set is *OR*'d with the *PM* fuzzy set. Figure 8.11 shows the final throttle action output fuzzy set. When the centroid defuzzification method is applied, an injector throttle action of +23 cm/sec is found as the expected value.

The centroid defuzzification is commonly the only defuzzification method used in control engineering applications (although this not a hard and fast rule). Figure 8.12 shows how the composite maximum technique produces a clearly different value for the throttle action. However, since we want a continuous and smoothly varying throttle action, the centroid is the preferred method of defuzzification.

The defuzzified value produces a change in the throttle injector at some instant in time. Moving the throttle changes the current temperature and pressure in the plant. Thus,

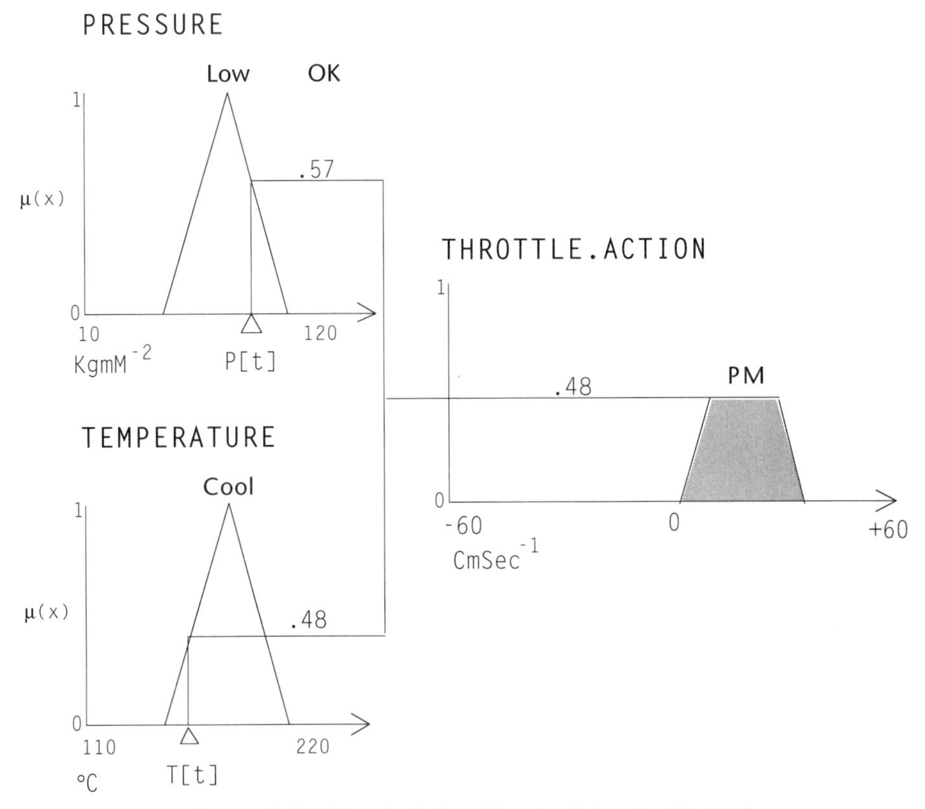

Figure 8.10 *Throttle.Action* After Applying the First Rule

in the next time period, the current values for temperature and pressure are used by the fuzzy control to produce another throttle movement. In this model, we have used the actual temperature and pressure values, but many control applications use the change in temperature and pressure (see Figure 8.4). These rules look something like this:

```
[R1] if ΔTemp is SN and ΔPressure is SN
   then throttle action is PS

[R2] if ΔTemp is SN and ΔPressure is LN
   then throttle action is PM
```

Δp indicates the amount of change in the parameter (which can be either a positive or negative amount). In any case, fuzzy control models are generally fairly simple (although quite powerful). Business models, on the other hand, involve much more complex processing.

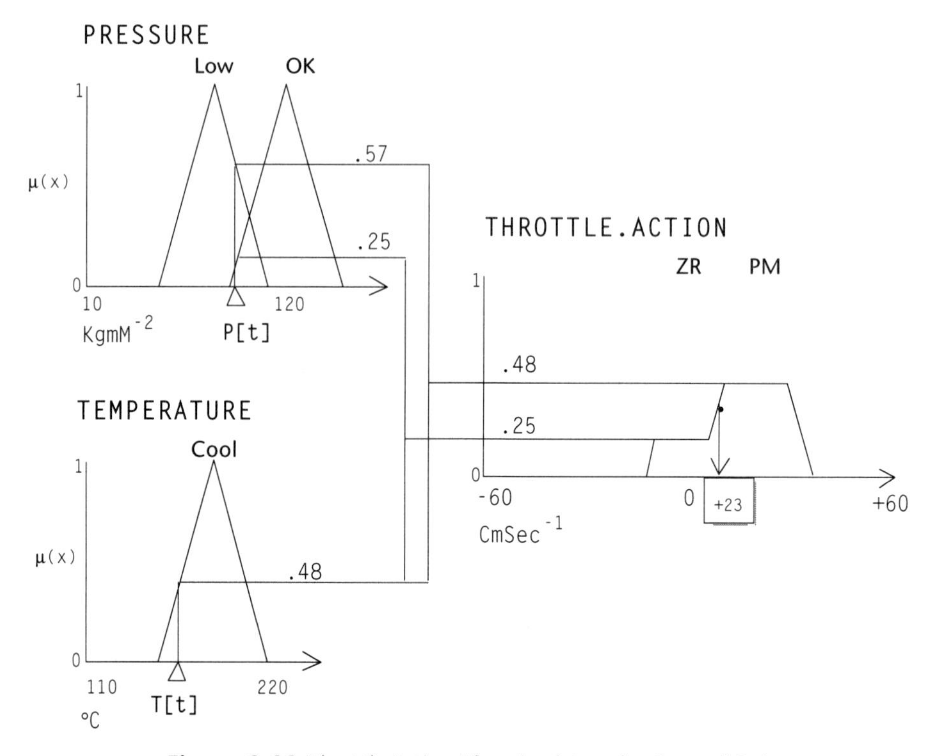

Figure 8.11 *Throttle.Action* After Applying the Second Rule

We now turn our attention to some actual fuzzy models in new product pricing and project risk assessment.

The New Product Pricing Model (Version 1)

A classical example of the way approximate reasoning is used in decision support is found in the price estimation model for new products constructed in the mid-1980s for a leading British retail firm. Developing a price for a new product involves a critical mix of many imprecise and uncertain factors: the estimated demand for the product; the competition's pricing and price strategies; the sensitivity of the market to price; the costs of manufacturing, transportation, storing, and replenishing standing stock; spoilage rates; seasonality of demand; probable life cycle of the product; capitalization requirements; lead time to market; product uniqueness; and its window of opportunity—just to name a few. Nearly all these constraints involve a degree of imprecision. And our model must

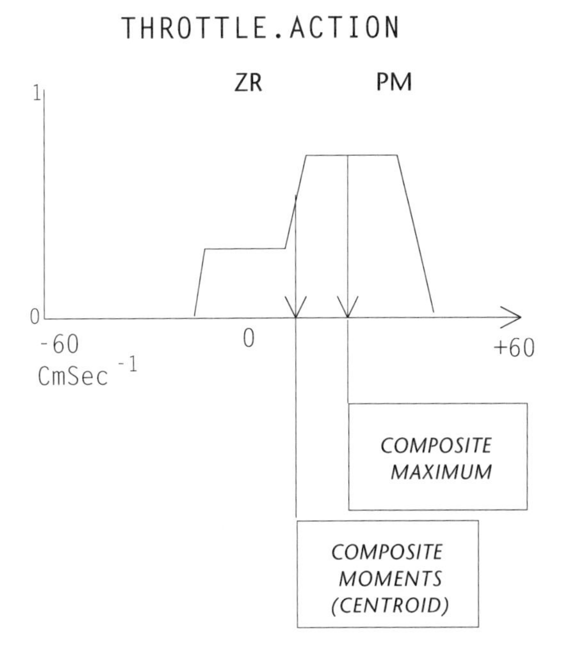

Figure 8.12 Alternative Ways of Interpreting the Output Fuzzy Region

respect not only constraints in pricing, but also the corporate objectives in setting prices relative to a wide number of equally imprecise considerations.

In this first version of the model, we want to establish a price based on just a few factors: our need to be profitable while sustaining high sales volumes, the average price of the competition's product in our market place, and the cost to manufacture the product. Figure 8.13 shows, schematically, how the model is organized.

Given these constraints and objectives, we run a fuzzy model that integrates the corporation's pricing position to yield an estimated price for the product. By changing these parameters, we can perform sensitivity tests on the price function as well as test alternate approaches to constraining the acceptable price.

Model Design and Objectives

The core fuzzy model has only four rules:

```
[R1] our price must be high
[R2] our price must be low
[R3] our price must be around 2*costs
[R4] if the competition.price is not very high
          then our price should be near the competition.price
```

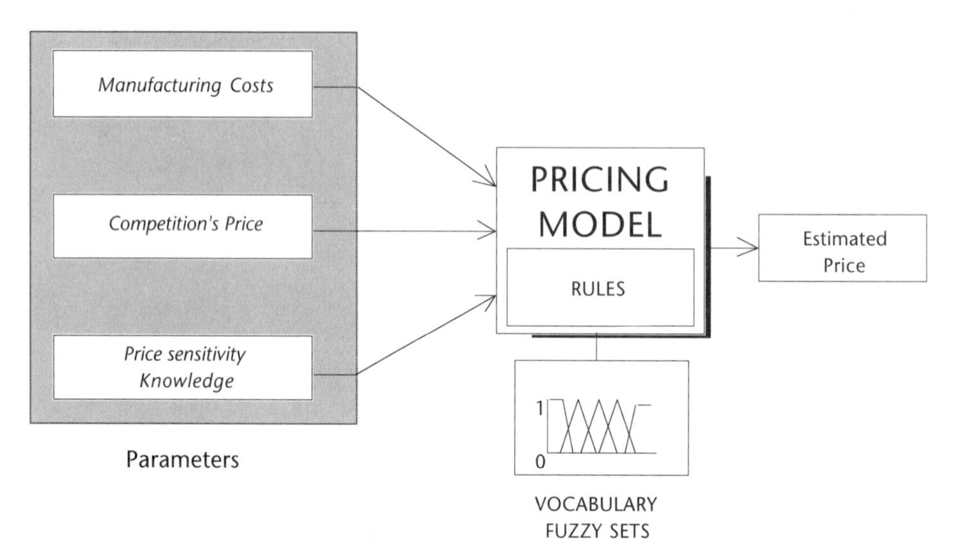

Figure 8.13 The Structure of the Basic New Product Pricing Model

This model introduces another seldom-appreciated feature of fuzzy systems—their ability to model often conflicting expert knowledge. The first rule, proposed by the finance director, ensures that we will have adequate margins and thus ensures profitability. The second rule, proposed by the sales and marketing vice president, ensures that we will have sufficient volume in our market area. The third rule, offered by the manufacturing director, ensures that the price will cover the direct costs of manufacturing. And the fourth rule, also entered by the marketing group, says that as long as the competition's price is not considered very high, our price should be close to that of the competition (thus preventing a price war).

THE MODEL EXECUTION LOGIC

We now follow the execution of the model. This is the FSHPRC1.CPP model. The following model activity log file extract (see Figure 8.14) shows how the base fuzzy sets are combined with fuzzy regions created from the current data points. When the program is executed, the model requests the manufacturing costs and competition's price. The actual pricing policy is executed and an estimated price is returned. Figure 8.14 shows how the model execution appears.

```
Enter the Manufacturing Costs:12
Enter the Competition's Price:26

The Recommended Price is: 24.75 with CompIdx of 0.597
```

Figure 8.14 Execution of the Basic Pricing Model

Create the Basic Price Fuzzy Sets

The fuzzy sets *High* and *Low*, shown in Figure 8.15, represent price sensitivity information. They indicate what points are considered for this kind of product to be a high price and a low price. The sets are used in the model for fundamental positioning. In the absence of any other information, we can place the pricing scheme somewhere in either the *High* or the *Low* set for price fuzzy sets—or a combination of the two.

```
---Policy 'PRICING' created.
---Default Hedges installed in Policy 'PRICING'
---Linear Increasing FuzzySet 'HIGH' created.
---Linear Decreasing FuzzySet 'LOW' created.

PRICE
Domain [UofD]:      16.00 to      36.00

        1.00*                                                                    .
        0.90 *******                                                        ......
        0.80        ******                                           .......
        0.70              ******                               ......
        0.60                    *******                 .......
        0.50                           ******.....
        0.40                          ...... *******
        0.30                    .......             ******
        0.20              ......                          *******
        0.10        .......                                      ******
        0.00......                                                    ******
          0---|---|---|---|---|---|---|---|---|---|---|---|---|---|---|---|---0
          16.00   18.50   21.00   23.50   26.00   28.50   31.00   33.50   36.00
        .  FuzzySet:    HIGH
          Description: High for Price
          Domained:         16.00 to      36.00
        *  FuzzySet:    LOW
          Description: Low for Price
          Domained:         16.00 to      36.00
```

Figure 8.15 The Price Sensitivity Fuzzy Sets: *High* and *Low* for Price

Create the Run-Time Model Fuzzy Sets

Figures 8.16 and 8.17 are model-base fuzzy sets that depend on the actual run-time data. Each new value of manufacturing costs and competition prices produces new versions of these fuzzy sets. The difference in the width of the fuzzy sets reflects the model semantics. The *Around 2*MfgCosts* has a 25% diffusion to account for a basic uncertainty, at this point, about the full manufacturing costs for the product, as well as the degree to which we want this factor to contribute to the default price value. The *Near.CompPrice* (*around the competition price*) fuzzy set has a narrower diffusion (15%) to account for the model's assumption that a proposed price would need to be quite close to the competition price to be regarded as being near the competition (see rule [*R4*]). Changing the width of these dynamically created fuzzy sets changes how the underlying default price is calculated and is one way of fine-tuning the model.

```
---Empty FuzzySet 'NULL' created.
---PI FuzzySet 'Near.2*MfgCosts' created.

        FuzzySet:    Near.2*MfgCosts
        Description: Near (Around) Twice MfgCosts
      1.00                         ..
      0.90                      ...  ....
      0.80                    ..         .
      0.70                   .            ..
      0.60                 ..               .
      0.50                .                  .
      0.40               .                    .
      0.30              .                      .
      0.20             .                        .
      0.10           ..                          ..
      0.00...........                              ........................
          0---|---|---|---|---|---|---|---|---|---|---|---|---|---|---|---0
        16.00   18.50   21.00   23.50   26.00   28.50   31.00   33.50   36.00
        Domained:        16.00 to      36.00
```

Figure 8.16 The Cost of Manufacturing Fuzzy Set (*Around 24 [2*MfgCosts]*)

```
---PI FuzzySet 'Near.CompPrice' created.

        FuzzySet:    Near.CompPrice
        Description: Near (Around) Competition Price
      1.00                                .
      0.90                              .. ..
      0.80                            ..     ..
      0.70
      0.60                           .         .
      0.50                          .           .
      0.40
      0.30                         .             .
      0.20                        .               .
      0.10                       .                 .
      0.00.......................                   .......................
          0---|---|---|---|---|---|---|---|---|---|---|---|---|---|---|---0
        16.00   18.50   21.00   23.50   26.00   28.50   31.00   33.50   36.00
        Domained:        16.00 to      36.00
```

Figure 8.17 The Near to Competiton's Price Fuzzy Set (*Around 26*)

Execute the Price Estimation Rules

The messages at the first part of Figure 8.15 show how the model's main program creates the *pricing* policy and then loads the default hedges and allocates the linear *High* and

Low price fuzzy sets. Listing 8.1 is the start of the actual pricing policy (*FSHPOL1.CPP*), where the parameter data is recorded and the fuzzy work area is initialized. An empty fuzzy set for *Price* is created as the formal solution variable is added to the work area.

```
Fuzzy Pricing Model. (c) 1992 Metus Systems Group.
[Mean] Competition Price:      26.00
Base Manufacturing Costs:      12.00
Price Estimation Policy Begins....
---SystemNote(003): Fuzzy Work Area Initialized
---Empty FuzzySet 'PRICE' created.
---SystemNote(005): Output Variable 'PRICE' added to Fuzzy Model.

The Rules:
R1   our price must be high
R2   our price must be low
R3   our price must be around 2*mfgcosts
R4   if the competition.price is not very high
        then our price should be near the competition.price
```

Listing 8.1 Execution of the Initial Pricing Policy

After evaluating and applying rule [*R1*], the solution fuzzy set appears as shown in Figure 8.18. In applying this unconditional fuzzy proposition, the fuzzy model makes the value of *Price* as true as possible in the fuzzy set *High*. Since the solution fuzzy set is empty, you must copy the *High* fuzzy set directly to the solution fuzzy set. Since unconditional sets are *AND*'d with the output fuzzy sets, we cannot perform this operation on an empty fuzzy set because $min(\mu(x),0)$ is always zero.

```
R1   our price must be high
---Premise Truth= 1.0000

       FuzzySet:     PRICE
        Description:
      1.00                                                                    .
      0.90                                                              ......
      0.80                                                       .......
      0.70                                                 ......
      0.60                                          .......
      0.50                                    ......
      0.40                              ......
      0.30                       .......
      0.20                 ......
      0.10          .......
      0.00......
          0---|---|---|---|---|---|---|---|---|---|---|---|---|---|---|---0
        16.00   18.50   21.00   23.50   26.00   28.50   31.00   33.50   36.00
        Domained:        16.00 to      36.00
```

Figure 8.18 The *Price* Solution Fuzzy Set After Executing Rule [*R1*]

When the second unconditional rule *[R2]* is applied, the solution fuzzy set is updated with the maximum truth for the *Low* proposition (the consequent fuzzy region). In applying an unconditional proposition, the solution is generated at the intersection of the two sets (they are effectively *AND*'d). Figure 8.19 shows the results of executing *[R2]*. Equations 8.1 and 8.2 describe the formal actions for handling unconditional rules in a fuzzy model.

```
R2   our price must be low
---Premise Truth= 1.0000

       FuzzySet:     PRICE
       Description:
     1.00
     0.90
     0.80
     0.70
     0.60
     0.50                                 .
     0.40                           ...... .......
     0.30                    .......             ......
     0.20              ......                          .......
     0.10        .......                                      ......
     0.00......                                                     ......
         0---|---|---|---|---|---|---|---|---|---|---|---|---|---|---|---0
       16.00   18.50   21.00   23.50   26.00   28.50   31.00   33.50   36.00
       Domained:         16.00 to       36.00
```

Figure 8.19 The *Price* Solution Fuzzy Set After Executing Rule *[R2]*

$$if(\forall \mu_{solution}[x]) = 0) \quad \mu_{solution}[x_i] \; = \mu_{consequent}[x_i]$$

Equation 8.1 Updating an Empty Solution Fuzzy Set with an Unconditional Proposition

$$\mu_{solution}[x_i] = \min(\mu_{solution}[x_i], \mu_{consequent}[x_i])$$

Equation 8.2 Updating a Working Solution Fuzzy Set with an Unconditional Proposition

After the execution of rules *[R1]* and *[R2]*, the model contains a triangular fuzzy region at the μ[.5] height bounded by the intersection of the *High* and *Low* fuzzy regions. If the model consisted solely of these two rules, we could find a value for price by simply defuzzifying the output. Both the composite moment and the composite maximum would yield a value of $26.00. This is an important and significant difference between fuzzy models and traditional symbolic expert systems. Fuzzy assertions (unconditional propositions) are not simply data assignments; they are active components of the model's underlying logic. These contradictory assertions illustrate fuzzy logic's unique ability to handle apparently contradictory propositions, stemming in part from its failure to follow the Boolean laws of the excluded middle and of noncontradiction.

After the application of the unconditional rule [*R3*], Figure 8.20 shows the state of the *Price* solution fuzzy region. This rule overlays the current working fuzzy region with the bell-shaped fuzzy region created by approximating twice the stated manufacturing costs (see Figure 8.16). Since this is an unconditional rule, we take the minimum of the solution fuzzy set and this consequent set. Figure 8.21 shows the fuzzy sets produced by these rules, and Figure 8.22 is the final fuzzy region produced by applying the minimum operator.

```
R3   our price must be around 2*mfgcosts
---Premise Truth= 1.0000

      FuzzySet:    PRICE
      Description:
    1.00
    0.90
    0.80
    0.70
    0.60
    0.50                                    .
    0.40                              ...... ....
    0.30                       .......         .
    0.20                 ......                  .
    0.10               ..                          ..
    0.00...........                                  ........................
        0---|---|---|---|---|---|---|---|---|---|---|---|---|---|---|---|---0
      16.00   18.50   21.00   23.50   26.00   28.50   31.00   33.50   36.00
      Domained:          16.00 to       36.00
```

Figure 8.20 The *Price* Solution Fuzzy Set After Executing Rule [*R3*]

```
    1.00                              ..
    0.90                           ...  ....
    0.80                         ..        .
    0.70                        .           ..
    0.60                      ..              .
    0.50                     .               * .
    0.40                    .         ****** *******
    0.30                   .   *******             .   ******
    0.20                 ******                     .       *******
    0.10          *******                            ..            ******
    0.00******.....                                    ..................******
        0---|---|---|---|---|---|---|---|---|---|---|---|---|---|---|---|---0
      16.00   18.50   21.00   23.50   26.00   28.50   31.00   33.50   36.00
  .    FuzzySet:    Near.2*MfgCosts
      Description: Near (Around) Twice MfgCosts
      Domained:          16.00 to       36.00
  *    FuzzySet:    PRICE
      Description:
      Domained:          16.00 to       36.00
```

Figure 8.21 The *Price* Solution Set Overlaid by *Around 2*MfgCosts*

As Figure 8.22 shows, when the *Around 2*MfgCosts* fuzzy set is actually applied to the solution region it restricts the left and right edges of the existing solution fuzzy region but has no effect on the height of the existing middle region.

Rule [*R4*] is a complicated conditional fuzzy proposition. The predicate fuzzy set is also used as a consequent constraint fuzzy set. This rule also uses a hedged linguistic variable in the predicate. In executing this rule, we must create and evaluate the predicate linguistic variables in order to determine how much truth is in the following proposition:

```
the competition.price is not very high
```

As Figure 8.23 shows, the first step in evaluating linguistic variables is the creation of the fuzzy region *very High*. This is done by applying the built-in hedge *very* to the *High* fuzzy set. The hedge operation creates a new fuzzy set containing *very High*, and the original fuzzy set *High* is not altered.

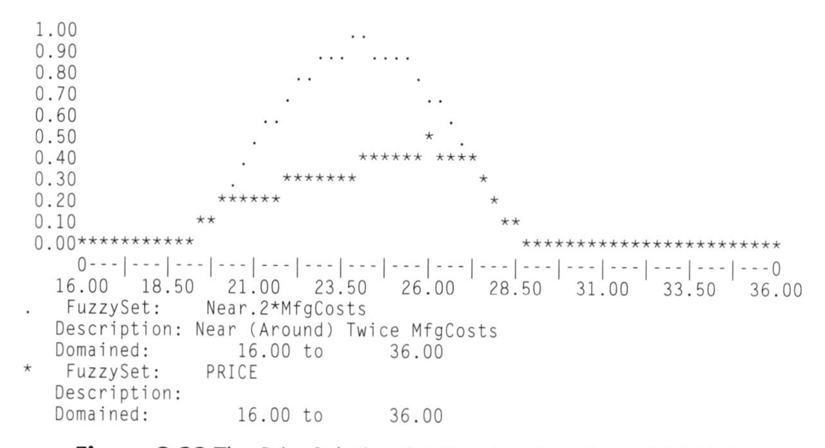

Figure 8.22 The *Price* Solution Set Restricted by *Around 2*MfgCosts*

```
R4   if the competition.price is not very high
       then our price must be around the competition.price
OK. Fuzzy set 'HIGH' copied into Fuzzy set 'HIGH'
---Hedge 'very' applied to Fuzzy Set "HIGH"
       FuzzySet:     HIGH
       Description: very HIGH
   1.00
   0.90                                                                        .
   0.80                                                                     ...
   0.70                                                                  ...
   0.60                                                              ....
   0.50                                                          ....
   0.40                                                     .....
   0.30                                                 ....
   0.20                                           ......
   0.10                                     ......
   0.00....................       .........
       0---|---|---|---|---|---|---|---|---|---|---|---|---|---|---|---|---0
     16.00   18.50   21.00   23.50   26.00   28.50   31.00   33.50   36.00
     Domained:           16.00 to       36.00
```

Figure 8.23 Creating the *very High* Fuzzy Region

Figure 8.24 shows the fuzzy region representing the final linguistic variable *not very High*. We create this region by applying the standard Zadeh *not* $(1-\mu_A(x))$ to the temporary fuzzy set created by the *very* hedge. The *FzyApplyNOT* function modifies the actual fuzzy set. (Since we will discard its memory after this rule, we need not create a third fuzzy set. If we wanted to preserve the *very High* fuzzy set, then the *FzyApplyHedge* function for the *not* hedge would create a new fuzzy set.)

```
---Hedge 'ZADEH' NOT applied to Fuzzy Set "HIGH"
        FuzzySet:     HIGH
        Description: ZADEH NOT HIGH
      1.00.......
      0.90       ...............
      0.80                      ........
      0.70                              ......
      0.60                                    ......
      0.50                                          ....
      0.40                                              .....
      0.30                                                   ....
      0.20                                                       ...
      0.10                                                          ....
      0.00                                                              ...
          0---|---|---|---|---|---|---|---|---|---|---|---|---|---|---|---0
        16.00   18.50   21.00   23.50   26.00   28.50   31.00   33.50   36.00
        Domained:         16.00 to        36.00
```

Figure 8.24 Creating the *not very High* Linguistic Variable

When this linguistic space is created, we can determine the truth of the predicate proposition by finding the grade of membership for the competition price (`$26.00`). This has a truth function value of [`0.75`]. The consequent proposition

```
our price must be around the competition price
```

must be evaluated, and the *Price* solution fuzzy set updated. Figure 8.25 shows the *Price* solution fuzzy set after applying rule [*R4*]. Using the predicate truth value [0.75], the correlation minimum process is applied to the fuzzy set *Near.CompPrice* (see Figure 8.17). The height of the consequent is reduced to this truth value. Conditional fuzzy propositions update the solution fuzzy set by applying the union of the consequent and the solution set (in effect *OR*ing the two sets). Equations 8.3 and 8.4 are the formal methods for applying conditional fuzzy rules.

```
---Premise Truth= 0.7519
        FuzzySet:     PRICE
        Description:
      1.00
      0.90
      0.80
      0.70                                .....
      0.60                              ..     ..
      0.50                             .         .
      0.40                            .           .
      0.30                     .......
      0.20               ......                    ..
      0.10             ..                            .
      0.00...........                                 ......................
          0---|---|---|---|---|---|---|---|---|---|---|---|---|---|---|---0
        16.00   18.50   21.00   23.50   26.00   28.50   31.00   33.50   36.00
        Domained:         16.00 to        36.00
```

Figure 8.25 The *Price* Solution Fuzzy Set After Executing Rule *[R4]*

$$\mu_{consequent \otimes Pt}[x_i] = \mu_{consequent}[x_i] \times \mu_{premise}$$

Equation 8.3 The Conditional Rule Correlation Process

$$\mu_{solution}[x_i] = \max(\mu_{solution}[x_i], \mu_{consequent \otimes Pt}[x_i])$$

Equation 8.4 Updating a Working Solution Fuzzy Set with a Conditional Proposition

Defuzzify to Find Expected Value for Price

The new product pricing policy is now complete. The *Price* fuzzy set contains a space representing the combined knowledge of the four rules. In order to elicit the expected value for price, we must defuzzify the solution fuzzy set.

EVALUATING DEFUZZIFICATION STRATEGIES

The composite moments or centroid technique is used to defuzzify the *Price* fuzzy set. Figure 8.26 shows the final fuzzy set and the results of the defuzzification, which include a recommended price of $24.75 with a fairly high degree of underlying compatibility. The choice of defuzzification methods in the model depends on how we want the result calculated. The composite moments technique blends both the conditional and unconditional into a result that tends to move smoothly around the output space as the model parameters change. The composite maximum bases its results on the point in the output space with the highest truth value. As long as rule [*R4*] has a truth greater than the unconditionals, the value of *Price* is based on this one rule. If the truth of rule [*R4*] falls below the unconditionals, then the value of price is based on the combined unconditionals (rules [*R1*] through [*R3*]). We can see the difference in the comparison chart in Figure 8.27.

```
  FuzzySet:     PRICE
  Description: FINAL Price Estimate
1.00
0.90
0.80
0.70                                    .....
0.60                                  ..     ..
0.50                                 .         .
0.40                                .           .
0.30                         .......
0.20                   ......                    ..
0.10                 ..                            .
0.00...........                                      .......................
    0---|---|---|---|---|---|---|---|---|---|---|---|---|---|---|---0
  16.00   18.50   21.00   23.50   26.00   28.50   31.00   33.50   36.00
  Domained:          16.00 to        36.00

---'CENTROID'     defuzzification. Value:     24.750, [0.5975]
Model Solution:
  Price        =        24.75
  CompIdx      =         0.60
  SurfaceHght  =         0.75
```

Figure 8.26 Defuzzifying the *Price* Solution Set Using the Centroid Technique

No.	Comp. Price	DoM	CENTROID Price	CENTROID CIX	MAXIMUM Price	MAXIMUM CIX
1.	23.50	.860	24.05	.819	23.50	.860
2.	24.00	.842	24.20	.837	23.97	.842
3.	24.50	.823	24.28	.818	24.52	.823
4.	25.00	.800	24.44	.764	24.98	.800
5.	25.50	.778	24.52	.675	25.53	.778
6.	26.00	.752	24.75	.597	26.00	.752
7.	26.50	.728	25.06	.538	26.47	.728
8.	27.00	.703	25.30	.463	27.02	.703
9.	27.50	.673	25.53	.475	27.48	.673
10.	28.00	.646	25.77	.486	28.03	.646
11.	28.50	.612	26.00	.498	28.50	.612
12.	29.00	.583	26.23	.490	28.97	.583
13.	29.50	.552	26.47	.479	29.52	.552
14.	30.00	.515	26.62	.471	29.98	.515
15.	30.50	.482	26.78	.463	26.08	.498
16.	31.00	.442	26.86	.459	26.08	.498
17.	31.50	.406	26.94	.455	26.08	.498
18.	32.00	.370	27.02	.451	26.08	.498
19.	32.50	.326	26.94	.455	26.08	.498
20.	33.00	.287	26.86	.459	26.08	.498
21.	33.50	.240	26.62	.471	26.08	.498
22.	34.00	.199	26.31	.486	26.08	.498
23.	34.50	.157	25.92	.494	26.08	.498
24.	35.00	.106	25.38	.467	26.08	.498
25.	35.50	.061	24.91	.444	26.08	.498

Figure 8.27 Comparison of Composite Moments and Maximum Defuzzification

Figure 8.27 (from the FSHPRC9.CPP model) shows the difference in estimated price between composite moments (centroid) and composite maximum for differing values of competition price (the manufacturing cost has been set to a constant value of $12.00). The column labeled DoM represents the competition price's degree of membership in the *not very High* fuzzy set.

From this chart we can see some interesting defuzzification behavior. As the competition price increases, the truth of rule [*R4*]'s predicate gradually decreases. (At $23.50 it is reasonably considered *not very High*, but at $35.00 it is almost certainly a *very High* value as indicated by the associated degree of membership in the *not very High* fuzzy set.) The composite moments technique "feels" the effects of the unconditionals and holds the price at a region toward the center of the unconditionals. The composite maximum method, in lines 1 through 14 of Figure 8.27, tracks the competition price rather closely. As soon as the strength of the output fuzzy region defined by rule [*R4*] falls below the strength of the region formed by the unconditionals, rule [*R4*] no longer controls the composite maximum result (the maximum region is defined by the intersection of the unconditionals that have a constant height of [.498] ≈ [.5] in the output fuzzy set). This switch happens at line 15 in Figure 8.27—the price jumps from $29.98 back to $26.08. The composite maximum price remains at this price as the competition price continues to rise.

How is this behavior expressed in the fuzzy model itself? Figure 8.28 (a very long figure) is the execution of the pricing model associated with the chart shown in Figure 8.27. This execution trace shows the final solution fuzzy set for *Price* at the point of defuzzification. We also display the associated competition price (along with its degree of membership in the not very *High* fuzzy set). The values for the centroid and composite maximum are displayed for each fuzzy set.

```
  CompPrice    =       23.50 [DoM: 0.860]

       FuzzySet:     PRICE
       Description:
     1.00
     0.90
     0.80                             .....
     0.70                            .     .
     0.60                           .       .
     0.50                          .         .  .
     0.40                                     .. ....
     0.30                         .                 .
     0.20                     .....                  .
     0.10                   ..                        ..
     0.00...........                                    ........................
        0---|---|---|---|---|---|---|---|---|---|---|---|---|---|---|---|---|---0
       16.00   18.50   21.00   23.50   26.00   28.50   31.00   33.50   36.00
       Domained:          16.00 to        36.00

Model Solution:
CENTROID:          Price        =       24.05   CIX          =       0.82
MAXIMUM:           Price        =       23.50   CIX          =       0.86
  SurfaceHght =         0.86

CompPrice    =       24.00 [DoM: 0.842]

       FuzzySet:     PRICE
       Description:
     1.00
     0.90
     0.80                              ....
     0.70                             .    ..
     0.60                            .       .
     0.50                           .         .
     0.40                          .         . ....
     0.30                         .               .
     0.20                     ......               .
     0.10                   ..                      ..
     0.00...........                                   .........................
        0---|---|---|---|---|---|---|---|---|---|---|---|---|---|---|---|---|---0
       16.00   18.50   21.00   23.50   26.00   28.50   31.00   33.50   36.00
       Domained:          16.00 to        36.00

Model Solution:
CENTROID:          Price        =       24.20   CIX          =       0.84
MAXIMUM:           Price        =       23.97   CIX          =       0.84
  SurfaceHght =         0.84
```

Figure 8.28 Execution of the Pricing Models for Various Competition Price Values

```
CompPrice    =      24.50 [DoM: 0.823]

      FuzzySet:    PRICE
      Description:
    1.00
    0.90
    0.80                              ...
    0.70                            ..   ..
    0.60                           .       .
    0.50                          .         .
    0.40                                     ....
    0.30                       ...               .
    0.20                 ......                   .
    0.10               ..                          ..
    0.00...........                                  ........................
       0---|---|---|---|---|---|---|---|---|---|---|---|---|---|---|---0
      16.00   18.50   21.00   23.50   26.00   28.50   31.00   33.50   36.00
      Domained:        16.00 to      36.00

Model Solution:
CENTROID:          Price       =       24.28   CIX          =       0.82
MAXIMUM:           Price       =       24.52   CIX          =       0.82
  SurfaceHght  =       0.82

CompPrice    =      25.00 [DoM: 0.800]

      FuzzySet:    PRICE
      Description:
    1.00
    0.90
    0.80                                ..
    0.70                              ..  ..
    0.60                             .      ..
    0.50                            .         .
    0.40                           .           ..
    0.30                       ....              .
    0.20                 ......                   .
    0.10               ..                          ..
    0.00...........                                  .........................
       0---|---|---|---|---|---|---|---|---|---|---|---|---|---|---|---0
      16.00   18.50   21.00   23.50   26.00   28.50   31.00   33.50   36.00
      Domained:        16.00 to      36.00

Model Solution:
CENTROID:          Price       =       24.44   CIX          =       0.76
MAXIMUM:           Price       =       24.98   CIX          =       0.80
  SurfaceHght  =       0.80
```

Figure 8.28 Execution of the Pricing Models for Various Competition Price Values *(continued)*

```
CompPrice    =       25.50 [DoM: 0.778]

        FuzzySet:    PRICE
        Description:
      1.00
      0.90
      0.80
      0.70                                ......
      0.60                               .      .
      0.50                              .        .
      0.40                             .          .
      0.30                       ......            .
      0.20                 ......                   .
      0.10               ..                          ..
      0.00...........                                  ........................
         0---|---|---|---|---|---|---|---|---|---|---|---|---|---|---|---0
        16.00   18.50   21.00   23.50   26.00   28.50   31.00   33.50   36.00
        Domained:        16.00 to       36.00

Model Solution:
CENTROID:          Price       =      24.52   CIX         =       0.68
MAXIMUM:           Price       =      25.53   CIX         =       0.78
  SurfaceHght =        0.78


CompPrice    =       26.00 [DoM: 0.752]

        FuzzySet:    PRICE
        Description:
      1.00
      0.90
      0.80
      0.70                                 .....
      0.60                               ..     ..
      0.50                              .         .
      0.40                             .           .
      0.30                       .......
      0.20                 ......                   ..
      0.10               ..                           .
      0.00...........                                  .......................
         0---|---|---|---|---|---|---|---|---|---|---|---|---|---|---|---0
        16.00   18.50   21.00   23.50   26.00   28.50   31.00   33.50   36.00
        Domained:        16.00 to       36.00

Model Solution:
CENTROID:          Price       =      24.75   CIX         =       0.60
MAXIMUM:           Price       =      26.00   CIX         =       0.75
  SurfaceHght =        0.75
```

Figure 8.28 Execution of the Pricing Models for Various Competition Price Values *(continued)*

```
CompPrice    =       26.50 [DoM: 0.728]

        FuzzySet:     PRICE
        Description:
      1.00
      0.90
      0.80
      0.70                                     ....
      0.60                                   ..     ..
      0.50                                  .         .
      0.40                               ...           .
      0.30                        .......               .
      0.20                  ......                       .
      0.10                ..                              ..
      0.00...........                                       .....................
          0---|---|---|---|---|---|---|---|---|---|---|---|---|---|---|---0
        16.00   18.50   21.00   23.50   26.00   28.50   31.00   33.50   36.00
        Domained:        16.00 to       36.00

Model Solution:
CENTROID:          Price        =       25.06   CIX          =       0.54
MAXIMUM:           Price        =       26.47   CIX          =       0.73
  SurfaceHght =         0.73

CompPrice    =       27.00 [DoM: 0.703]

        FuzzySet:     PRICE
        Description:
      1.00
      0.90
      0.80
      0.70                                      ...
      0.60                                    ..   ..
      0.50                                   .       ..
      0.40                              .....          .
      0.30                       .......                .
      0.20                 ......                        .
      0.10               ..                               .
      0.00...........                                       .....................
          0---|---|---|---|---|---|---|---|---|---|---|---|---|---|---|---0
        16.00   18.50   21.00   23.50   26.00   28.50   31.00   33.50   36.00
        Domained:        16.00 to       36.00

Model Solution:
CENTROID:          Price        =       25.30   CIX          =       0.46
MAXIMUM:           Price        =       27.02   CIX          =       0.70
  SurfaceHght =         0.70
```

Figure 8.28 Execution of the Pricing Models for Various Competition Price Values *(continued)*

```
CompPrice    =       27.50 [DoM: 0.673]

        FuzzySet:    PRICE
        Description:
      1.00
      0.90
      0.80
      0.70
      0.60                                      .......
      0.50                                    ..       .
      0.40                                .......       .
      0.30                            .......            .
      0.20                        ......                  ..
      0.10                      ..                          .
      0.00...........                                        ..................
          0---|---|---|---|---|---|---|---|---|---|---|---|---|---|---|---0
        16.00   18.50   21.00   23.50   26.00   28.50   31.00   33.50   36.00
        Domained:         16.00 to       36.00

Model Solution:
CENTROID:           Price       =       25.53   CIX         =       0.47
MAXIMUM:            Price       =       27.48   CIX         =       0.67
  SurfaceHght =         0.67


CompPrice    =       28.00 [DoM: 0.646]

        FuzzySet:    PRICE
        Description:
      1.00
      0.90
      0.80
      0.70
      0.60                                       ......
      0.50                                    . ..      ..
      0.40                                ...... .        .
      0.30                            .......              .
      0.20                        ......                    .
      0.10                      ..                            ..
      0.00...........                                           ................
          0---|---|---|---|---|---|---|---|---|---|---|---|---|---|---|---0
        16.00   18.50   21.00   23.50   26.00   28.50   31.00   33.50   36.00
        Domained:         16.00 to       36.00

Model Solution:
CENTROID:           Price       =       25.77   CIX         =       0.49
MAXIMUM:            Price       =       28.03   CIX         =       0.65
  SurfaceHght =         0.65
```

Figure 8.28 Execution of the Pricing Models for Various Competition Price Values *(continued)*

```
CompPrice   =     28.50 [DoM: 0.612]

       FuzzySet:    PRICE
       Description:
     1.00
     0.90
     0.80
     0.70
     0.60                                     ...
     0.50                              .    ...    ...
     0.40                         ...... ...          .
     0.30                  .......                     ..
     0.20           ......                               .
     0.10         ..                                      .
     0.00...........                                       ................
         0---|---|---|---|---|---|---|---|---|---|---|---|---|---|---|---0
       16.00   18.50   21.00   23.50   26.00   28.50   31.00   33.50   36.00
       Domained:        16.00 to      36.00

Model Solution:
CENTROID:          Price       =      26.00   CIX        =      0.50
MAXIMUM:           Price       =      28.50   CIX        =      0.61
  SurfaceHght =        0.61


CompPrice   =     29.00 [DoM: 0.583]

       FuzzySet:    PRICE
       Description:
     1.00
     0.90
     0.80
     0.70
     0.60
     0.50                              .    ........
     0.40                         ...... .....         ..
     0.30                  .......                       .
     0.20           ......                                .
     0.10         ..                                       ..
     0.00...........                                         .............
         0---|---|---|---|---|---|---|---|---|---|---|---|---|---|---|---0
       16.00   18.50   21.00   23.50   26.00   28.50   31.00   33.50   36.00
       Domained:        16.00 to      36.00

Model Solution:
CENTROID:          Price       =      26.23   CIX        =      0.49
MAXIMUM:           Price       =      28.97   CIX        =      0.58
  SurfaceHght =        0.58
```

Figure 8.28 Execution of the Pricing Models for Various Competition Price Values *(continued)*

```
CompPrice    =      29.50 [DoM: 0.552]

        FuzzySet:    PRICE
        Description:
      1.00
      0.90
      0.80
      0.70
      0.60
      0.50                                   .        .......
      0.40                             ...... .... ..        ..
      0.30                      .......            .           ..
      0.20               ......                                  .
      0.10             ..                                         ..
      0.00...........                                               ...........
          0---|---|---|---|---|---|---|---|---|---|---|---|---|---|---|---0
        16.00   18.50   21.00   23.50   26.00   28.50   31.00   33.50   36.00
        Domained:         16.00 to      36.00

Model Solution:
CENTROID:        Price       =      26.47   CIX        =      0.48
MAXIMUM:         Price       =      29.52   CIX        =      0.55
  SurfaceHght =       0.55

CompPrice    =      30.00 [DoM: 0.515]

        FuzzySet:    PRICE
        Description:
      1.00
      0.90
      0.80
      0.70
      0.60
      0.50                                   .         .....
      0.40                             ...... ....   ...     ..
      0.30                      .......            . .         ..
      0.20               ......                     .            ..
      0.10             ..                                          .
      0.00...........                                               ..........
          0---|---|---|---|---|---|---|---|---|---|---|---|---|---|---|---0
        16.00   18.50   21.00   23.50   26.00   28.50   31.00   33.50   36.00
        Domained:         16.00 to      36.00

Model Solution:
CENTROID:        Price       =      26.62   CIX        =      0.47
MAXIMUM:         Price       =      29.98   CIX        =      0.51
  SurfaceHght =       0.51
```

Figure 8.28 Execution of the Pricing Models for Various Competition Price Values *(continued)*

```
CompPrice    =      30.50 [DoM: 0.482]

        FuzzySet:    PRICE
        Description:
      1.00
      0.90
      0.80
      0.70
      0.60
      0.50                             .
      0.40                      ...... ....     .........
      0.30                .......          .  ..         ..
      0.20          ......                  ..             ..
      0.10        ..                                         ..
      0.00...........                                          ........
         0---|---|---|---|---|---|---|---|---|---|---|---|---|---|---|---0
        16.00   18.50   21.00   23.50   26.00   28.50   31.00   33.50   36.00
        Domained:        16.00 to       36.00

Model Solution:
CENTROID:        Price      =      26.78   CIX        =       0.46
MAXIMUM:         Price      =      26.08   CIX        =       0.50
  SurfaceHght =        0.50

CompPrice    =      31.00 [DoM: 0.442]

        FuzzySet:    PRICE
        Description:
      1.00
      0.90
      0.80
      0.70
      0.60
      0.50                             .
      0.40                      ...... ....     .......
      0.30                .......          .   ...         ...
      0.20          ......                  . .              .
      0.10        ..                         ..                ...
      0.00...........                                              ......
         0---|---|---|---|---|---|---|---|---|---|---|---|---|---|---|---0
        16.00   18.50   21.00   23.50   26.00   28.50   31.00   33.50   36.00
        Domained:        16.00 to       36.00

Model Solution:
CENTROID:        Price      =      26.86   CIX        =       0.46
MAXIMUM:         Price      =      26.08   CIX        =       0.50
  SurfaceHght =        0.50
```

Figure 8.28 Execution of the Pricing Models for Various Competition Price Values *(continued)*

```
CompPrice   =      31.50 [DoM: 0.406]

        FuzzySet:    PRICE
        Description:
      1.00
      0.90
      0.80
      0.70
      0.60
      0.50
      0.40
      0.30
      0.20
      0.10
      0.00...........
         0---|---|---|---|---|---|---|---|---|---|---|---|---|---|---|---0
        16.00   18.50   21.00   23.50   26.00   28.50   31.00   33.50   36.00
        Domained:        16.00 to       36.00

Model Solution:
CENTROID:          Price       =      26.94   CIX        =       0.46
MAXIMUM:           Price       =      26.08   CIX        =       0.50
  SurfaceHght =          0.50

CompPrice   =      32.00 [DoM: 0.370]

        FuzzySet:    PRICE
        Description:
      1.00
      0.90
      0.80
      0.70
      0.60
      0.50
      0.40
      0.30
      0.20
      0.10
      0.00...........
         0---|---|---|---|---|---|---|---|---|---|---|---|---|---|---|---0
        16.00   18.50   21.00   23.50   26.00   28.50   31.00   33.50   36.00
        Domained:        16.00 to       36.00

Model Solution:
CENTROID:          Price       =      27.02   CIX        =       0.45
MAXIMUM:           Price       =      26.08   CIX        =       0.50
  SurfaceHght =          0.50
```

Figure 8.28 Execution of the Pricing Models for Various Competition Price Values *(continued)*

```
CompPrice   =      32.50 [DoM: 0.326]

        FuzzySet:    PRICE
        Description:
      1.00
      0.90
      0.80
      0.70
      0.60
      0.50                                .
      0.40                          ...... ....
      0.30                   .......           .            .......
      0.20             ......                   .       ....       ...
      0.10           ..                          ..  ...              ...
      0.00...........                              ..                    ..
          0---|---|---|---|---|---|---|---|---|---|---|---|---|---|---|---0
        16.00   18.50   21.00   23.50   26.00   28.50   31.00   33.50   36.00
        Domained:        16.00 to      36.00

Model Solution:
CENTROID:        Price        =     26.94   CIX         =       0.46
MAXIMUM:         Price        =     26.08   CIX         =       0.50
   SurfaceHght =      0.50


CompPrice   =      33.00 [DoM: 0.287]

        FuzzySet:    PRICE
        Description:
      1.00
      0.90
      0.80
      0.70
      0.60
      0.50                                .
      0.40                          ...... ....
      0.30                   .......           .
      0.20             ......                   .         .............
      0.10           ..                          ..    ...             ...
      0.00...........                              ....                   .
          0---|---|---|---|---|---|---|---|---|---|---|---|---|---|---|---0
        16.00   18.50   21.00   23.50   26.00   28.50   31.00   33.50   36.00
        Domained:        16.00 to      36.00

Model Solution:
CENTROID:        Price        =     26.86   CIX         =       0.46
MAXIMUM:         Price        =     26.08   CIX         =       0.50
   SurfaceHght =      0.50
```

Figure 8.28 Execution of the Pricing Models for Various Competition Price Values *(continued)*

```
CompPrice   =      33.50 [DoM: 0.240]

        FuzzySet:    PRICE
        Description:
      1.00
      0.90
      0.80
      0.70
      0.60
      0.50                                    .
      0.40                              ...... ....
      0.30                       .......           .
      0.20               ......                     .          ...........
      0.10             ..                            ..    ....           ...
      0.00...........                                  ......
        0---|---|---|---|---|---|---|---|---|---|---|---|---|---|---|---0
       16.00   18.50   21.00   23.50   26.00   28.50   31.00   33.50   36.00
       Domained:        16.00 to      36.00

Model Solution:
CENTROID:        Price       =     26.62   CIX        =      0.47
MAXIMUM:         Price       =     26.08   CIX        =      0.50
  SurfaceHght =      0.50


CompPrice   =      34.00 [DoM: 0.199]

        FuzzySet:    PRICE
        Description:
      1.00
      0.90
      0.80
      0.70
      0.60
      0.50                                    .
      0.40                              ...... ....
      0.30                       .......           .
      0.20               ......                     .             .....
      0.10             ..                            ..      ......     ....
      0.00...........                                  .........
        0---|---|---|---|---|---|---|---|---|---|---|---|---|---|---|---0
       16.00   18.50   21.00   23.50   26.00   28.50   31.00   33.50   36.00
       Domained:        16.00 to      36.00

Model Solution:
CENTROID:        Price       =     26.31   CIX        =      0.49
MAXIMUM:         Price       =     26.08   CIX        =      0.50
  SurfaceHght =      0.50
```

Figure 8.28 Execution of the Pricing Models for Various Competition Price Values *(continued)*

```
CompPrice   =      34.50 [DoM: 0.157]

       FuzzySet:    PRICE
       Description:
     1.00
     0.90
     0.80
     0.70
     0.60
     0.50                                .
     0.40                          ...... ....
     0.30                   .......        .
     0.20             ......                .
     0.10          ..                        ..            .............
     0.00..........                            ...........
         0---|---|---|---|---|---|---|---|---|---|---|---|---|---|---|---0
       16.00   18.50   21.00   23.50   26.00   28.50   31.00   33.50   36.00
       Domained:        16.00 to       36.00

Model Solution:
CENTROID:        Price       =       25.92   CIX        =        0.49
MAXIMUM:         Price       =       26.08   CIX        =        0.50
  SurfaceHght =        0.50


CompPrice   =      35.00 [DoM: 0.106]

       FuzzySet:    PRICE
       Description:
     1.00
     0.90
     0.80
     0.70
     0.60
     0.50                                .
     0.40                          ...... ....
     0.30                   .......        .
     0.20             ......                .
     0.10          ..                        ..                  ........
     0.00..........                            ................
         0---|---|---|---|---|---|---|---|---|---|---|---|---|---|---|---0
       16.00   18.50   21.00   23.50   26.00   28.50   31.00   33.50   36.00
       Domained:        16.00 to       36.00

Model Solution:
CENTROID:        Price       =       25.38   CIX        =        0.47
MAXIMUM:         Price       =       26.08   CIX        =        0.50
  SurfaceHght =        0.50
```

Figure 8.28 Execution of the Pricing Models for Various Competition Price Values *(continued)*

```
CompPrice    =     35.50 [DoM: 0.061]

       FuzzySet:    PRICE
       Description:
     1.00
     0.90
     0.80
     0.70
     0.60
     0.50                                   .
     0.40                             ...... ....
     0.30                      .......          .
     0.20               ......                   .
     0.10             ..                          ..
     0.00..........                                 ........................
        0---|---|---|---|---|---|---|---|---|---|---|---|---|---|---0
       16.00   18.50   21.00   23.50   26.00   28.50   31.00   33.50   36.00
       Domained:         16.00 to       36.00

Model Solution:
CENTROID:        Price      =     24.91   CIX        =      0.44
MAXIMUM:         Price      =     26.08   CIX        =      0.50
  SurfaceHght =       0.50
```

Figure 8.28 Execution of the Pricing Models for Various Competition Price Values *(continued)*

The New Product Pricing Model (Version 2)

Version 1 of the pricing model reflected a conservative price estimating approach. In Version 2, the strategies for finding a default price reflect a more aggressive attitude toward market positioning. The unconditional rules in this model employ more sophisticated linguistic variables. The core fuzzy model contains the following four rules:

```
[R1] our price must be very high
[R2] our price must be somewhat low
[R3] our price must be above around 2*mfgcosts
[R4] if the competition.price is not very low
     then our price must be around the competition.price
```

Model Design Strategies

The first rule, taking into account the costs of spoilage, distribution, and transportation, specifies that the product price must be very high, if lacking any other evidence. We might accomplish the same objective either by breaking out this criteria in additional supporting rules such as

```
[R5] if spoilage.rate is excessive
   then price must be quite high

[R6] if distribution.costs are rather high
   then price should be high

[R7] if inventory.costs are high
   then price should be somewhat high
```

or by formulating another policy that calculates and returns the probable support costs for the product. In any case, the first rule [*R1*]encodes the finance manager's view that, to cover the ancillary costs of supporting the product, the price must be toward the upper limit of the price sensitivity scale. The second rule [*R2*] relaxes the constraint that the price should be definitely low. The hedge *somewhat* permits the price to rise up the price sensitivity range.

Rule [*R3*] more realistically alters the way the manufacturing costs constrain the model. In this model, the price must be above twice the manufacturing costs, not simply around the costs. This reflects a more common sense view that the manufacturing costs set a minimum price level but are not intricately tied to the price's acceptable values.

The final rule [*R4*] takes a different view of the competition's pricing strategy. As long as our competitors are not engaging in a price war—dropping their price close to costs in order to gain volume sales—we want our price to be close to that of the competition. Once more, the linguistic variable not *very Low* could be adjusted to set the actual precision of this strategy (such as simply *not Low* or *positively not very Low*).

THE MODEL EXECUTION LOGIC

This version of the pricing model (found in *FSHPRC3.CPP*) has the same logical flow as the first version.[3] The difference lies in the organization of the fuzzy sets and the knowledge of the production rules. The model example here is executed with slightly different parameters: a manufacturing cost of $12.00 and a competition price of $29.00. Execution logic that is essentially the same as the first model version has generally been omitted.

Create the Basic Fuzzy Sets

The vocabulary fuzzy sets in Figure 8.29 are formed by applying the built-in *very* and *somewhat* hedges to the base *High* and *Low* fuzzy sets, respectively. The *very* hedge intensifies the fuzzy set *High*, reducing the truth membership for any value that would normally be *High*. The *somewhat* hedge dilutes the fuzzy set *Low*, increasing the truth membership for any value that would normally be *Low*. For a detailed discussion of hedge intensification and dilution see Chapter 5, *Fuzzy Set Hedges,* Intensifying and Diluting Fuzzy Regions on page 230.

```
---Policy 'PRICING' created.
---Default Hedges installed in Policy 'PRICING'
---Linear Increasing FuzzySet 'VERY.HIGH' created.
---Linear Decreasing FuzzySet 'SOMEWHAT.LOW' created.

PRICE
Domain [UofD]:      16.00 to      36.00

      1.00**                                                                  .
      0.90  ************                                                   ...
      0.80              ***********                                     ...
      0.70                         *********                        ....
      0.60                                  ********            ....
      0.50                                          *******.
      0.40                                         ....    ******
      0.30                                   ......              ****
      0.20                             ......                        ***
      0.10                    .........                                 **
      0.00....................                                            *
         0---|---|---|---|---|---|---|---|---|---|---|---|---|---|---|---0
       16.00   18.50   21.00   23.50   26.00   28.50   31.00   33.50   36.00
     .  FuzzySet:     VERY.HIGH
        Description: Very High for Price
        Domained:          16.00 to      36.00
     *  FuzzySet:     SOMEWHAT.LOW
        Description: Somewhat Low for Price
        Domained:          16.00 to      36.00
```

Figure 8.29 The Price Constraint Fuzzy Sets: *very High* and *somewhat Low*

Create the Run-Time Model Fuzzy Sets

Figure 8.30 illustrates another significant change we make in Version 2 of the pricing model. The twice manufacturing costs window is much more narrow than the fuzzy region generated in the first model. We are using a 10% diffusion instead of the previous 25% diffusion value. Since we will actually be looking at the region that is above the approximated space for this value, we want to narrow its horizon so the final fuzzy space is not quite so broad.

```
        FuzzySet:     Near.2*MfgCosts
        Description:
      1.00                           .
      0.90                        ... .
      0.80                             .
      0.70                        .
      0.60                              .
      0.50                       .
      0.40
      0.30                      .        .
      0.20
      0.10                     .          .
      0.00....................            ..................................
         0---|---|---|---|---|---|---|---|---|---|---|---|---|---|---|---0
       16.00   18.50   21.00   23.50   26.00   28.50   31.00   33.50   36.00
        Domained:          16.00 to      36.00
```

Figure 8.30 A Narrower Cost of Manufacturing Fuzzy Set (*Around 24 [2*MfgCosts]*)

Figure 8.31 illustrates the final linguistic variable for above around twice the manufacturing costs. The slope of the fuzzy set is zero until we reach the midpoint of the *Around 2*MfgCosts* base fuzzy region when it starts increasing its truth membership until it is at absolute truth at the point where *Around 2*MfgCosts* has zero membership. Figure 8.32 shows how the curves appear when overlaid.

```
---Hedge 'above' applied to Fuzzy Set "Near.2*MfgCosts"
       FuzzySet:     Near.2*MfgCosts
       Description: Above Near (Around) Twice MfgCosts
     1.00                                   ................................
     0.90                                  .
     0.80                                 .
     0.70
     0.60                                .
     0.50
     0.40                               .
     0.30
     0.20                              .
     0.10
     0.00............................
         0---|---|---|---|---|---|---|---|---|---|---|---|---|---|---|---0
       16.00   18.50   21.00   23.50   26.00   28.50   31.00   33.50   36.00
       Domained:          16.00 to        36.00
```

Figure 8.31 The *Above Around 2*MfgCosts* Linguistic Variable

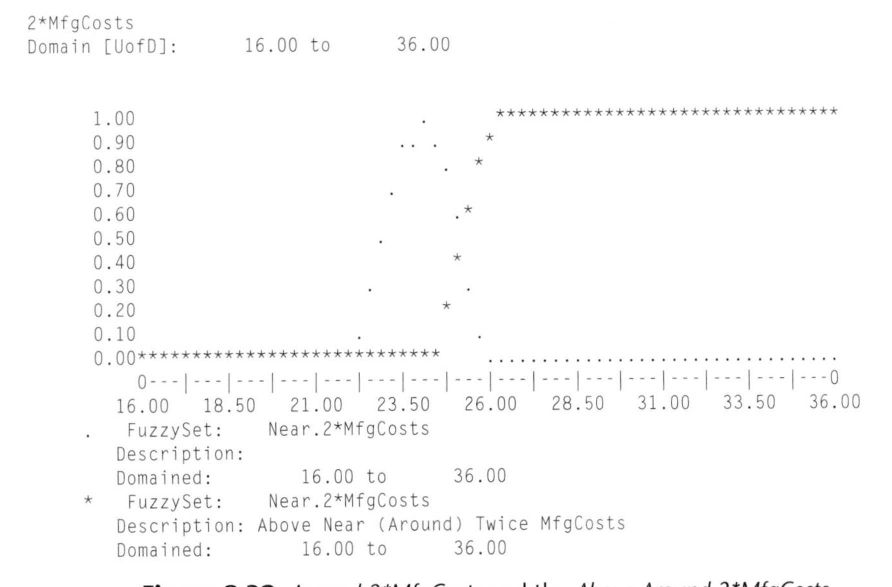

Figure 8.32 *Around 2*MfgCosts* and the *Above Around 2*MfgCosts*

The *above* hedge (like the *below* hedge) restricts a fuzzy region to a space that is above (or below) the unit membership point of a bell-shaped or triangular fuzzy set. As you can see in Figure 8.32, the generated fuzzy set, on the right side of the unit membership point, is the standard Zadeh complement of the fuzzy set.

As Figure 8.33 indicates, the fuzzy set for near the competitor's price remains unchanged in this model version (a little stability from one model to the next is always useful).

```
---PI FuzzySet 'Near.CompPrice' created.
        FuzzySet:     Near.CompPrice
        Description: Near (Around) Competition Price
      1.00                                .
      0.90                              .. ..
      0.80                            ..     ..
      0.70
      0.60                           .         .
      0.50                          .           .
      0.40
      0.30                         .             .
      0.20                        .               .
      0.10                       .                 .
      0.00.......................                   .......................
          0---|---|---|---|---|---|---|---|---|---|---|---|---|---|---|---0
        16.00   18.50   21.00   23.50   26.00   28.50   31.00   33.50   36.00
        Domained:        16.00 to      36.00
```

Figure 8.33 *Near.CompPrice* Fuzzy Set (*Around 26*)

Execute the Price Estimation Rules

The first unconditional rules are applied in the same fashion as the rules in the first model version. Unconditionals are combined using the *And* operation. Figures 8.34 and 8.35 show how the output fuzzy space is updated as each rule is fired.

The solution fuzzy set for *Price*, shown in Figure 8.35, represents the intersection of the *very High* and *somewhat Low* linguistic variables. Since these sets are nonlinear and no longer completely symmetrical, in this case *somewhat LOW* is not the complement of *very High*; the intersection point is shifted to the right.

```
Rule Execution....
R1   our price must be very high

        FuzzySet:     PRICE
        Description:
      1.00                                                                .
      0.90                                                             ...
      0.80                                                          ...
      0.70                                                      ....
      0.60                                                  ....
      0.50                                             .....
      0.40                                         ....
      0.30                                   ......
      0.20                             ......
      0.10                    .........
      0.00....................
          0---|---|---|---|---|---|---|---|---|---|---|---|---|---|---|---0
        16.00   18.50   21.00   23.50   26.00   28.50   31.00   33.50   36.00
        Domained:        16.00 to      36.00
```

Figure 8.34 The *Price* Solution Fuzzy Set After Executing Rule [*R1*]

```
R2    our price must be somewhat low

     FuzzySet:     PRICE
     Description:
  1.00
  0.90
  0.80
  0.70
  0.60
  0.50                                                ....
  0.40                                            ....     ......
  0.30                                        ......           ....
  0.20                                    ......                   ...
  0.10                              .........                        ..
  0.00......................                                           .
     0---|---|---|---|---|---|---|---|---|---|---|---|---|---|---|---|---0
     16.00   18.50   21.00   23.50   26.00   28.50   31.00   33.50   36.00
     Domained:        16.00 to       36.00
```

Figure 8.35 The *Price* Solution Fuzzy Set After Executing Rule [*R2*]

The change in the model's properties through the use of linguistic variables is an important feature of fuzzy systems. In the initial pricing model, the application of the first two unconditional rules create a triangular region centered over the midpoint of the domain. You might wonder why we need a fuzzy model to find this value since it can just as easily be calculated as shown in Equation 8.5.

$$price = low + \frac{1}{2}(high - low)$$

Equation 8.5 Locating Price as the Domain's Midpoint

However, this is an artifact of *High* and *Low*'s linear nature. If these were sigmoid (S-curved) fuzzy sets, such a calculation would not be quite so simple. In fact, by adding the *very* and *somewhat* hedges to the first two rules, the morphology of the model is altered in a semantic and predictable way. A strictly mathematical model that attempts to mirror the action of complex linguistic variables would prove extremely difficult (consider the changes to Equation 8.5 to accommodate the nonlinear shape of the new intersection point for these two rules). Fuzzy models can quickly and easily approximate a wide spectrum of complex nonlinear spaces simply through changes in their linguistic variables.

When rule [*R3*] is applied, the left-hand side of the *Price* solution fuzzy set is clipped by the slope of the manufacturing costs linguistic variable. Figure 8.36 shows the solution after this clipping is performed. This truncation of the *Price* fuzzy sets is caused by the way unconditionals are applied: through the intersection of the two sets.

```
R3   our price must be above around 2*mfgcosts

       FuzzySet:    PRICE
       Description:
     1.00
     0.90
     0.80
     0.70
     0.60
     0.50                                                         ....
     0.40                                                    ....     ......
     0.30                                             ......               ....
     0.20                                       ......                         ...
     0.10                                      .                                  ..
     0.00............................                                                .
         0---|---|---|---|---|---|---|---|---|---|---|---|---|---|---|---0
       16.00   18.50   21.00   23.50   26.00   28.50   31.00   33.50   36.00
       Domained:        16.00 to      36.00
```

Figure 8.36 The *Price* Solution Fuzzy Set After Executing Rule [*R3*]

The following two graphs, in Figures 8.37 and 8.38, illustrate how the solution is reduced along the left edge of the solution fuzzy set by applying rule [*R3*]. You can see that this restriction slices through the current solution fuzzy region, clipping off any area to the left of the slope. This action ensures that, if no other evidence is accumulated, the final price will always be above the twice manufacturing costs value.

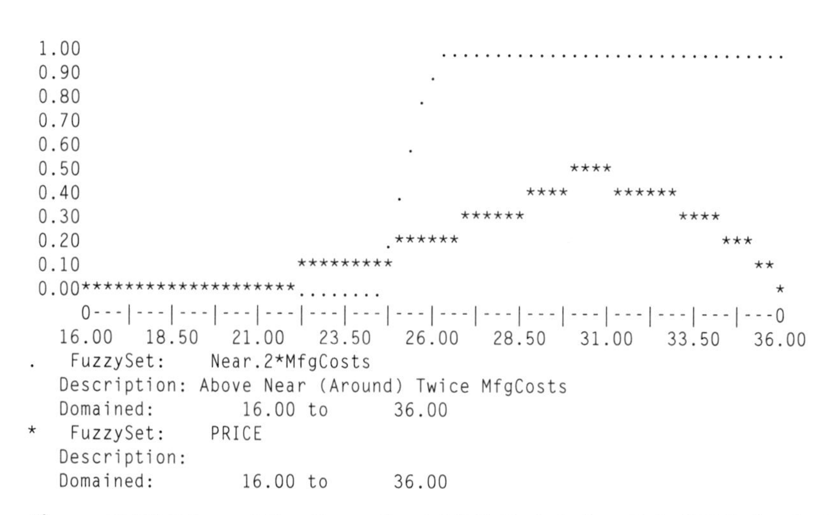

Figure 8.37 *Price* and the *Above About 2*MfgCosts* Before Rule [*R3*] Is Applied

```
1.00                                        ................................
0.90                                       .
0.80                                      .
0.70
0.60                                     .
0.50                                                    ****
0.40                                    .           ****    ******
0.30                                          ******              ****
0.20                                   .******                        ***
0.10                                   *                                 **
0.00****************************                                           *
     0---|---|---|---|---|---|---|---|---|---|---|---|---|---|---|---|---0
   16.00   18.50   21.00   23.50   26.00   28.50   31.00   33.50   36.00
 .   FuzzySet:     Near.2*MfgCosts
   Description: Above Near (Around) Twice MfgCosts
   Domained:         16.00 to      36.00
 *   FuzzySet:     PRICE
   Description:
   Domained:         16.00 to      36.00
```

Figure 8.38 *Price* and the *Above About 2*MfgCosts* After Rule [*R3*] Is Applied

Conditional rule [R4] is processed in the same manner as conditional rule [R4] in the first version of the pricing model. In this instance, however, we are interested in determining the truth of the proposition:

```
the competition.price is not very low
```

As we see in Figure 8.39, the initial fuzzy region for *very Low* is created by applying the built-in hedge *very* to the *Low* fuzzy set. In this model, we initially created a base fuzzy set, *somewhat Low* by hedging the base set *Low*. At this point, we retrieve the base *Low* set and hedge it once more to produce the *very Low* fuzzy region.

```
R4   if the competition.price is not very low
        then our price must be around the competition.price
---Hedge 'very' applied to Fuzzy Set "LOW"

   FuzzySet:     LOW
   Description: very LOW
 1.00.
 0.90 ...
 0.80     ....
 0.70         ...
 0.60            ....
 0.50                .....
 0.40                     .....
 0.30                          .....
 0.20                               .......
 0.10                                      ........
 0.00                                              .....................
     0---|---|---|---|---|---|---|---|---|---|---|---|---|---|---|---|---0
   16.00   18.50   21.00   23.50   26.00   28.50   31.00   33.50   36.00
   Domained:         16.00 to      36.00
```

Figure 8.39 Creating the *very Low* Fuzzy Region

The final *not very Low* linguistic variable is shown in Figure 8.40. It is created by applying the standard Zadeh *not* $(1\text{-}\mu_A(x))$ to the temporary fuzzy set created by the *very* hedge. When the *not very Low* linguistic variable is complete, we can determine the truth of the predicate proposition by finding the grade of membership for the competition price ($29.00). Since the competition's price of $29.00 is well to the right of the *not very Low* curve, the proposition has a truth function value of [0.89].

```
---Hedge 'ZADEH' NOT applied to Fuzzy Set "LOW"
FuzzySet:     LOW
        Description: ZADEH NOT LOW
      1.00                                                       .......
      0.90                                               ................
      0.80                                       ........
      0.70                                 ......
      0.60                            .....
      0.50                       .....
      0.40                  ....
      0.30              ....
      0.20          ....
      0.10      ....
      0.00...
          0---|---|---|---|---|---|---|---|---|---|---|---|---|---|---|---0
        16.00   18.50   21.00   23.50   26.00   28.50   31.00   33.50   36.00
        Domained:         16.00 to        36.00
```

Figure 8.40 Creating the *not very Low* Linguistic Variable

At this point in the model process, the consequent proposition

```
our price must be around the competition price
```

must be evaluated and the *Price* solution fuzzy set updated.

The first step in this process is correlating (through correlation minimum) the consequent fuzzy set (*Around.CompPrice*) with the truth of the predicate ([.87]). Figure 8.41 shows the state of the *Price* solution variable overlaid with the correlated consequent fuzzy set. As usual, the solution fuzzy region is updated by applying the union of the adjusted consequent and the solution set.

```
---PremiseTruth=        0.87

      1.00
      0.90                                             ......
      0.80                                           .      ..
      0.70                                          .         .
      0.60                                         .
      0.50                                        .         ****
      0.40                                       .     ****     ******
      0.30                                      . ******          .     ****
      0.20                                 ******                  .        ***
      0.10                                *  .                      .          **
      0.00****************************...                            ..........*
          0---|---|---|---|---|---|---|---|---|---|---|---|---|---|---|---0
        16.00   18.50   21.00   23.50   26.00   28.50   31.00   33.50   36.00
   .    FuzzySet:     Near.CompPrice
        Description: Near (Around) Competition Price
        Domained:         16.00 to        36.00
   *    FuzzySet:     PRICE
        Description:
        Domained:         16.00 to        36.00
```

Figure 8.41 *Price* and the Correlated *Near.CompPrice* Before Completing Rule *[R4]*

Figure 8.42 shows the solution fuzzy set for *Price* after the last rule is executed. The model is now complete, and the expected value for price can be determined by defuzzifying this region (see Figure 8.43).

Using the composite moments technique to defuzzify, we find an expected value for *Price* of \$29.83. Note that this value is almost squarely on top of the current competition price, which is in keeping with the working of fuzzy models involving conditional and unconditional propositions. The unconditionals (when they are issued before the conditionals) describe the default state of the system. If none of the conditionals fires with a strength greater than the highest truth of the unconditionals, then the unconditionals define the model's value. In this case, conditional rule [*R4*] said, in effect, "to the degree that the competition's price is not very low, make our price close to the competition's price." Since the antecedent proposition

the competition price is not very low

had a significant degree of truth, the truth of the consequent action was also very high. A large solution fuzzy region was centered almost directly over the competitor's price of \$29.00. Only the influence of the unconditionals, felt by the centroid defuzzification method, pulled the value slightly to the right. If composite maximum defuzzification had been used, the expected value for *Price* would be the same as the competition's price.

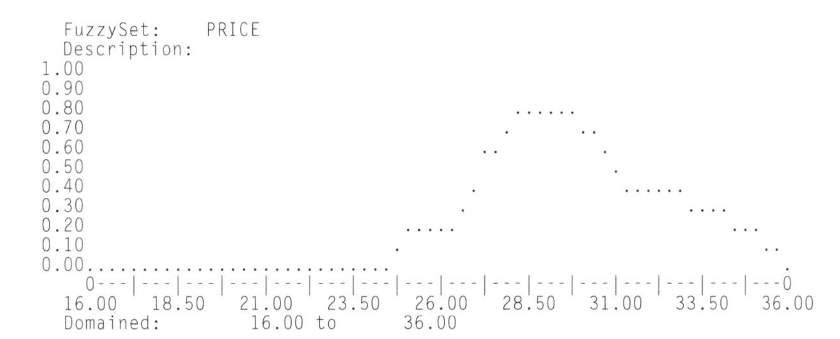

Figure 8.42 The *Price* Solution Fuzzy Set After Executing Rule [*R4*]

```
     FuzzySet:     PRICE
     Description:
   1.00
   0.90
   0.80                                          ......
   0.70                                         .      ..
   0.60                                       ..         .
   0.50                                                   .
   0.40                                      .             ......
   0.30                                     .                    ....
   0.20                                .....                         ...
   0.10                               .                                 ..
   0.00............................                                       .
       0---|---|---|---|---|---|---|---|---|---|---|---|---|---|---|---|---0
     16.00   18.50   21.00   23.50   26.00   28.50   31.00   33.50   36.00
     Domained:          16.00 to       36.00
---'CENTROID'     defuzzification. Value:     29.828, [0.8112]
Model Solution:
  Price       =       29.83
  CompIdx     =        0.81
  SurfaceHght =        0.87
```

Figure 8.43 Defuzzifying the *Price* Solution Set Using the Centroid Technique

The New Product Pricing Model (Version 3)

In Version 3 of the pricing model, we explore the effects of moving the unconditional propositions from the front to the end of the model. In a conventional symbolic expert system, the order in which the rules appear in the knowledge base is immaterial—the inference engine is responsible for ordering the conflict agenda.The same philosophy holds for fuzzy systems that contain only *if–then* rules or only unconditionals. When the model (or policy) contains a mixture of conditional and unconditional rules, then the order in which they are executed determines the final solution fuzzy space. The model has the same fuzzy sets and rules as the pricing model in Version 1 except the order of rule execution has been modified as follows:

```
[R1] if the competition.price is not very high
        then our price must be near the competition.price
[R2] our price must be high
[R3] our price must be low
[R4] our price must be around 2*mfgcosts
```

The Model Execution Logic

The conditional rule is executed first, followed by all the unconditionals. Other than this change, the parameters to the model are the same as those for Version 1: a manufacturing cost of $12.00 and a competition price of $26.00. The basic and run-time fuzzy sets are not shown but are also the same as the Version 1 pricing model. Whenever the execution logic is essentially the same as Version 1, the fuzzy sets are shown without any extended commentary.

Execute the Price Estimation Rules

As noted previously, when the *not very High* linguistic variable is created (Figures 8.44 and 8.45), we can find the truth of the predicate proposition by finding the membership of the competition price value ($26.00) in this fuzzy space. This process results in a truth function value of [.75]. Figure 8.46 shows how the correlation minimum is performed by restricting the consequent fuzzy set to the maximum height of the predicate

```
R1   if the competition.price is not very high
        then our price must be around the competition.price
---Hedge 'very' applied to Fuzzy Set "HIGH"

        FuzzySet:     HIGH
        Description: very HIGH
      1.00                                                                        .
      0.90                                                                     ...
      0.80                                                                  ...
      0.70                                                              ....
      0.60                                                          ....
      0.50                                                     .....
      0.40                                                 ....
      0.30                                           ......
      0.20                                     ......
      0.10                             .........
      0.00.....................
         0---|---|---|---|---|---|---|---|---|---|---|---|---|---|---|---0
        16.00   18.50   21.00   23.50   26.00   28.50   31.00   33.50   36.00
        Domained:        16.00 to      36.00
```

Figure 8.44 Creating the *very High* Fuzzy Region

```
---Hedge 'ZADEH' NOT applied to Fuzzy Set "HIGH"

       FuzzySet:     HIGH
       Description: ZADEH NOT HIGH
     1.00.......
     0.90          ...............
     0.80                         ........
     0.70                                 ......
     0.60                                       ......
     0.50                                             ....
     0.40                                                 .....
     0.30                                                      ....
     0.20                                                          ...
     0.10                                                             ....
     0.00                                                                ...
         0---|---|---|---|---|---|---|---|---|---|---|---|---|---|---|---0
       16.00   18.50   21.00   23.50   26.00   28.50   31.00   33.50   36.00
       Domained:        16.00 to       36.00
```

Figure 8.45 Creating the *not very High* Linguistic Variable

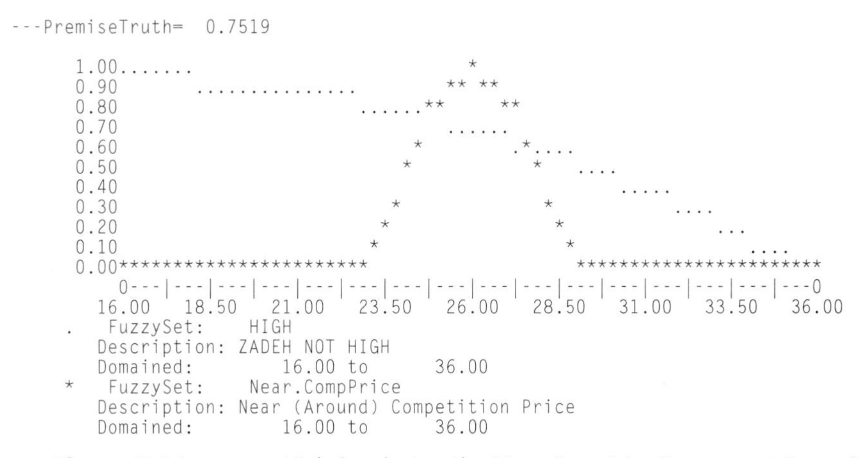

```
---PremiseTruth=  0.7519

     1.00.......                                *
     0.90          ...............            ** **
     0.80                         ......**          **
     0.70                                 ......
     0.60                              *        .*....
     0.50                             *           *   ....
     0.40                                                 .....
     0.30                            *             *          ....
     0.20                           *               *             ...
     0.10                          *                 *               ....
     0.00***********************                   ***********************
         0---|---|---|---|---|---|---|---|---|---|---|---|---|---|---|---0
       16.00   18.50   21.00   23.50   26.00   28.50   31.00   33.50   36.00
   .    FuzzySet:     HIGH
       Description: ZADEH NOT HIGH
       Domained:        16.00 to       36.00
   *    FuzzySet:     Near.CompPrice
       Description: Near (Around) Competition Price
       Domained:        16.00 to       36.00
```

Figure 8.46 *not very High* Overlaying the *Near.CompPrice* Consequent Fuzzy Set

We now execute [*R2*], the first unconditional rule. As Figure 8.47 shows, the current *Price* fuzzy set is the space defined by the near the competition price reduced by the truth of the rule [*R4*] predicate. Figure 8.48 shows how the *High* fuzzy space overlays the current *Price* solution set.

```
FuzzySet:     PRICE
       Description:
     1.00
     0.90
     0.80
     0.70                                       .....
     0.60                                     ..     ..
     0.50                                    .         .
     0.40                                   .           .
     0.30
     0.20                                 ..             ..
     0.10                                .                 .
     0.00.......................                   ......................
         0---|---|---|---|---|---|---|---|---|---|---|---|---|---|---|---0
       16.00   18.50   21.00   23.50   26.00   28.50   31.00   33.50   36.00
       Domained:        16.00 to       36.00
```

Figure 8.47 The *Price* Solution Fuzzy Set After Executing Rule [*R4*]

```
R2   our price must be high

     1.00                                                                      .
     0.90                                                                 ......
     0.80                                                          .......
     0.70                                 *****                ......
     0.60                               **     **   .......
     0.50                              *    .....*
     0.40                             *.....      *
     0.30                       .......
     0.20                 ......     **             **
     0.10        .......            *                 *
     0.00**********************                        **********************
         0---|---|---|---|---|---|---|---|---|---|---|---|---|---|---|---0
        16.00   18.50   21.00   23.50   26.00   28.50   31.00   33.50   36.00
    .  FuzzySet:    HIGH
       Description: High for Price
       Domained:         16.00 to       36.00
    *  FuzzySet:    PRICE
       Description:
       Domained:         16.00 to       36.00
```

Figure 8.48 The *Price* Solution Set Overlaid by *High*

Like all unconditional propositions, the solution space is updated by taking the minimum (the intersection) of the solution set's membership function and the membership function of the consequent fuzzy set. Figure 8.49 shows how the solution space is clipped across the top and the left by the *High* fuzzy set, resulting in the final solution fuzzy set.

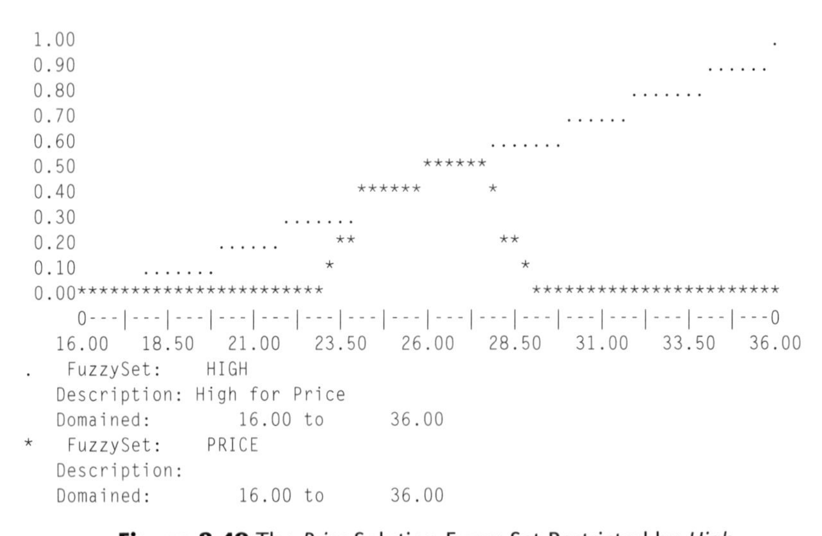

```
     1.00                                                                      .
     0.90                                                                 ......
     0.80                                                          .......
     0.70                                                      ......
     0.60                                               .......
     0.50                                   ******
     0.40                             ******      *
     0.30                       .......
     0.20                 ......     **             **
     0.10        .......            *                 *
     0.00**********************                        **********************
         0---|---|---|---|---|---|---|---|---|---|---|---|---|---|---|---0
        16.00   18.50   21.00   23.50   26.00   28.50   31.00   33.50   36.00
    .  FuzzySet:    HIGH
       Description: High for Price
       Domained:         16.00 to       36.00
    *  FuzzySet:    PRICE
       Description:
       Domained:         16.00 to       36.00
```

Figure 8.49 The *Price* Solution Fuzzy Set Restricted by *High*

We now execute [R3], the second unconditional rule. As Figure 8.50 shows, the current *Price* fuzzy set is the space defined by the *High* fuzzy set and the near the competition price reduced by the truth of the rule [*R2*] predicate. Figure 8.51 shows how the *Low* fuzzy space overlays the current *Price* solution set.

```
FuzzySet:     PRICE
        Description:
      1.00
      0.90
      0.80
      0.70
      0.60
      0.50                                        ......
      0.40                                  ......      .
      0.30
      0.20                                ..              ..
      0.10                               .                  .
      0.00.......................                            ......................
          0---|---|---|---|---|---|---|---|---|---|---|---|---|---|---|---0
        16.00    18.50    21.00    23.50    26.00    28.50    31.00    33.50    36.00
        Domained:         16.00 to       36.00
```

Figure 8.50 The *Price* Solution Fuzzy Set After Executing Rule [*R2*]

```
R3   our price must be low

Domain [UofD]:       16.00 to       36.00

      1.00.
      0.90 .......
      0.80        ......
      0.70              ......
      0.60                    .......
      0.50                           .....******
      0.40                          ******  .....*.
      0.30                                        ......
      0.20                        **              **      .......
      0.10                       *                  *           ......
      0.00***********************                    ***********************
          0---|---|---|---|---|---|---|---|---|---|---|---|---|---|---|---0
        16.00    18.50    21.00    23.50    26.00    28.50    31.00    33.50    36.00
     .   FuzzySet:     LOW
        Description: Low for Price
        Domained:         16.00 to       36.00
     *   FuzzySet:     PRICE
        Description:
        Domained:         16.00 to       36.00
```

Figure 8.51 The *Price* Solution Set Overlaid by *Low*

Rule [*R3*] is also an unconditional, so the solution space is updated by taking the minimum (the intersection) of the solution set's membership function and the membership function of the consequent fuzzy set. Figure 8.52 shows how the solution space is clipped across the top and the right by the *Low* fuzzy set, resulting in the final solution fuzzy set.

```
Domain [UofD]:      16.00 to      36.00

      1.00.
      0.90 .......
      0.80       ......
      0.70            ......
      0.60                  .......
      0.50                         .....*
      0.40                        ****** ******.
      0.30                                       ......
      0.20                      **             **      .......
      0.10                     *                 *           ......
      0.00***********************                   ***********************
          0---|---|---|---|---|---|---|---|---|---|---|---|---|---|---|---0
        16.00   18.50   21.00   23.50   26.00   28.50   31.00   33.50   36.00
     .   FuzzySet:    LOW
        Description: Low for Price
        Domained:        16.00 to      36.00
     *   FuzzySet:    PRICE
        Description:
        Domained:        16.00 to      36.00
```

Figure 8.52 The *Price* Solution Fuzzy Set Restricted by *Low*

We now execute rule [R4], the last unconditional rule. As Figure 8.53 shows, the current *Price* fuzzy set is the space defined by the *High* fuzzy set, the *Low* fuzzy set, and the near the competition price reduced by the truth of the rule [*R1*] predicate. We now want to further restrict the solution space by the manufacturing costs rule. Figure 8.54 shows how the *Around.2*MfgCosts* fuzzy space overlays the current *Price* solution set.

```
        FuzzySet:    PRICE
        Description:
      1.00
      0.90
      0.80
      0.70
      0.60
      0.50                                .
      0.40                        ...... ......
      0.30
      0.20                      ..             ..
      0.10                     .                 .
      0.00.......................                   .......................
          0---|---|---|---|---|---|---|---|---|---|---|---|---|---|---|---0
        16.00   18.50   21.00   23.50   26.00   28.50   31.00   33.50   36.00
        Domained:        16.00 to      36.00
```

Figure 8.53 The *Price* Solution Fuzzy Set After Executing Rule *[R3]*

```
R4   our price must be around 2*mfgcostss
Domain [UofD]:       16.00 to       36.00

      1.00                                ..
      0.90                            ...   ....
      0.80                          ..        .
      0.70                         .           ..
      0.60                       ..              .
      0.50                      .           *     .
      0.40                     .       ****** ******
      0.30                    .                    .
      0.20                   .         **           .**
      0.10                 ..         *              ..*
      0.00***********************                      .***********************
         0---|---|---|---|---|---|---|---|---|---|---|---|---|---|---|---0
        16.00   18.50   21.00   23.50   26.00   28.50   31.00   33.50   36.00
      .   FuzzySet:    Near.2*MfgCosts
        Description: Near (Around) Twice MfgCosts
        Domained:          16.00 to      36.00
      *   FuzzySet:    PRICE
        Description:
        Domained:          16.00 to      36.00
```

Figure 8.54 The *Price* Solution Fuzzy Set Overlaid by *Around.2*MfgCosts*

Executing unconditional rule [*R4*] updates the solution fuzzy set by taking the minimum of the solution set's membership function and the membership function of the consequent fuzzy set. Figure 8.55 shows how the solution space is clipped along the right edge by the *Around.2*MfgCosts* curve.

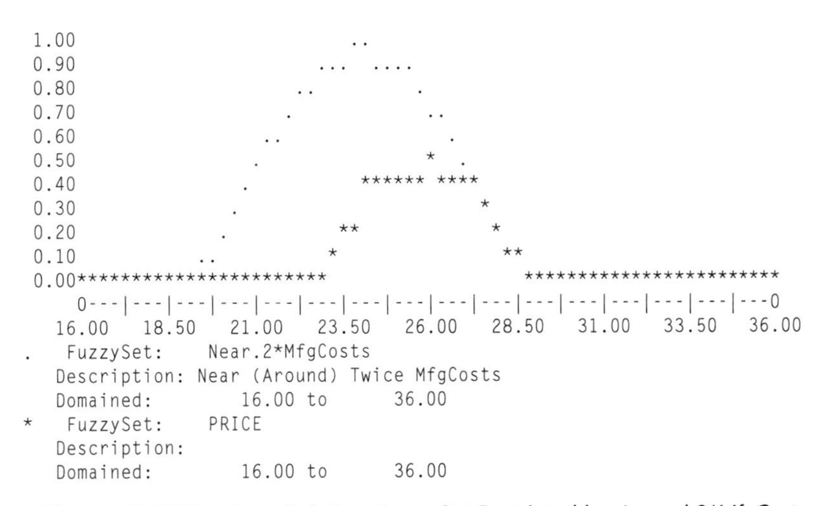

Figure 8.55 The *Price* Solution Fuzzy Set Restricted by *Around.2*MfgCosts*

Defuzzify to Find Expected Value for Price

The model is now complete. Figure 8.56 shows the final *Price* fuzzy set. Using the composite moments (centroid) defuzzification method, we find the expected value for *Price* (which is $25.84).

```
  FuzzySet:    PRICE
  Description:
1.00
0.90
0.80
0.70
0.60
0.50                                    .
0.40                              ...... ....
0.30                                         .
0.20                             ..           .
0.10                            .              ..
0.00......................                       ........................
    0---|---|---|---|---|---|---|---|---|---|---|---|---|---|---|---|---0
  16.00   18.50   21.00   23.50   26.00   28.50   31.00   33.50   36.00
  Domained:        16.00 to      36.00

---'CENTROID'     defuzzification. Value:     25.844, [0.4903]
Model Solution:
  Price        =       25.84
  CompIdx      =        0.49
  SurfaceHght  =        0.50
```

Figure 8.56 Final *Price* Fuzzy Set from Model Execution

While the result, in this case, of applying the unconditionals last is not remarkably different from Version 1, this is an artifact of the model rather than a general statement about the differences between the two approaches. In general, applying the unconditionals last will restrict any sets formed by the conditionals, while applying unconditionals first simply provides a minimum default solution set. For any real-world applications, the two approaches would, and usually do, produce significantly different results.

The New Product Pricing Model (P&L Version)

We conclude the new product pricing models with a look at how fuzzy policies can be used to create a simple profit and loss analysis for the product introduction program. Until now, the pricing policies have been analyzed *in vacuo*—as though they were independent entities. In fact, pricing, inventory, sales forecast, risk assessment, and similar fuzzy policies are almost always embedded in larger applications.

DESIGN FOR THE P&L MODEL

In this product profit and loss (P&L) model, there are two fuzzy policies: one to calculate a recommended price and one to use this price in predicting a product sales volume. Figure 8.57 shows how the profit and loss planning model is organized.

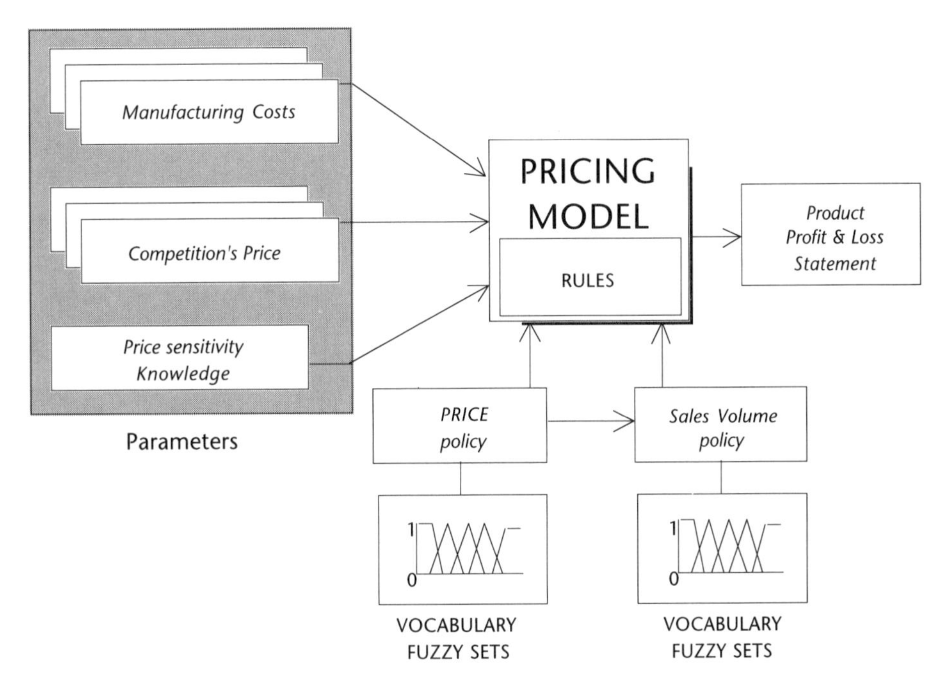

Figure 8.57 The Structure of the New Product Profit and Loss Analysis Model

This model estimates gross profits (before taxes, extraordinary expenses, write-offs, and inventory adjustments and discounts) based on monthly direct costs and competitor prices in the same marketplace. For this study we examine the rather simple model shown in Figure 8.58.

```
SalesMargins=Price-DirectCosts;
GrossSales=SalesVolume*Price;
GrossProfit=SalesMargins*SalesVolume;
```

Figure 8.58 The New Product P&L Model

Figure 8.58 is a time series model covering twelve months of data, so there is an implied time subscript attached to each variable. This time bias becomes evident when the model code itself is examined. Listing 8.2 shows a segment of the actual P&L model. (*FSGPRC5.CPP* contains the actual model code.)

```
for(i=0;i<MaxMonths;i++)
    {
     compprice=competitionPrice[i];
     mfgcosts=directCosts[i];
     Price[i]=fshpr5policy(
       Rules,mfgcosts,compprice,&cindex,&status);
     salesVolume[i]=SVolpolicy(Price[i],&cindex,&status);
     salesMargin[i]=Price[i]-directCosts[i];
     grossSales[i] =salesVolume[i]*Price[i];
     grossProfit[i]=salesMargin[i]*salesVolume[i];
    }
```

Listing 8.2 The Code Representation of the Pricing P&L Model

Model Execution and Logic

The values for direct costs and competition price are stored in arrays as part of the main model program although, in an actual application, the values would most probably be read from an external file or database system. When the model is executed,[4] a competition price and direct cost value are selected. The pricing policy (*fshpr5policy*) is used to find the recommended price. This price is then passed to the sales volume policy (*SVolpolicy*) to produce a new sales volume based on the price. The sales volume policy uses a monotonic logic scheme to predict a sales volume based on the premise that the higher the price the lower the sales volume. From price and sales volume, we go on to calculate margins, gross sales, and gross profits. Figure 8.59 shows the profit and loss statement generated by the model.

	Volume	Our Price	Direct Costs	Sales Margin	Gross Sales	Gross Profit	Comp. Price	UCIdx
Jan	94.92	17.09	8.25	8.84	1622.57	839.47	16.25	0.760
Feb	94.14	17.25	8.50	8.75	1623.93	823.73	16.50	0.816
Mar	92.97	17.48	8.75	8.73	1625.50	812.02	16.75	0.828
Apr	87.11	18.66	9.00	9.66	1625.13	841.15	18.50	0.978
May	85.55	18.97	9.25	9.72	1622.72	831.41	18.75	0.970
Jun	75.39	21.00	9.50	11.50	1583.20	866.99	21.25	0.921
Jul	72.27	21.62	9.75	11.88	1562.74	858.15	22.00	0.889
Aug	63.67	23.42	10.00	13.42	1491.31	854.60	24.50	0.682
Sep	56.64	24.83	10.25	14.58	1406.28	825.71	26.75	0.387
Oct	52.73	25.53	10.50	15.03	1346.37	792.66	28.25	0.162
Nov	48.83	26.39	10.75	15.64	1288.60	763.70	30.50	0.016
Dec	47.66	26.62	11.00	15.62	1268.85	744.63	31.75	0.051

Figure 8.59 The P&L Statement for the New Product Pricing Model (*FSHPRC5*)

There are quite simple causal relationships at work in this model: as the competition price increases, our recommended price also increases. As our recommended price increases, the sales volume steadily decreases. There is a point, in June, when the decrease in the projected sales volume begins to affect gross sales (and, consequently, profitability[5]). In this P&L statement, the unit compatibility index (*UCIdx*) reflects the strength of the estimated price. To simplify matters, the compatibility index for the sales volume is not reflected in the model.

> The *SVolpolicy* uses simple fuzzy propositional or monotonic logic to predict a new sales volume based on the current recommended price. Consider the implicit fuzzy proposition:
>
> *if price is low then sales.volume is high*
>
> This is obviously quite naive. Assuming our competition has adequate sales volumes at its current price, a better approach might be to examine the positive or negative distance between our price and the competition's price and make decisions based on how rapidly we are closing on or pulling away from that price.

```
double SVolpolicy(double Price,float *CIX,int *statusPtr)
 {
  PDB     *prcPDBptr;
  FDB     *svFDBptr,*LoFDBptr;
  double   volume;
  float    PremiseTruth;
  FILE    *mdlout;

  *statusPtr=0;
  prcPDBptr=XSYSctl.XSYScurrPDBptr;
  mdlout=MtsGetSystemFile(SYSMODFILE);
  fprintf(mdlout,"%s\n","Sales Volume Estimate Policy Begins..");
//--Find the fuzzy sets in the policy dictionary
  svFDBptr=MdlFindFDB("HIGH.VOLUME",    prcPDBptr,statusPtr);
  LoFDBptr=MdlFindFDB("LOW",            prcPDBptr,statusPtr);
//
  volume=FzyMonotonicLogic(
      LoFDBptr,svFDBptr,Price,&PremiseTruth,statusPtr);
  return(volume);
 }
```

Listing 8.3 The Fuzzy Sales Volume Estimating Policy

The sales volume policy uses the basic *Low* vocabulary fuzzy set (see Figure 8.60) along with a new fuzzy set that defines the semantics of high volumes based on the number of orders (see Figure 8.61). It is the relationship between these two fuzzy regions that we model in monotonic reasoning.

```
PRICE
Domain [UofD]:       16.00 to       36.00


   1.00*                                                                     .
   0.90 *******                                                        ......
   0.80        ******                                           .......
   0.70              ******                               ......
   0.60                    *******                  .......
   0.50                           ******.....
   0.40                          ...... *******
   0.30                    .......             ******
   0.20              ......                          *******
   0.10       .......                                       ******
   0.00......                                                     ******
       0---|---|---|---|---|---|---|---|---|---|---|---|---|---|---|---0
     16.00   18.50   21.00   23.50   26.00   28.50   31.00   33.50   36.00
  .   FuzzySet:     HIGH
     Description: High for Price
     Domained:          16.00 to       36.00
  *   FuzzySet:     LOW
     Description: Low for Price
     Domained:          16.00 to       36.00
```

Figure 8.60 The *High* and *Low* Price Fuzzy Sets

```
     FuzzySet:     HIGH.VOLUME
     Description: High Sales Volumes
   1.00                                                                      .
   0.90                                                                ......
   0.80                                                         .......
   0.70                                                   ......
   0.60                                            .......
   0.50                                      ......
   0.40                                ......
   0.30                         .......
   0.20                   ......
   0.10            .......
   0.00......
       0---|---|---|---|---|---|---|---|---|---|---|---|---|---|---|---0
      0.00    12.50   25.00   37.50   50.00   62.50   75.00   87.50  100.00
     Domained:          0.00 to       100
```

Figure 8.61 The *High.Volume* Fuzzy Set for Sales Orders

Using Policies to Calculate Price and Sales Volume

In Figure 8.62 we see the final *Price* solution fuzzy set for one of the monthly price estimation policy executions. The underlying rule structure is the same as the Version 1 pricing policy. Each new direct cost (*MfgCost* in the pricing policy) and competition price yields a new recommendation for the price parameter. In this instance, a value of $23.42 is found through the composite moments defuzzification method.

```
  FuzzySet:    PRICE
  Description:
1.00
0.90
0.80                            ...
0.70                          ..   ..
0.60                         .       .
0.50                        .         .
0.40                                   .
0.30                    ...             .
0.20               ......
0.10        .......                      ..
0.00......                                 .............................
   0---|---|---|---|---|---|---|---|---|---|---|---|---|---|---|---0
  16.00   18.50   21.00   23.50   26.00   28.50   31.00   33.50   36.00

---'CENTROID'     defuzzification. Value:     23.422, [0.6817]
Model Solution:
  Price       =     23.42
  CompIdx     =      0.68
  SurfaceHght =      0.82
```

Figure 8.62 The Final *Price* Solution Fuzzy Set and Its Expected Value (*$23.42*)

This recommended value for price is used by *SVolpolicy* to find an estimated sales volume. The new price's degree of membership in the fuzzy set *Low* is found. We use this degree of membership to find a corresponding value in the *High Volume* fuzzy set by moving horizontally across the fuzzy set until we find the curve. We use the scalar value from the domain as the value for sales volume. This implies that a relationship exists between the shape of *Low* price and *High Volume* sales. Figure 8.63 shows the implication relationship between the two fuzzy sets.

Each row in the implication matrix defines a fuzzy implication relationship set between *Low* price and *High Volume* sales. Figure 8.64 shows this fuzzy region for the current recommended price value.

Approximate Reasoning System
Implication Matrix for 'LOW' and 'HIGH.VOLUME'

LOW		HIGH.VOLUME 0.00	1.25	2.50	3.75	5.00	6.25	7.50	8.75	10.00
16.00	1.00	0.00								1.00
18.50	0.88	0.00							0.88	0.88
21.00	0.75	0.00						0.75	0.75	0.75
23.50	0.63	0.00					0.63	0.63	0.63	0.63
26.00	0.50	0.00				0.50	0.50	0.50	0.50	0.50
28.50	0.38	0.00			0.37	0.38	0.38	0.38	0.38	0.38
31.00	0.25	0.00		0.25	0.25	0.25	0.25	0.25	0.25	0.25
33.50	0.13	0.00	0.13	0.13	0.13	0.13	0.13	0.13	0.13	0.13
36.00	0.00	0.00	0.00	0.00	0.00	0.00	0.00	0.00	0.00	0.00

Scaling:HIGH.VOLUME.....x 10.00

Figure 8.63 The Implication Maxtrix for *Low* Price and *High.Volume* Sales

```
'HIGH.VOLUME' when 'LOW' is      23.42

FuzzySet:    HIGH.VOLUME
Description: High Sales Volumes
1.00
0.90
0.80
0.70
0.60                                                   ..........................
0.50                                             ......
0.40                                       ......
0.30                                ....... 
0.20                          ......
0.10                ....... 
0.00......
    0---|---|---|---|---|---|---|---|---|---|---|---|---|---|---|---0
    0.00    12.50    25.00    37.50    50.00    62.50    75.00    87.50   100.00
    Domained:            0.00 to       100.00
```

Figure 8.64 The Implication Fuzzy Space Between *Low* and *High.Volume*

In this profit and loss model, we have used a simple linear relationship: as *Low* decreases, *High Volume* increases. Thus, a high truth membership in *Low* will find a high truth membership value in *High Volume*. Modeling different curve dependencies will have a significant impact on model performance. The implication matrix and a fuzzy relationship set for a set of values from either of the fuzzy set domains can be produced through the *FzyImplicationMatrix* function (which is inserted in the *FSHPRC5.CPP* pricing model in this book).

A Project Risk Assessment Model

A business organization is often faced with a wide spectrum of service, technology, and research opportunities in the shape of formally defined projects. A manager must decide how to allocate limited financial, staffing, and equipment resources to projects in order to satisfy the organization's functional responsibilities as well as corporate strategic business, political, and image constraints. While both a broad technology assessment of the project pool and a financial analysis of project payback are common practices in industry, too little attention is given to assessing the risks associated with assuming an individual project. This risk is measured in starkly simply terms: the organization terminates a project without meeting the project's objectives. Curtailing a project involves wasted resources, the postponement of other projects, adverse political exposure, and a decrease in organizational morale. A model that ranks the project pool in terms of completion risk provides management with a valuable tool in deciding, among the top x projects, which should be allocated resources and which should undergo further study.

The Model Design

Figure 8.65 shows the architecture of a model that evaluates the critical factors underlying a project's success and makes a recommendation on whether or not to accept the project.

The factors considered by the model represent just a few of the actual data points in the production model. For the sake of clarity, we have omitted rules covering such topics as the organization's prior experience with similar projects; the experience and age of the project manager; the number of different skills required during the project; the degree of resource sharing across multiple projects; the percent of staff comprised of consultants; part-time, and full-time employees; the internal project complexity (number of activities, number of different resources, etc.); the level of return on investment; the sponsoring organization's perceived commitment; and similar factors. The core risk assessment model has four rules:

```
[R1] if project.duration is Long then risk is increased;
[R2] if project.staffing is Large then risk is increased;
[R3] if project.funding is Low then risk is increased;
[R4] if project.priority is High then risk is increased;
```

We use these rules in a special way. The risk assessment model takes a different approach to fuzzy reasoning than models discussed in the previous case studies. As Figure 8.66 illustrates, the model employs a special kind of reasoning known as *scalable monotonic chaining.*

Scalable monotonic chaining does not create and then defuzzify a solution fuzzy set. Rather, it uses monotonic chaining to map the risk specified in individual rules to an intermediate risk measuring fuzzy set (in this case, *Increased.Risk*). The result of this mapping

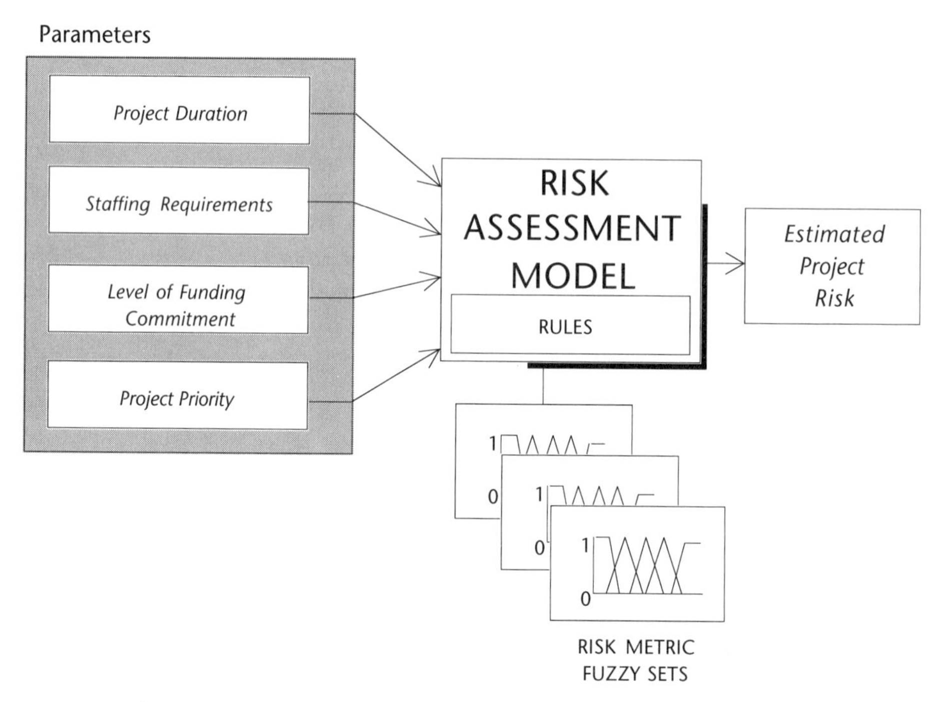

Figure 8.65 The Structure of the Project Risk Assessment Model

is a scalar value from the domain of the risk metric indicating the degree of risk for this particular model factor. The monotonic reasoning results for each rule are summed to produce a final risk value. This value, a scalar, called the *TotalRisk*, is used to find the actual project risk. We find *TotalRisk*'s degree of membership in a controlling fuzzy set (the *High.Risk* set in our model). This process scales the risk. The degree of membership is the overall project risk. In our model, we simply multiply the truth function by 1000 to produce a risk factor within the range of the *Increased.Risk* domain.

Model Application Issues

The scalable monotonic chaining approach to fuzzy reasoning addresses a persistent problem in fuzzy models that have a large number of rules with the same consequent fuzzy set.[6] The solution fuzzy set quickly becomes completely saturated. In the min/max inference technique, the solution fuzzy set is updated (for conditional *if–then* rules) by taking the maximum of the consequent fuzzy set and the solution fuzzy set. This means that after a few rules, the "high water mark" in the solution set will move quickly toward [1.0]. It also means that rules whose truths are less than the current truth level of the solution fuzzy region will not contribute to the solution. In risk assessment models, for example, this pro-

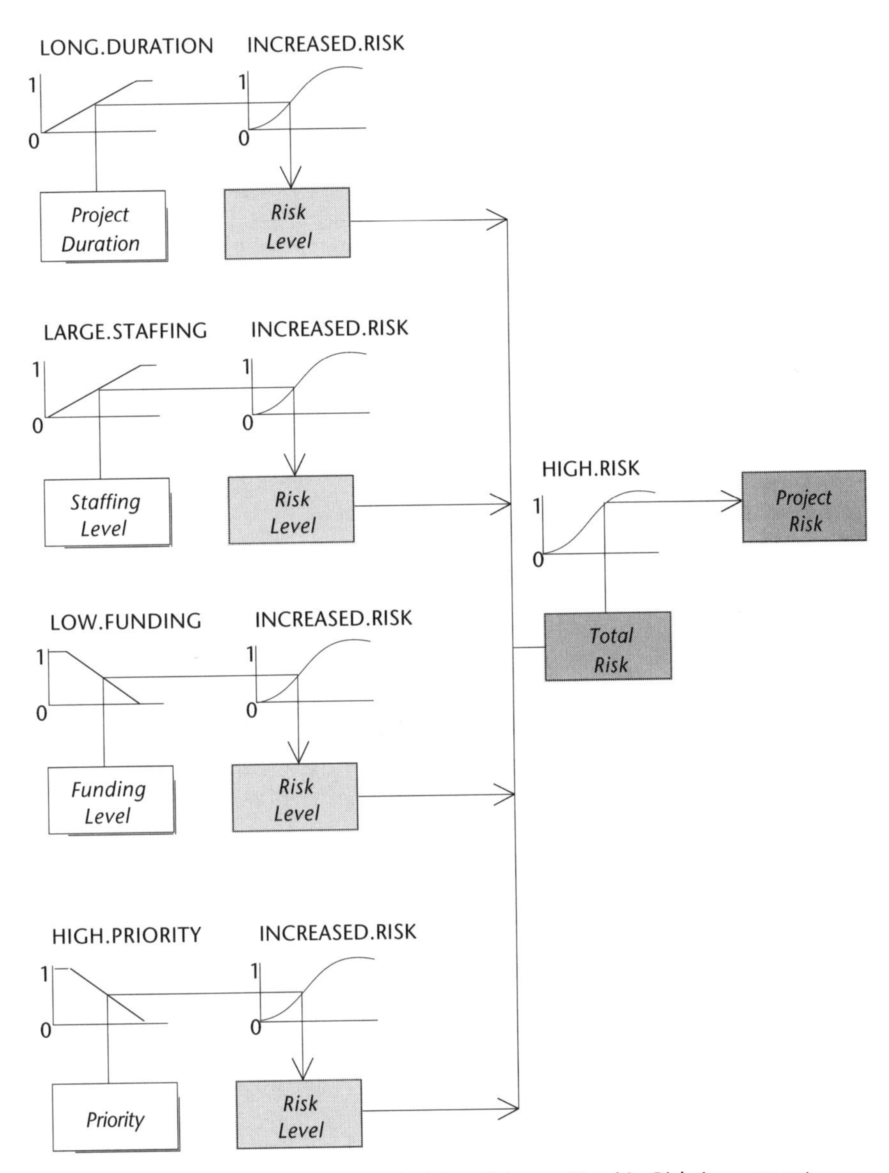

Figure 8.66 The Monotonic Chaining Scheme Used in Risk Assessment

cess is unsatisfactory. We want to accumulate as much possible evidence as we can. As an example, consider a more extensive set of rules (although the phenomenon of saturation can occur even in a two-rule system):

```
[R1] if project.duration is Long then risk is Increased;
[R2] if project.staffing is Large then risk is Increased;
[R3] if project.funding is Low then risk is Increased;
[R4] if project.priority is High then risk is Increased;
[R5] if project.complexity is High then risk is Increased;
[R6] if project.resources are Large then risk is Increased;
[R7] if project.experience is Low then risk is Increased;
[R8] if project.size is Big then risk is Increased;
```

If rule [*R1*] has a predicate truth of [.82] then the solution fuzzy set *Risk* contains the consequent fuzzy set *Increased* to a degree of [.82]. If none of the other rules has a truth greater than [.82], none will affect the *Risk* solution space. The risk of the project is determined by a single rule. Yet this does not seem reasonable. The cumulative effect of the other rules should influence the determination of risk since if any of these rules also have a significant truth (say, [.68], [.80], or [.52]) then the evident risk of assuming this project is much higher than indicated by the state of *Risk*.

Using the additive form of fuzzy inferencing helps but does not solve this problem. After executing two rules with predicate truths of [.62] and [.48], the solution fuzzy set will have a combined truth of [1.0] and will be saturated. You should recall that the maximum truth of a fuzzy set is bounded at [1.0] through the relationship $\mu A[x]=min(1,\mu_A[x])$. Actually, this is a problem of the semantics associated with the way *Increased* is used. For a large number of rules that map a consequent to a fuzzy solution set, if many of these rules are true and the solution effect is arithmetically cumulative rather than morphological, then we must turn to another method of representing this effect. Chained monotonic scaling is such a method.

In this technique, we introduce two major ideas. First, we discuss the idea of determining the risk associated with each factor based on an associated fuzzy region and its mapping to a risk measuring fuzzy set. Since these fuzzy sets can have different topologies, this process allows a highly nonlinear relationship on a factor-by-factor basis. Second, we develop the idea of taking the cumulative risks by simple addition. (However, in the production model, the risk value was weighted by its grade of membership in the consequent fuzzy set and by finding this number in another fuzzy set that calibrates this number to a final risk determination.) This last step also introduces a high degree of nonlinearity into our model. These kinds of fuzzy models can handle any number of rules and still maintain the important relationships between the underlying rules and the final risk assessment.

In building the model, we need only maintain a domain correspondence between the unit risk fuzzy set (*Increased*) and the scaling fuzzy set (*High.Risk*). The domain of the scaling fuzzy set must extend from the lower domain of the unit fuzzy set to the maximum possible value if each rule's truth was [1.0].

Model Execution Logic

The model execution requests the project parameters and then evaluates each for its degree of risk. This is the FSHPRA1.CPP model. When the model completes, the project risk assessment is displayed along with the individual risks for each project property. The first parts of the model (Figures 8.67 through 8.70) show the fuzzy sets associated with each risk property, measured by the model.

```
 FuzzySet:     LONG.DURATION
 Description: Long Duration (in weeks)
1.00                                                         .....
0.90                                                  ...........
0.80                                              .....
0.70                                         .....
0.60                                      ...
0.50                                  ....
0.40                               ...
0.30                           ....
0.20                      .....
0.10                .......
0.00..............
    0---|---|---|---|---|---|---|---|---|---|---|---|---|---|---|---0
   0.00    3.75    7.50   11.25   15.00   18.75   22.50   26.25   30.00
 Domained:           0.00 to        30.00
```

Figure 8.67 The *Long* Duration Fuzzy Set

```
 FuzzySet:     LARGE.STAFFING
 Description: Large Staffing
1.00                                                              .
0.90                                                          ......
0.80                                                   .......
0.70                                             ......
0.60                                      .......
0.50                                ......
0.40                          ......
0.30                   .......
0.20             ......
0.10       .......
0.00......
    0---|---|---|---|---|---|---|---|---|---|---|---|---|---|---|---0
   0.00    3.00    6.00    9.00   12.00   15.00   18.00   21.00   24.00
 Domained:           0.00 to        24.00
```

Figure 8.68 The *Large* Project Staffing Fuzzy Set

```
  FuzzySet:     LOW.FUNDING
  Description: Low for Funding
1.00.
0.90 .......
0.80        ......
0.70              ......
0.60                    .......
0.50                           ......
0.40                                 .......
0.30                                        ......
0.20                                              .......
0.10                                                     ......
0.00
                                                              ......
     0---|---|---|---|---|---|---|---|---|---|---|---|---|---|---|---0
     0.00   12.50   25.00   37.50   50.00   62.50   75.00   87.50  100.00
  Domained:           0.00 to      100.00
```

Figure 8.69 The *Low* Percent of Required Funding Fuzzy Set

```
  FuzzySet:     HIGH.PRIORITY
  Description: High for Priority
1.00.
0.90 .......
0.80        ......
0.70              ......
0.60                    .......
0.50                           ......
0.40                                 .......
0.30                                        ......
0.20                                              .......
0.10                                                     ......
0.00
                                                              ......
     0---|---|---|---|---|---|---|---|---|---|---|---|---|---|---|---0
     0.00    1.25    2.50    3.75    5.00    6.25    7.50    8.75   10.00
  Domained:           0.00 to      10.00
```

Figure 8.70 The *High* Priority Fuzzy Set

```
  FuzzySet:     INCREASED
  Description: Increased Risk
1.00                                                               .....
0.90                                                      ..........
0.80                                                 .....
0.70                                            .....
0.60                                         ...
0.50                                     ....
0.40                                  ...
0.30                              ....
0.20                          ....
0.10                    ......
0.00................
     0---|---|---|---|---|---|---|---|---|---|---|---|---|---|---|---0
     0.00  125.00  250.00  375.00  500.00  625.00  750.00  875.00 1000.00
  Domained:           0.00 to      1000.00
```

Figure 8.71 The *Increased* Risk Consequent Fuzzy Set

Figure 8.71 is the unit risk measuring fuzzy set. Each rule's predicate is evaluated for its degree of truth. This degree of truth is used to find a point on the *Increased* fuzzy set. Dropping a plumb line to the domain, a value is found corresponding to the risk for this rule. This can be a value between 0 and 1000 (which is an arbitrary scale, but one that works well for a large number of rules). This scaling means that if all the rules have a maximum degree of truth, then the total cumulative risk would be 4000.

Figure 8.72 is the scaling fuzzy set used to find the actual project risk. The domain goes from 0 to 5000 so that it can map the cumulative unit risk to its proper position on the domain. If the cumulative risk is close to 4000, then the rules carry a large amount of risk. If the cumulative risk is toward the lower end of the domain, then the rules carry very little risk.

You should note that the shape of this fuzzy set (and the *Increased* consequent fuzzy set) allows us to adjust the way a risk value is interpreted. Changing the shape of the scaling fuzzy set, in particular, provides an easy method of mapping cumulative risk values to individual perspectives on how they relate to project risk.

```
  FuzzySet:     HIGH.RISK
  Description: Increased Risk
1.00                                                                   .....
0.90                                                         ...........
0.80                                                    .....
0.70                                              .....
0.60                                          ...
0.50                                     ....
0.40                                ...
0.30                            ....
0.20                      .....
0.10               ......
0.00..............
    0---|---|---|---|---|---|---|---|---|---|---|---|---|---|---|---0
   0.00  625.00 1250.00 1875.00 2500.00 3125.00 3750.00 4375.00 5000.00
  Domained:            0.00 to     5000.00
```

Figure 8.72 The *High.Risk* Rescaling Fuzzy Set

Executing the Risk Assessment Rules

Figures 8.73 and 8.74 are executions of the risk assessment model for different project description properties. For each risk factor, its value in the *Increased* set is shown along with the unit compatibility index (in this case the grade of membership). The *Cumulative* project risk is the simple arithmetic sum of the individual factors, and the membership value represents the average of the factor membership grades. Finally, the actual project risk assessment ([.712] in Figure 8.74) is the degree of membership of the cumulative fuzzy set multiplied by 1000 to bring its value into line with the individual factor ranges.

```
Fuzzy Project Risk Assessment Model.
     (c) 1998 Metus Systems Group.
Duration                :      8.00
Staff Load Level        :     12.00
Funding Commitment      :     40.00
Prioritization          :      3.00
The Rules:
R1   if project.duration is Long then risk is increased;
R2   if project.staffing is Large then risk is increased;
R3   if project.funding is Low then risk is increased;
R4   if project.priority is High then risk is increased;


Duration Risk           :  292.97 [0.141]
Staffing Risk           :  515.62 [0.498]
Funding  Risk           :  574.22 [0.603]
Priority Risk           :  632.81 [0.704]
CUMULATIVE Risk         : 2015.62 [0.487]
Project Risk Assessment :  323.76
```

Figure 8.73 A *Low* Risk Assessment Model Execution

```
Duration                :     25.00
Staff Load Level        :     19.00
Funding Commitment      :     10.00
Prioritization          :      1.00

The Rules:

R1   if project.duration is Long then risk is increased;
R2   if project.staffing is Large then risk is increased;
R3   if project.funding is Low then risk is increased;
R4   if project.priority is High then risk is increased;


Duration Risk           :  839.84 [0.944]
Staffing Risk           :  687.50 [0.786]
Funding  Risk           :  792.97 [0.903]
Priority Risk           :  792.97 [0.903]
CUMULATIVE Risk         : 3113.28 [0.884]
Project Risk Assessment :  712.86
```

Figure 8.74 A *High* Risk Assessment Model Execution

The FSHPRA1.CPP risk assessment model is an example of two concepts: the scalable monotonic chaining method and how a fuzzy model can be encoded in a single, simple, main C++ program. This model does not use the fuzzy system policy feature. Fuzzy sets are stored as local program variables.

```
#include <stdio.h>
#include <string.h>
#include <math.h>
#include "FDB.hpp"
#include "fuzzy.hpp"
#include "mtypes.hpp"
#include "mtsptype.hpp"
//
//--These are the "logical" rules.
static const char *Rules[]=
 {
  "R1   if project.duration is Long then risk is Increased;",
  "R2   if project.staffing is Large then risk is Increased;",
  "R3   if project.funding is Low then risk is Increased;",
  "R4   if project.priority is High then risk is Increased;"
 };
static const Rulemax=4;
void main(void)
 {
  FDB     *ldFDBptr,
          *lsFDBptr,
          *lfFDBptr,
          *hpFDBptr,
          *irFDBptr,
          *hrFDBptr;
  int      i,
           status,
           Idxpos,
           TryCtl[2];
  float    totgrade,
           memgrade,
           totDofRisk;
  double   prjDUR,
           prjSLL,
           prjFUN,
           prjPRI,
           durRisk,
           sllRisk,
           funRisk,
           priRisk,
           cumRisk;

  double   Domain[2],Parms[4];
  FILE    *mdlout;

  MdlConnecttoFMS(&status);
  mdlout=MtsGetSystemFile(SYSMODFILE);
```

```
//
//--Create the basic fuzzy sets
//
  Domain[0]=0;Domain[1]=30;
  ldFDBptr=FzyCreateSet(
     "LONG.DURATION",GROWTH  ,Domain,Parms,0,&status);
  strcpy(ldFDBptr->FDBdesc,"Long Duration (in weeks)");

  Domain[0]=0;Domain[1]=24;
  lsFDBptr=FzyCreateSet(
     "LARGE.STAFFING",INCREASE,Domain,Parms,0,&status);
  strcpy(lsFDBptr->FDBdesc,"Large Staffing");

  Domain[0]=0;Domain[1]=100;
  lfFDBptr=FzyCreateSet(
     "LOW.FUNDING",DECREASE,Domain,Parms,0,&status);
  strcpy(lfFDBptr->FDBdesc,"Low for Funding");

  Domain[0]=0;Domain[1]=10;
  hpFDBptr=FzyCreateSet(
     "HIGH.PRIORITY",DECREASE,Domain,Parms,0,&status);
  strcpy(hpFDBptr->FDBdesc,"High for Priority");

  Domain[0]=0;Domain[1]=1000;
  irFDBptr=FzyCreateSet(
     "INCREASED",GROWTH,Domain,Parms,0,&status);
  for(i=0;i<VECMAX;i++)
     irFDBptr->FDBvector[i]=
       (float)(pow(irFDBptr->FDBvector[i],1.1));
  strcpy(irFDBptr->FDBdesc,"Increased Risk");

  Domain[0]=0;Domain[1]=5000;
  hrFDBptr=FzyCreateSet(
    "HIGH.RISK",GROWTH,Domain,Parms,0,&status);
  strcpy(hrFDBptr->FDBdesc,"Increased Risk");
 //
//--Prompt for the model Data
//
  TryCtl[0]=8;
  TryCtl[1]=0;

  Domain[0]=0;Domain[1]=30;
  prjDUR=MtsAskforDBL(
    "duration",
    "Enter the duration (in weeks):",
    Domain,TryCtl,&status);
  if(status!=0) return;
```

```
  Domain[0]=0;Domain[1]=24;
  prjSLL=MtsAskforDBL(
    "staffing",
    "Enter the staffing load level:",
    Domain,TryCtl,&status);
  if(status!=0) return;

  Domain[0]=0;Domain[1]=100;
  prjFUN=MtsAskforDBL(
    "funding",
    "Enter the funding commitment :",
    Domain,TryCtl,&status);
  if(status!=0) return;

  Domain[0]=0;Domain[1]=10;
  prjPRI=MtsAskforDBL(
    "priority",
    "Enter the project priority   :",
    Domain,TryCtl,&status);
  if(status!=0) return;
//
  fprintf(mdlout,"%s\n%s\n",
    "Fuzzy Project Risk Assessment Model.",
    "       (c) 1992 Metus Systems Group.");
  fprintf(mdlout,"%s%10.2f\n","Duration             : ",prjDUR   );
  fprintf(mdlout,"%s%10.2f\n","Staff Load Level   : ",prjSLL   );
  fprintf(mdlout,"%s%10.2f\n","Funding Commitment : ",prjFUN   );
  fprintf(mdlout,"%s%10.2f\n", "Prioritization     : ",prjPRI   );
//
  fprintf(mdlout,"%s\n","The Rules:");
  for(i=0;i<Rulemax;i++) fprintf(mdlout,"%s\n",Rules[i]);
  fputc('\n',mdlout);

  totgrade=0;
  durRisk=FzyMonotonicLogic(
   ldFDBptr,irFDBptr,prjDUR,&memgrade,&status);
  fprintf(mdlout,"%s%7.2f%s%5.3f%s\n",
    "Duration Risk             : ",durRisk," [",memgrade,"]");
  totgrade+=memgrade;

  sllRisk=FzyMonotonicLogic(
   lsFDBptr,irFDBptr,prjSLL,&memgrade,&status);
  fprintf(mdlout,"%s%7.2f%s%5.3f%s\n",
    "Staffing Risk             : ",sllRisk," [",memgrade,"]");
  totgrade+=memgrade;

  funRisk=FzyMonotonicLogic(
```

```
  lfFDBptr,irFDBptr,prjFUN,&memgrade,&status);
 fprintf(mdlout,"%s%7.2f%s%5.3f%s\n",
   "Funding  Risk           : ",funRisk," [",memgrade,"]");
 totgrade+=memgrade;

 priRisk=FzyMonotonicLogic(
  hpFDBptr,irFDBptr,prjPRI,&memgrade,&status);
 fprintf(mdlout,"%s%7.2f%s%5.3f%s\n",
   "Priority Risk           : ",priRisk," [",memgrade,"]");
 totgrade+=memgrade;

 cumRisk=(durRisk+sllRisk+funRisk+priRisk);
 fprintf(mdlout,"%s%7.2f%s%5.3f%s\n",
   "CUMULATIVE Risk         : ",cumRisk," [",totgrade/4,"]");
 totDofRisk=FzyGetMembership(hrFDBptr,cumRisk,&Idxpos,&status);
 fprintf(mdlout,"%s%7.2f\n",
   "Project Risk Assessment : ",totDofRisk*1000);
 return;
}
```

Listing 8.4 The Project Risk Assessment Model (FSHPRA1.CPP)

After it defines the fuzzy sets and reads the model parameters, the main part of the model calls the *FzyMonotonicLogic* function for each parameter, adds their risk values, and uses *FzyGetMembership* to find the degree of membership for the cumulative value in the scaling fuzzy set. This example model has been broken down for clarity; in a production model, the data and fuzzy sets are stored in arrays so that the entire body of the model collapses to the code sown in Listing 8.5:

```
for(i=0;i<MaxParms;i++)
  {
   Risk[i]=FzyMonotonicLogic(
     parmFDBptr[i],irFDBptr,parmData[i],parmMem[i],&status);
   cumRisk+=Risk[i];
  }
totDofRisk=FzyGetMembership(hrFDBptr,cumRisk,&Idxpos,&status);
```

Listing 8.5 The Production Version of Project Risk Assessment

Notes

1. Kosko, Bart, "Addition as Fuzzy Mutual Entropy," *Information Sciences*, vol. 73, pp. 273–284, Oct. 1993.

2. In control applications, even very small degrees of memberships are usually processed since alpha-cut thresholds are generally not used.

3. For those astute readers wondering about FSHPRC2—this is the same program as FSHPRC1 except that it contains an embedded explanatory facility and explores the final solution fuzzy set using multiple defuzzification methods.

4. The actual execute profile of the model is not included for the P&L model due to the volume of output produced by even a simple trace. The model activity log file from FSHPRC5 occupies 129,418 bytes on disk (that's 2075 lines of information).

5. This would be an ideal model for a simple feedback protocol. Although not explored in this book, an optimization policy following the sales volume estimate could look at whether the forecast is an increase or decrease from last period. If it's a decrease, then we need to readjust the recommended price to account for this purchase sensitivity threshold.

6. I owe this approach to an idea for monotonic scaling developed by Roger Stein, a vice president at Moody's Investors Service in New York City. The case study itself is part of an actual risk analysis system developed by the author for a large investment management company.

Nine

Building Fuzzy Systems

A System Evaluation and Design Methodology

In this chapter, we bring together the conceptual and functional ideas of fuzzy systems, as well as the ideas underlying the physical implementation of models, into both a criteria for evaluating fuzzy projects and a broad methodology for designing, implementing, and validating fuzzy models. In the evaluation section, we will analyze the properties of business and technical "systems" in light of their relationship to fuzzy logic, and we will develop criteria for evaluating the feasibility of applying fuzzy systems methodologies. This discussion is followed by a general (and easily broken) methodology for designing fuzzy systems (based on the author's "this is the way I do things" procedure).

Evaluating Fuzzy System Projects

How do we recognize an opportunity to employ fuzzy logic (or, more properly, approximate reasoning)? In some cases the answer is obvious and intuitive—at least to knowledge engineers accustomed to evaluating systems in terms of continuous variables, intrinsic imprecision, and level of complexity. In other instances, the use of fuzzy logic is not as apparent. A few of these less obvious cases involve the transformation of crisply dynamic systems into fuzzy process–based systems so that they can be effectively and cleanly modeled. This use of fuzzy logic (actually a consequence of Zadeh's *principle of incompatibility*) is not so easily visualized by system builders accustomed to working with conventional logic, Bayesian probabilities, or mathematical modeling systems.

The Ideal Fuzzy System Problem

The *ideal* is the template against which we compare and contrast reality. The ideal manufacturing company provides the background for the idealized linear programming solutions to inventory, trans-shipment, bill-of-materials, and assembly problems. The ideal salesperson is used to illustrate the ideal traveling salesman problem. The ideal business organization is addressed by the ideal enterprise model that ideally integrates management decisions with bottom-line visibility and a complete (omniscient) understanding of the complex cause-and-effect dynamics of the supply and demand dichotomy (not to mention the effects of opportunity costs, random staff attrition, and macro/microeconomic fluctuations). Consequently, we nearly always view idealizations as representations of the perfect physical state rather than as actual models of the "way things work." This same caveat holds for fuzzy systems.

Fuzzy Model Characteristics

The following characteristics apply to the prototypical (or perhaps *archetypal*) fuzzy model. The demarcation of the characteristics' behavior in any system may be difficult to recognize. (As an example, only a few of the system control parameters are marginally fuzzy—how do we then evaluate the effectiveness of applying fuzzy set theory to the whole system?)

Fuzzy Control Parameters

Fuzzy control parameters provide the most straightforward and clear-cut indication that a fuzzy approach will be fruitful. Models whose control parameters[1] have intrinsically imprecise properties are well suited to fuzzy modeling. The trick here, of course, is familiarity with the semantics and representational characteristics of fuzzy sets and hedges. Once a knowledge engineer gains expertise in "fuzzy visualization," his or her task is often reduced to mapping the conceptual framework of the expert's domain into a series of fuzzy sets that represent the elements in the expert's decision and judgment protocol.

Let's examine the way we might approach the development of a fuzzy company acquisition and credit analysis system for a brokerage house. A (very) brief knowledge acquisition session between a credit analyst (Expert) and the knowledge engineer (KE) might run something like this:

KE	Well, how do you decide which companies are candidates for the acquisition portfolio?
Expert	First, of course, I look for relatively high and consistent profits.
KE	What do you mean by *high profits*?
Expert	That, naturally, depends on the company's earnings, so I can't give you an absolute number, like five million dollars, but I'd say 15% on earnings would be fine.

KE So (going to the blackboard),

```
candidate=profits > 1.15*earnings
```

Expert That's right.

KE You would absolutely eliminate a company that was at 14% of earnings.

Expert (reflecting for a moment) Not necessarily. Profits are only one indicator. Anyway 14% isn't a bad profit picture.

KE I see, so you're actually looking for 15% as the (pause) point where you would unequivocally say that the company is highly profitable. Companies with less profitability might still be of interest but to a lesser degree. (The knowledge engineer returns to the blackboard and sketches out a rough fuzzy set.) Something like this (see Figure 9.1):

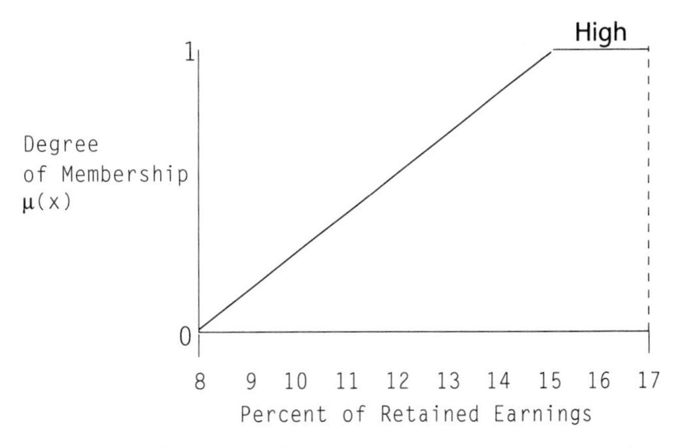

Figure 9.1 The Fuzzy Representation of Highly Profitable

I just chose 8% as an arbitrary low-end cutoff—we can come back to that later. But isn't this what you mean; as profits increase, the compatibility of the company with your criterion for highly profitable steadily increases?

Expert Yeah, sort of. I never really thought of it that way—but yes, that's generally what I mean. These companies (the expert takes a marker and circles the middle set of profitability numbers, as shown in Figure 9.2) are still important—hey, not every business is going to make the best numbers. There's not any particular point where I'd say this is a definite "no" and the next company is a definite "yes"—except maybe down there at the bottom end, maybe just a little below your 8%.

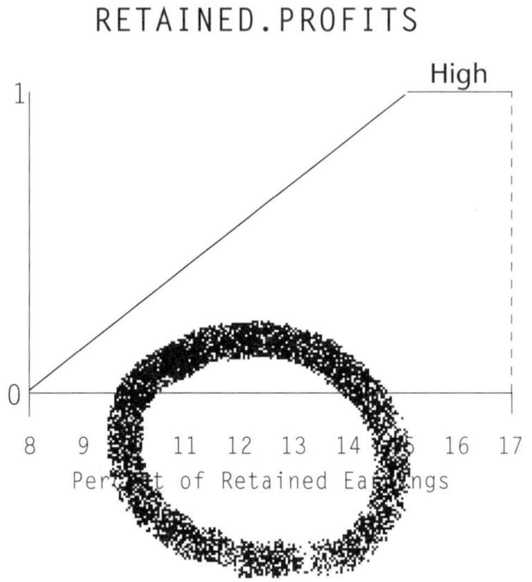

Figure 9.2 An Area of Special Interest in the Highly Profitable Region

KE — Fine. So we've generally established what you mean by high profits. And then?

Expert — I also try to find small- to medium-sized companies trading in the private sector.

KE — How do you look at company size? Is it just the number of employees? The number of products? Their gross or net revenues?

This approach to articulating the system control parameters requires a shift in knowledge representation from logical determinism and arithmetic formalism to a semantics and property-based representation expressed directly through the surface characteristics of fuzzy sets. The fuzzy vocabulary stores the underlying relationships between the expert and the decision process. Model design proceeds from changes in the number, the shape, and the overlaps of the fuzzy sets rather than solely through production rules. These fundamental changes in model representation also mean a change in the method of encoding the model knowledge. In a fuzzy system, we express the dynamics of the system as a series of set-level inferences held in a series of conditional and unconditional fuzzy rules.What this means is simple—a fuzzy model is constructed from a base vocabulary of fuzzy sets (and hedges) defining the semantic properties of the model. Rules manipulate fuzzy sets. A knowledge engineer must learn to think in this higher level of abstraction in order to properly codify expert experience.

Multiple Experts

Fuzzy models apply conditional and unconditional assertions to a problem solution space—one such solution space is developed for each consequent control parameter. The entire rule set is processed to determine the shape of this final space, thus, in effect, running all the rules in parallel. This final state is bounded by (if you follow Zadeh's laws) the minimum surface of the unconditional assertions and the maximum surface of the conditionals—in other words, the Cartesian product of the intersecting production rules. Since this product does not obey the law of the excluded middle or the law of non-contradiction,[2] it has some counterintuitive effects for the system designer—the most remarkable is the ability to explicitly encode mutually exclusive rules.

Contradiction, however, only exists in the Boolean idea of rigidly defined set membership limits. In a fuzzy model, contradiction takes on a different meaning (that of overlapping or partial sets). Consider the new product pricing model (see "The New Product Pricing Model (Version 1)" on page 428). This model contains two apparently contradictory rules:

```
our price must be high;
our price must be low;
```

These fuzzy assertions establish the limits on the possible outcome spaces for the model. The financial expert's intent is to maximize profits by setting the price at the highest possible level. The goal of the marketing and sales expert is to maximize volume by making the price as low as possible. Note that in fuzzy models, assertions are active, not passive—they contribute directly to the model's solution space.

This can seem like a trivial example of conflicting experts, but in fact the model dynamics are especially difficult to replicate with conventional non-fuzzy modeling tools. The difficulty becomes apparent when we change the underlying shape of the fuzzy sets or add hedges to the rules. As an example, consider the following pricing policy:

```
our price must be quite high
our price must be very low;
```

Changing the relationships expressed in these two rules by hedging the *High* and *Low* fuzzy sets changes the entire nature of the problem space. The unconditional assertions now specify an intersection of two nonlinear spaces. Yet this is the way multiple experts think about problems.

Elastic Relationships Among Continuous Variables

The normal implication function associated with fuzzy system models (that is, approximate reasoning) establishes a correspondence between two or more control variables in the model. This correspondence means that a change in one variable produces a pronounced effect on the state of the dependent variables. Fuzzy sets represent the semantics of continuous variables. By breaking a continuous variable into semantic "chunks," we can establish a relationship between a chunk in one fuzzy region and a chunk in another fuzzy region. The chunks are the fuzzy term sets underlying each of the model control and solution variables. As an example, the following rules from a sales analysis system establish a relationship between the number of orders in the accounts receivables and the rates of customer sales:

```
[R1] if Account.Receivables is LARGE and Sales.Rate is EVEN
   then Collection.Rate is SLOW;

[R2] if Account.Receivables is MODERATE and Sales.Rate is EVEN
   then Collection.Rate is below NORMAL;

[R3] if Account.Receivable is SMALL and Sales.Rate is EVEN
   the Collection.Rate is ACCEPTABLE;
```

The model diagnoses the problems in the sales organization. In this case, it observes in rule [*R1*] that since there is a relationship between sales and accounts receivable, if the rate of sales remains relatively constant (*Even* is a bell-shaped fuzzy set) but the number of receivables is *Large* (that is, increasing), then there must be a problem with the organization's collection process. The degree to which the model estimates that the process is slow is related to the depth of the backlog and the evenness of the sales rate.

Since the implication process is based on the set theoretic operations of fuzzy logic, the control and solution variables must be represented as either fuzzy sets or data elements that can be approximated as fuzzy spaces. In practical applications, this latter restriction is usually not severe, since the symmetry between fuzzy sets and numbers allows us to approximate any one-dimensional product point as a fuzzy product space. In fact, the ability to convert noncontinuous variables into continuous fuzzy spaces provides a powerful medium for system function modeling.

Complex, Poorly Understood, or Nonlinear Problems

In the previous section on modeling systems with multiple experts, we noted that the application of hedges to a fuzzy rule can induce a highly nonlinear nature in the model. This is only one aspect of approximate reasoning's ability to facilitate the representation of systems for which no (or no reasonable) mathematical model exists, systems which exhibit very complex and nonlinear behaviors, or systems for which a complete understanding of the system process is not available. Fuzzy systems are particularly well suited to modeling nonlinear systems. The nature of fuzzy rules and the relationship between fuzzy sets of differing shapes provides a powerful capability for incrementally modeling a system whose complexity makes traditional expert system, mathematical, and statistical approaches very difficult.

Uncertainties, Probabilities, and Possibilities in Data

As we have previously discussed and demonstrated, there is a fundamental mathematical and semantic difference between fuzziness and probability. While probabilities are concerned with the imprecision associated with the outcome of well-defined events, fuzzy logic is concerned with the imprecision that is an intrinsic part of the event (or concept) itself. You may consider this as the difference between fuzziness (or imprecision) and randomness. Fuzzy system modeling usually provides a more flexible and richer representational scheme than either certainty factors or Bayesian probabilities. This is due to the nature of fuzzy logic—it attempts to characterize imprecision and undecidability within the control structure itself rather than in the outcome of the model. Additionally, fuzzy systems can represent a much more complex "possibility" space than probabilities for the following reasons:

1. Their cumulative distributions can sum to less than or more than one. This is due to the interpretation of a fuzzy set as a "possibility" rather than as a "probability" distribution.
2. They are independent of *a priori* frequencies.
3. They can more nearly represent interdependent control variables.
4. They are able to reflect the imprecision in probabilities themselves, since fuzzy models can reduce contradictory solution states to a fuzzy surface.
5. They provide a more intuitive method of expressing concepts for which probability distributions are unknown or unattainable.
6. They provide a mathematically sound and semantic-based modeling capability at a high level of abstraction, where changes in the system are reflected through linguistic modifications.

Note, however, that fuzzy systems and probability representations are not mutually contradictory. Both can, and often are, employed in the same model. Statements that contain both probability and fuzzy terms look something like this:

```
there is a 30% chance the motor temperature is hot
   (or, the temperature is hot 30% of the time)

if our price is low,
   then our market share will increase 50%
```

The percentage represents the probability of the event. Probability constraints can be associated with fuzzy outcomes as well as with the mechanisms of fuzzy implications. But these are two distinct and separate issues, unrelated to the nature of expressing the imprecision of control variables in terms of probability spaces.

As we have seen, the characteristic or discriminant function of a fuzzy set resembles a probability distribution function. However, because the underlying phenomenon is associated with vagueness or imprecision rather than the expected occurrence of an event, they are more properly called *possibility distributions*. A possibility distribution defines how easily an event may occur rather than its probability of occurrence. Fundamentally, the possibility is a measure of the compatibility between an element in the domain and the truth of the fuzzy concept. This is the epistemological definition of a possibility distribution. A probability distribution follows the cumulative odds of an event's outcome, while possibility distributions follow the compatibility of the event state with the concept. Generally a possibility distribution will follow, but not necessarily coincide with, a probability distribution. And we have an intuitive feel that high probability implies high possibility, but the complement is not necessarily true.

In departing from the absolute frequentist assessment of probability, a Bayesian interpretation of subjective probability theory comes the closest to representing the kind of context-dependent frequency distributions usually associated with the possibility infrastructure of fuzzy models. The crucial differences, however, are still implicit in the nature of probability as a predictor of future events.

The expected state of a fuzzy control variable, on the other hand, ranges across the allowed possibility space for that variable's underlying domain. This means that the cumulative possibility value for a fuzzy system need not sum to one, nor is the cumulative interval—itself a fuzzy region—necessarily monotonic. To see how this can be the case, consider a software development project involving new technology (such as case-based reasoning or neural networks), but staffed with technically competent and aggressive people. Our estimate of project success might be stated as:

```
the level of failure is high
the level of success is fairly high
```

Clearly, we have mixed or fuzzy intuitions about the probable success of the project. To put this in a slightly more crisp context, we might say that there is a 90% chance the project will fail, but we wouldn't automatically feel that its chance of success is only 10%. We might intuitively feel that it has a 30% to 40% chance of succeeding. This corresponds to a fuzzy region at the intersection of success and failure. The degree of truth in the intersection, its truth membership in the fuzzy space, determines the strength of the implication and, consequently, the degree to which success or failure might be expected to occur.

Fuzzy Set and Data Representational Issues

In the section on fundamental design issues, we were concerned with the physical organization of a model, often at the C++ code level. That section covered several topics:

the ways in which fuzzy sets, hedges, formal variables, and external data elements are represented in model policies; how the explanatory facility is incorporated; the way rules are expressed physically and logically; and the methods used to defuzzify the results. In this section, we turn our attention to other areas, including: the ways in which fuzzy sets are derived; how fuzzy sets are related to data variables; the meaning and nature of fuzzy set groups; and the appropriate organization of the fuzzy infrastructure.

Variable and Parameter Decomposition

To write rules—fuzzy propositions—about the behavior of a model, we must establish a relationship between a collection of fuzzy spaces and our control and solution variables. In particular, it is necessary to decompose a variable into one or more fuzzy sets defining the semantic properties of the variable. Each fuzzy set describes some range of the variable's values and attaches a linguistic meaning to that range. The name of the fuzzy set becomes our handle for accessing this semantic property of the variable.

Variables define the data used and produced by our fuzzy model. An incoming variable's current value is either retrieved from sensors, found in a database or file record, entered by the user (through a dialog box, as an example), or generated by another software program. Outgoing variables are created by the fuzzy model. As an example, the concept of profitability could be modeled around a variable *Profit*. The values of the variable *Profit* can be positive (the corporation made money), negative (the corporation lost money), or zero (the corporation broke even). Across this range of values, we can view particular ranges as representing various ideas about the corporation's profitability. You might slice up this range in such sections as a big loss, a small loss, break even, a small profit, and a big profit. Figure 9.3 illustrates how *Profit* is decomposed into a set of fuzzy regions.

These are not crisply defined segments; however, as the value of profits moves up or down, we feel it is more or less compatible with one concept than one of the neighboring concepts. For example, as you run deeper and deeper into the red, the compatibility of the negative value with *SL* (a small loss) begins to decline, and the compatibility with the idea of *BL* (a big loss) begins to increase. Each of these semantic partitions becomes a fuzzy set associated with the variable *Profit*.

Semantic Decomposition of Profit

In Figures 9.4 through 9.8, the variable *Profit* is decomposed into a collection of underlying fuzzy sets. The universe of discourse for the variable *Profit* goes from –150 (a loss of $150,000,000) to +150 (an after-tax profit of $150,000,000). Each individual fuzzy set represents a particular idea or concept associated with the overall idea of corporate profits. These figures represent only one possible decomposition scheme for the variable *Profit,* and, in fact, an actual representation would depend on the model context.

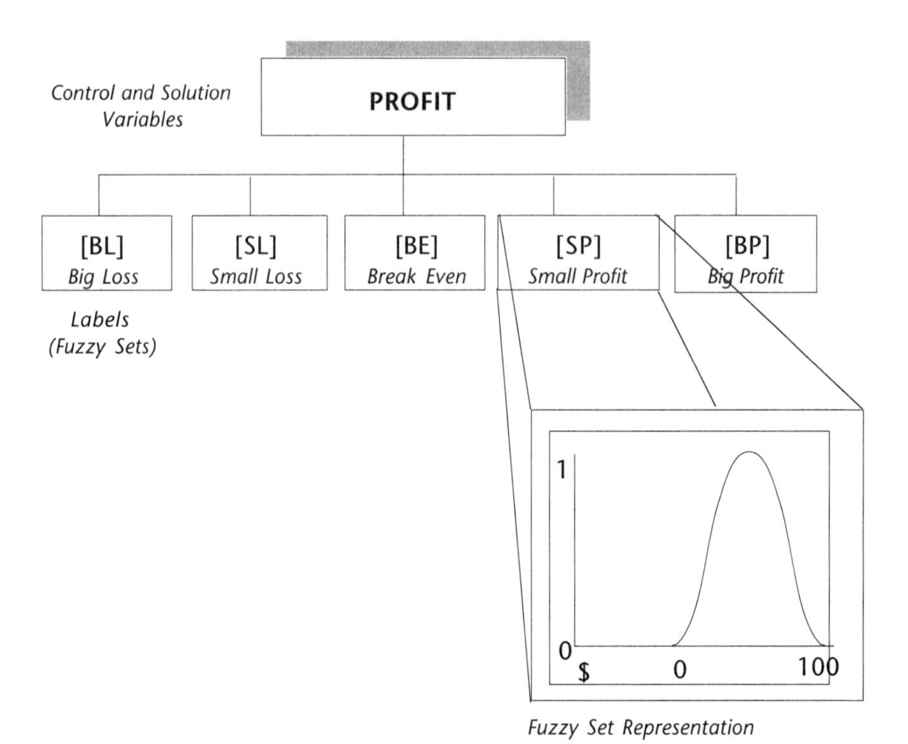

Figure 9.3 Term Set Decomposition of the *Profit* Variable

```
  FuzzySet:     BL
  Description: A Big Loss
1.00.....
0.90      ...........
0.80                 .....
0.70                      .....
0.60                           ...
0.50                              ....
0.40                                  ...
0.30                                     ....
0.20                                         .....
0.10                                              ......
0.00                                                    ..............
    0---|---|---|---|---|---|---|---|---|---|---|---|---|---|---|---0
-150.00 -137.50 -125.00 -112.50 -100.00  -87.50  -75.00  -62.50  -50.00
  Domained:       -150.00 to     -50.00
```

Figure 9.4 The Idea of a Big Loss

```
  FuzzySet:    SL
  Description: A Small Loss
1.00                                   .....
0.90                              .....     .....
0.80                           ...               ...
0.70                         ..                    ..
0.60                       ..                        ..
0.50                     ..                            ..
0.40                    .                                .
0.30                  ..                                  ..
0.20               ...                                      ...
0.10            ...                                            ...
0.00.......                                                       .......
    0---|---|---|---|---|---|---|---|---|---|---|---|---|---|---|---0
-100.00  -87.50  -75.00  -62.50  -50.00  -37.50  -25.00  -12.50    0.00
  Domained:       -100.00 to       0.00
```

Figure 9.5 The Idea of a Small Loss

```
  FuzzySet:    BE
  Description: Break Even
1.00                                   .....
0.90                              .....     .....
0.80                           ...               ...
0.70                         ..                    ..
0.60                       ..                        ..
0.50                     ..                            ..
0.40                    .                                .
0.30                  ..                                  ..
0.20               ...                                      ...
0.10            ...                                            ...
0.00.......                                                       .......
    0---|---|---|---|---|---|---|---|---|---|---|---|---|---|---|---0
 -50.00  -37.50  -25.00  -12.50    0.00   12.50   25.00   37.50   50.00
  Domained:        -50.00 to       50.00
```

Figure 9.6 The Idea of Breaking Even

```
  FuzzySet:    SP
  Description: A Small Profit
1.00                                   .....
0.90                              .....     .....
0.80                           ...               ...
0.70                         ..                    ..
0.60                       ..                        ..
0.50                     ..                            ..
0.40                    .                                .
0.30                  ..                                  ..
0.20               ...                                      ...
0.10            ...                                            ...
0.00.......                                                       .......
    0---|---|---|---|---|---|---|---|---|---|---|---|---|---|---|---0
   0.00   12.50   25.00   37.50   50.00   62.50   75.00   87.50  100.00
  Domained:          0.00 to      100.00
```

Figure 9.7 The Idea of A Small Profit

```
  FuzzySet:    LP
  Description: A Large Profit
1.00                                                        .....
0.90                                                  ...........
0.80                                             .....
0.70                                        .....
0.60                                     ...
0.50                                 ....
0.40                              ...
0.30                          ....
0.20                     .....
0.10                .......
0.00..............
    0---|---|---|---|---|---|---|---|---|---|---|---|---|---|---|---0
  50.00   62.50   75.00   87.50  100.00  112.50  125.00  137.50  150.00
  Domained:         50.00 to      150.00
```

Figure 9.8 The Idea of A Large Profit

When the individual fuzzy sets are complete, the structure of *Profit* can be viewed as a collection of overlapping fuzzy regions (this view is gained with the *FzyPlotVar* function). Figure 9.9 shows how *Profit* appears after decomposition.

```
PROFIT
Domain [UofD]:    -150.00 to      150.00

      1.00..                  ***         ++          ::               ---
      0.90 .....             ** **      +++ ++      :: :::            ----
      0.80     ..           *     *     ++     ++     ::     ::          --
      0.70       ..         *       *  +      ++ ::       :          ---
      0.60        ..        **       **+       + :        :        --
      0.50         ..        *        ++        :        ::     --
      0.40           .      *         +*        :        :    --
      0.30            ..  *          ++*       : +       :: --
      0.20             .*           ++   *      :: ++      --
      0.10             *  ..       ++     *    ::    ++     ---::
      0.00          ***      ....+++     **:::     ++------    :::
          0---|---|---|---|---|---|---|---|---|---|---|---|---|---|---|---0
      -150.00 -112.50  -75.00  -37.50    0.00   37.50   75.00  112.50  150.00
      .  FuzzySet:    BL
         Description: A Big Loss
         Domained:      -150.00 to     -50.00
      *  FuzzySet:    SL
         Description: A Small Loss
         Domained:      -100.00 to       0.00
      +  FuzzySet:    BE
         Description: Break Even
         Domained:       -50.00 to      50.00
      :  FuzzySet:    SP
         Description: A Small Profit
         Domained:         0.00 to     100.00
      -  FuzzySet:    LP
         Description: A Large Profit
         Domained:        50.00 to     150.00
```

Figure 9.9 The Variable *Profit* Decomposed into Its Fuzzy Regions

Proper decomposition of a variable into a complete set of fuzzy terms is a crucial aspect of creating a robust and resilient model. Only named fuzzy regions can be addressed in your rule set. As an example, from this decomposition we can write rules such as:

```
if profits are BL, then expansion.program is very reduced;
if profits are BE, then expansion.program is slighlty advanced;
if profits are BP, then expansion.program is highly advanced;
```

But, given the current decomposition of *Profit*, a rule triggering some action based on the degree to which profits are between 25 and 80 cannot be written. This range of values does not have an associated fuzzy set definition. It is not semantically meaningful to the model. So, all semantically meaningful regions of a variable must be represented by a fuzzy set.

Fuzzy Set Naming Conventions

Naming the fuzzy set components of a model has important implications for understandability, maintenance, and model validation. In control engineering, variables commonly represent states of change: change of velocity (Δv), change in the error (Δe), and change in pressure (Δp) or temperature (Δt). The fuzzy sets associated with a variable, called the *term set*, are generally labeled to indicate positive or negative changes. Figure 9.10 illustrates the fuzzy sets associated with a steam turbine throttle action control variable.

The fuzzy sets are given labels descriptive of the action rather than the semantics associated with the action itself. The labels of the *Throttle.Action* variable include *LN* (large negative), *MN* (medium negative), *SN* (small negative), and *ZR* (zero change or nothing). This contrasts with the way the *Profit* variable was decomposed. Instead of large negative (*LN*)—which accurately reflects the magnitude of the negative *Profit* value—the label *BL* (big loss) was used.

As much as possible, the name of a fuzzy set should suggest *what it means* rather than *what it is*. Understandably, in control applications, there is not always a recognizable semantic label attached to changes in sensor errors, temperatures, and pressures, so that *MN* is just about as descriptive in the engineering world as something like *MediumNegativeChange*.

Although our previous example on decomposition (see Figure 9.9) used rather terse labeling conventions for the underlying term set—*BL, SL, BE*—these cryptic names should, in general, be avoided in information models. Since a fuzzy model is based on a linguistic approach to model representation, fuzzy set names should reflect, as closely as possible, the natural language meaning of the term. From a model maintenance and validation viewpoint this is an important property of fuzzy models. As an example, the following rules impart relatively clear semantic meaning to their actions:

```
if profits are a Big Loss, then the expansion.program is
    very REDUCED;
if profits are at Break.Even, then the expansion.program is
    slightly ADVANCED;
```

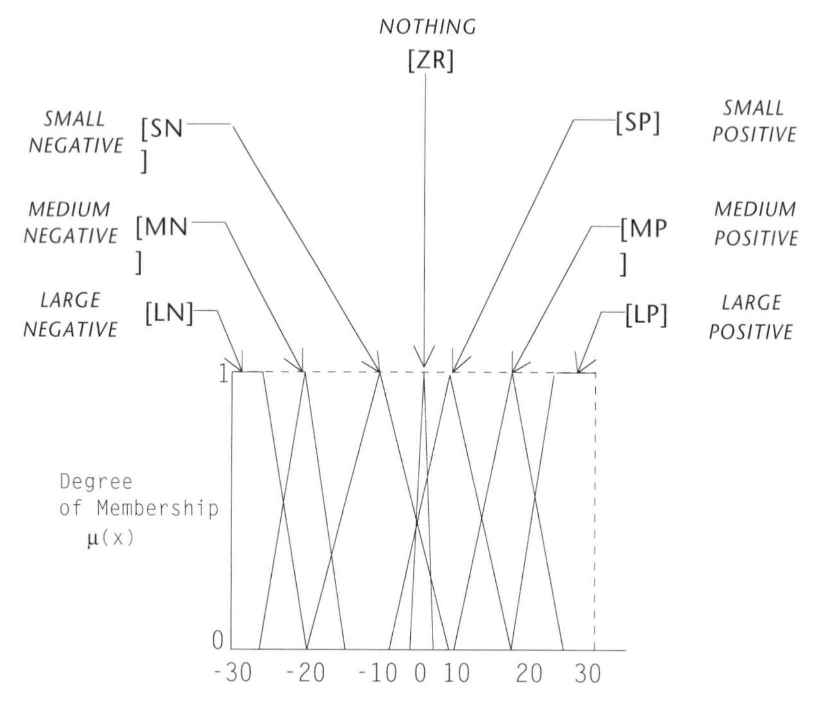

Figure 9.10 The Fuzzy Term Set for the *Throttle.Action* Variable

This is preferable to the following rules written with abbreviated fuzzy set names, which make understanding the rules' meanings difficult:

```
if profits are BL, then expansion.program is VR;
if profits are BE, then expansion.program is SA;
```

The naming conventions for fuzzy sets are a more specific case of naming conventions for linguistic variables. The semiotics associated with combining multiple hedges with a fuzzy set to produce informative strings requires a careful analysis of the underlying meanings of the fuzzy set vocabulary. In general, a fuzzy set should be a class description (similar to a noun) and a hedge should function like an English modifier (similar to an adverb or adjective). Much of how we name our sets and hedges depends on whether the fuzzy rules are **classifying** (*if Height is Tall then Weight is Heavy*) or **prescribing** (*if RateofLoss is High then Risk is somewhat Increased*).

THE MEANING AND DEGREE OF FUZZY SET OVERLAP

To convert a series of individual fuzzy regions into one continuous and smooth surface, each fuzzy set must, to some degree, overlap its neighboring set. There is no precise algorithm for determining the minimum or maximum degree of overlap, but this interference pattern should reflect the semantics of the associated control or solution variable. This overlap is not an artifact of fuzzy reasoning but reflects the actual nature of the fuzzy sets underlying the variable's domain. An overlap is the natural consequence of the fuzziness and ambiguity associated with the segmentation and classification of a continuous space. As an example, in the *Profit* (see Figure 9.11) variable decomposition, we can see that the overlap reflects the definite semantics of each concept.

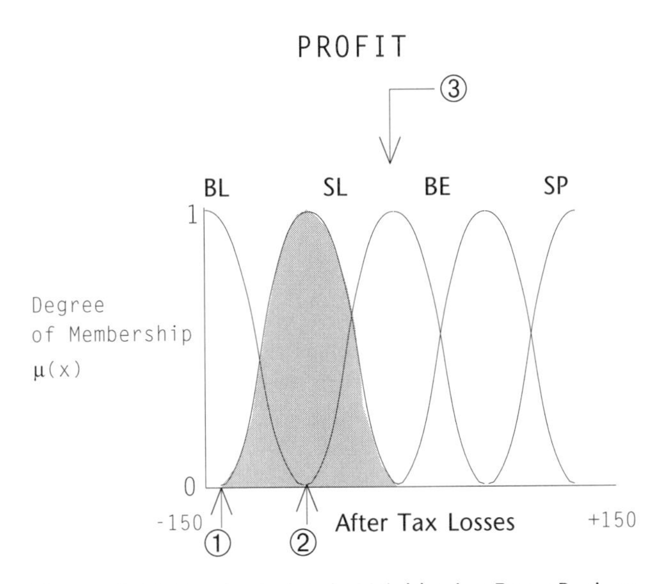

Figure 9.11 Overlap Points in Neighboring Fuzzy Regions

The extreme left edge of the fuzzy term *SL* (*Small.Loss*) begins inside the *BL* (*Big.Loss*) term region (①), showing that there is some ambiguous loss where we might consider the profitability, to some degree, to be both a big loss and a small loss. The unity point for *Small.Loss* (②) is at the zero point for *BL* and also the zero point for the *BE* (*Break.Even*) region. We find the same unambiguous unity membership point for the *Break.Even* region (③). As we gradually move to the right of the small loss region, its truth steadily declines and the truth for the break even region increases.

Experience dictates that the overlap for midpoint-to-edge for neighboring fuzzy regions averages somewhere between 25% and 50% of the fuzzy set base. Figure 9.12 illustrates the conventional midpoint overlap scheme for trapezoidal, triangular, and bell-shaped fuzzy sets.

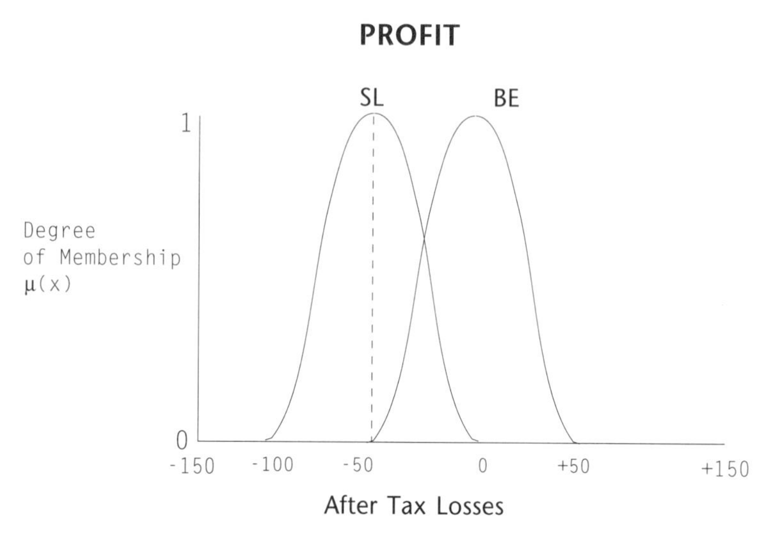

Figure 9.12 Conventional 50% Fuzzy Set Overlap

The maximum membership point for one fuzzy set represents the minimal membership point for each of its neighbors. This makes some semantic sense. After a fuzzy region reaches it unit membership point ($\mu[1.0]$), its membership value begins to decrease as the domain values increase (it is no longer a perfect representative of the set). At the same time, the domain values move into the neighboring fuzzy region to its right with greater and greater strength.

While the degree of actual overlap depends on the underlying concept and the intrinsic degree of imprecision associated with the two neighboring states, there are certain representations that create problems in fuzzy system models. Figure 9.13 shows one problem state. The neighboring fuzzy regions are completely or very weakly disjoint.

Fuzzy sets specified in this manner cause abrupt changes in the rule behavior. The control surface becomes extremely choppy or ragged. Fuzzy sets without overlap also fail to represent properly the semantics underlying the model; in fact, these disjoint sets create model states that are somewhat akin to interval arithmetic. Figure 9.14 illustrates a less clear-cut example of a decomposition problem. Here the fuzzy regions overlap to such a degree that even when a value is completely in *Small.Loss* [*SL*], it also has a significant membership in *Break.Even* [*BE*].

Such a design leads to problems in the model rule processing as well as in the conceptual nature of the system itself. Although, at the edges of a fuzzy set, fuzzy logic does not obey strictly the classical *Law of the Excluded Middle* and the *Law of Noncontradiction* (for details see "Counterintuitives and the Law of Noncontradiction" on page 186), we run into a fundamental epistemological problem when a domain value can be perfectly representative of one set and also have a membership in another set. Even fuzzy logic obeys the *Law of Noncontradiction* at the unit ($\mu[1.0]$) membership point. There are some cases

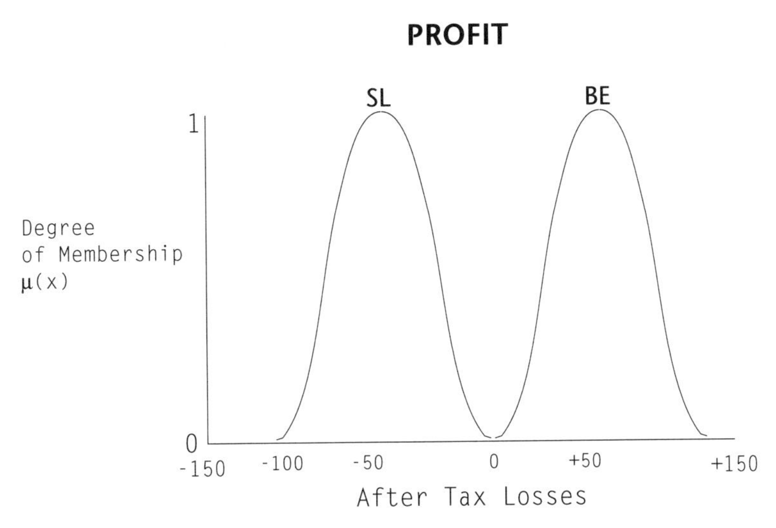

Figure 9.13 Completely Disjoint Neighboring Fuzzy Regions

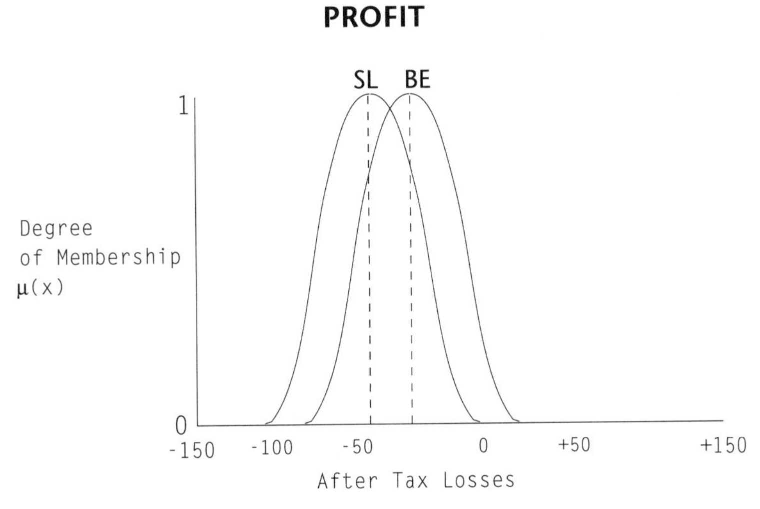

Figure 9.14 Excessive Overlap in Neighboring Fuzzy Regions

where this excessive kind of overlap is necessary, usually in control problems. In modeling information systems, however, we should examine carefully the meaning of such excessive overlaps. The fuzzy sets in Figure 9.14 represent a special problem for rules like:

```
[R1] if profit is SL, then bonus is very small;
[R2] if profit is at BE, then bonus is moderate;
```

If the value of profit in *SL* (*Small.Loss*) is *μ[1.0]*, then the bonus should be *very small*. At this point, however, profit also has a significant truth value in the *BE* (*Break.Even*) fuzzy set, so rule [*R2*] also fires, combining the linguistic variables *very small* and *moderate*. To avoid this kind of semantic ambiguity, some general rules of thumb have been developed for setting the overlap among neighboring fuzzy sets. Figure 9.15 illustrates one general rule of thumb for overlapping fuzzy sets.

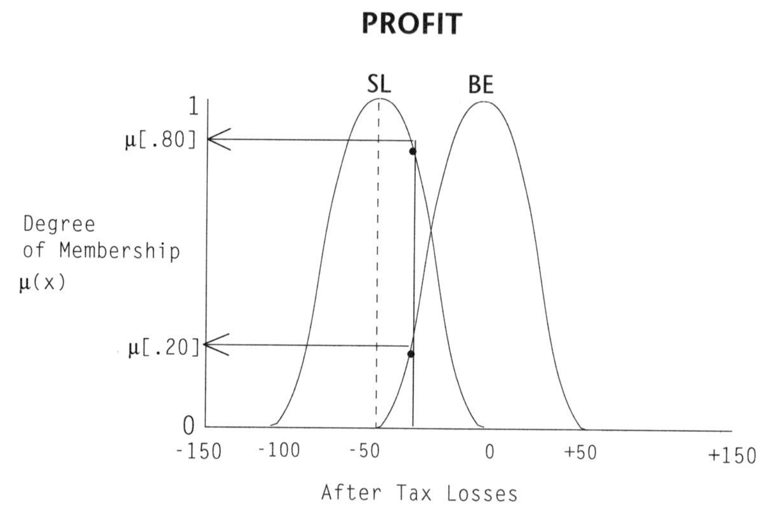

Figure 9.15 The "Sum-to-One [or Less]" Rule of Thumb for Set Overlaps

As Equation 9.1 states, the sum of all points through the overlapping fuzzy sets should be equal to or less than one. Thus is the *sum-to-one [or less] rule.*

$$\sum_{i=1}^{n} \mu_i[x] \leq 1$$

Equation 9.1 Truth Function Formula for Overlapping Fuzzy Sets

In almost all cases, n should equal 2 with the sum-to-one rule setting a general upper limit on n of 3. Broadly, this means that a particular region of the variable's domain should not have more than two simultaneous interpretations, with a third in some special cases. Usually, the third region represents a definition of the space between the other two. Figure 9.16 illustrates this point with the addition of the *WL* (*Weak.Loss*) region between the *SL* (*Small.Loss*) and the *BE* (*Break.Even*) fuzzy sets.

This exception to the maximum coverage heuristic violates the sum-to-one [or less] rule since a plumb line dropped from the *μ[1.0]* membership point of the *WL* (weak loss)

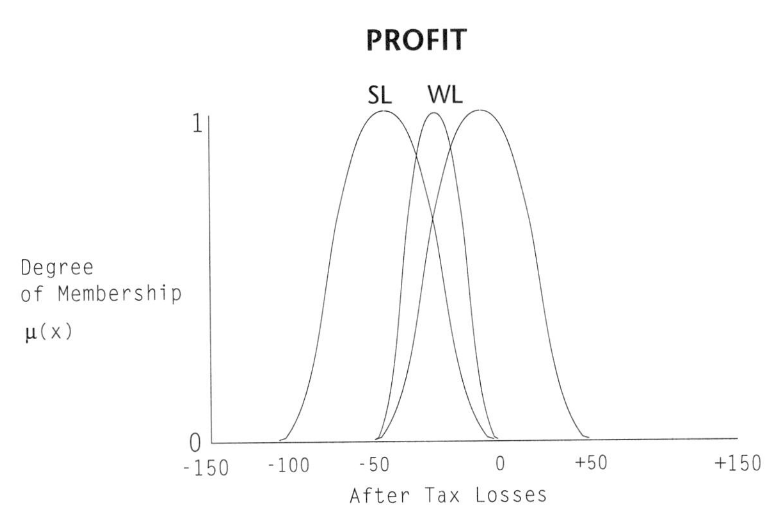

Figure 9.16 The "Maximum Coverage" Rule of Thumb for Set Overlap

fuzzy set crosses the *SL* and *BE* sets at the *μ[.5]* level. This means the cumulative truth membership at this point is:

$$\sum_{i=1}^{n} \mu_i[x] = \mu_{SL}[.5] + \mu_{BE}[.5] + \mu_{WL}[1.0] = 2.0$$

Figure 9.17 illustrates a conceptual and permissible violation of the maximum overlap and the sum-to-one rules when the additional sets are hedged versions of the base fuzzy set.

Care must still be exercised in combining hedged fuzzy sets that reference overlapping pieces of the domain, due to the effect on rule execution. Consider the following set of rules describing an action based on profits within *SL* (small loss):

```
[R1] if profit is a somewhat SL
   then bankruptcy.risk is small;
[R2] if profit is a SL
   then bankruptcy.risk is nominal;
[R3] if profit is a very SL
   then bankruptcy.risk is minimal;
```

The logical problem with these rules stems from the fact that these hedged fuzzy sets represent special cases of truth function overlap since they are related through subset membership (see Equation 9.2).

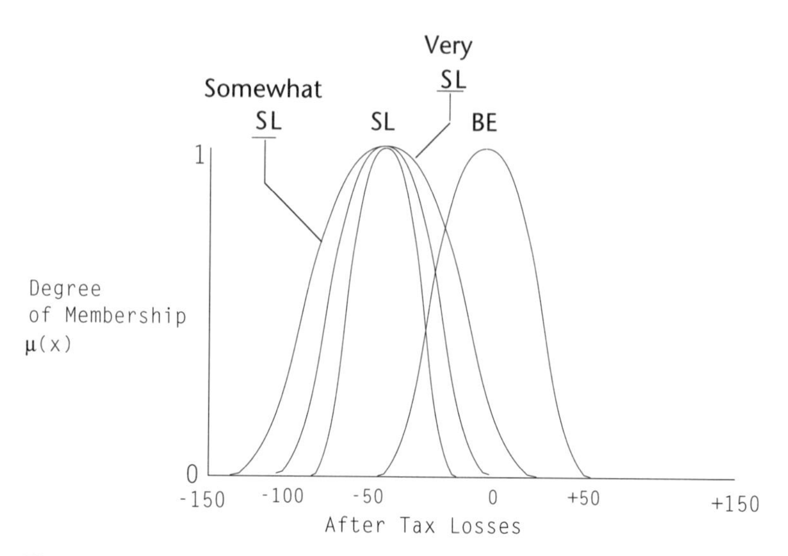

Figure 9.17 Exceeding the Maximum Coverage with Hedged Fuzzy Sets

$$SL_{very} \in SL \in SL_{somewhat}$$

Equation 9.2 Hedged Fuzzy Sets as Subsets of Larger Fuzzy Regions

Several of these rules are correlated. Whenever rule [*R3*] executes, rules [*R2*] and [*R1*] will always execute; whenever [*R2*] executes, [*R1*] will always execute. And since the area encompassed by the subset fuzzy sets is very large, we also have a high probability that [*R2*] and [*R3*] will execute whenever [*R1*] is executed. In most instances, this kind of dependency among rules based on the relationship of their fuzzy sets will cause anomalous behavior in the model. In a robust fuzzy model—robust in the engineering sense that the model can tolerate changes in its operating state—we want multiple rules to fire when a value from the domain exists in multiple, neighboring fuzzy regions. This corresponds to the degree to which the system is in transition from one state to the next. When rules execute due to subset membership, the model is not in a true transition but is reflecting a reinforcement of truth values from the superposition of related fuzzy sets.

Control Engineering Perspectives on Overlap and Composition

Fuzzy systems in control engineering have a different orientation and purpose than fuzzy information models. In a control system, the fuzzy system usually operates in real-time with an objective of finding optimal (or near optimal) performance values for the device. The current state of the device is sensed and fed back through the fuzzy engine as a set of difference measurements. Fuzzy rules ensure that the operating state of the device remains within its proper range. The nature of the term set associated with control system variables differs somewhat from those normally used in information modeling. To see these differences, consider a fuzzy system that controls a steam turbine. The fuel injector rate is determined by two sensor readings: the current temperature and the current pressure. Figure 9.18 is a schematic of the turbine plant.

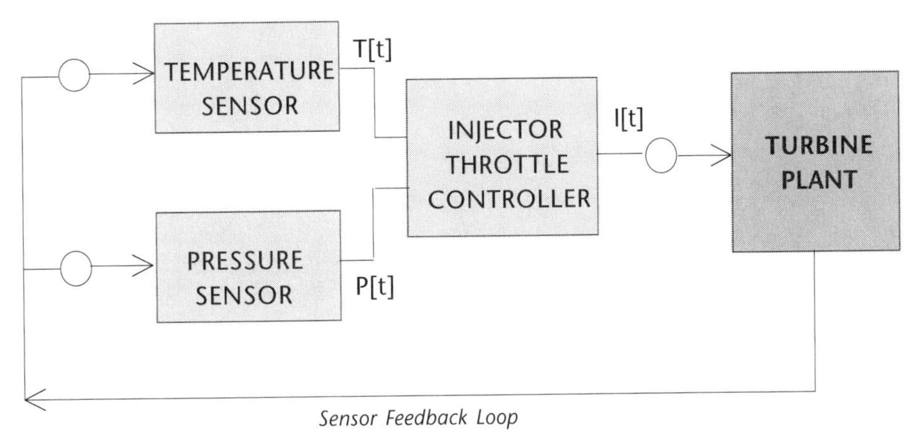

Figure 9.18 A (Very) Simplified Steam Turbine Plant

The steam turbine injector should remain at the normal or zero movement position as long as the temperature and pressure in the turbine are within normal operating ranges. As temperature or pressure increases or decreases, the throttle opens or closes to increase or decrease the amount of fuel reaching the turbine. Figure 9.19 shows the term sets underlying the *THROTTLE.ACTION* solution variable.

From Figure 9.19, we can make several observations about the term set decomposition for control engineering systems. In a general way, a fuzzy control or solution variable has the following properties:

- The fuzzy sets in the variable are ordered into regions of differing areas. The area covered by the fuzzy sets decreases as the domain converges on the variable's desired operating region. This means that a value in one of the outer regions will induce a general rule action that brings the system back toward the region of acceptable behavior. The effect here is to reduce the number of rules necessary to control the plant, since a single rule can handle all device states in these edge regions.

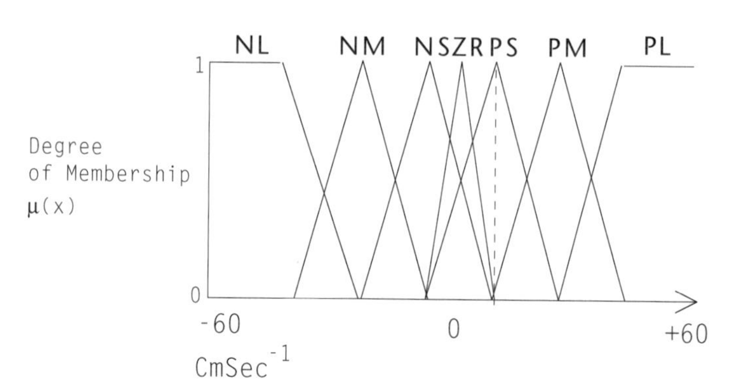

Figure 9.19 The Term Set for the *THROTTLE.ACTION* Control Variable

- The degree of overlap at the region of preferred performance often exceeds the "sum-to-one [or less]" heuristic. This provides for both fine-tuning and control by the rules as well as a bracketing of the region by narrowly defined neighboring fuzzy regions (see the previous point about the decrease in fuzzy set areas). Although this violates our rules about the degree of overlap, the increase of overlapping fuzzy sets at this point ensures that multiple rules will execute as the problem state moves to the left or right of the preferred operating region. In some mechanical devices such as those in the pendulum balancing class, the high degree of overlap ensures that, when the system is in its optimal state (the pendulum is balanced), any small changes in this state are immediately detected and handled.
- The number of fuzzy sets within the variable is almost always an odd number between 3 and 11, with 5 or 7 occurring most frequently. Figure 9.20 illustrates why this is the case.
- If a variable, such as *temperature*, is decomposed into its operating ranges with the center region as the state of preferred behavior, then the regions on the left and right must be labeled and used in the rule set to indicate a state requiring some action. Consequently, this idea of decomposition is recursive. If one of the regions on the left or right is decomposed into a fuzzy number (a triangular or bell-shaped fuzzy set), then the region to its left or right must be labeled also. This process invariably leads to an odd number of fuzzy sets.

Control problems are most often defined in terms of triangular and trapezoidal fuzzy sets since the representation of these shapes can be handled with minimum storage, and

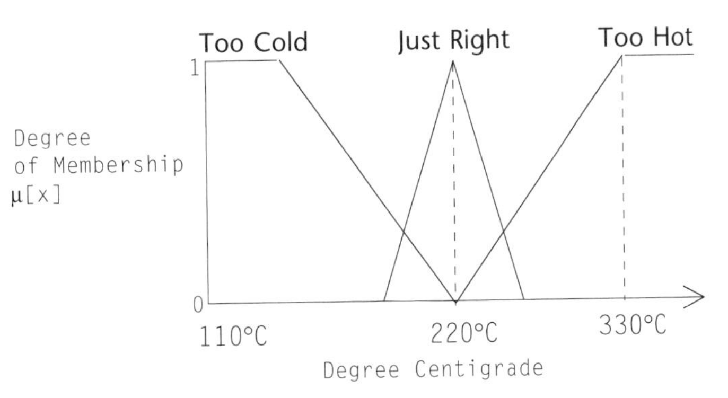

Figure 9.20 A Minimal Controlling Fuzzy Term Set

the degree of membership determination can be performed with a small amount of cycle time. This attention to representation is driven by the target environment—4–, 8–, and 16–bit microprocessors.

Highly Overlapping Fuzzy Regions

Another family of important representational constructs involves highly overlapping fuzzy regions. Unlike term sets that are decomposed into a series of adjacent semantic fuzzy sets, these highly overlapping fuzzy regions are normally restricted to just a few fuzzy sets. The most common representation involves a fuzzy set and its complement. Figure 9.21 illustrates this idea with the redefinition of *Profit* using two fuzzy sets: *Profitable* and *Unprofitable*.

```
PROFIT
Domain [UofD]:    -150.00 to     150.00
                                                                     *****
     1.00.....                                                 ***********
     0.90     ...........                                 *****
     0.80                .....                       *****
     0.70                     .....                ***
     0.60                          ...         ****
     0.50                             ...****
     0.40                             ***  ...
     0.30                         ****        ....
     0.20                    *****                .....
     0.10              ******                          ......
     0.00**************                                      ..............
         0---|---|---|---|---|---|---|---|---|---|---|---|---|---|---|---0
     -150.00 -112.50 -75.00 -37.50   0.00  37.50  75.00 112.50 150.00
   .  FuzzySet:    Unprofitable
      Description: Loss Money
      Domained:       -150.00 to     150.00
   *  FuzzySet:    Profitable
      Description: Made Money
      Domained:       -150.00 to     150.00
```

Figure 9.21 *Profit* Defined by Two Highly Overlapping Fuzzy Regions

This form of decomposition is used quite often in fuzzy information modeling. It plays an important role in risk assessment in such fuzzy concepts as *Increased* risk and *Decreased* risk, where intermediate concepts are less important than a general movement toward higher or lower states of risk. Highly overlapped sets are also important in models that employ unconditional fuzzy rules or where the model is attempting to represent the decision space of several conflicting experts. In these models, the rules are often contradictory, or they specify a model state that lies somewhere (as yet undetermined) along the entire domain of the variable.

Designing and Eliciting Fuzzy Sets

Now that we have examined some of the important elements in decomposing variables into their component fuzzy sets, the next logical step is exploring the ways in which these fuzzy are created and properly interrelated. Much speculative literature, mostly from the academic sector, has gone into the techniques for generating fuzzy sets, as though this process was a critical stumbling block in model formation. While eliciting the proper number, shape, and overlap of the fuzzy sets is a crucial component of a model, the actual derivation of fuzzy sets is rather straightforward (much more difficult issues involve the choice implication and defuzzification methods).

Knowledge Engineering

The fundamental extract of fuzzy set information begins with the knowledge acquisition phase of the overall knowledge engineering process. Subject matter experts (SMEs) provide the basic source of intelligence in an expert system. Other sources include articles, procedure manuals, reference documents, repair manuals, and so forth. However, knowledge from articles and other documents should be used carefully since they are potentially subject to both practice and procedural errors. (Manuals, sometimes written before a process is actually implemented, often describe how something should be done, not how it is actually done.) The basic foundation of knowledge engineering involves the extraction of the actual decision making actions underlying a process. Processes identification—*BPM*, or business process modeling—involves the functional decomposition of an expert's activities into a set of tasks (sometimes these tasks are called policies, contexts, or functional units). At the core of knowledge acquisition process is the *narrative*:

> **Narrative** (NAR'A-TIV) Giving an account of any occurrence; an orderly account of a series of events; story telling [L. *narrare*, *-atum*, fr. *gnarus*, knowing, fr. Greek, γνοΘων, to know].

Knowledge acquisition begins with a narrative. This narrative is extracted from the subject matter expert through gentle, low interference, but consistent guidance from the knowledge engineer. It is important to capture the expert's sense of how he or she first perceives the task's nature and then makes decisions, as well as to determine the degree of difficulty and complexity of the task itself. The preliminary interview should capture as much of the process as possible in the expert's own voice. From this narrative, you can extract the basic processes underlying an activity. The process itself consists of activities that interact with events. It is, fundamentally, these events that are the variables and parameters in the fuzzy model and must be decomposed into their component fuzzy sets.

A Knowledge Acquisition Methodology

The following is a simple, low-technology, iterative methodology for extracting variables and fuzzy sets from the expert's narrative. Much of it is common sense, but it has provedto be a useful technique in building even highly complex systems over the past seventeen years.

First, you should extract all the objects (variables) and their properties (fuzzy sets) in the narrative. You are going to create a separate 3x5 index card for each variable. On the back of the card, you will draw the collection of fuzzy sets that define the variable. As you go through the narrative, use one color highlighter to identify each variable and another color highlighter to identify each fuzzy set (or potential fuzzy set). Bear in mind that variables are often implied by the fuzzy set. As an example, consider the statement:

> When the elderly are asked about dental pain, they have difficulty specifying its place of origin.

The term *elderly* is a fuzzy set associated with the implied variable *Age*. Thus we create an entry for *Age* (depending on the context, we might actually create the variable *PatientAge* or *SubjectAge*.) I usually use all uppercase letters for the variable and mixed case for fuzzy sets. I write the variable name at the top left of the index card in block letters. On the back of the card, we draw the fuzzy sets as they are identified. In this case, *Elderly* is the first fuzzy set. Generally the domain of the fuzzy set is derived from the narrative, but more likely than not, the expert simply used the term without bothering to define its exact boundaries. When the narrative is refined, the width, shape, and overlap of the fuzzy sets comprise the bulk of the first round questions back to the expert. If we do not have access to the expert (when knowledge is extracted from an article, as an example), we must make an estimate of the fuzzy sets topology from its use in the narrative. Thus, on the back of the index card, we draw the fuzzy set (see Figure 9.22). Sketch out the general range of values for the variable, and put the fuzzy set where it belongs on the range.

Draw the vertical axis (with the [0,1] scales and the "Degree of Membership" label) and horizontal axis in dark ball-point ink. But always draw the horizontal range and the actual fuzzy set in pencil. You should also jot down the fuzzy set name and its explicit domain under the graph—trying to figure out what you meant and where boundary points start and end after you have sketched out three or more fuzzy sets is extremely difficult and very error-prone.

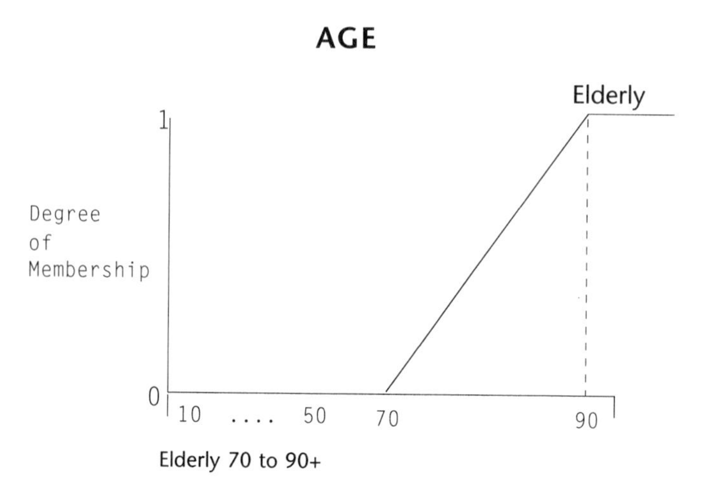

Figure 9.22 The *Elderly* Fuzzy Set

In addition to variables and their fuzzy sets, we also need to collect any constants or numbers, text strings, dates, and Boolean switches used in the narrative. These are named and sorted in with the variables. Pay particular attention to numeric constants that might become fuzzy numbers. As an example, the constant "minimum potassium blood level" (*MINKBLVL*) is determined, from the clinical blood management expert, as 10.5. You may return to this constant at a later time to indicate its degree of fuzziness when it appears, as an example, that some rule is expressed in terms of an approximate constraint:

if BloodPotassium is near MINKBLVL then...

The term *near* applied to a number implies that the constant *MINKBLVL* is being treated like a fuzzy number (that is, a bell or triangular-shaped fuzzy set spread out around the central value). If so, then we want to draw the fuzzy set around this number so we can visualize how it will work in the model. Figure 9.23 illustrates how this process might be done.

Idealizing the shape and width of a fuzzy number is not as crucial as decomposing a variable into its individual, overlapping fuzzy sets, but it is good practice. Visualizing the form of a fuzzy number helps us understand the meaning of a statement that uses such fuzzy numbers.

You repeat this identification process as you move through the narrative. When you're done, each variable in the narrative should be identified, and, for each numeric variable, the preliminary fuzzy sets should be isolated. Now go back to each variable, and on the front of the card, list each fuzzy set in left-to-right order along with the set's domain. You should now have something like the following:

(MINKBLVL)
Miniumum Blood Potassium

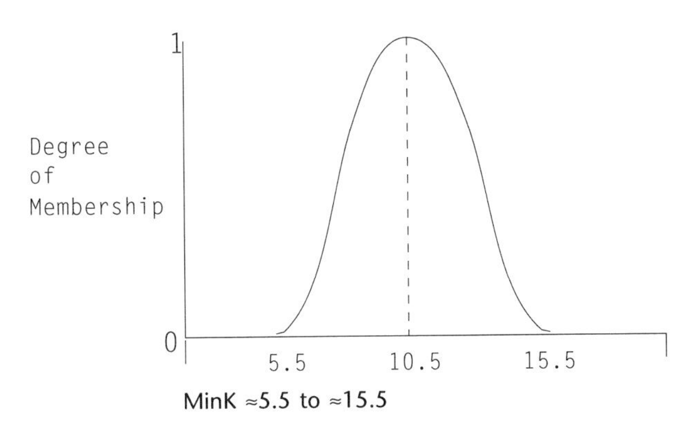

Figure 9.23 Capturing a Fuzzy Number

PATIENTAGE
Number
0 to 110

Child	0 to 18
YoungAdult	16 to 30
MiddleAged	25 to 45
Mature	40 to 75
Elderly	70 to 90+

On the back of each card is the complete fuzzy set collection, pulled from the narrative, defining the semantics of the variable. Note that this technique immediately highlights any variable that has a partially defined fuzzy set collection.

VOTED-FOR DISTRIBUTIONS

We can also view a fuzzy set's membership function as a representation of how a population of expert voters classified each value (or class interval) in the data population. When, as an example, we want to construct a fuzzy set for the concept *Tall*, our experts are asked to vote on whether they would consider a particular height as *Tall*. Given a handful of experts, Figure 9.24 illustrates how the voters judge a specific height as *Tall*.

Such a frequency distribution generates a membership function point by taking the percentage of voters who assigned a degree of compatibilty between a height and the concept *Tall*. As an example, assume we have ten experts. We ask who would consider four

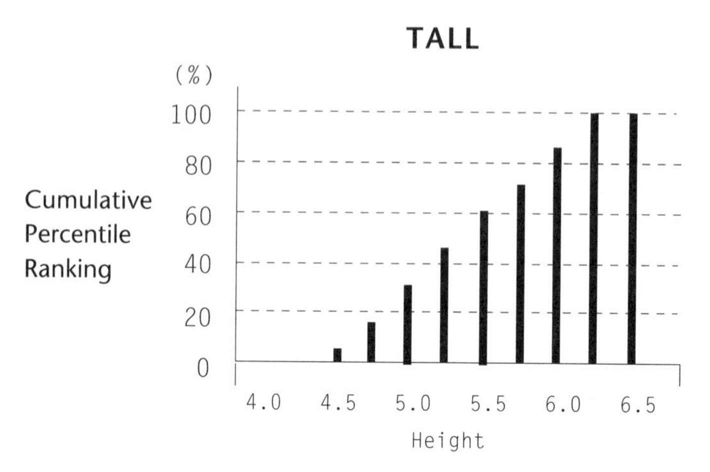

Figure 9.24 Voting Frequency to Determine the Concept *Tall*

feet as *Tall*. No one raises his or her hand, so the membership function is zero. We move to four feet, six inches (4.5 feet). Two experts raise their hands. Thus we have:

$$\frac{2}{10} = .2$$

And this becomes the membership function value at the domain value of 4.5 feet. Eventually, we reach a height where everyone, including those who voted before, will vote that this height is definitely *Tall*. Figure 9.25 shows how the fuzzy set is actually constructed.

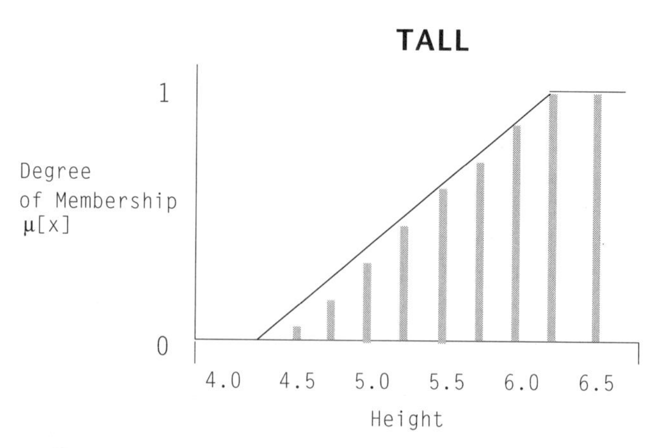

Figure 9.25 Fuzzy Set *Tall* from a Voting Distribution

The voters can be a group of subject matter experts or a group of end-users (as an example). When voters are experts, we can use their relative peer rankings to influence the topology of the fuzzy set. The total vote for the *k*th domain point becomes the weighted average of the votes by all the experts who recognized a compatibilty between a domain value and the fuzzy concept. This vote is expressed as Equation 9.3.

$$V_k = \frac{\sum_{i=1}^{n} v_i \times w_i}{\sum_{i=1}^{n} w_i}$$

Equation 9.3 The Weighted Vote Ranking

A weighted voting protocol assigns a higher count to experts with greater peer ranking. In general, a voting methodology can create fuzzy sets for any set of real, observable and (usually) physical concepts. Voting is another method of extracting subjective information about model parameters from a pool of experts. It has the virtue of fusing the opinions of multiple experts into a single integrated concept. Where there is wide disagreement among experts about the nature of a phenomenon, however, the voting method can produce unworkable results. Knowledge engineers using a voting approach must also insure that the actual semantics of the concept under discussion are well understood by each expert. Failure to narrow the conceptual framework can produce unpredictable results. As a simple example, when developing a fuzzy set for *Tall,* it is important that the experts agree on the underlying domain: are we discussing tall males, tall females, tall basketball players, or tall as a general demographic feature?

Statistical Properties of the Data

Another mechanism for deriving fuzzy sets rests on the properties of the domain data discovered through standard statistical analysis. While this is not the place to delve into the nature of statistical methods, a brief review of the central measures of random populations and their relations to the properties of fuzzy sets is in order. If the data is normally distributed, and we know the mean and standard deviations, then fuzzy sets can be constructed from the data's instrinsic characteristics. Consider the generation of *Tall* from the distribution of heights in a general population. Figure 9.26 shows both a population with a mean height of 5.5 feet and the intervals representing one and two standard deviations.

Deriving a measure for *Tall* from this population distribution involves making some assumptions about the semantics of a *Tall* person. One simple way to generate the fuzzy set representation is to map the frequency distribution of the Gaussian curve to a rough fuzzy set approximation. We then start at the average or median domain point with a membership

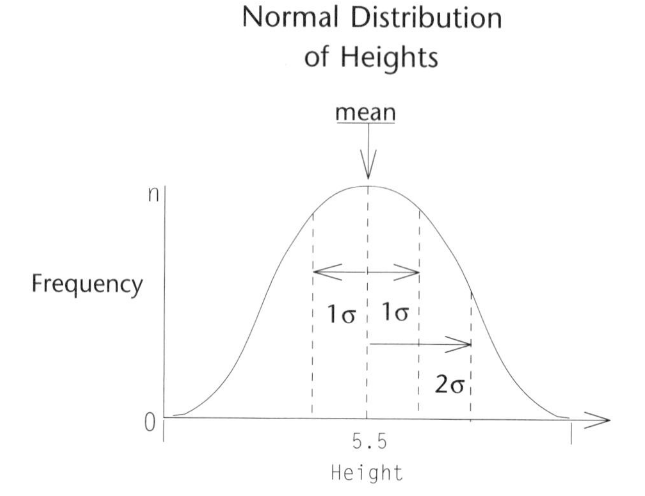

Figure 9.26 Normal Distribution of Heights

value of zero, and create a surface that moves outward. We draw an upward sloping line to the domain value at one standard deviation, giving this point a membership of [.5]. The line is then continued to the value at two standard deviations, giving this a membership of [.75]—that is, one half the interval [.5,1.0])—and then linearly to the [1.0] membership. In this way, the generated fuzzy set is biased toward the heights that lie beyond one and a half standard deviations from the average. Figure 9.27 illustrates schematically how the fuzzy set *Tall* is generated.

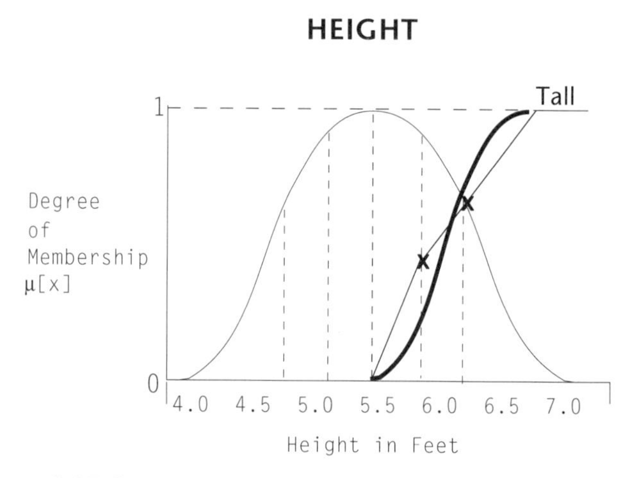

Figure 9.27 Generating the *Tall* Fuzzy Set by Linear Segmentation

Once the points have been found, the actual fuzzy set is easily generated using the fuzzy coordinate method (see the *FzyCoordSeries* procedure). However, other techniques can be applied. The bold line in Figure 9.27 is one possible shape for *Tall* using the growth S-curve. In any case, the shape of the fuzzy set is generally smoothed to roughly follow the data's statistical properties. Figure 9.28 shows a possible decomposition of height into short, medium, and tall, based on the distribution of data.

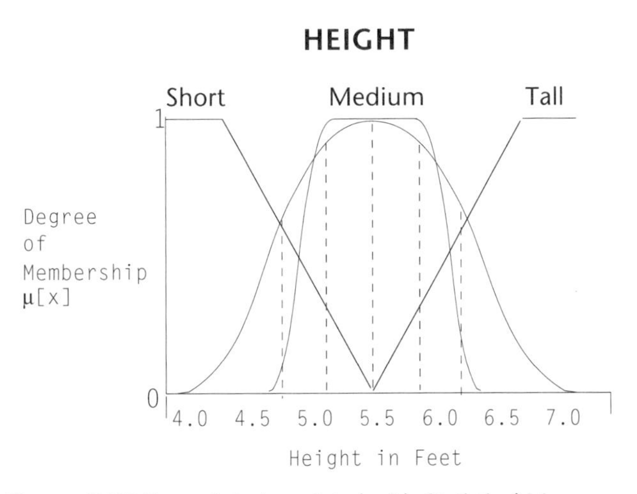

Figure 9.28 Fuzzy Sets Associated with Statistical Measures

If you suspect (or know that) your data is not normally distributed, the statistical measures can still be used to derive fuzzy set representations. But, in this case, you would be well advised to use nonparametric analysis to probe the actual distribution associated with your data. Fundamentally, this means examining the third and fourth statistical moments. The third moment, *kurtosis,* describes the shape of the distribution. Figure 9.29 shows the three basic forms.

The kurtosis of a curve is related to its variance. Leptokurtic curves have a small variance while platykurtic curves have a high or wide variance. The fourth statistical moment, *skew,* defines the distribution bias of the data. Skew indicates whether or not the data has a pronounced tail toward the left or the right. In the process of deriving fuzzy sets from data characteristics, skew is a very important property when the population is not normally distributed. Figure 9.30 shows the orientation of left and right skewed data.

A measurement of skew is often important because it indicates that the mean value may be shifted left or right in the underlying population. As an example, Figure 9.31 illustrates schematically how the mean in a skewed distribution (A) differs from the mean of a normal distribution (B).

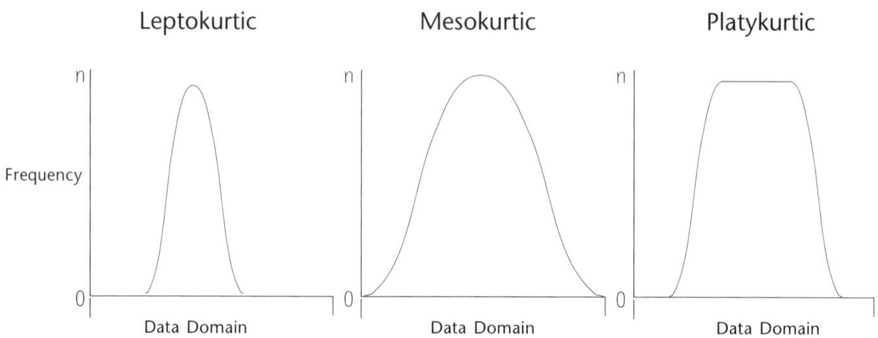

Figure 9.29 Types of Distributions (Third Statistical Moment)

Distribution Shapes

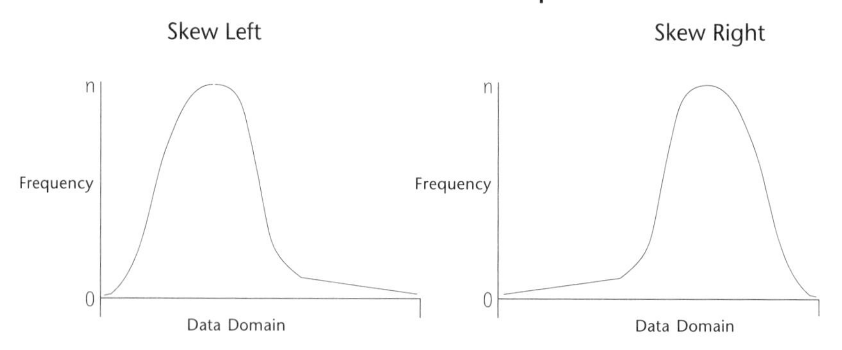

Figure 9.30 Types of Skewed Data Distributions (Fourth Statistical Moment)

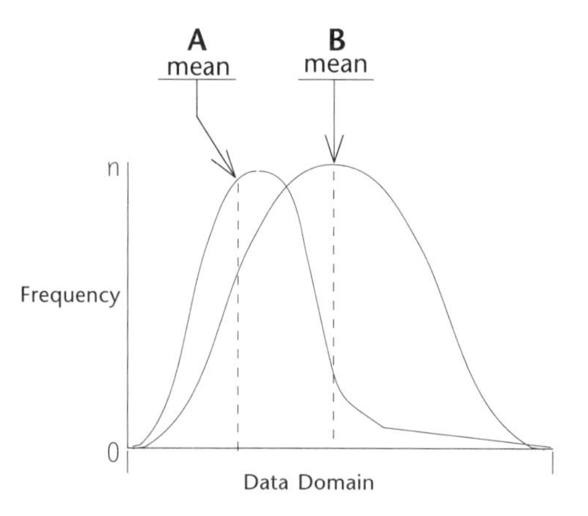

Figure 9.31 Shift in Means Among Skewed and Normal Distributions

Normally distributed curves are not necessarily mesokurtic (as Figure 9.29 might imply); however, they do have some common properties. The highest point in a normally distributed curve is at the center and is the average (or arithmetic mean). The curve slopes down from this point until it reaches a point where it begins to move outward. This is the inflexion point and is precisely one standard deviation from the mean. The area from the mean out to one standard deviation on both sides contains 68% of the observations. Roughly 90% of the observations are contained within two standard deviations from the mean. A particular observation's distance from the mean, measured in units of standard deviation, is called its *z-score,* or *z-statistic*.

Interestingly, these proportions are the same for all normally distributed curves whether they are leptokurtic (a population with low variance) or platykurtic (a population with high variance). Thus, from the viewpoint of fuzzy set generation, a knowledge of the skew for non-normal populations is more important than a knowledge of the kurtosis (since, for normal curves, we can convert all observations in the candidate population to z-statistic intervals.)

Extracting fuzzy sets from statistical data is by no means a straightforward process. First, unlike academic examples, data in such real-world areas as financial analysis, marketing, portfolio evaluation, pharmaceutical drug discovery, clinical diagnosis, urban planning, project management, risk assessment, and production scheduling is seldom normally distributed, clean, and noise-free. Further, data in most business environments is highly volatile, time-varying, and subject to changes in corporate operating and accounting policies. System builders and knowledge engineers often have very little idea of the data's true stability and underlying statistical properties. Second, there is only a loose relationship between the statistical properties of data and the semantics of the model. Even if we know the mean, standard deviation, kurtosis, and skew of our data, these factors are often more important in understanding the boundaries of a model than they are in evolving the actual fuzzy sets.

Psychometrics of Fuzzy Set Evolution

Many (perhaps most) fuzzy researchers, knowledge engineers, and academics are surprised to find the underlying concepts of fuzzy logic buried in the psychometric and psychophysics literature dating back to the early and middle 1800s. During this time, Ernst Heinrich Weber (1795–1878) pioneered work on the concept of the "just noticeable difference" (*jnd*). This theory was refined by his student, Gustav Theodor Fechner (1801–1887), producing two axioms of experimental psychology: *Weber's Law* and *Fechner's Law.* Both of these rules deal with our ability to perceive changes in stimuli against a background of an initial stimulus. Weber found that in every case the magnitude of the *jnd* varied as a well-defined ratio with the magnitude of the background or base-line stimulus (the one with which a second stimulus was compared). Weber's Law can be expressed as a simple formula:

$$k = \frac{\Delta R}{R}$$

The formula says that the ratio between the change in stimulus to the *jnd* point (ΔR) and the magnitude of the base-line stimulus, R, is some constant k for any sensory system. Weber's Law was the first statement of its kind, establishing a quantitatively precise relationship between the physical and the psychological worlds.

Gustav Fechner extended Weber's work by noticing that while the absolute difference between two stimuli increases with the magnitude of the stimuli, the perceiver's sensation of a *jnd* remains the same. Fechner integrated and recast Weber's Law into a new mathematical form:

$$S = n \log R$$

This formula means, in plain terms, that incremental increases in intensity are the results of doubling the stimulus intensity (scaled by some ratio or coefficient factor). This is called Fechner's Law.

The "just noticeable difference" and its relationship to Weber's Law have important implications for the design and use of fuzzy systems. Why? The first hint can be found early in the post–World War II literature, when psychologists began the process of regularizing cognitive science and psychiatry in an all-out attempt to bring its underpinnings into the same mathematical foundations as the "hard" sciences of chemistry and physics. An excellent summary of these psychometric laws and the *jnd* appeared in the article entitled "On the *Psychological Law*," by S. S. Stevens of Harvard University (published in *The Psychological Review*, Vol. 64, No. 4, May 1957, *pp.* 153–180). Interestingly, on page 156 are two diagrams showing the interlaced curves for lower and higher pitch and softer and louder loudness.

Using fifty subjects, the psychometric functions (shown in Figure 9.32) were associated with the test population's ability to distinguish the pitch and loudness for bands of noise. Each test contained 110 items consisting of a standard noise (lasting two seconds) immediately followed by a comparison noise (also lasting two seconds). The band of noise used in the pitch test was shifted up and down the frequency scale.

These graphs look suspiciously like a rudimentary form of a fuzzy set. The diagrams graphically show the cross-over at the *jnd* point, where subjects can tell the difference between the lower and higher pitch. The plots indicate how many subjects could discriminate differences in pitch and loudness. The same kind of discrimination applies to a wide range of physical phenomena, many of which appear regularly in business, scientific, and government agency planning models. As an example, consider a graph (Figure 9.33) that shows the transition between yellow and red in terms of the underlying wavelengths.

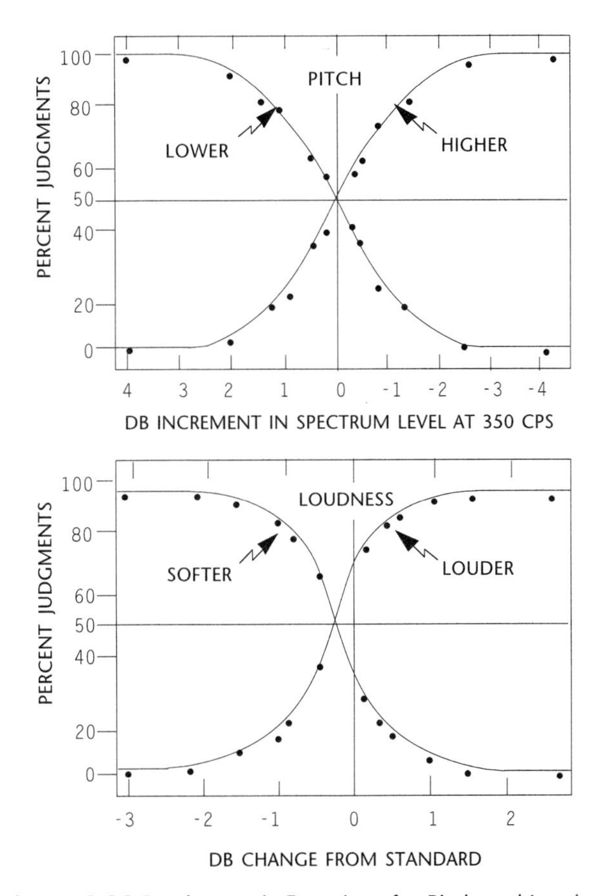

Figure 9.32 Psychometric Functions for Pitch and Loudness

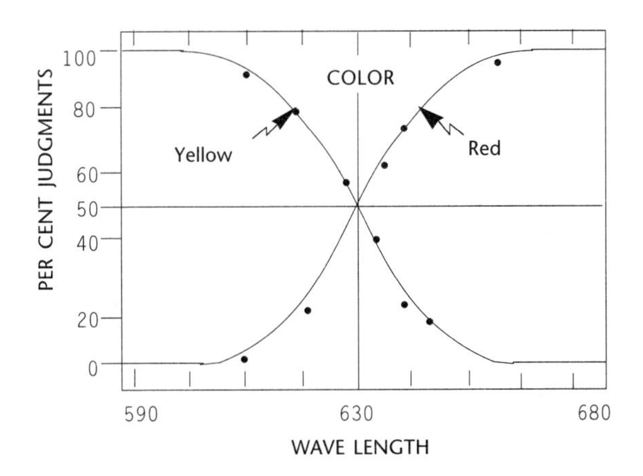

Figure 9.33 Noticeable Difference in Color Transitions

FUZZY SET IMPLICATIONS: CROSS-OVER POINT AND THE VOTING OF POPULATIONS.

The psychometrics underlying the *jnd* provide an important construction and management tool for knowledge engineers creating fuzzy systems. The measurement of an expert's ability to distinguish between changes in a phenomenon is a critical factor in his or her ability to vote on the relationship between a state of the phenomenon and a particular semantic category (which is the essense of a fuzzy set). As an example, consider *Tall* and *Short* in Figure 9.34.

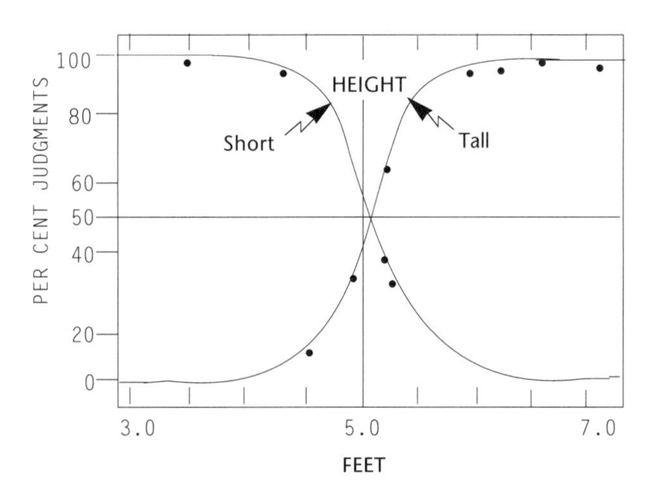

Figure 9.34 Noticeable Differences in *Tall* and *Short*

We might very well show a population of experts a medium-sized individual from the target population (the source stimulus) and then a series of individuals whose heights are either greater than or less than the source stimulus. Through such a process, the sensitivity of the experts to changes in height, and their ability to discriminate between different degrees of shortness and tallness, can be discovered. Such a base-line approach to voting, incorporating the psychometric functions associated with the process, can be used to adjust the vote score and produce better estimates for our fuzzy sets (for details on inducing a fuzzy set through expert voting see "Voted-For Distributions" on page 515).

The overlap of fuzzy sets, in particular, reflects nothing more than the *jnd* between neighboring sets. We can extend the idea of the *jnd* to fuzzy sets by considering what a cross-over point means: it means that we can now tell the difference between two commingled concepts. At the point where the *Tall* and *Short* set surfaces cross, we have the just noticeable difference between *Tall* and *Short*. Within the boundaries of the sets to the left and right, ambiguity exists, in the sense that a small change in the height does not elicit a strong response to label the new value as *Tall* or *Short*.

But the *jnd* concept can be used for more than simply deciding where two overlapping fuzzy sets should intersect. The actual form of a fuzzy set—its shape, width, and slope—can then be induced from a population of experts that votes on the difference between a series of examples. We can elicit the fuzzy sets for *Tall* and *Short* by showing a set of experts a series of base examples and asking them to define the *jnd* for a change in height from that example. Using Fechner's Law, a sufficiently robust set of observations, we can then extrapolate how the population as a whole would be represented for that particular group of experts. We must remember that fuzzy sets are always context-dependent. Their shapes are based on the semantics of the model, and the semantics of the models are derived ultimately from the subject matter experts.

Best Estimate of the Variable's Semantics (SWAG)

We now briefly consider one of two approaches that can be used when expert knowledge is unavailable (recall that even in the statistical evolution of fuzzy sets, we are relying on some form of expert knowledge to guide our selection of the statistic measurements used to produce the fuzzy sets). Known in the knowledge engineering industry (and endemic throughout the military) as a Scientific Wild-Ass Guess (SWAG), we can often approximate the shape of a fuzzy set from both a common sense understanding of the problem, and informal conversations with the end-user. Having said this, however, we should bear in mind Renè Descartes' warning:

> Le bon sens est la chose du monde la mieux partagee, car chacun pense en etre bien pourvu. (from *Le Discours de la Methode*)
>
> Common sense is the most widely shared commodity in the world, for every man is convinced that he is well supplied with it.

Rough or approximate estimates about the shape of a fuzzy set usually work well due to the overall tolerance to set membership functions exhibited by fuzzy models. We can often estimate the semantics of a model parameter and encode the fuzzy set based on a guess about its probable semantics. Figure 9.35 illustrates how a variety of different (but related) fuzzy sets can represent the concept *MiddleAged*.

From practical experience, I have seen that triangular and bell-shaped fuzzy sets generally provide extremely close data representations in a fuzzy model, while trapezoidal fuzzy sets, depending on the width of the plateau, can occasionally introduce "defuzzification wells" in the output. In any case, from our initial shape estimates, we can often develop more accurate fuzzy sets by observing the system behavior. Combining model performance feedback (such as the ongoing values of the statistical compatibility index) with the sensitivity of the model to changes in the vocabulary fuzzy sets, a knowledge engineer can often improve the shape of the fuzzy sets.

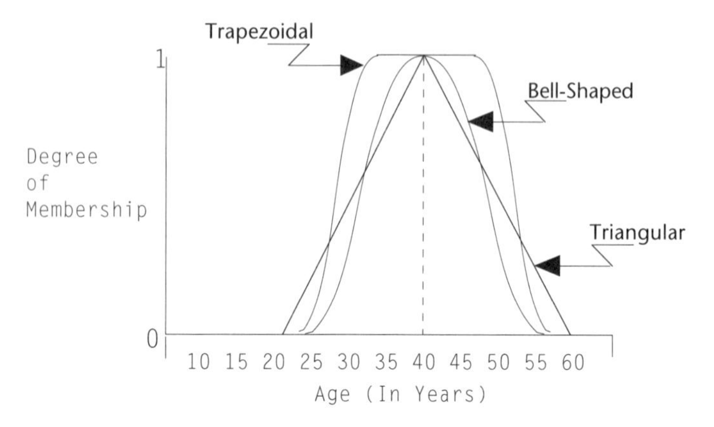

Figure 9.35 Fuzzy Set Representations of MiddleAge

Automatic Variable Decomposition

Although the generation of fuzzy sets from expert knowledge is an important aspect in fuzzy modeling, in recent years fuzzy systems have been generated through various forms of supervised and unsupervised data mining using sophisticated rule induction techniques. In a knowledge discovery or data mining approach, fuzzy sets are automatically generated by the system. The domains of the underlying variables are decomposed into arbitrary collections of fuzzy sets. While this approach does, indeed, elicit rules from the data, the rules do not actually reflect any real model semantics. There are two general techniques for automatically decomposing variables into fuzzy sets. The first, illustrated in Figure 9.36, places a bell-shaped (or triangular or trapezoidal) fuzzy set at the center of the domain and then generates a collection of fuzzy sets to the left and right.

Center-based generation normally labels the middle fuzzy set as *CE* (the center), and then sets to the left as *Sxx* (small) and sets to the right as *Bxx* (big). The center method always produces an odd number of fuzzy sets. For many control systems where the fuzzy sets measure changes (Δx), this procedure makes some sense. The center is often the *ZR* (zero) set, with sets to the left indicating negative changes and sets to the right indicating positive changes. An important property in automatic fuzzy set generation is the width of the overlap between sets, known as the expectancy level of the fuzzy set. This is generally expressed as a percentage of the neighboring fuzzy set's base. Figure 9.37 shows how this degree of overlap is an important property. The overlap is usually set to 50%, but the actual overlap often depends on the nature of the expected control surface.

A variation on this, of course, is linear decomposition. In this process, the entire range of a variable is decomposed into a set of overlapping fuzzy sets, starting at the left-most edge and moving right until the entire range is covered. Both odd and even numbers of

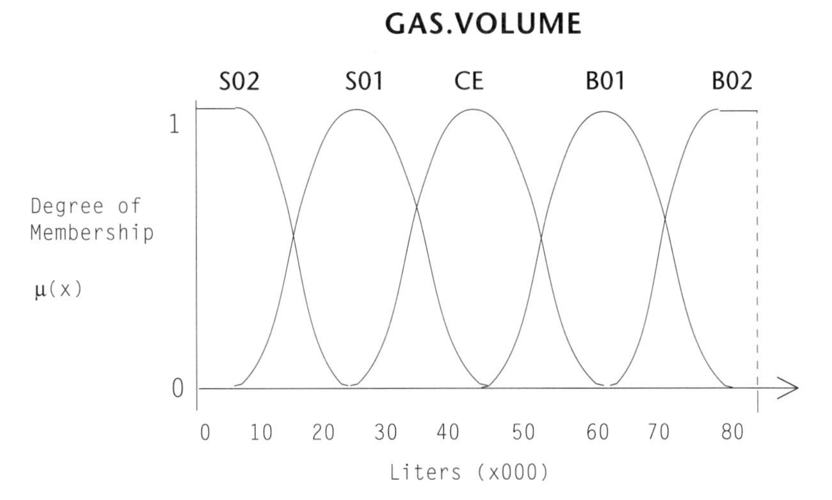

Figure 9.36 Automatic Generation of Fuzzy Sets (Center-Based)

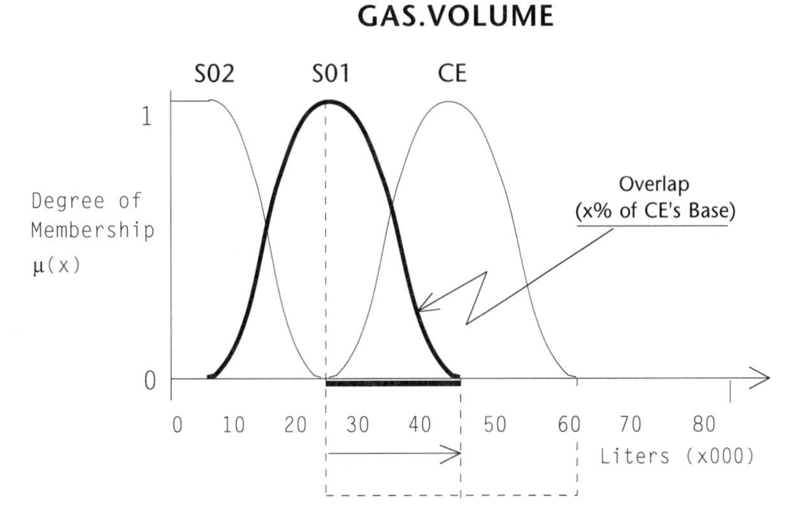

Figure 9.37 Specifying the Generated Overlap (Expectancy)

fuzzy sets can be generated using this method. Figure 9.38 illustrates the linear decomposition of the *ROI* variable.

Neither of the previous automatic fuzzy set generation methods include knowledge about the underlying data distributions. The "around (or near) mean" methodology, on the other hand, positions a bell-shaped (or triangular) fuzzy set over the mean of the distribution

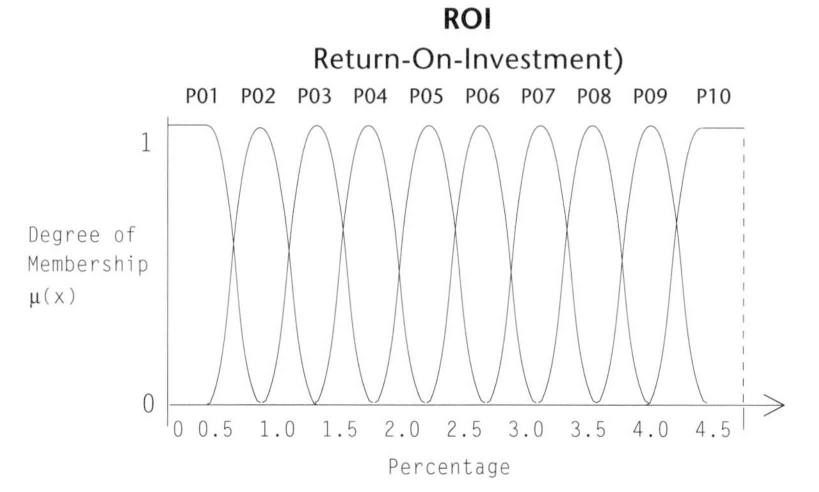

Figure 9.38 Automatic Generation of Fuzzy Sets (Linear)

with a width of one standard deviation. This set is bracketed by narrow fuzzy numbers covering the area out to two standard deviations. Additional fuzzy sets are added with wider domains to cover the entire variable's range. Figure 9.39 shows a mean-based decomposition over the *Gas.Volume* variable.

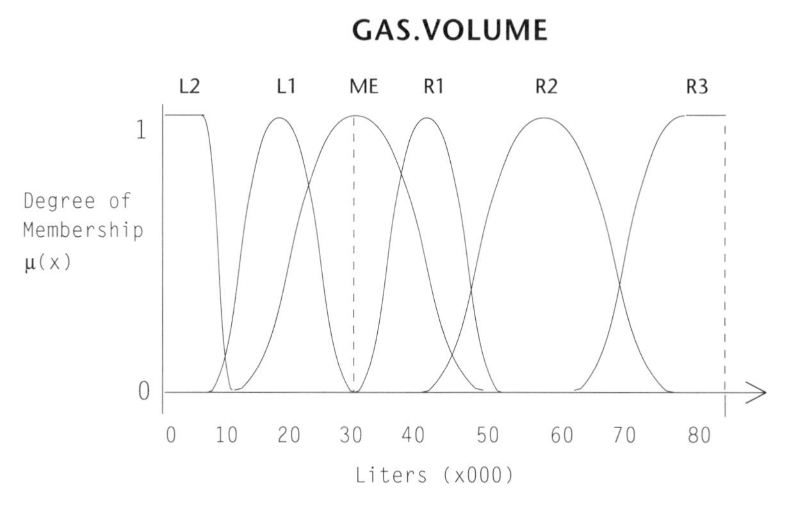

Figure 9.39 Automatic Generation of Fuzzy Sets (Around Mean)

Automatic generation of fuzzy sets plays an important role in the emerging technologies of data mining and knowledge discovery. These two technologies partition the variable's domain into a set of overlapping regions so that cluster analysis can find related masses of knowledge. However, they are not a satisfactory way of generating fuzzy sets for fuzzy expert and decision-support systems. In order to create truly effective and maintainable fuzzy models, the fuzzy sets must reflect the actual semantics of the system parameters.

Boolean and Semi-Fuzzy Variables

Fuzzy inferencing uses the degree of truth in a predicate proposition to determine the degree of compatibility in the consequent. When the variables used in the model are continuous and mapped to a fuzzy term set, this process works well. Consider the following rules from a vehicle underwriting risk assessment model in the insurance industry:

```
[R1] if age is young
   then risk is increased;
[R2] if distance.to.work is far
   then risk is slightly increased;
[R3] if accidents.last2years is excessive
   then risk is very increased;
[R4] if DWI.convictions is significant
   then risk is totally increased;
```

The degree of risk is determined by evaluating each fuzzy proposition and building a representation of a risk fuzzy set. This result is defuzzified to find the expected value of risk. Now we add another rule:

```
[R5] if sex="male" then risk is increased;
```

The predicate of this rule involves a Boolean expression. Sex can only be male or female (assuming we are discussing human beings and not something like slime molds, myxomycetes, which have eighteen sexes). This causes the implication function to assign a truth value of [1.0] to the predicate. In our risk assessment model, the result is a saturation of the solution space. The Boolean rule preempts all the fuzzy rules.[8]

Using Boolean Filters

If the fuzzy rule can be made conditional on a Boolean filter, this kind of modeling problem can usually be avoided. An explicit Boolean filter acts as a qualifying predicate on the rule. If this predicate is true (in the Boolean sense), then the fuzzy rule is evaluated.

The truth of the qualifying predicate is discarded. Using this kind of explicit filter, [*R5*] might appear as:

```
[R5] if sex="male"
   then if age is young
       then risk is highly increased
              else risk is somewhat increased;
```

Another approach for rules with less complex predicates is simply to filter the rule evaluation by using the Boolean expression in the predicate. As an example, another version of [*R5*] with an implicit Boolean filter appears as:

```
[R5] if sex="male" and age is young
   then risk is highly increased;
```

If the expression *sex="male"* is true ($\mu[1.0]$) and the truth of age is young is $\mu[.38]$, then, since this is an *and* condition, the minimum of (1.0,.38) is ".38". So the rule is evaluated, and the degree of truth in the Boolean expression does not contribute to the outcome. When the expression *sex="male"* is false ($\mu[0]$), the minimum is zero, so this rule is not executed (which is our intent in the first place). In both of these examples the Boolean expression has been used to indicate whether or not a particular rule should execute; we have not addressed the main issue: how does a Boolean expression contribute to a conditional fuzzy proposition?

Applying Explicit Degrees of Membership

In attempting to decide how a Boolean state can contribute meaningfully to a fuzzy proposition, we need to return to the ideas underlying a fuzzy set. In our definition of fuzzy sets, we mapped a continuous, numeric space to a truth membership function. But fuzzy sets can also be non-numeric, and they can assume the form of lattices. In a non-numeric fuzzy set, members are explicitly ranked according to their subjective mapping to the fuzzy set semantics. An example of this is illustrated in Figure 9.40, the set of all American states that are square-like.

Each state is placed in the fuzzy set according to its degree of "squareness." There is no proper horizontal axis in this kind of fuzzy set (the states can appear horizontally in any order), only a vertical mapping of each state to its degree of membership. This forms a proper fuzzy set. Figure 9.41 shows the membership ranking of each state in the fuzzy set.

The square states fuzzy set is a special case of a more general family of lattice fuzzy sets. A lattice set contains members arrayed in an MxN space. Figure 9.42 illustrates one such lattice fuzzy region with its members ranked in space according to their specified membership values. Ranking along the other axes is model-dependent (such as alphanumeric lexical ordering).

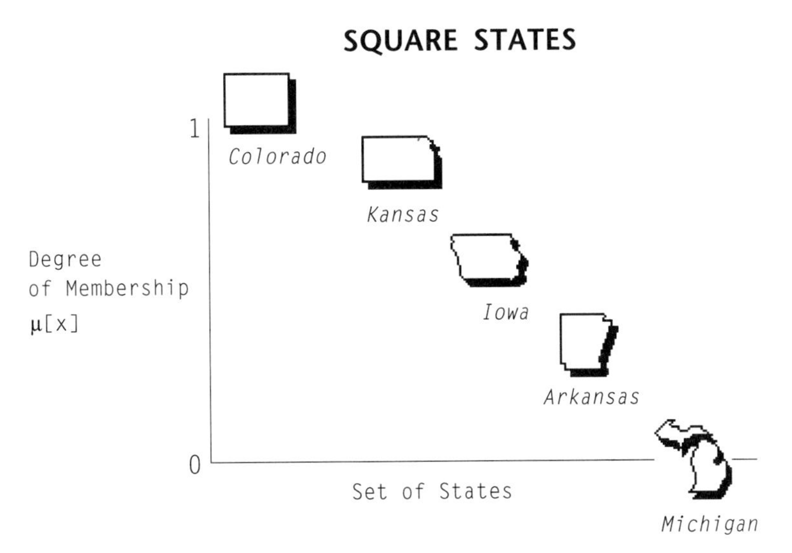

Figure 9.40 The Fuzzy Set of Square States

$$SquareStates = \begin{bmatrix} Colorado & 1.0 \\ Kansas & .8 \\ Iowa & .6 \\ Arkansas & .3 \\ Michigan & 0 \end{bmatrix}$$

Figure 9.41 The Membership Roster for the Square States Fuzzy Set

How does a lattice fuzzy set help us mix fuzzy and Boolean expressions? It gives us the facility for explicitly placing the Boolean expression in a fuzzy region. From this point, we can modify the way fuzzy rules are formulated. A Boolean expression is viewed as a point in a lattice fuzzy set with an explicit membership assignment. The risk assessment rule now becomes:

```
if sex="male"[.60] then risk is increased;
```

If the predicate of this rule is true, then its truth membership is $\mu[.60]$. This is a reasonable approach to handling Boolean variables, since it accords well with the underlying intent of the rule. When sex is "male," we mean that the overall risk is increased to some degree independent of any other factors. A simple Boolean expression fails to map with this intuition—the risk is unconditionally increased to its maximum value.

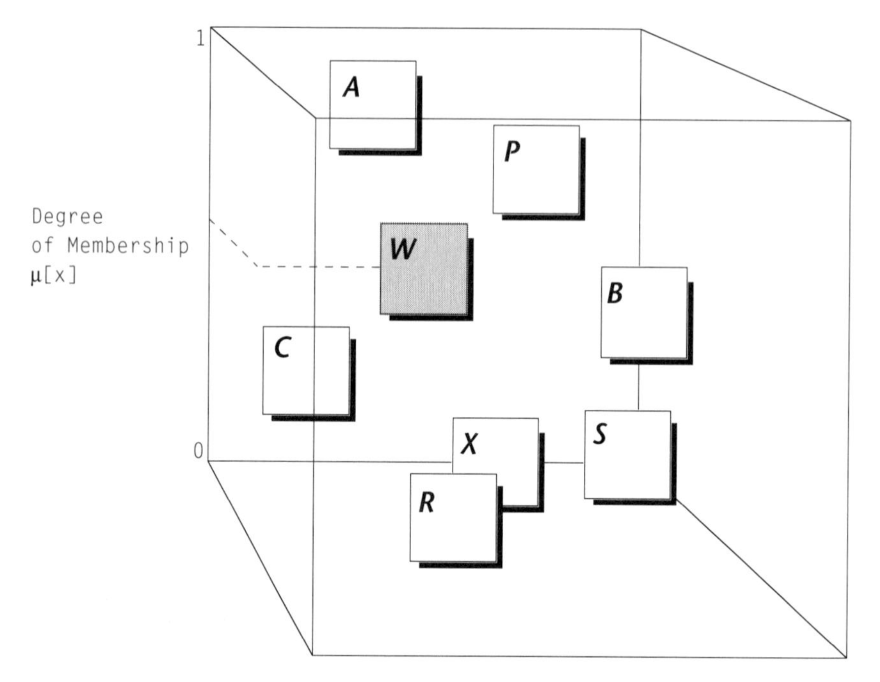

Figure 9.42 An Arbitrary Lattice Fuzzy Set

Uncertain and Noisy Data

Both information and control systems must deal with questionable data. An item of data carries with it a level of uncertainty related to our confidence in the measuring device's accuracy and precision, the transmission medium's clarity and stability, and, fundamentally, the properties of the data itself. When a model receives its data, there are two kinds of uncertainty: confidence in the nature of the data and confidence in the accuracy of the data. Confidence in the nature of data must be resolved outside the model. It is an exogenous control condition. On the other hand, uncertainty and noise in the data stream can be addressed, in many cases, by viewing the data as fuzzy numbers.

Understanding Fuzzy Numbers

All numbers are fuzzy sets. We seldom make the connection between our confidence in a data point and the nature of fuzzy numbers. In effect, the numbers that we use without qualification in a model—the manufacturing price of a product, the last quarter's after-tax profit, a client's age—are *singleton* fuzzy sets: fuzzy spaces that have no horizontal dimension. It is this horizontal dimension that indicates the degree of ambiguity or noise in the data.

Understanding Fuzzy Numbers (continued)

A Fuzzy Variable

As an example, you go to a warehouse bin containing four-dimensional widgets and count them, and you find there are exactly twelve widgets on hand. There is no noise or lack of certainty in this number. Figure 9.43 shows the fuzzy set "12"—we would ordinarily consider this a simple number.

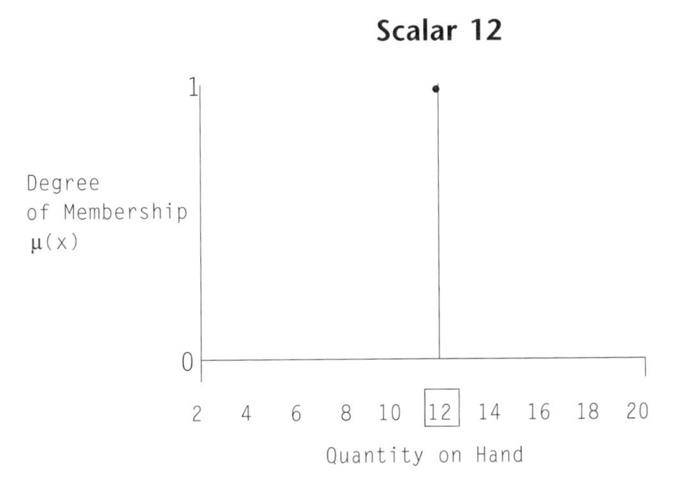

Figure 9.43 The Number 12

On the other hand, you send Daryl, your chief assistant, to count the number of four-dimensional widgets. Sitting behind your desk, you watch Daryl look in the bin, rapidly move the widgets around, and make a very quick count. Daryl reports that there are twelve widgets. This number carries less certainty because of Daryl's cursory count. The actual on-hand quantity might be one or two widgets more or less than Daryl reported. Figure 9.44 shows how a less confident scalar 12 is represented. In fuzzy terms, we would say this number of widgets is *about*, *near*, or *around* 12.

Understanding Fuzzy Numbers (continued)

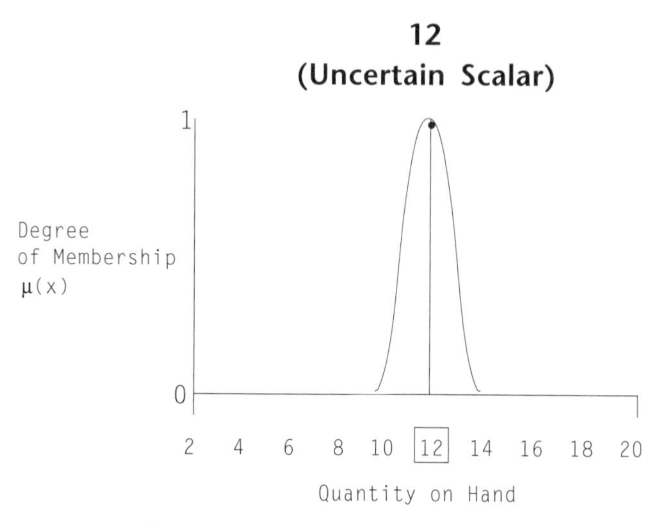

Figure 9.44 The Uncertain Number 12

Thus, the horizontal diffusion of the number, its loss of crispness, reflects our belief in its degree of certainty. Figure 9.45 illustrates a highly uncertain fuzzy quantity centered around 12. This uncertainty can also be interpreted as ambient noise in the data.

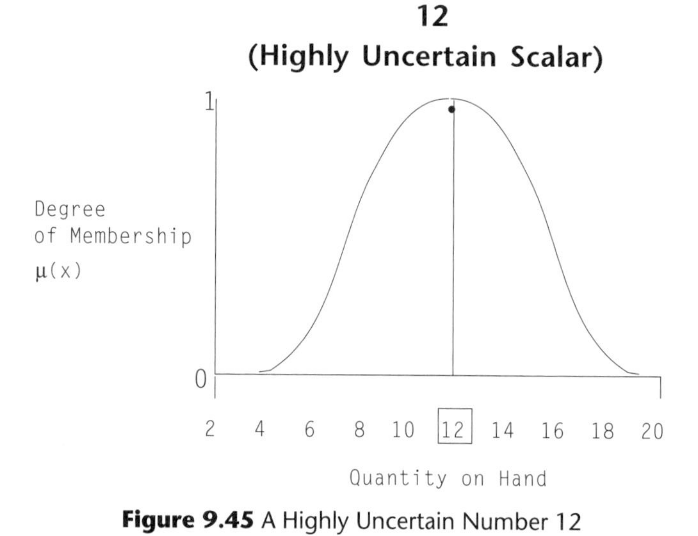

Figure 9.45 A Highly Uncertain Number 12

A Fuzzy Class or Quantity

A fuzzy class or quantity is also a fuzzy number and parallels the ideas underlying the uncertainty of external fuzzy numbers. These are generally represented as intrinsic fuzzy sets within the model (while fuzzy variables are primarily used as input to the model). The horizontal diffusion of the fuzzy number indicates the degree of ambiguity or uncertainty existing in the set concept. Figure 9.46 illustrates the fuzzy set *MiddleAged*.

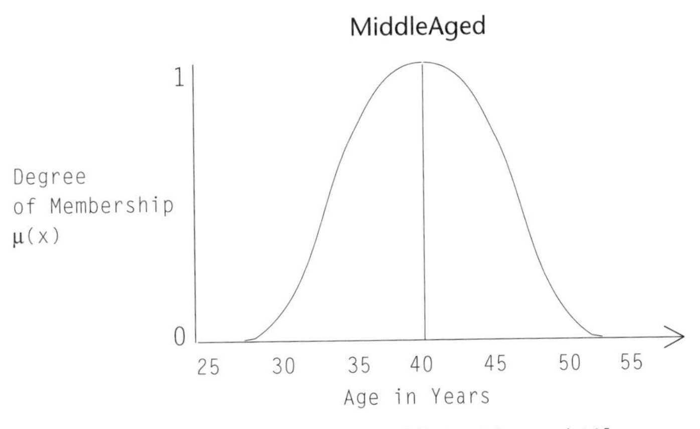

Figure 9.46 The Concept *MiddleAged [around 40]*

The idea of *MiddleAged* is expressed as a fuzzy number *around 40*. We know that 40 is definitely middle aged, but we do not know a single point on the left where middle aged begins and on the right where middle age ends (because these points do not exist). The width of the bell curve is a measure of this ambiguity. If we improved our estimate of the onset of middle age, the degree of uncertainty about what constitutes middle agedness might result in a new, narrower, fuzzy quantity. If we believed that middle agedness started and disappeared after reaching 40, then this fuzzy number reduces to a singleton.

Fuzzy quantities of this kind are used extensively throughout the fuzzy modeling environment. They represent as bell curves, triangles, or trapezoids the fuzzy sets underlying the decomposition of control and solution variables.

Handling Uncertain and Noisy Data

With the exception of statistical and explicitly probabilistic data, information systems are not generally designed to deal effectively with approximate or uncertain information. Fuzzy information systems, on the other hand, often view their data spaces as collections of points with varying degrees of certainty. In our previous discussions, we have viewed fuzziness as a property of the conceptual knowledge structures in the model, that is, as attributes associated with the vocabulary fuzzy sets. In this instance, however, the fuzziness resides in the incoming data. The degree to which we are uncertain about a data point because of noise or the lack of quantitative or qualitative information induces a fuzzy membership function around the data point.

Consider the following two rules from an (admittedly hypothetical) credit assessment system:

```
[R1] if income is MODERATELY.HIGH
   then credit.risk is MEDIUM;
[R2] if income is HIGH
   then credit.risk is LOW;
```

Figure 9.47 shows the fuzzy sets associated with the model variable `Income`.

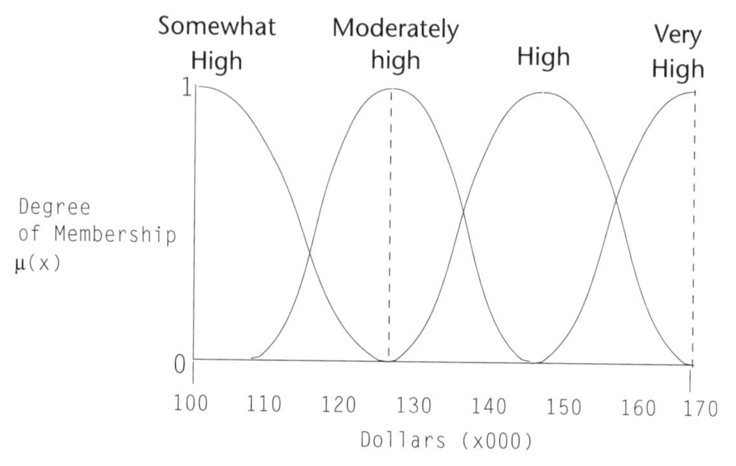

Figure 9.47 The Fuzzy Sets for the *Income* Variable

If we know the value of *Income* exactly (within the granularity required by the system), the execution of rules [*R1*] and [*R2*] is straightforward. When our knowledge about the candidate's income is less precise—the figures on the credit application are blurred, the figure in the database is approximate, etc.—we need some way to evaluate the rule's predicate with less than crisp data.

A possible system–user interaction from the hypothetical credit assessment system might request the individual's annual income. An amount prefixed with the "≈" symbol indicates an estimate.

```
Enter the candidate's income: ≈135000
Enter confidence in this figure[0 to 1]: .8
```

The figures in bold are entered by the credit analyst. The confidence factor is a measure of how much noise we believe is associated with the data (the lower the number, the broader the fuzzy number's bell curve). Figure 9.48 shows how the approximate income of $135,000 is represented as a fuzzy number.

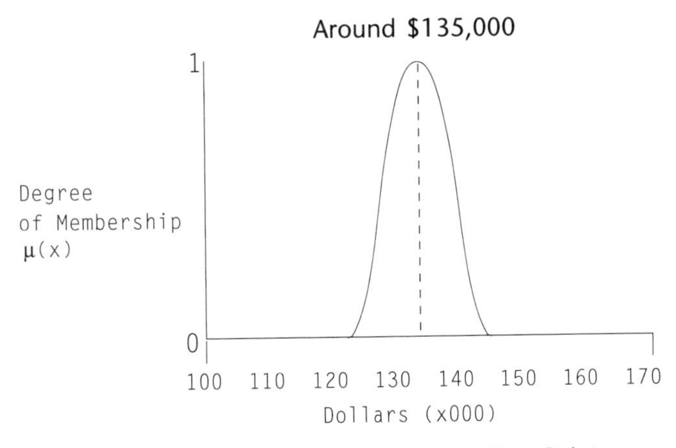

Figure 9.48 Evaluating a Fuzzy Input Data Point

How do we evaluate a rule using fuzzy numbers? We must find the point of maximum correlation between the predicate fuzzy set and the fuzzy region associated with the data value. This is the maximum truth membership point that is cojoint in each set—an operation equivalent to *AND*ing the sets and finding the maximum truth of the resulting fuzzy region. Figure 9.49 shows this intersection point for rule [*R1*].

Inferencing with Fuzzy Data

The approximate reasoning process with uncertain data is essentially the same as the process involving crisp input data points with one exception: the degree of membership in the predicate fuzzy sets must be found by combining the fuzzy data with the premise term set. Figure 9.50 shows the fuzzy sets associated with the *Risk* solution variable. For simplicity, these are simply *Low*, *Medium*, and *High*.

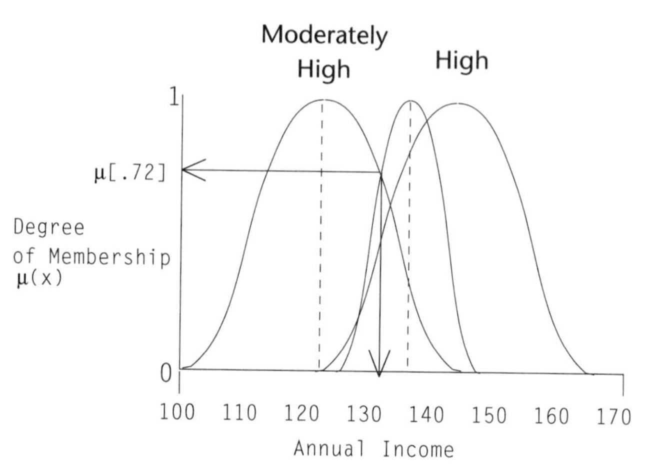

Figure 9.49 An Approximate *Income* Data Point

The implication process for the credit assessment rules involves finding the truth of the predicate fuzzy proposition, correlating the *Risk* term, and updating the *Risk* output fuzzy region. Figure 9.51 illustrates the way rule [R1] is processed.

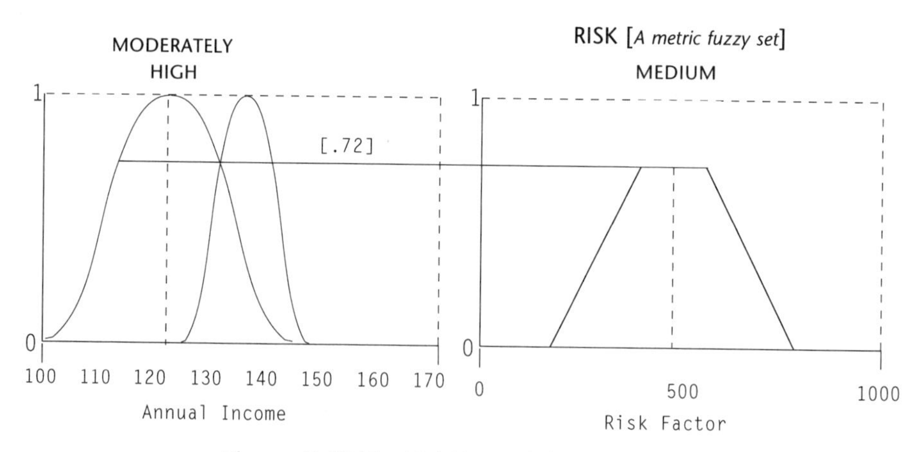

Figure 9.50 The Risk Factor Solution Fuzzy Sets

The inherent fuzziness of the income value comes into play when we select all the rules that can fire for a single input data point. With crisp data, this is determined directly

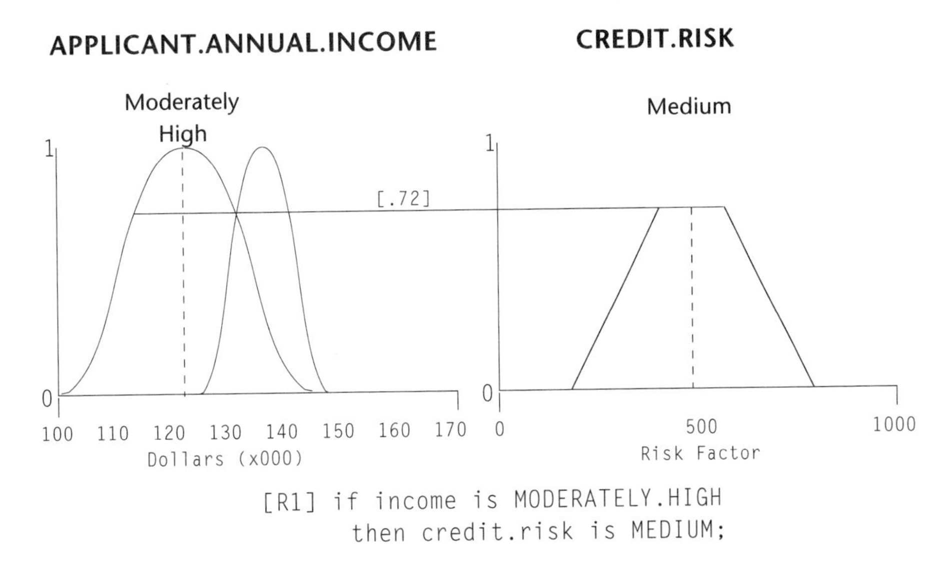

Figure 9.51 Implication Process for Credit Assessment Rule [R1]

by the degree of overlap among neighboring fuzzy sets. Using fuzzy data, the overlap of the term sets and the overlap of the fuzzy data point are both involved. Figure 9.52 illustrates how a value for the fuzzy income is used to correlate the *Low* risk term.

The correlation maximum process truncates the *Low* fuzzy set at the [.90] truth level. Figure 9.53 illustrates how the final credit *Risk* solution fuzzy set appears after the two rules are executed. The dotted line indicates the expected value after a composite moments defuzzification.

Building Fuzzy System Models

In this section, we examine a fuzzy system development methodology. The methodology prescribes a loosely structured technique for building and delivering a fuzzy system model, but it is not intended as anything more than a guideline. This is a *protocycling methodology*; it describes a cyclical method for defining a base system that is refined and extended as you apply different performance measurements. Figure 9.54 is a high-level schematic of the methodology.

The methodology itself represents an attempt to formalize and add structure to an interactive two-part building process. (Yes, there are three blocks, but the first is generally not iterative!) In this process, the initial conceptual design is done on paper, not on the computer. This is a critical part of the fuzzy model evolution: understanding the mechanics

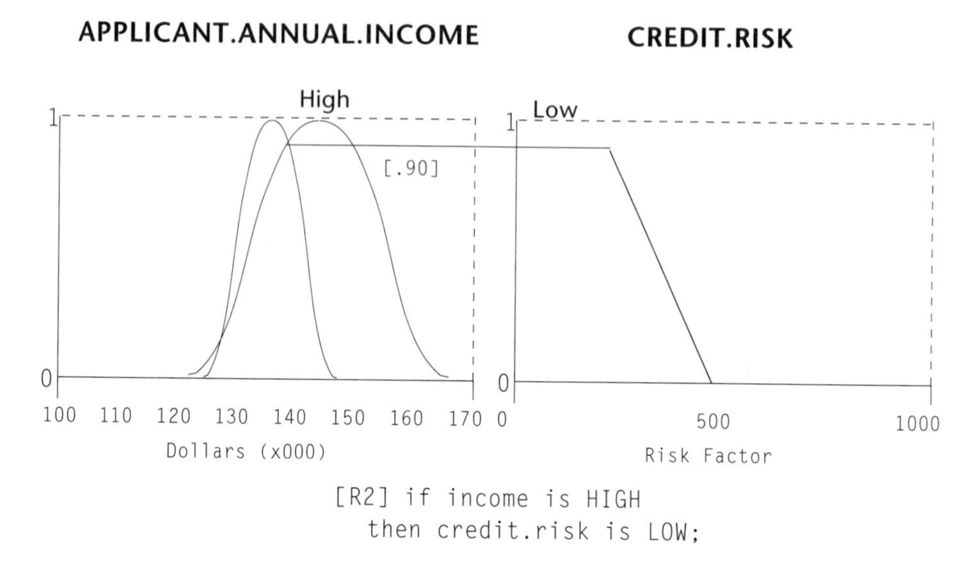

Figure 9.52 The Correlation of *Low* to *High* Income in Credit Assessment Rule *[R2]*

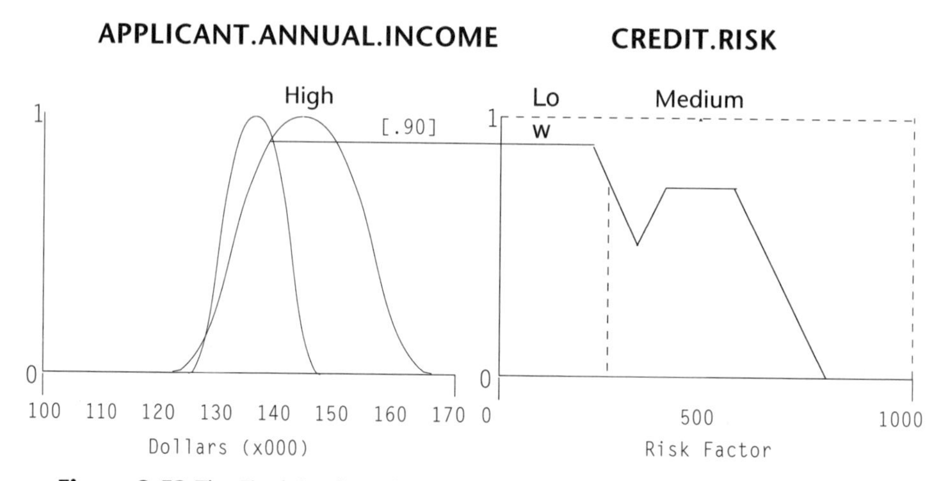

Figure 9.53 The Final Credit *Risk* Solution Fuzzy Region After Applying Rule *[R2]*

behind your system's behavior and identifying the system dynamics in terms of the conventional input–process–output model (Figure 9.55).

Because fuzzy systems are predominantly verified through visual means, a simulation and surface display capability is critical to the model acceptance process. (In the fuzzy modeling library, the *FzyDrawSet*, *FzyPlotVar*, and *FzyExamineSet* functions provide extensive

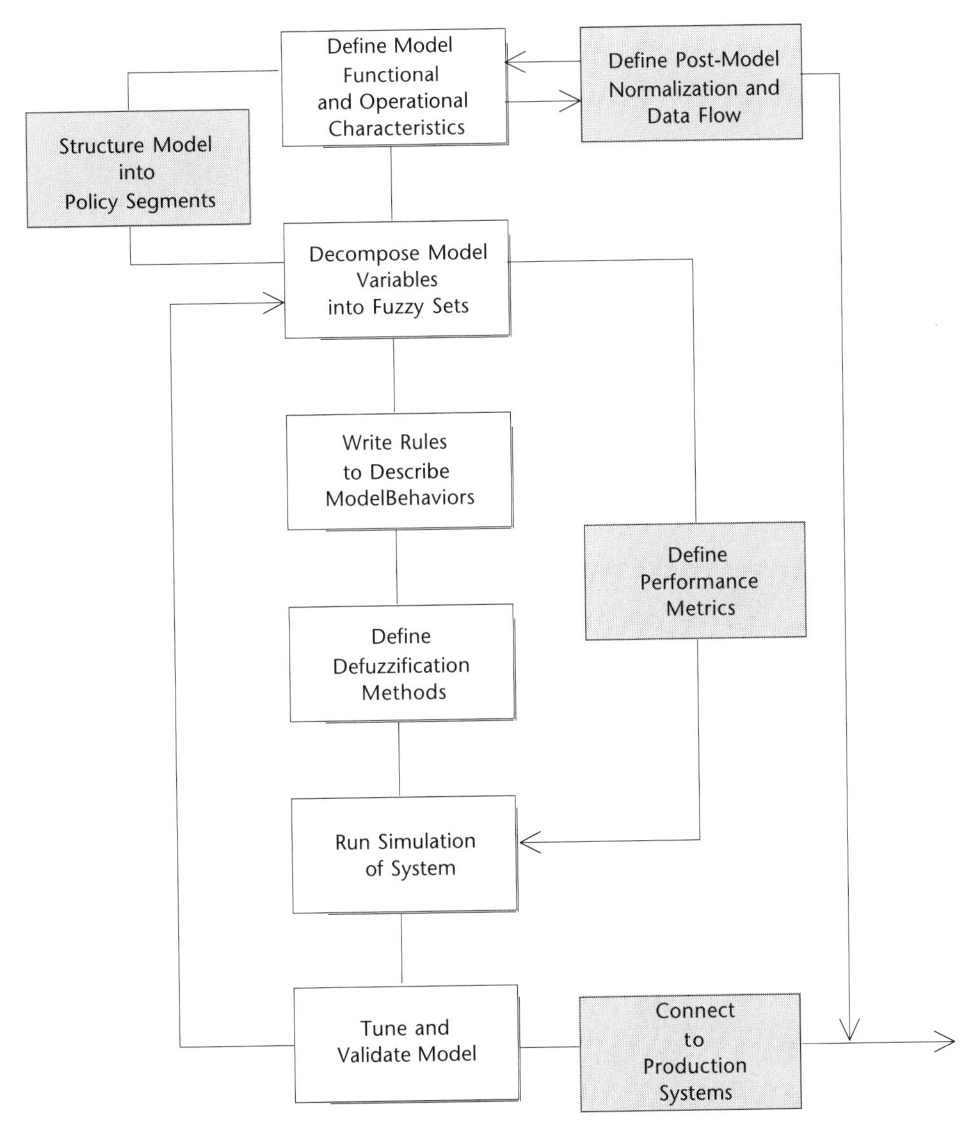

Figure 9.54 The Fuzzy System Development Methodology

fuzzy surface display capabilities.) This simulation can be either control-surface or data-driven. In control-surface simulation, the development tool cycles over a *k(n)* dimensional space defined by the Cartesian products of the input and output variable spaces. In a data-driven simulation (a more common and generally a much better approach), the design engineer prepares a file of probable sample instances for each variable. The model reads values

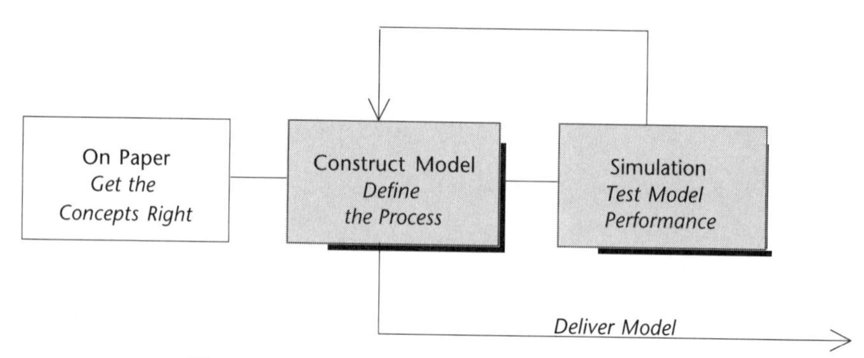

Figure 9.55 The Meta-Level Protocycling Flow

from this data file (or generates random values, usually following a Poisson or beta distribution) and examines the model output performance at each instance.

The Fuzzy Design Methodology

Essentially, there are five fundamental stages in the construction of fuzzy models, and these steps are incorporated implicitly in our methodology. The stages are:

- Selecting the performance (control and solution) variables
- Defining the control surfaces (the fuzzy sets)
- Constructing the relationship between input and output spaces (rules)
- Running a simulation of the system model
- Interpreting the output (defuzzifying to get expected values)

The fuzzy modeling of information systems involves a relatively simple process—a control surface is designed that responds correctly to a set of expected data points. This control surface (the system response) is made to react correctly to data (database records, spread sheet cells, etc.) by changing either the shape of the underlying fuzzy sets and/or the meaning of the rules (the system relationships).

Define the Model's Functional and Operational Characteristics

In this first stage of a fuzzy model design, we want to establish the architectural characteristics of the current system and, at this time, define the specific operating characteristics of the proposed fuzzy model. This is the point in the design where the functional decomposition of the system into multiple models, and models into multiple policies, should be considered. Isolating the system's operation and responsibility into several models allows a much more structured approach to system design and development. The basic vocabulary fuzzy sets, hedges, and variable definitions can be isolated either in models or policies, or shared among functional units.

Define the System in Terms of an Input–Process–Output Model

The first step in designing a fuzzy system follows the analysis techniques used by commercial application systems analysts and intelligent-systems knowledge engineers. Our first task is defining three elements: what information (data points) flows into the system, what basic or fundamental transformations are performed on the data, and what data elements are eventually output from the system (Figure 9.56).

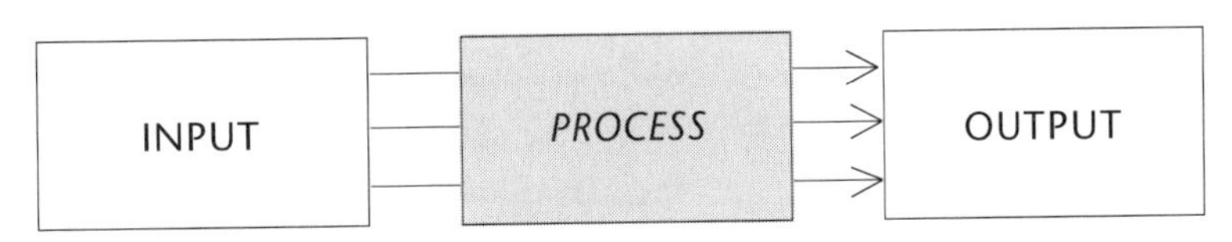

Figure 9.56 The Conventional System Flow Process

Even if we do not have a deep mathematical model of the system process, we must have a deep understanding of these three phenomena. Fuzzy modeling involves developing the best estimation surface for a formal system. The *Process* component of the model centers around a set of rules describing the behavior of the model output, given a set of inputs that interact with the fuzzy set vocabulary. Unlike purely mathematical models and symbolic expert systems in general, fuzzy models work from the outside in; that is, we want to simulate a given system behavior by using a collection of fuzzy regions. Each process block is a candidate for encapsulating this activity into a separate fuzzy model or, for functional components of a model, into a set of policies.

Localize the Model in the Production System

At the time we sit down to design a fuzzy model and determine the flows into and out of the system, we should also position the model in relation to the total system architecture. This architecture provides a clear picture of how inputs and outputs flow through the system (Figure 9.57).

Localization will help the designer make decisions about the expected number of sensor inputs, the required number and mode of the outputs, and the range and strength of the output. Localization provides reinforcement for the input–process–output design step.

Depending on the nature and complexity of your overall system, the order of model design can change. You might find it more appropriate to localize your fuzzy component

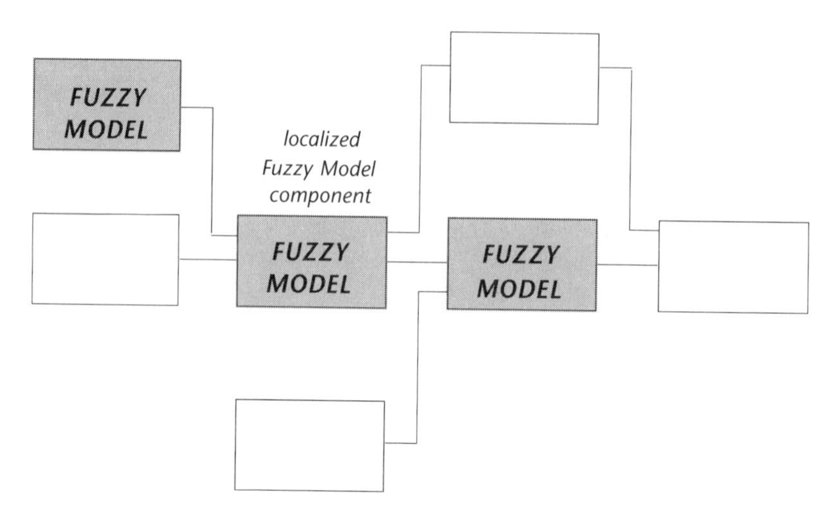

Figure 9.57 Localization of the Fuzzy Model Components

before tackling the details of input and output (since these might be determined by the overall system architecture).

Segment the Model into Functional and Operational Components

In some ways, this step is a refinement of the localization process, but we consider it as a separate issue to highlight its importance. Fuzzy models, like ordinary models, need to handle a wide variety of input and output data representations. There are two related considerations here. First, if a fuzzy model receives information from or sends information into a statistical filter, aggregator, or profiler, then the designer must consider both the certainty of data coming into the model and the required characteristics of data leaving the model. Second—and this is somewhat philosophical—you should not confuse fuzziness or imprecision with statistical measures. Where statistical and probabilistic models are appropriate, they should be used.

Isolate the Critical Performance Variables

This is undoubtedly the most difficult part of the initial model formulation. From a universe of data that is continuous and discrete, qualitative and quantitative, sparse and dense, as well as certain and uncertain, we must identify those variables that directly control the operation of the fuzzy system and those that lead to an output value from the model. These are control and solution variables, respectively. Performance variables can be both endogenous to the model (in the sense that we have arbitrary control over their values) and exogenous to the model (such as the duration of daylight in October, the mean temperature in Ohio, the required maximum speed of a motor, the impedance and capacitance of a resistor set, the price of a competitor's product in our market area). An important part of performance variable identification is consolidation (aggregation), separability, subordination, and pruning. By this we mean identifying the variables that actually affect the model performance out of a population of possible variables. Similar variables can be consolidated. Macro-variables—those at a higher level of granularity than the model—can be decomposed into individual variables. Control variables that take effect only at boundary points or only when primary variables reach specific state levels are subordinated in the rule reference set. Unimportant variables must be pruned from the model.

Choose the Mode of Solution Variables

A fuzzy model produces output in the form of consequent fuzzy regions. These are the solution variables that are used independently or fed into other components of the overall system. In describing the internal transformation process, we need to identify the path of implication steps leading to the final output set. These are the solution variables generated by the fuzzy model. There exists a dichotomous partition in models between the fuzzy nature of the control surfaces (which are quantitative and continuous) and the scalar measurements output by the model. The physical control or decision space that receives output from the model may need to respond at discrete points rather than across a continuous range of values. This means that the designer must decide on an output mode for each of the solution variables—either continuous or discrete. This process of converting a continuous scale to a discrete scale is called *discretization*.

Depending on the capabilities of the modeling tool, this step should also establish—or at least make an initial estimate of—how the consequent (output) solution spaces will be represented. In general, there are two options: as a set of discrete singletons and as a fuzzy region. This selection will affect the nature of defuzzification and the precision of the model. Usually, singleton output representations provide much faster defuzzification (since the entire area need not be integrated), but since the output is represented by a set of sparse points, the precision can often be low. In some process control applications, the requirement for high speed will automatically dictate the output representation (singletons).

Resolve Basic Performance Criteria

Decide on a Level of Granularity

How fine a control do you need to exercise on the model performance surface? How fine are the measurement capabilities on the data space? (This reflects the capabilities of the sensors in physical systems and the aggregation level of financial, manufacturing, and project data in business and information systems.) Intrinsic data space granularity is not only a measure of the model's predictor capability, but is also a powerful limiting factor on the control capabilities of the model. Granularity, however, works in fuzzy models in another direction. Since a fuzzy model can interpolate control and solution spaces through

rough approximations of the actual value space (as if it were sampling the model's current steady state), a designer can build fuzzy models at a much higher level of abstraction (or metaphor) than is possible in conventional mathematical, Boolean, probabilistic, or statistical models. As an example, the pricing model rule

```
if the competition.price is not very high
   then our price must be near the competition;
```

adequately expresses a constraint on the solution variable price even when the actual value of the competitor's price in the market is only approximated. In process engineering models, the high level of granularity accepted in motor, electromechanical, and automotive systems, exemplified by such rules as

```
if temperature is high and pressure is low
   then rotor.speed is a decrease;
```

allows the definition of a workable control surface without a high level of detail and mathematical precision.

Determine Domain of the Model Variables

After identifying each control and solution variable, we need to determine their ranges of applicability. This range is called the *domain* of the variable, and when the fuzzy sets are defined, this domain is transferred directly to the characteristic function. You must note that the conformal mapping between control variables and their dependent solution variables requires a translation along similar domains. Pay particular attention to units of measure. Variables corresponding to the same domain should be measured in the same units. If we are modeling the skid of an anti-lock braking system on an automobile, we might want a system that recognizes the coefficient of friction, the force dynamics of the car, its speed, its impulse and momentum characteristics, etc. If mass is measured in ergs or dynes or joules, then acceleration should not be measured in feet per second but in centimeters or meters. A decohesion between measurement or calibration among control and solution variables is a leading cause of model failure (this is often reflected in unpredictable or inconsistent model predictions).

Determine the Degree of Uncertainty in the Data

More often applicable to process engineering systems, we need to understand the crispness of the incoming data elements in order to accommodate sensor insensitivity, noise, vibration effects, thermodynamic (heating) instabilities, sparse data, and intrinsically fuzzy data points. In a fuzzy model, the quantized input value in a dirty or imprecise space is determined by finding the maximum intersection of the term (fuzzy) space defining the input control variable and the fuzzy region associated with the sensor.

When a sensor reading[3] is received by a fuzzy system model, its topology determines the degree of confidence we have in the central measure of the data space. If the sensor device is high quality and has little noise, the reading approaches a singleton somewhere in the range of the fuzzy set term range. As noise and other factors of uncertainty increase, the input data point becomes a diffuse fuzzy space. The amplitude and period of the sensor input defines a new fuzzy region whose space must be measured (matched) against the input variable's fuzzy set.

Define the Limits of Operability

Although related to the specification of conformal domains for the term set, the limit of system operability is a higher and more restrictive metric. In this measurement, we want to find the optimal (or comfortable) region of operation for the system under its expected input data spaces. What would be considered allowable response values for this model? Such a region represents the middle of the possible density function (or the Cartesian product) for a series of input and related output variables.

Thus, the contour surface for the (X,Y) manifold defines the operating characteristics of your model. The exact nature of the contour will be explored when the fuzzy term sets are defined and tuned during simulation. At this step in model development, we simply want to define the way in which normal (X,Y) values are related. At this point, the designer should also define the extreme limits of operation (called the *pathological state*).

Establish Metrics for Model Performance Requirements

As the model architecture is formalized, a list of required performance characteristics should be developed. These will be used later to measure how well the final system meets our design criteria. Response measures include switching and rise time, minimum input to maximum output, signal (activation) strength, discretization classes, feedback, and feed-forward requirements.

Define the Fuzzy Sets

The next step in designing a fuzzy model involves the actual definition and construction of the fuzzy sets.[4] Fuzzy sets describe the actual control surfaces over which the process is mapped by the rules. Since the basic vocabulary of a production rule is the fuzzy set, these must be built before we can write instructions on how they are related to the incoming data. A more detailed examination of the issues surrounding the design of fuzzy sets and the use of fuzzy numbers is found earlier in this chapter; see "Fuzzy Set and Data Representational Issues" on page 496.

Determine the Type of Fuzzy Measurement

What does the fuzzy set measure? Fuzzy regions can provide orthogonal mappings between domain values and their membership in the set—this is called an "ordinary" fuzzy set. The fuzzy sets for the steam turbine are ordinary fuzzy sets since they map a degree of steam temperature to its membership in one or more fuzzy sets.

A fuzzy set can also define differential surfaces corresponding to a control space that would be expressed as a differential or difference equation in a mathematical model. These fuzzy sets correspond to a change in some control variable and could be used in a rule such as:

```
if speed is very high
   then throttle.movement is a BigNegative;
```

Generally these fuzzy sets represent the first derivative of some action, the degree of change between model states (especially in time-series or time-dependent systems) or the force of a control that must be applied to bring a system back into equilibrium.

Another type of fuzzy region is the *proportional metric*. This set reflects a degree of proportional compatibility between a control state and a solution state. Proportional and metric fuzzy sets play an important role in risk analysis, policy assessment, and other systems that rank the solution value. The sample rules in the following (alternative) project risk assessment model will eventually rank the risk of the current project on an arbitrary scale between zero and one hundred:

```
if project.duration is somewhat long
   then risk is an increase;
if project.budget is above avg(budget)
   then risk is a positive increase;
```

By itself, this fuzzy set is semantically empty. Its only meaning is assigned by the rule behaviors that populate the solution fuzzy region with degrees of *Increase* corresponding to the level of truth in the antecedent (premise) expression.

Another kind of measurement fuzzy set is the *degree of proportionality set*. This set is generally a logistic or sigmoid curve reflecting a degree of proportionality between a control state and a solution state, which can be seen in the following examples of *Usually* and *Most*. *Usually* has its inflection point at the [.50] membership point so that it becomes more and more true after that point. In this case, the domain can be interpreted as a percentage of the population so that at 70% the degree of membership in `Usually` is [.90]. The fuzzy set *Most* has its degree of proportionality skewed toward the 50% population level. That is, in a fuzzy proposition about the occurrence of event *[s] in [y]*, it is only true when it happens more than 50% of time. You should note that these classifications address the purpose and meaning of the fuzzy term; they do not connote any particular shape for the characteristic surface function.

Choose the Shape of the Fuzzy Set (Its Surface Morphology)

This step is undoubtedly the most critical part of building a fuzzy model, since the shape of the fuzzy set determines the correspondence between data and the underlying concept. If the fuzzy set is not properly shaped, then the degree of membership truth will be either too high or too low. Experimental evidence, however, points to a high tolerance for fuzzy shape approximation in most models; that is, a fuzzy model will still behave well even when the fuzzy shapes are not precisely drawn.

In process engineering models, the conventional space representations are either trapezoidal or triangular. The trapezoid usually maps membership functions at the domain extremes, while the triangles slice up the variable's operating space into a series of smaller but well-defined fuzzy regions. The idea here is to capture the linguistic nature of each subregion—as we move left to right across the domain, the compatibility with the fuzzy region increases until it reaches unity; after this, it begins to fall off toward zero.

Related to the idea of triangular surfaces is the PI curve or bell-shaped fuzzy set. Bell curves are more sensitive to the distribution of membership functions in large populations. The concept *Some* could be modeled with such an idea.

There are *no* fixed topologies for fuzzy sets, since the shape of a fuzzy set must map the underlying domain back to the set membership function, and this mapping can take any form. Typical fuzzy sets in the information and social sciences area include S-curves (exponential or logistic functions), linear surfaces, PI curves, and general multimodal surfaces. In deciding how to represent a fuzzy space, you must understand how each element out of the domain is mapped into the concept. The domain of the fuzzy region should also be elastic rather than restrictive; that is, it should extend into the absolute zero and absolute member ranges of the data. Every base fuzzy set must be normal. A normal fuzzy set has at least one membership value at [1.0] and at least one membership value at [0.0].

Elicit a Fuzzy Set Shape

Under most usual circumstances, a design engineer or systems analyst will construct a fuzzy set from the intuitive understanding of the target control surface. This is a natural part of knowledge engineering. Is there a better way of deriving the fuzzy surface than simply estimating the surface characteristics? Some extended research has gone into this topic. At the beginning of this chapter, we discussed some important ways to derive fuzzy sets (see "Designing and Eliciting Fuzzy Sets" on page 512); the following four tasks are additional approaches to automating, to some degree, the generation of fuzzy sets:

Probability frequency distributions. If the control variable is drawn from a continuous random population, or can be easily viewed as a continuous variable, then a probability density function or a probability mass function (for noncontinuous variables) can be used to describe the variable's behavior. We can take the distribution curve produced in this fashion and view it as a possibility set, that is, a fuzzy set. Probability models for one or more model variables, when available, should be used to help construct a fuzzy control surface.

Neural network models. A neural network can provide sophisticated nonlinear separability analysis on large quantities of data. Neural systems have been used to find natural membership functions in the data and thus directly create fuzzy regions.

Mathematical surface sampling. When a complete (or, perhaps, a partial) mathematical model exists for the process, you can derive a fuzzy surface by fuzzifying the mathematical surface. By simulating the process and randomly sampling the surface, a relationship between the performance variables and one or more control surfaces can be determined. Such techniques as the root locus method, frequency response plots, Bode diagrams, and polar plots provide sampling of the active model surface.[5]

Subjective approximations of the control surface. This is the "best guess" theory and corresponds to the way experts and experienced design engineers develop fuzzy systems. Since, as we have noted, fuzzy models seem very tolerant of approximate fuzzy sets, this method remains the best (in terms of opportunity costs) and the easiest way to design and implement a fuzzy system. In developing fuzzy sets from empirical data through a best estimate of fit, you should view the fuzzy region as a measure of compatibility and elasticity between elements in the domain and the underlying concept. To what degree is the temperature 220°C compatible with the concept *Warm* or *Hot* in the current context?

Select an Appropriate Degree of Overlap

To convert a series of individual fuzzy controls into one continuous and smooth surface, each fuzzy set must overlap its neighboring set to some degree. There is no precise algorithm for determining the minimum or maximum degree of overlap, but this interference pattern should reflect the semantics of the associated control or solution variable. Experience dictates that the overlap for triangle-to-triangle and trapezoid-to-triangle fuzzy regions averages somewhere between 25% and 50% of the fuzzy set base. The degree of overlap depends on the concept modeled and the intrinsic degree of imprecision associated with the two neighboring states.

Decide on the Space Correlation Metrics

For advanced fuzzy systems using Delphi policy analysis, adaptive feedback protocols, or m-dimensional fuzzy regions, we need to make a mapping between the n-dimensional model space and the k-dimensional process space. For adaptive and Delphi fuzzy systems, this means that we must establish both goal states and goodness-of-fit objective functions for the system, as well as weights for each of the fuzzy sets (that is, we need to build a "peer" weighting among the fuzzy regions). The situation is slightly more complex for m-dimensional fuzzy models.

Ensure that the Sets are Conformally Mapped

You must ensure that the domains among the fuzzy sets associated with the same control or solution variables are *conformal,* that is, they share the same universe of discourse.

It is also important to ensure that fuzzy sets employing monotonic fuzzy reasoning, proportional inference, and associative implication functions are mapped into the same configuration space. Thus, for a weight estimating model, the rule

```
if height is TALL then weight is HEAVY
```

requires that the fuzzy regions for *Tall* and *Heavy* have a conformal topology. Conformal means that the manifolds of the control surfaces behave in a manner that is related to the correlation between the two surfaces. If weight is roughly proportional to the cube of the height, then as *Tall* increases, *Heavy* must also increase in a shape that is representative of the *Tall* concept. Failure to map related fuzzy regions conformally is a significant reason for the failure of otherwise simple fuzzy models.

The fuzzy library does *not* contain a rule execution procedure; rather, it provides the low-level functions necessary to construct a rule object with a rule execution function. To build a generalized rule processor, you should understand how rules are constructed and maintained as data structures. As Figure 9.58 illustrates, an *if–then–else* rule is really a conditional binary tree.

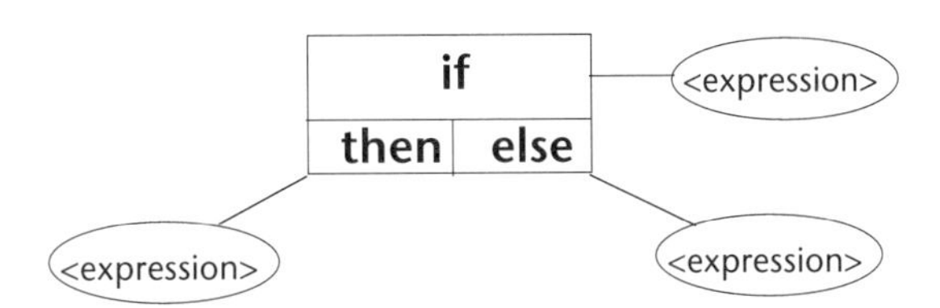

Figure 9.58 The Structure of an *if–then–else* Rule

The *<expression>* associated with the **if** (the rule predicate) is evaluted. If it is true, the **then** branch expression is evaluated; otherwise, the **else** branch expression is evaluated. These expressions are converted to early-operator, reverse polish notation and then transformed into a binary expression tree. As an example, in the rule, *if w is A and x is B then y is C else z is D* the predicate or premise is converted into the reverse polish notation form, *AND.w.A.is.x.B.is*

An expression in the early operator form of reverse polish notation has its elements organized in the form *<term><term><operator>*. We then convert the expression into a binary tree as illustrated schematically in Figure 9.59.

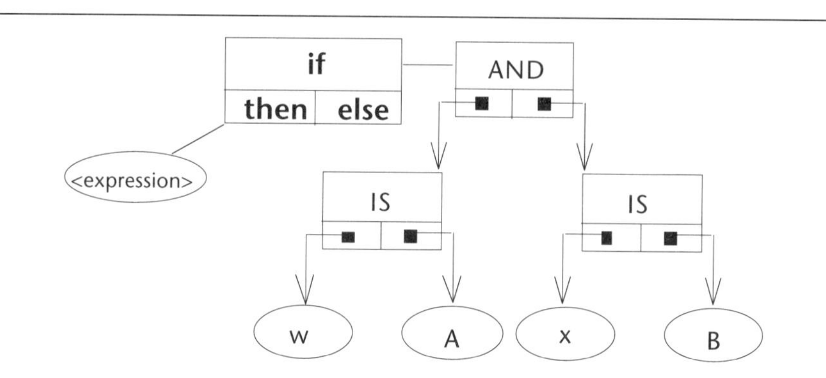

Figure 9.59 The Rule Premise as an Expression Tree

We can now resolve the expression by performing a post-order traversal of the binary tree, replacing each node with the value of the node's operation on the two leaf elements. For pure fuzzy rules, this means finding the membership of the left-hand element in the right-hand fuzzy set. Of course, some expressions imply the insertion of implicit operators. As an example, consider the rule premise expression *if A is near 2*MfgCosts*. The binary execution tree for this appears in Figure 9.60.

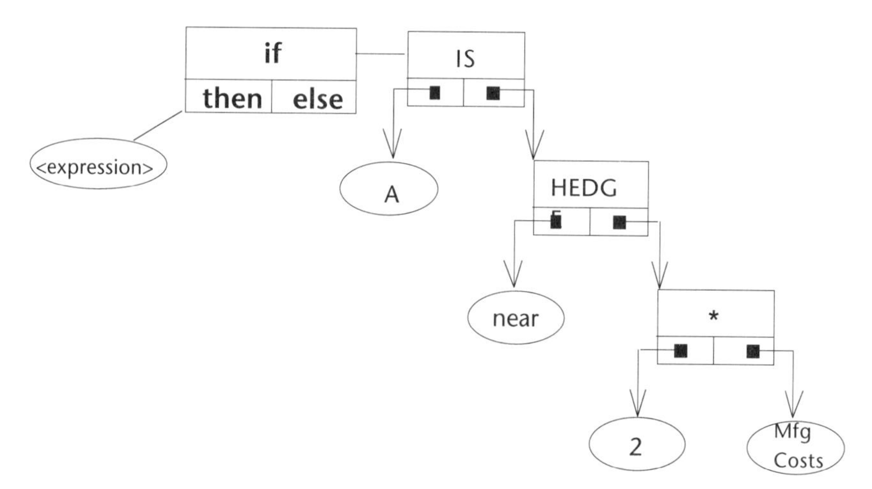

Figure 9.60 Expression Tree for Fuzzy Hedge Operations

> We must insert the *Hedge* operator into our expression in order to properly cast this predicate into both reverse polish and a proper expression tree. The hedge operator also generates a new fuzzy set, which must be entered into the system and then removed after the rule has been processed. Any good book on compiler design will provide the necessary background in converting expressions into reverse polish and then into an evaluation tree.

WRITE THE RULES

Writing the rules that describe how the system operates is the final compositional step in building a fuzzy model. When the rules have been written and compiled, you have specified a set of actions against the underlying term set. The rules are written in the form

```
if x is S then y is T;
```

where *x* and *y* are scalar expressions (or semi-fuzzy scalars[6] from the physical sensor systems), and *S* and *T* are linguistic variables in the model context. For systems that do not support hedges, the linguistic variable is exactly equivalent to a fuzzy set.

When you write rules, consider the following practices:

- Group together all rules that set the same solution variable.

```
if temp is cold and pressure is low
      then throttle is a BigPlus
if temp is cold and pressure is high
      then throttle is a MediumPlus
```

- Arrange the rules for easy readability.
- Use indentation to show the structure of the rule.
- Use a naming convention to identify different classes of performance variables. You might consider labeling control variables with "*Ctl*" or "*Inp*" (for input) and solution variables with "*Slt*" or "*Otp*" (for output). A naming convention can also be used to indicate the source of the variable, its use, and its dimensionality or cardinality.
- Use comments to describe the purpose of the rule, including any method and philosophical approaches. You should also use comments for a block of rules so its overall purpose is evident.

- Use white space around and between rules.
- If possible, use mixed-case rules—all variables are in uppercase and lowercase (the first character is uppercased), all fuzzy sets and hedges are in uppercase, and all language elements are in lowercase.

These practices encourage a structured way of encoding the model logic. At the same time, they will improve model maintainability, quality, and expandability. Since the rules support the functional mechanics of your fuzzy model, their clarity and comprehension is very important.

Write the Ordinary Conditional Rules

Ordinary conditional rules are the rules that are in the standard form, such as "if <fuzzy expression> then <fuzzy action>," and the degree to which the fuzzy action is taken depends on the degree of truth in the antecedent proposition. Each conditional rule examines the compatibility between a control variable group and one or more fuzzy sets. These comparisons can be defined through *AND* and *OR* operators (of various operator classes). The truth of the result is used to create a fuzzy region corresponding to one of the solution variables. Most fuzzy models, especially in process control applications, will consist predominantly of conditional rules.

Enter Any Unconditional Rules

An unconditional rule establishes a boundary constraint on the consequent fuzzy space. These rules function as limiters in the linear programming sense by constraining the permissible shape of the solution fuzzy set. Mixing conditional and unconditional rules in the same model segment, without understanding the dynamics of the control surface, can result in counter-intuitive results.

Select Compensatory Operators for Special Rules

The correct performance of some models depends on changing the basic min–min compositional rules specified by Zadeh for the *AND* and *OR* operators. Classes of alternative operators exist that will apply different compositional spaces to conjunctive and disjunctive fuzzy propositions. These operators include algebraic transformers such as the weighted average, the bounded sum, and the truth product. Other operators are special classes defined by such researchers as Yager and Zimmerman. As a matter of refining and describing the model, a practical approach to implementing rules with compensatory operators (when they comprise a small subset of the total rules) consists of deferring these rules until the standard rules have been written.

Review the Rule Set and Add Any Hedges

If the fuzzy modeling system supports surface transformers or hedges, this is the time to examine the rule set for opportunities to write hedged rules. These rules generally capture pathological behavior patterns (operations at the extreme edges of the control surface) and also use the base fuzzy vocabulary to create new rules.

```
if temperature is very hot
   then throttle is a very BigNegative;
```

In any event, hedges are important prototyping facilities since they allow the generation of nonlinear fuzzy regions that would be difficult to describe correctly through a coordinate system.

Add Any Alpha Cuts to Individual Rules

An alpha cut establishes the minimum truth threshold for a rule. If the truth of a premise evaluation falls below the alpha cut, the rule is not executed. As an example, if we specify

```
(Alpha=0.2) if temperature is cold
   then throttle is a BigPlus;
```

then if the truth of the fuzzy proposition *temperature is cold* is less than [0.2], this rule will not fire. Thus an alpha cut is an important control in managing the execution of fuzzy rules. Note that, from experience, you will find the alpha cut is more important in rules that perform unions and intersections on fuzzy sets, since these operations tend to produce "residual" fuzzy spaces. A *residual space* is one that has a very small but finite truth value (such as [.002] or [.09]). Remember that a fuzzy rule will fire if any truth (no matter how small) exists in the premise.

Enter the Rule Execution Weights

In most fuzzy models, you can weight the importance of rules by supplying a weight multiplier. All rules have a default weight of [1.0] (so the weight is invisible). As an example, if we specify

```
(Wght=0.8) if temperature is cold
   then throttle is a BigPlus;
```

then the truth of the premise proposition *temperature is cold* is multiplied by [.8], thus, in effect, reducing its force by 20%. Rule execution weights provide the model designer with a way of concentrating force in the rule set.

Define the Defuzzification Method for Each Solution Variable

This is a critical control factor in completing the design of a fuzzy model. Defuzzification selects the expected value of the solution variable from the consequent fuzzy region. How this is done will affect the predictive performance of your model. Unless you have reason to suspect that your model will behave better under other defuzzification techniques, the composite moments (centroid) method appears to provide a consistent and well-balanced approach. The centroid method is sensitive to the height and width of the total fuzzy region. It is also well suited to handling solution regions that are maintained as sparse singletons. Expected values from the centroid defuzzification also tend to move smoothly from one observation to the next (unlike composite maximum or the average heights method that tends to jump erratically on widely noncontiguous and nonmonotonic input values).

Notes

1. For those unaccustomed to dynamic system modeling, a control parameter is a variable in the model state whose value either determines the next state of the model or is established by one or more (and presumably final) states of the model. In a vehicle insurance underwriting model, as an example, control variables might include the applicant's age, driving record, and residential location. Another control variable, *risk*, is the final model solution variable—but it is also a control variable in the broadest sense.

2. This is a fundamental property of fuzzy set theory. Since the complement of a set [A] is defined as 1-μ[x], and the membership function gradient is a continuous vector between [0,1], then the Boolean complement conditions are only obeyed at the set extremes (where the truth values are ordinarily Boolean, zero, and one). Thus the intersection of [A] and ~[A] is some real number representing the difference between the set characteristics at that point in the fuzzy surface. We can see that the intersection operation on a fuzzy set provides a "generally true" state, on the average, by observing that μ[x]=1-μ[x]. Solving for μ[x] yields μ[x]=1/2.

3. Note that "sensor reading" means any input into a fuzzy system. For physical system controllers this is the standard device electrical, optical, or mechanical sensor; for information systems a "sensor" is any I/O port (such as a database gateway, a spreadsheet connection, or a desktop window).

4. The population of fuzzy sets in a model is also called the *term set.*

5. If we have a mathematical model, why would we want to create a fuzzy control system? The answer lies in two directions: flexibility and extensibility. The ordinary differential equations and partial differential equations (for time-phased models) are generally difficult to engineer, sometimes unpredictable outside the narrow domains of their applicability, and for systems with *k*-degrees of interlocking freedom, the *k*-order equations cannot be solved or implemented at the process-chip level. Thus fuzzy models are much more flexible since we can change the control surface heuristically, and they are generally more extensible since we need not develop a deep mathematical model for the new problem space.

6. A semi-fuzzy expression is a scalar that has been fuzzified for input.

Ten

Using the Fuzzy Code Libraries

This chapter provides an overview of the fuzzy modeling system and a description of the major fuzzy modeling library software. We also examine the various structures used by the modeling software. The 16- and 32-bit dynamic link libraries for Windows 3.1 and Windows 95 are introduced, and we examine how they are used in the Microsoft Visual Basic environment. Each of the major fuzzy functions is described. However, the tool and utility software accompanying the book (such as the work entokeners and system file interfaces) is not described in detail.

The fuzzy logic service routines are the building blocks for all the fuzzy modeling facilities. These C++ functions handle the actual management of fuzzy set and hedge operations. You can use these routines for various tasks to construct your own fuzzy reasoning system, to build and deliver sophisticated approximate reasoning models, and to enhance the operation of existing information technology systems by the incorporation of fuzzy logic. These functions will tightly integrate a wide spectrum of fuzzy logic services into your own applications by managing fuzzy sets, hedges, and system variables.

The Code and Interface Libraries

Each library is stored in its own directory. There are three C/C++ code libraries with source code, two Visual Basic interface libraries with module definitions, and a directory of demonstration programs and all the programs used in this book. There is also a directory containing the code examples associated with Bill Comb's rapid inferencing method. The following libraries are included with the book:

met16dos The 16-bit DOS C/C++ fuzzy logic code. This is the code library used throughout the book to illustrate the various fuzzy models. The code has been compiled with the Microsoft Visual C++ compiler (version 1.52) to generate a series of DOS executables (*.exe*).

met16dem The 16-bit DOS C/C++ demonstration and book example programs. These programs have been compiled and tested under the 16-bit Microsoft Visual C++ compiler (version 1.52).

met16win The 16-bit Windows 3.1 fuzzy logic C/C++ code. This library has been compiled with the Microsoft Visual C++ compiler (version 1.52) and generates a dynamic link library (DLL) containing all the fuzzy logic functions. It can be used with the Visual Basic 3.0 interface (see *met16vb3* below).

met32win The 32-bit Windows 95 fuzzy logic C/C++ code. This library has been compiled with the Microsoft Visual C++ compiler (version 4.0) and generates a dynamic link library (DLL) containing all the fuzzy logic functions. It can be used with the Visual Basic 5.0 interface (see *met32vb5* below).

met16vb3 The 16-bit Visual Basic module definitions for the 16-bit dynamic link library (DLL) generated by the *met16win* project file. With this module included in your Visual Basic project, you can access any of the fuzzy logic functions through a standard CALL statement. Visual Basic 3.0 examples are also included, illustrating how the dynamic link libraries are connected to your application and how the fuzzy logic functions are called.

met32vb5 The 32-bit Visual Basic module definitions for the 32-bit dynamic link library (DLL) generated by the *met32win* project file. As with the *met16vb3* definitions, including this module in your Visual Basic 5.0 project provides access to all the fuzzy logic functions through a standard CALL statement. Visual Basic 5.0 examples are also included, illustrating how the dynamic link libraries are connected to your application and how the fuzzy logic functions are called.

combsrim The modules, Excel spreadsheets, and examples for the Comb's rapid inferencing method. This newly discovered technique reconfigures a fuzzy system so that the exponential explosion of rules due to the dimensionality of a problem is reduced to a near linear expansion.

Each library contains a *read.me* file containing technical details on how the code is organized, how to incorporate the code into your application framework, and other important notices that are specific to the library.

General Software Issues

All the library code shares a common design and implementation philosophy. The philosophy is reflected in the software architecture and encompasses a set of code-specific functions as well as client-interface–specific functions. The following topics are of general interest to anyone concerned with compiling and using the fuzzy logic code libraries.

System and Client Error Diagnostics

The system error diagnostic facility is tightly integrated with the software. When a significant error—usually a fatal error—occurs, the library routines call a special error diagnostic function and return a nonzero status code (see the next section). Standard error diagnostics are generated by the *MtsSendError* function. This function selects the proper error message from a file, formats it in a standard way, and displays the message on your terminal. The error message is also written on the system activity log (*syslog.fil*).

The system error diagnostic file is created in a two-step process:

1. The error diagnostic message, along with its class, is added to the *fmssys.sel* (source error listing) file using any conventional editor. The line number of the message in this file is the diagnostic number associated with the message used by *MtsSendError* to select a particular message.
2. The *generrdf.exe* (generate error diagnostic file) program is executed. This reads the source file, serially numbers each diagnostic message within its class, and writes the *fmssys.edf* file used by the actual error diagnostic function.

Figure 10.1 shows how the error diagnostic file is produced from the source file. When an error condition is raised in the code, a call to *MtsSendError* is made indicating the line number (in *fmssys.sel*) of the required message.

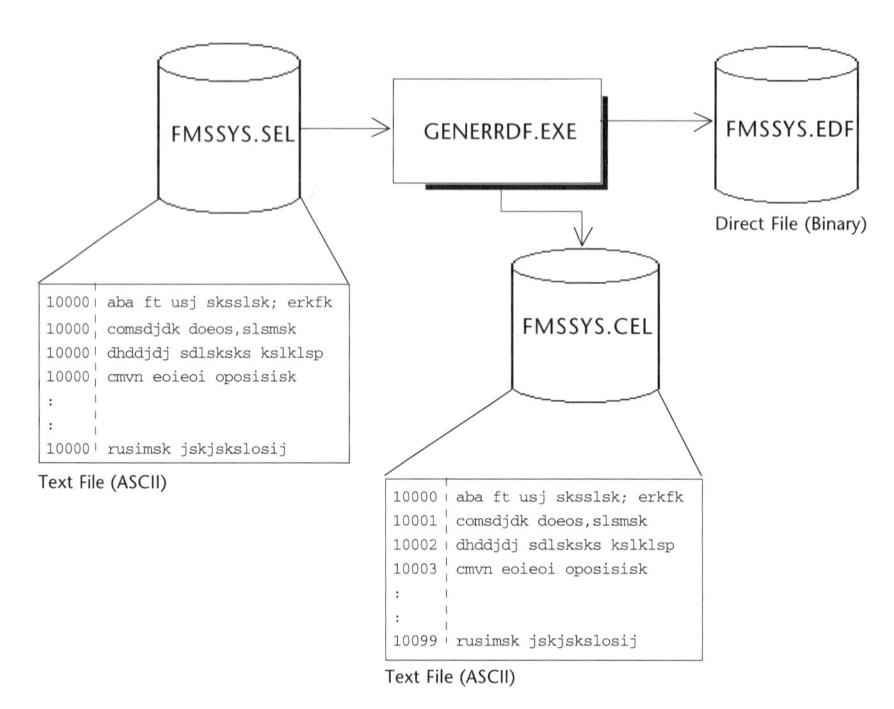

Figure 10.1 Generating the System Error Diagnostic File

If *XSYSctl.XSYSerrmsgs* is turned off ('0'), then error messages are not displayed at the terminal. Messages are always written to the system activity log.

The error diagnostic provides information about the exact software module that generated the error. Along with the error message, the system displays the identity of the software module, and a unique diagnostic number that includes the error class and the sequence number within the class. For example:

```
ERROR--(mtfzdfz.18.004) Unrecognized Defuzzification: 41
```

The ERROR prefix identifies this message as an error. This is followed by *mtfzdfz*, the actual physical file name of the module. The module name is followed by a two-part number. The first part (18) is the class—in this case, eighteen is fuzzy logic. The second part (004) is the sequence number within the class. The next element is an error message, which is, optionally, followed by the offending item.

An error diagnostic message generally means the interruption of whatever module is currently executing. This interrupt can be propagated upward so that the entire application is affected.

Software Status Codes

The fuzzy modeling software design incorporates a more traditional type of error status detecting than is found in many C/C++ programs. The standard *errno* external variable is not used, since this variable is also set by C/C++ service routines and is not reinitialized on subsequent module calls. This provides an unreliable method of communicating actual errors back to the calling program. The software also avoids the ambiguous, although more traditional, method of returning an error indicator as the function value. Combining actual values with error values in a program leads to increased program complexity and means that the programmer must constantly be aware of the signal used to indicate an error status. Instead, all library routines carry, as their last parameter, a pointer to an integer status variable. As an example, consider the function that creates a fuzzy set:

```
FDB *FzyCreateSet(char *SetId,
     int SetType,double Domain[],double Parms[],
       int ParmCnt,int *statusPtr);
```

When this function is called, the status parameter (**statusPtr*) is automatically set to zero. If an error occurs in the description of the fuzzy set or if insufficient memory exists to allocate the fuzzy set, then the software will issue an error diagnostic message and set the status parameter (**statusPtr*) to a nonzero number. The value of the status code indicates which error actually occurred.

Information and Warning Messages

In addition to error diagnostics, the fuzzy modeling system also issues formal explanatory and warning messages during its processing. Each message begins with the prefix "FMS" and contains an environment indicator, a unique message number, and a message level indicator. For example:

```
FMSsys0002I--Available RAM=620K
```

This message is organized as follows: the prefix "FMS" indicates that it is a formal system message; the "sys" indicates that it is issued by the system, "0002" is the message number, and "I" means that the message is intended for information only. Information messages are the most common and provide details about the current environment, confirm default option settings, and so forth. These messages are for your information only and generally have no effect on software operation.

Warning messages—those with a message level indicator of "W"—alert you to possible problems in the way the system is working. A warning indicates some problem that is not necessarily fatal to the operation of the software. For example, a drive is not ready:

```
MTSfil0036W--Device is not Ready: "B:"
```

In this case, the system either defaults to the *C:* drive or prompts for the correct device address. Conditions that are more extreme than a warning are handled by the error diagnostic processor.

Using Dynamic Link Library (DLL) Files

The CD-ROM contains Microsoft Visual C++ projects that create both 16-bit and 32-bit dynamic link libraries (DLLs) for use with rapid application development environments in both Windows 3.1 and Windows 95. A DLL exports the name and code for each of the fuzzy logic and utility functions. You can then use these entry points in any software system that supports a DLL linkage. As Figure 10.2 illustrates, you can integrate fuzzy logic operations into a Visual Basic application simply by including the module definitions.

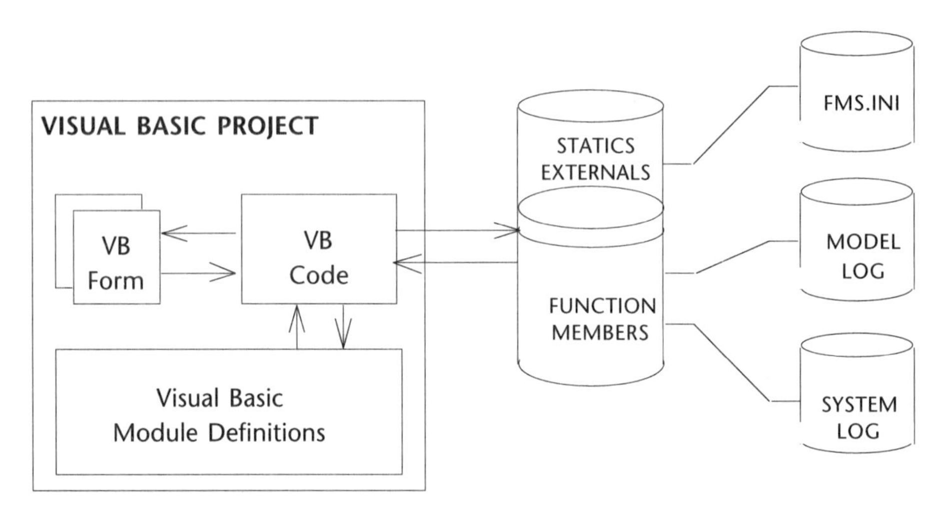

Figure 10.2 The DLL Interface to Microsoft Visual Basic

The dynamic link libraries are created in the project directories, or in sub-directories depending on your compiler and which version of Visual C++ you are using. These DLL files must be placed in the currently active path before they can be found.

THE VISUAL BASIC MODULE DEFINITIONS

The module definition file in Visual Basic contains program constants (corresponding to the C/C++ symbolics), the variable and array definitions used by local software services, and all the dynamic link library member definitions. This last component, the member definitions, is the crucial link between the DLL and the application environment. You can see this in the following fragment from the fuzzy logic module; Listing 10.1 is the set of constants used to create fuzzy sets with the *FzyCreateSet* program.

```
Global Const gcEMPTYSET = 0
Global Const gcINCREASE = 1
Global Const gcDECREASE = 2
Global Const gcGROWTH = 3
Global Const gcDECLINE = 4
Global Const gcPI = 5
Global Const gcBETA = 6
Global Const gcGAUSS = 7
Global Const gcTRIANGLE = 8
Global Const gcLEFTSHDR = 9
Global Const gcRIGHTSHDR = 10
Global Const gcTRAPEZOID = 11
Global Const gcTRANSFORM = 12
Global Const gcCOORDINATES = 13
Global Const gcMEMSERIES = 14
Global Const gcTRUTH = 15
```

Listing 10.1 Constants in the Fuzzy Visual Basic Module File

Although you can rename this constants and add your own, you cannot change their value without a corresponding change in the C/C++ source code. A module file also contains global variable definitions used with the DLL entry names. Listing 10.2 shows a small segment of these variable definitions.

```
'
Global gdParms(1 To 8)          As Double
Global giParmCnt                As Integer
Global giCurvetype              As Integer
Global gdDomain(1 To 2)         As Double
Global gsCurveNames(0 To 28)    As String
Global gsHedgeNames(1 To 28)    As String
Global gsHedgeTypes(1 To 28)    As String
Global giHedgeModes(1 To 28)    As Integer
```

```
Global gdValues(1 To 128)        As Double
Global gfGrades(1 To 128)        As Single
Global giValCnt                  As Long
Global gsParmString              As String
```

Listing 10.2 Global Variables in the Fuzzy VB Module File

The core of the module is the definition of each DLL member. You cannot use a DLL function in your Visual Basic program until it has been defined in the module file. The definition, as shown in Listing 10.3 contains the name type (subroutine or function), the name and location of the DLL physical file, and the calling parameter definitions.

```
Declare Sub SysOpenFuzzy Lib "met32fz.dll" (Status As Integer)
Declare Function FzyAboveAlfa Lib "met32fz.dll" (ByVal TruthValue
    As Single, ByVal Alfacut As Single, ByVal AlfaType As Inte-
    ger) As Integer
Declare Function FzyAddFZYblk Lib "met32fz.dll" (ByVal VDB As
    Long, FDB As Long, FSV As Long, Status As Integer) As Long
Declare Function FzyAND Lib "met32fz.dll" (ByVal TruthValue As
    Single, ByVal TruthValue As Single) As Single
Declare Sub FzyApplyAlfa Lib "met32fz.dll" (ByVal FDB As Long,
    ByVal TruthValue As Single, ByVal AlfaType As Integer)
```

Listing 10.3 DLL Member Name Definitions in the Fuzzy Visual Basic Module File

The specifications for each parameter must be done carefully (if you make any changes to the C/C++ source code) and there are some significant differences in the way parameters are handled in the 16- and 32-bit versions of Visual Basic (most importantly in the way default integer values are handled).

USING THE DLL NAMES IN VISUAL BASIC

When the DLL names have been defined in your Visual Basic project, you can use them through a conventional CALL (as long as you respect the proper parameter data specifications). As an example, the following code is drawn from the demonstration fuzzy set editor. Calls to DLL members are indicated in **bold**. It combines Visual Basic and C/C++ code into an integrated framework for creating various kinds of fuzzy sets. Listing is used to create an empty fuzzy set.

```
'------------------------------------------------------------
'--Now we create an empty fuzzy set as a working area to be used
'--by the editor. Each current fuzzy set populates this working
'--set. This will alsoupdate the graphic image of the set in the
'--editor window
'------------------------------------------------------------
'
    dDomain(0) = 1
```

```
        dDomain(1) = 100
        gsCurrFuzzyset = "Empty"
        lFuzzyset = FzyCreateSet(_
          gsCurrFuzzyset, gcEMPTYSET, dDomain(0), dParms(0), 0, iSta-
    tus)
        gsCurrFuzzyset = ""
```

Listing 10.4 Creating an Empty Fuzzy Set in Visual Basic

The pointer (or handle) to the fuzzy set's *FDB* is returned in the Visual Basic variable *lFuzzyset*. Although Visual Basic cannot process this structure directly, it can pass the address of the fuzzy set descriptor block (FDB) to any of the other DLL members. Listing 10.5 shows part of the case structure used to create fuzzy sets of various shapes.

```
gsCurrFuzzyset = FEnames.Text
Call DDsetidtoFDB(gsCurrFuzzyset, lFuzzyset, iStatus)
iGenType = FEgenerators.ListIndex
'
'-----------------------------------------------------------------------
'--Get the high and low domain values off the screen. Make these the universe
'--of discourse (full domain) for the fuzzy set. Also put these values into the
'--local domain() array.
'-----------------------------------------------------------------------
'
dLo = Val(FElodom.Text)
dHi = Val(FEhidom.Text)
Call DDdomaintoFDB(dLo, dHi, lFuzzyset, iStatus)
dDomain(0) = dLo
dDomain(1) = dHi

Select Case iGenType
    Case gcEMPTYSET
        Call FzyEmptySet(lFuzzyset)
    Case gcINCREASE, gcDECREASE
        If giParmCnt > 0 Then
            dLo = gdParms(1)
            dHi = gdParms(2)
        End If
        Call FzyLinearCurve(lFuzzyset, dLo, dHi, iGenType, iStatus)
    Case gcGROWTH, gcDECLINE
        dFzyFlexPt = dLo + ((dHi - dLo) / 2)
        If giParmCnt > 0 Then
            dLo = gdParms(1)
            dFzyFlexPt = gdParms(2)
            dHi = gdParms(3)
        End If
        Call FzySCurve(lFuzzyset, dLo, dFzyFlexPt, dHi, iGenType, iStatus)
```

```
Case gcPI
    dFzyWidthPt = (dHi - dLo) / 2
    dFzyCenterPt = dLo + dFzyWidthPt
    If giParmCnt > 0 Then
        dFzyCenterPt = gdParms(1)
        dFzyWidthPt = gdParms(2)
    End If
    Call FzyPiCurve(lFuzzyset, dFzyCenterPt, dFzyWidthPt, iStatus)
```

Listing 10.5 Creating Various Fuzzy Sets in Visual Basic

The DLL members encapsulate the fuzzy logic library in a way such that the Visual Basic environment has complete access to all its functional capabilities. However, the library works by allocating and manipulating pointers to fuzzy sets, hedges, and other related structures. These pointers can't be handled directly by Visual Basic but are simply stored in local variables and then passed to the DLL routines when required. Pointers are always stored in long (32-bit) integers in both the 16- and 32-bit versions of Visual Basic. In the previous example, *lFuzzyset* is a local Visual Basic variable containing the address of the FDB structure created by the *FzyCreateSet* function. It is passed to other functions, such as *FzyPicurve*, where the contents of the fuzzy membership array are modified.

Modeling and Utility Software

The fuzzy code library is arranged in three sections. The first collection handles all the fuzzy logic functions—creating and using fuzzy sets and hedges. The second collection contains the general fuzzy modeling functions—programs for connecting to the fuzzy control blocks, for creating policies, and for using policy dictionaries. The third collection of functions contains miscellaneous tools and utilities used throughout the system by the fuzzy logic and modeling code.

Symbolic Constants, Global Data, and Prototypes

These *include* files contain the symbolic constants, static data elements, tables, and function prototypes used throughout the modeling system (see Table 10.1). The *mtsptype.hpp* file contains every function prototype as well as the name of the physical file where the function is located (*FzyApplyHedge* is located in *mtfzaph.cpp*, as an example).

Data Structures

The structure definitions for fuzzy sets, hedges, formal variables, compatibility measurement, and the system and fuzzy reasoning control blocks are maintained in a set of *include* files (see Table 10.2).

Include File	Use
fuzzy.hpp	Constants and types for fuzzy functions
mdltypes.hpp	Tables and static variables for general modeling
mtsptype.hpp	Prototype definitions for all functions
mtypes.hpp	The general pool of symbolics and constants

Table 10.1 Include Files for Constants and Prototypes

Include File	Use
ASSERT.hpp	The code-level assertion function
CIX.hpp	The statistical compatibility index structure
FDB.hpp	Fuzzy set descriptor block
FSV.hpp	Fuzzy solution variable structure
HDB.hpp	Hedge descriptor block
MDB.hpp	Model descriptor block
NDB.hpp	Noise word descriptor block
RXB.hpp	Rule explanatory block
PDB.hpp	Policy descriptor block
SSYSCTL.hpp	Top-level structure for *XSYSctl* structure
TOKENS.hpp	Token structure (used by entokening routines)
VDB.hpp	Variable descriptor block
XFZYCTL.hpp	The fuzzy reasoning work area (*XFZYctl*)
XSYSCTL.hpp	The system control block (*XSYSctl*)

Table 10.2 Data Structures and Control Blocks

Fuzzy Logic Functions

Table 10.3 shows the basic service routines supporting all the fuzzy logic and approximate reasoning functions. All these functions begin with the *Fzy* prefix and are located in files with physical names starting with *mtfz*.

The Fuzzy System Modeling Functions

The general modeling routines support the development of fuzzy systems using the policy concept. These routines manipulate model objects such as fuzzy sets and hedges in the policy dictionaries (see Tables 10.3 and 10.4).

Function	Use
FzyAboveAlfa	Checks whether truth value is above alpha cut
FzyAddFZYctl	Adds a variable to the fuzzy logic control block
FzyAND	Performs the standard Zadeh AND
FzyApplyAlfa	Applies an alpha cut to a fuzzy set
FzyApplyAND	Applies an AND operation to two fuzzy sets
FzyApplyHedge	Applies a hedge to a fuzzy set
FzyApplyNOT	Applies a Zadeh or compensatory `NOT` operator
FzyApplyOR	Applies an OR operation to two fuzzy sets
FzyAutoScale	Autoscaling factor for fuzzy sets
FzyBetaCurve	Creates a beta curve fuzzy set
FzyCloseFZYctl	Closes the fuzzy working area and releases memory
FzyCompAND	Applies a compensatory AND class
FzyCompOR	Applies a compensatory OR class
FzyCondProposition	Implements a conditional fuzzy proposition
FzyCoordSeries	Creates fuzzy set from a value membership array
FzyCopySet	Copies one fuzzy set into another fuzzy set
FzyCopyVector	Copies the truth membership vector of a fuzzy set
FzyCorrAverage	Performs a correlation average on a fuzzy set
FzyCorrMinimum	Performs a correlation minimum on a fuzzy set
FzyCorrProduct	Scales a fuzzy set by correlation product method
FzyCreateHedge	Creates a user-defined hedge
FzyCreateSet	Creates a fuzzy set
FzyDefuzzify	Resolves expected value of a fuzzy solution set
FzyDeleteSet	Removes a fuzzy set from a policy
FzyDisplayFSV	Displays the fuzzy logic control block
FzyDisplayMemVector	Displays contents of fuzzy set membership vector
FzyDrawSet	Draws a fuzzy set on the specified system file
FzyEquivalentScalar	Finds the scalar at a specified membership grade
FzyExamineSet	Generates *MxN* matrix of fuzzy set density
FzyFindFSV	Locates a variable's *FSV* in the fuzzy work area
FzyFindPlateau	Locates the maximum plateau region in a fuzzy set
FzyFuzzyScalar	Converts a scalar into a fuzzy set
FzyGaussianCurve	Creates a bell-shaped curve from Gaussian points
FzyGetCoordinates	Parses coordinate pairs from a text buffer
FzyGetHeight	Finds the maximum truth value for a fuzzy region
FzyGetHighPoint	Finds the height of set and returns vector cell index
FzyGetMembership	Returns the truth membership for a particular scalar
FzyGetScalar	Returns scalar associated with fuzzy truth value

Table 10.3 The Fuzzy Logic Functions

Function	Use
FzyImplMatrix	Draws implication space between two fuzzy regions
FzyInitCIX	Initializes the compatibility index statistics structure
FzyInitFZYctl	Initializes the fuzzy logic control block space
FzyInitHDB	Initializes a hedge descriptor block
FzyInitFDB	Initializes a fuzzy set
FzyInitVector	Initializes a fuzzy set truth membership vector
FzyInsertHedges	Inserts built-in system hedges into an HDB array
FzyInterpVec	Performs piecewise linear interpolation on a fuzzy set
FzyIsMemberof	Indicates whether or not number is in a fuzzy set
FzyIsNormal	Determines if a fuzzy set is normalized
FzyIsNormalType	Indicates if a fuzzy set is in specified normal form
FzyLinearCurve	Creates a linear increase/decrease fuzzy set
FzyMemSeries	Creates a fuzzy set from a truth membership series
FzyMonotonicLogic	Implements the monotonic implication strategy
FzyNormalizeSet	Normalizes a fuzzy set
FzyOR	Performs the standard Zadeh OR
FzyPiCurve	Creates a bell-shaped or PI curve fuzzy set
FzyPlateauCnt	Counts plateaus in a fuzzy set
FzyPlotSets	Plots a series of fuzzy sets across common domain
FzyPLotVar	Plots fuzzy sets associated with a model variable
FzySCurve	Creates a left- or right-facing S-curve fuzzy set
FzyShoulderedCurve	Creates a shouldered trapezoidal fuzzy set
FzyStatCompIndex	Calculates statistical compatibility index measures
FzySupportSet	Finds the strong "support set" of a fuzzy region
FzyTriangleCurve	Creates a triangular-shaped fuzzy set
FzyTrueSet	Generates a linear, diagonal *true* fuzzy set
FzyUnCondProposition	Implements an unconditional fuzzy proposition
FzyUnitCompIndex	Calculates the nonstatistical compatibility index
FzyWgtdBetaCurve	Creates a weighted beta curve
FzyXPSacctext	Associates textual description with rule number
FzyXPSaddRXC	Adds a new clause explanatory node to rule's list
FzyXPSandtext	Associates textual description with AND operator
FzyXPSconclusion	Displays model conclusion in textual format
FzyXPScreateRXB	Creates the top explanatory node for a rule
FzyXPScreateRXC	Creates a clause (proposition) node for a rule
FzyXPSdegtext	Associates textual description with degree of strength
FzyXPSexplain	Provides full textual explanation for a rule
FzyXPSInitRXB	Initializes the RXB structure
FzyXPSInitRXC	Initializes the RXC structure

Table 10.3 The Fuzzy Logic Functions (Continued)

Function	Use
FzyXPSinset	Calculates rule display inset based on premise truth
FzyXPSmemtext	Associates textual description with set membership
FzyXPSpostext	Associates textual description with scalar value
FzyXPSprftext	Associates textual description with proposition
FzyXPSshowrules	Displays all the rules
FzyXPStentext	Associates textual description with *then* option

Table 10.3 The Fuzzy Logic Functions (Continued)

Function	Use
MdlAddNoiseWord	Adds a noise word to the policy dictionary
MdlClosePolicy	Removes a policy and its contents from memory
MdlConnecttoFMS	Connects to the fuzzy modeling system
MdlCreateModel	Creates a model descriptor block
MdlCreatePolicy	Creates a policy descriptor block
MdlFindFDB	Returns a pointer to a fuzzy set in a policy
MdlFindHDB	Returns a pointer to a hedge descriptor in a policy
MdlFindNDB	Returns a pointer to a noise word in a policy
MdlFindPDB	Returns a pointer to a policy in the *XSYSctl* block
MdlFindVDB	Returns a pointer to a variable in a policy
MdlInitMDB	Initializes a model descriptor
MdlInitPDB	Initializes a policy descriptor
MdlInsertHedges	Inserts the built-in hedges into a policy
MdlInsertNWords	Inserts the built-in noise words into a policy
MdlLinkFDB	Adds a fuzzy set to a policy
MdlLInkHDB	Adds a hedge descriptor to a policy
MdlLinkMDB	Attaches a model to the *XSYSctl* block
MdlLinkVDB	Adds a variable to a policy
MdlRemoveFDB	Deletes a fuzzy set from a policy
MdlRemoveHDB	Deletes a hedge descriptor from a policy
MdlRemoveNDB	Deletes a noise word from a policy
MdlRemoveVDB	Deletes a variable from a policy

Table 10.4 The Fuzzy System Modeling Functions

The solution variable in a fuzzy model must be defined through a formal variable definition—a variable descriptor block (VDB). These functions create and initialize the VDB structure for use with the approximate reasoning facilities (Table 10.5).

Function	Use
VarInitVDB	Initializes a formal variable descriptor
VarCreateScalar	Creates a scalar for use in fuzzy reasoning

Table 10.5 Variable Management Functions

MISCELLANEOUS TOOLS AND UTILITIES

In addition to the fuzzy logic functions and the model-building functions, the code library also contains a number of programs that support these functions. The following utilities and tools of Table 10.6 are found in the library. Some of these, such as those that handle the error diagnostic processor, are used widely throughout the system.

Function	Use
MtsAskforBOOL	Prompts for a Boolean value
MtsAskforDBL	Prompts for a double precision value
MtsAskforINT	Prompts for a signed long integer value
MtsAskforSTR	Prompts for a text string
MtsAskforVAL	Prompts for a string that is member of a table
MtsCenterString	Centers a string inside a specified field length
MtsEntoken	Breaks string content down into individual words
MtsFactorial	Computes *n!* for reasonable value of *n*
MtsFillString	Fills blank and tabs in string with indicated byte
MtsFindsubstr	Returns offset position of substring
MtsFormatDate	Formats a date in `yyyymmdd` in several ways
MtsFormatDbl	Formats a double value into a string
MtsFormatFlt	Formats a float value into a string
MtsFormatLong	Formats a signed long value into a string
MtsFormatInt	Formats a signed int value into a string
MtsFormatTime	Formats system time into several display formats
MtsGetDate	Retrieves current system date
MtsGetErrorMsg	Returns text of error message for error number
MtsGetDate	Retrieves current system date
MtsGetErrorMsg	Returns text of error message for error number
MtsGetsubstr	Extracts a substring
MtsGetsystemFile	Returns file pointer to one of the system log files
MtsGetTime	Retrieves the current system time
MtsHashString	Applies hashing function to string contents
MtsIsBoolean	Indicates whether string contains a Boolean value
MtsIsNumeric	Indicates whether string contains a number

Table 10.6 Tools and Utility Service Functions

Function	Use
MtsIsPrime	Indicates whether a number is a prime
MtsLeftJustify	Left justifies a string in a specified field width
MtsLogFileName	Returns identifier of file associated with system log
Mtspause	Initiates a pause until key is pressed
MtsPutsubstr	Inserts a string into another string
Mtsprimes	Generates a table of prime numbers
MtsRandomNumber	Produces a long series of pseudo-random numbers
MtsReduceLine	Justifies and removes multiple white spaces from line
MtsSearchTable	Performs a serial search on a table
MtsSendError	Writes an error diagnostic to terminal and log file
MtsSendClientError	Writes an error that is not part of the modeling code
MtsSendWarning	Writes a warning message of different levels
MtsStrIndex	Finds offset of one string in another
MtsWritetoLog	Writes a string to a system file

Table 10.6 Tools and Utility Service Functions

Demonstration and Fuzzy Model Programs

In addition to the fuzzy logic, modeling, and utilities code, this book also contains a large number of demonstration programs that explore the use of hedges and alternate operators. The source code, inventory backorder, new product pricing, project risk assessment, and weight estimation are also included. Table 10.7 lists these support programs.

Function	Use
alfacut.cpp	Effects of alpha threshold cuts on fuzzy set shape
altands.cpp	Truth tables from alternate AND operators
altors.cpp	Truth tables from alternate OR operators
cosnnot.cpp	Truth tables for cosine OR operator
dilhedge.cpp	Effects of dilution hedge on fuzzy sets
fshinv1.cpp	Inventory backorder model based on sales orders
fshinv2.cpp	Inventory backorders using time-series changes
fshinv3.cpp	Single rule inventory model showing undecidability
fshinv4.cpp	Inventory model showing saturation of fuzzy region
fshinv5.cpp	Inventory model with backorders scaled by CIX
fshpra1.cpp	The project risk assessment model
fshprc1.cpp	Standard new product pricing model

Table 10.7 Stand-Alone Programs Containing Demonstrations and Fuzzy Models

Function	Use
fshprc2.cpp	New product pricing with explanatory facility
fshprc3.cpp	New product pricing model with hedged price sets
fshprc4.cpp	New product pricing model with unconditionals last
fshprc5.cpp	Profit and loss (P&L) using fuzzy pricing model
fshprc6.cpp	Pricing model with spoilage and distribution costs
fshprc7.cpp	Pricing model with defuzzification comparisons
fzycoord.cpp	Shows fuzzy sets created with coordinate points
fzyird.cpp	Temperature fuzzy sets for infrared detector system
fzyprfv.cpp	Profitability fuzzy sets from corporate asset model
fzytrbn.cpp	Fuzzy sets for steam turbine controller model
inthedge.cpp	Effects of intensification hedge on fuzz sets
lihedge.cpp	Effects of hedges on linear fuzzy sets
pihedge.cpp	Effect of hedges on PI bell-shaped fuzzy set
propfst.cpp	Illustrates various proportional fuzzy sets
schedge.cpp	Effects of hedges on S-curve fuzzy sets
sugennot.cpp	Truth tables for the Sugeno NOT operator
sugennot.cpp	Truth tables for the Sugeno NOT operator
thldnot.cpp	Truth tables for the threshold NOT operator
weight1	The basic monotonic logic weight estimator model
weight2	Multiple parameter weight estimator model
yagerand.cpp	Truth tables for the Yager AND operator
yagernot.cpp	Truth tables for the Yager NOT operator
yageror.cpp	Truth tables for the Yager OR operator
zadehand.cpp	Truth tables for the standard Zadeh AND operator
zadehor.cpp	Truth tables for the standard Zadeh OR operators

Table 10.7 Stand-Alone Programs Containing Demonstrations and Fuzzy Models (Continued)

Description of Fuzzy Logic Functions

The following is a summary description of the major fuzzy logic functions (a few of the lesser-used or redundant functions have been omitted, but are included in the source library). Each function description includes the calling format, the parameter specifications, error conditions, a general section about how the function is used, any notes about usage, the names of related functions, the name of the function's source file (also obtained through the *mtsptype.hpp* file), and, when appropriate, a brief example of how the function is used.

FzyAboveAlfa

Format:

`bool FzyAboveAlfa(float` *memval*, `float` *alfacut*, `int` *AlfaType*`);`

Parameters:

float `memval`	A truth membership value in the range [0,1]
float `alfacut`	The alpha-cut threshold, also a corresponding truth membership value in the range [0,1]
int `AlfaType`	The type of alpha-cut enforcement. This can be either `STRONG` or `WEAK`.

Returns:

bool	`TRUE` if the truth membership is above the alpha cut, otherwise `FALSE`.

Status/Errors:

0	(Boolean `FALSE`) returned if the AlfaType is unrecognized or if the `memval` is not within the correct range.

Use:

You use this function to determine whether or not a particular membership value is above the current alpha-cut threshold. The function provides a test for both the `STRONG` and the `WEAK` alpha-cut levels. Also see the description of the *FzyApplyAlfa* function.

Related Processes:

`FzyApplyAlfa`	Truncates a truth membership function at the indicated alpha-cut threshold. This function uses *FzyAboveAlfa* to determine which values should be set to zero.
`FzyAlfaCut`	Returns the alpha-cut threshold attached to a specified fuzzy set. This differs from the global set alpha cut associated with the *XSYSctl* and the *XFZYctl* blocks.

Example:

```
//--truncate at the alpha cut threshold
for(i=0;i<VECMAX;i++)
   if(!(FzyAboveAlfa(FDBptr->FDBvector[i],AlfaCut,WEAK))
      FDBptr->FDBvector[i]=0;
```

FzyAddFZYctl

Format:

```
bool FzyAddFZYctl(
   VDB *VDBptr,FDB **FDBptr,FSV **FSVptr,int *statusPtr);
```

Parameters:

VDB *VDBptr — A pointer to a variable descriptor block (VDB). You must allocate and initialize a standard variable in order to use the fuzzy logic and approximate reasoning facilities.

FDB **FDBptr — The address of the pointer to the fuzzy set de-scriptor block (FDB). This consequent or solution fuzzy set is created and initialized by the *FzyAddFZYctl* function, thus you need to pass a pointer to the fuzzy set pointer so that the actual location of the fuzzy set can be specified.

FSV **FSVptr — The address of the pointer to the fuzzy solution variable structure. The structure is allocated by the function and then inserted into the *XFZYctl* control block array. Its address is returned for your inspection (see the *FzyDisplayFSV* function) but should not otherwise be directly manipulated.

int *statusPtr — This is the address of an integer variable that will receive any error codes. If no error occurs, then its value on exiting from this function will be zero.

Returns:

bool — TRUE is the variable that was successfully added to the fuzzy logic control block, otherwise FAILED.

Status/Errors:

0 No error.

1 Too many solution variables in the model. The *XFZYctl* block can hold a maximum of thirty-two (32) simultaneous undergeneration variables.

3 Insufficient memory to allocate the FSV structure.

5 Unable to create the FSV fuzzy set. (Check with the system log file *[syslog.fil]* to find the actual error generated by *FzyCreateSet*).

Use:

Each solution variable in your fuzzy model must be stored in the *XFZYctl* control block. This function inserts a new variable into the control block and performs several housekeeping chores:

- Ensures that no duplicates are entered.
- Allocates storage for a new solution variable (FSV).
- Copies the default defuzzification method, geometry, implication technique, and alpha cut from the associated variable descriptor.
- Creates a new, empty fuzzy set with the same name as the solution variable to hold the results of the fuzzy implication process.
- Constructs the correct array space, when required, for a preponderance of evidence geometry.
- Attaches a compatibility index block (CIX) to the output for use in statistical compatibility measurement.

In developing and exploiting a fuzzy model at the low-level services platform, you must explicitly allocate and initialize a new variable descriptor block (VDB) for each solution variable and then store these in the *XFZYctl* control block. The fuzzy proposition processors—such as *FzyUnCondProposition*, *FzyCondProposition*, and *FzyMonotonicLogic*—all use the contents of this control region to manage and update the current fuzzy set space associated with the solution variable.

Notes:

1. In your main application program you must include the *SFZYctl.hpp* header file. This is the storage allocation version of the external control block used by the fuzzy reasoning facility to manage each of the output solution variables. This main program

must also call *FzyInitFZYctl* to establish addressability to the fuzzy reasoning facilities.

2. You must also use part of the general model building facilities in fuzzy reasoning. In particular, an output solution variable must be represented as a variable, that is, defined as a variable descriptor block (VDB). The solution variable must be a double precision floating-point scalar.

Related Processes:

`FzyInitFZYctl`	Initializes the *XFZYctl* control block.

Example:

```
FDB     *PriceFDBptr;
FSV     *FSVptr;
•
•
//--Now add the PRICE variable's VDB
//--to the fuzzy solution variable control block
if(!(FzyAddFZYctl(VDBptr,&PriceFDBptr,&FSVptr,statustr)))
   {
      *statusPtr=1;
      return(NoPrice);
   }
```

FzyAND

Format:

float `FzyAND (float` *truth1*,`float` *truth2*`);`

Parameters:

float `truth1`	A truth value from a membership function in the range [0,1]
float `truth2`	A truth value from a membership function in the range [0,1]

Returns:

float	A truth value within the range [0,1]

Status/Errors:

-1	The first truth value is not within bounds.
-2	The second truth value is not within bounds.

Use:

This function implements the standard Zadeh AND operator. The AND of two truth values is found by taking the arithmetic minimum of their values. The AND operation is used in the predicate fuzzy evaluation and is also used in the consequent space of a solution variable when unconditional fuzzy propositions are implemented.

Related Processes:

FzyCompAND	This form of the AND process implements a variety of compensatory operators. The ZADEH compensatory class is equivalent to this basic fuzzy intersection process.

Example:

```
//--AND two fuzzy sets into a temporary set

FzyGetTruthVec(HIGH,Fzytruth1);

FzyGetTruthVec(FAST,Fzytruth2);

FzyInitVec(WrkVector,0);

for(i=0;i<VECMAX;i++)

                    WrkVec-
                    tor[i]=FzyAND(Fzytruth1,Fzytruth2);
```

FzyApplyAlfa

Format:

```
void FzyApplyAlfa(
        const FDB *FDBptr,float Alfacut,int AlfaType,int *status);
```

Parameters:

FDB *FDBptr*	A pointer to the fuzzy set whose truth membership function will be modified by the alpha-cut process.
float *Alfacut*	A single precision floating-point number. The value is the alpha-cut threshold.
int *AlfaType*	The type of alpha cut. This can be either `STRONG` or `WEAK`.

Status/Errors:

-1	(Warning) The alfa-cut value is equal to one.
0	Alpha-cut was applied successfully.
1	The alpha-cut threshold value was less than zero.
3	The alpha-cut threshold value was greater than one.
5	The alfa-type parameter was not recognized.

Use:

A technical concept closely related to the support set is the alpha-level set or the "α–cut." An alpha-level is a threshold restriction on the domain based on the membership grade of each domain value. This set contains all the domain values that are part of the fuzzy set at a minimum membership value of α. There are two kinds of α-cuts, `STRONG` and `WEAK`. The strong α-cut is defined as,

$$\mu_A(x) > \alpha$$

and the weak alpha-cut is defined as,

$$\mu_A(x) \geq \alpha$$

But why should we care about α-cuts? There are two reasons. First, the strong α-cut at zero [0] defines the support set for a fuzzy set. You can see this easily by comparing the α-level set produced by this slicing function and the support set for heavy. Second, the alpha-level set describes a power or strength function that is used by fuzzy models to decide whether or not a truth value should be considered equivalent to zero [0]. This is an important facility that controls the execution of fuzzy rules as well as the intersection of multiple fuzzy sets.

Related Processes:

`FzyAboveAlfa`	This Boolean function indicates whether or not a membership function is above a specified `STRONG` or `WEAK` alpha-cut threshold.
`FzySupportSet`	This procedure returns the left and right edges of the nonzero portion of the fuzzy set. It also considers the current alpha-cut level in locating the support set.

Example:

```
float AlfaLevel=.40;
FzyApplyAlfa(HiFDBptr,AlfaLevel,STRONG,&status);
```

FzyApplyAND

Format:

```
FDB *FzyApplyAND(
   FDB *FDBptr1,FDB *FDBptr2,
      int ANDClass,double ANDCoeff,int *statusptr);
```

Parameters:

FDB **FDBptr1*	A pointer to the first fuzzy set
FDB **FDBptr2*	A pointer to the second fuzzy set
int *ANDClass*	The class of `AND` operation required. This can be any of the compensatory `AND` classes, including the standard Zadeh `AND`.
double *ANDCoeff*	Used for several of the compensatory `AND` classes. This is the class coefficient used with the specified `AND` compensatory class.
int **statusptr*	The status return code.

Returns:

FDB *	A pointer to a new fuzzy set that is allocated and initialized by the function. The truth membership function of this fuzzy set is the ANDed product of the two incoming fuzzy sets.

Status/Errors:

0	No error, new fuzzy set is created.
1	Insufficient memory to allocate the fuzzy set.
n	Other errors are produced from the FzyCompAND routine, which is called by this function.

Use:

This function performs an AND operation on a complete fuzzy set (the other low-level functions *FzyAND* and *FzyCompAND* act only on individual truth membership function values). When the AND is complete, a new fuzzy set is created. A pointer to this fuzzy set is returned. The new fuzzy set is initialized with the values from *FDBptr1*.

Related Processes:

FzyApplyAND	Performs an AND operation on two fuzzy sets.
FzyCompAND	Performs a compensatory AND operation on two truth membership values.
FzyAND	Performs a standard Zadeh (minimum) AND on two truth membership values.

Example:

```
//--Take the AND of High and Low
FDB     *HiLoFDBptr;
HiLoFDBptr=FzyApplyAND(
   HiFDBptr,LoFDBptr,ZADEHAND,0,&status);
if(status!=0) return;
```

FzyApplyHedge

Format:

```
void FzyApplyHedge(
   FDB *inFDBptr,HDB *hedgeid,FDB *outFDBptr, int *statusptr);
```

Parameters:

FDB *inFDBptr — A pointer to the fuzzy set that will be modified by the hedge operation.

HDB *hedgeid — A pointer to the hedge descriptor.

FDB *outFDBptr — A new fuzzy set that is created and returned by the hedging process. This pointer is automatically set to `NULL` and then set to the address of the new fuzzy set.

int *statusptr — The procedure completion status. A nonzero value indicates that the hedge operation was not performed.

Status/Errors:

0 — Hedge was successfully performed. *outFDBptr* now contains an exact copy of *inFDBptr* except that the truth membership function has been modified by the hedge.

1 — Unrecognized user hedging operation.

3 — Unrecognized built-in hedge.

Use:

A hedge modifies the shape of a fuzzy set's surface causing a change in the membership truth profile. Hedges play the same role in fuzzy production rules that adjectives and adverbs play in English sentences and are processed in a similar manner. The application of a hedge changes the behavior of the logic, in the same way that adjectives change the meaning and intention of an English sentence. Accordingly, the number and order of hedges are significant, so that the expressions,

```
not very high
very not high
```

are subject to two different interpretations—just as we would interpret them differently in English or other natural languages. Since a hedge is linguistic in nature, multiple hedges can be applied to a single fuzzy region in a manner that corresponds to a restriction and a fine definition of the region's semantic characteristics. As an example,

```
very very high
positively not very high
generally around the median costs
```

are hedged fuzzy expressions. Because the hedge and its fuzzy set constitute a single semantic entity in the modeling language, they are often collectively called *linguistic variables*. An important property of a hedge is closure—that is, the application of a hedge to a fuzzy set produces another distinct fuzzy region. The built-in hedge identifiers are located in the `fuzzy.hpp` header, which recognizes the following built-in hedges:

```
ABOUT         general approximation
ABOVE         fuzzy set restriction
BELOW         fuzzy set restriction
CLOSE         narrow approximation
EXTREMELY     fuzzy set intensification (strong)
GENERALLY     contrast diffusion
NOT           fuzzy set complement
POSITIVE      contrast intensification
SLIGHTLY      fuzzy set intensification (weak)
SOMEWHAT      fuzzy set dilution
VERY          fuzzy set intensification
VICINITY      broad approximation
```

Notes:

1. Hedges are not applied by name, but through the use of a hedge descriptor block (HDB) that defines the properties of a hedge. Both built-in and user-defined hedges use the HDB as their means of communicating with the *FzyApplyHedge* function. To apply a hedge follow the following steps:

- Allocate an HDB structure.
- Fill out its properties, indicating the name of the hedge, whether it is a built-in hedge or a user hedge.
- Call *FzyApplyHedge* with the address of this HDB.
- Release the storage for the HDB if necessary.

The *HDBmode* field of the HDB structure indicates the kind of hedge. If this field is zero (0) then this is a user-defined field; otherwise it is a built-in hedge and this field contains the value of the hedge constant from the *fuzzy.hpp* header. For a user hedge, the *HDBop* field contains the type of transformation. For complete details see the *FzyCreateHedge* function.

2. The input and output fuzzy sets cannot be the same. As a first operation, the hedge function issues the *FzyCopySet* routine to move the incoming fuzzy set into the outgoing fuzzy set.

3. When the fuzzy set has been hedged, the alpha cut associated with the source fuzzy set is applied to the new set. If this does not exist (*FDBalfa==0*), then the global alpha-cut limit is applied.

4. When the hedge is complete an annotation is made in the system model log (*mdllog.fil*). Applying the hedge `VERY` to the fuzzy set "*HighforPrice*" produces this message:

```
Hedge 'VERY' applied to Fuzzy Set "HighforPrice"
```

The name (*FDBid*) of the new fuzzy set is not changed (that is, it will be same as the incoming fuzzy set), however, the description of the new fuzzy set (`FDBdesc`) will be changed to reflect the application of the most recent hedge. Each hedge operation will add to the fuzzy set description.

5. All the built-in hedges are context and domain independent. That is, the application of these hedges is not dependent on some special knowledge of or restriction to the domain of the underlying fuzzy set. User fuzzy sets that employ the `ADD`, `SUBTRACT`, `MULTIPLY`, or `DIVIDE` operations are generally not context and domain independent.

6. You should be aware that the user hedge operations are performed without regard to possible truncation or minimization of the truth vector. This means that a shift in the truth membership values may cause an unexpected "clipping" of the membership table.

Related Processes:

`FzyApplyNOT`	Applies standard as well as compensatory `NOT` operators. The standard Zadeh `NOT` is also available as a hedge operation in the *FzyApplyHedge* function.

Example:

```
if(!(HDBptr=new HDB))
   {
      *statusPtr=1;
```

```
        return;
    }
FzyInitHedge(HDBptr);
strcpy(HDBptr->HDBid,"VERY");
HDBptr->HDBmode=VERY;
HDBptr->HDBop=0;
HDBptr->HDBvalue=2.0
FzyApplyHedge(HiFDBptr,HDBptr,VeryHiFDBptr,statusPtr);
if(*statusPtr!=0) return;
```

FzyApplyNOT

Format:

```
void FzyApplyNOT(
    int NOTclass,float Classweight,FDB *FDBptr,int *statusptr)
```

Parameters:

int *NOTclass*	The class of NOT operator.
float *NOTweight*	For non-Zadeh (1-μ[*x*]) classes, you may need to specify a class "weight" for this function. Such *Classweights* control the intensity or degree of the NOT operator. Not every NOT class requires a weight, in which case the value is ignored.
FDB **FDBptr*	A pointer to the incoming fuzzy set.
int **statusptr*	The status return code.

Status/Errors:

0	No errors
1	Incorrect *Classweight* for Yager Class
3	Incorrect *Classweight* for Sugeno Class
5	Incorrect *Classweight* for Threshold Class
7	Unrecognized NOT class

Use:

This function implements an extended set of compensatory NOT operators as well as the standard Zadeh (min-max) NOT also available in the *FzyApplyHedge* function. In many instances the Zadeh NOT is too restrictive or insufficiently restrictive. This function implements several classes of NOT operators that provide a wide range of functional and algebraic operations.

Related Processes:

FzyApplyHedge	Applies a variety of fuzzy set surface transformers to a fuzzy set. You can use this function to apply the standard Zadeh NOT to a fuzzy set.

Example:

```
//--apply a moderate Yager NOT to a fuzzy set
float ClassWeight=4.0;
FzyApplyNot(YAGERNOT,ClassWeight,HiPriceFDB,statusPtr);
if(statusPtr!=0) return;
```

FzyApplyOR

Format:

```
FDB *FzyApplyOR(
    FDB *FDBptr1,FDB *FDBptr2,
        int ORClass,double ORCoeff,int *statusptr);
```

Parameters:

FDB *FDBptr1*	A pointer to the first fuzzy set.
FDB *FDBptr2*	A pointer to the second fuzzy set
int *ORClass*	The class of OR operation required. This can be any of the compensatory OR classes, including the standard Zadeh OR.
double *ORCoeff*	Used for several of the compensatory OR classes. This is the class coefficient used with the specified OR compensatory class.

int *`statusptr`	The status return code.

Returns:

FDB *	A pointer to a new fuzzy set that is allocated and initialized by the function. The truth membership function of this fuzzy set is the `OR`'d product of the two incoming fuzzy sets.

Status/Errors:

0	No error, new fuzzy set is created.
1	Insufficient memory to allocate the fuzzy set.
n	Other errors are produced from the *FzyCompOR* routine, which is called by this function.

Use:

This function performs an `OR` operation on a complete fuzzy set (the other low-level functions *FzyOR* and *FzyCompOR* act only on individual truth membership function values). When the `OR` is complete, a new fuzzy set is created. A pointer to this fuzzy set is returned. The new fuzzy set is initialized with the values from *FDBptr1*.

Related Processes:

`FzyApplyAND`	Performs an `AND` operation on two fuzzy sets.
`FzyCompOR`	Performs a compensatory `OR` operation on two truth membership values.
`FzyOR`	Performs a standard Zadeh (maximum) `OR` on two truth membership values.

Example:

```
//--Take the OR of High and Low
FDB*HiLoFDBptr;
HiLoFDBptr=FzyApplyOR(
   HiFDBptr,LoFDBptr,ZADEHOR,0,&status);
if(status!=0) return;
```

FzyAutoScale

Format:

```
float FzyAutoScale(double lo,double hi);
```

Parameters:

double *lo*	The required low domain of the fuzzy set
double *hi*	The required high domain of the fuzzy set

Returns:

float	A single precision floating-point number that can be used to scale automatically the fuzzy set domain values.

Use:

This procedure will produce a scaling factor for fuzzy set domain values based on their range of values and the required size of the output field. You can call this function to find the proper scaling automatically. If the value is within limits, a scale of 1.0 is returned.

Example:

```
Scale=FzyAutoScale(HiDomain,LoDomain);
OutX=X/Scale
```

FzyBetaCurve

Format:

```
Void FzyBetaCurve(
    FDB *FDBptr,double Center,double Inflexion,int *statusptr);
```

Parameters:

FDB **FDBptr*	A pointer to a fuzzy set. The beta curve generator only creates the truth membership function; it does not alter any other properties of the fuzzy set.

Since the bell-shaped topology generator references the domain in the FDB, this structure must be completed before invoking this function (see the example).

double `Center` — The center value of the beta curve. This is the value from the domain where the truth membership function is one ([1.0]).

double `Inflexion` — The inflection or crossover point of the beta curve. This value is added to and subtracted from the `Center` value to give the conical diameter of the fuzzy set.

int `*statusptr` — The status return code.

Status/Errors:

0	No error, beta curve was produced.
1	Center lies outside domain of fuzzy set.
3	Improper inflection point.

Use:

This low-level surface topology generator creates a bell-shaped fuzzy set truth membership function in the shape of a beta curve. Generally, you will not use this function directly since it is incorporated into the *FzyCreateSet* function; however, the beta curve surface generator can be used to change the current surface of an existing fuzzy set or where the overhead of including the fuzzy set creation function is not necessary.

The beta curve differs from the PI-curve in its general topology. The beta curve is usually used to represent a concept such as

```
Y is close to W
```

where `W` would be the center of the fuzzy region. This curve is defined using a center and an inflection point measured as the half-width of the curve at the crossover point. Notice that this curve goes to zero at infinity (a PI-curve goes to zero within the specified domain).

Notes:

1. You should also note that the width of a bell-shaped fuzzy set can be changed through the application of the approximation hedge, AROUND, as well as the intensification and dilution hedges, VERY and SOMEWHAT. The AROUND hedge will increase the fuzziness of the curve by broadening the curve.

Related Processes:

FzyPiCurve	This function also produces a bell-shaped curve, however, its properties are different from and controlled differently from the beta curve.

Example:

```
//--create a beta fuzzy set
if(!NearOurCostFDBptr=new FDB))
    {
        *statusptr=1;
        MtsSendClientError(606,PgmId,"NearOurCost");
        return;
    }
FzyInitSet(NearOurCostFDBptr);
strcpy(NearOurCostFDBptr->FDBid,"AvgCost");
NearOurCostFdbptr->FDBdomain[0]=16.0;
NearOurCostFDBptr->FDBdomain[1]=36.0;
double Center=24;
double Inflexion=8;
FzyBetaCurve(NearOurCostFDBptr,Center,Inflexion,&status);
```

FzyCompAND

Format:

```
float FzyCompAND(
    int ANDClass,double ANDCoeff,
        float truth1,float truth2,int *statusptr);
```

Parameters:

int *ANDClass*	The compensatory class designator
double *ANDCoeff*	The coefficient of the class algebra

float *truth1*	A truth membership value from a fuzzy set
float *truth2*	A truth membership value from a fuzzy set
int **statusptr*	The status return code

Returns:

float	The truth value from the compensatory AND operation applied to *truth1* and *truth2*

Status/Errors:

0	Compensatory AND applied successfully
1	ANDClass not recognized
3	ANDCoeff is not valid for this class
5	ANDCoeff is too large for this machine
7	truth1 is not a valid truth membership value
9	truth2 is not a valid truth membership value

Use:

This function performs a compensatory AND operation on two truth values, returning the resulting truth membership. Unlike the Zadeh AND, which simply takes the minimum of the two truth values, compensatory operators apply a wide variety of logical and arithmetic algebras to the truth functions.

Related Processes:

FzyAND	The standard Zadeh AND operation. If you specify a class of *ZADEHAND*, then the compensatory AND is equivalent to the Zadeh AND.

FzyCompOR

Format:

```
float FzyCompOR(
   int ORClass,double ORCoeff,float truth1,float truth2,
   int *statusptr);
```

Parameters:

int *ORClass*	The compensatory class designator
double *ORCoeff*	The coefficient of the class algebra
float *truth1*	A truth membership value from a fuzzy set
float *truth2*	A truth membership value from a fuzzy set
int **statusptr*	The status return code

Returns:

float	The truth value from the compensatory OR operation applied to *truth1* and *truth2*

Status/Errors:

0	Compensatory OR applied successfully
1	*ORClass* not recognized
3	*ORCoeff* is not valid for this class
5	*ORCoeff* is too large for this machine
7	*truth1* is not a valid truth membership value
9	*truth2* is not a valid truth membership value

Use:

This function performs a compensatory OR operation on two truth values, returning the resulting truth membership. Unlike the Zadeh OR, which simply takes the minimum of the two truth values, compensatory operators apply a wide variety of logical and arithmetic algebras to the truth functions.

Related Processes:

FzyOR	The standard Zadeh OR operation. If you specify a class of *ZADEHOR*, then the compensatory OR is equivalent to the Zadeh OR.

FzyCondProposition

Format:

```
void FzyCondProposition(
    FDB *inFDBptr,FSV *FSVptr,
       int CorrMethod,float PredTruth,int *statusptr);
```

Parameters:

FDB **inFDBptr*
A pointer to the fuzzy set that will be used to update the undergeneration solution variable's fuzzy region (see the FSV parameter).

FSV **FSVptr*
A pointer to the solution variable fuzzy solution variable (FSV) block stored in the *XFZYctl* structure. This pointer is returned by *FzyAddXFZYctl* when the solution variable is added to the model. You should store this pointer address for use in the *FzyCondProposition* function.

int *CorrMethod*
The correlation method that will be applied to the *inFDBptr* fuzzy set. This can be either of the two supported methods:

`CORRMINIMUM` Truncates the fuzzy set at the truth level of the *PredTruth* before the output is updated.

`CORRPRODUCT` Scales the fuzzy set using the *PredTruth* before the output is updated.

You can also indicate two other "placeholder" values for this parameter. There are two alternate parameters:

`CORRNONE` Indicates that no correlation process should be applied to this fuzzy set. The fuzzy set is applied to the output without any adjustment for the predicate truth.

`CORRDEFAULT` Indicates that the correlation method attached to the output variable's VDB should be used for the correlation process. This should generally be the default for the function.

	The correlation method symbolic constants are stored in the *fuzzy.hpp* header file.
float *PredTruth*	The aggregate truth of the fuzzy predicate. This value is used in the correlation method to adjust the height of the incoming fuzzy set.
int **statusptr*	The status return code from the conditional fuzzy proposition.

Use:

This function updates the fuzzy solution variable block for a model output variable using the standard Zadeh min-max implication method. For conditional fuzzy propositions, that is, a proposition that would be expressed in the rule formulation language as an `IF-THEN` statement, the undergeneration FSV is updated by taking the maximum of the current FSV fuzzy region and the incoming fuzzy set (*inFDBptr*). This is the same as `OR`ing the new fuzzy set with the FSV contents.

In a conditional fuzzy proposition the truth of the consequent fuzzy region is restricted to a function of the predicate or premise fuzzy proposition. Typically, this is the min-max functional relationship (the maximum consequent truth is limited by the minimum truth of the predicate). This function usually involves the application of a correlation method (such as correlation minimum) to the current consequent fuzzy set and then the update of the output variable's own fuzzy work region. From this adjusted consequent, some implication generating function (*g*) uses the fuzzy consequent (*fcadj*) and the current fuzzy solution variable space (*fsvcurr*) to generate a new fuzzy solution variable output fuzzy region.

The *FzyCondProposition* also increments the rule density array for all nonzero membership positions across the domain. In this case, "nonzero" means above the current alpha-cut threshold of the output fuzzy region.

Related Processes:

`FzyUnCondProposition`	Implements an unconditional fuzzy proposition. This is the same as a fuzzy assertion and establishes the default structure of the model when none of the conditional fuzzy propositions contributes.
`FzyFindFSV`	Finds the FSV for a particular VDB.

Example:

```
//--apply the rule:
//--IF competition.price is not very high
//--then our price should be near the competition.price
FzyApplyHedge(HiFDBptr,VeryHDBptr,WrkFDBptr,statusptr);
FzyApplyNOT(ZADEHNOT,0,inFDBptr,NotVryHiFDBptr,statusptr);
PremiseTruth=FzyGetMembership(
      NotVryHiFDBptr,CompPrice,&Indx,statusptr);
FzyCondProposition(
   NearCPFDBptr,PriceFSVptr,
      CORRDEAFULT,PremiseTruth,statusptr);
```

FzyCoordSeries

Format:

```
void FzyCoordSeries(
   FDB *FDBptr, double Values[],float Members[],
      int Valcnt,int *statusptr)
```

Parameters:

FDB *FDBptr*	A pointer to the FDB for the fuzzy set whose membership function will be created (or recreated) by the value and membership array.
double *Values[]*	An array of scalar values from the domain of the fuzzy set.
float *Members[]*	An array of truth membership values in the range [0,1], where *Members[i]* is the truth membership associated with *Values[i]*.
int *Valcnt*	The number of elements in the *Values[]/Members[]* array space.
int *statusptr*	The status return code. Set to nonzero if the fuzzy set's truth membership vector is not created.

Status/Errors:

-1 This is not a normal fuzzy set.

0	Fuzzy set's membership function is created.
1	Improper value and membership array. A *Value[i]* element is beyond the right (high) domain of the fuzzy set.
77	Improper value and membership array. The distance between two domain values exceeds the domain of the fuzzy set. Interpolation fails.

Use:

The *FzyCoordSeries* function constructs a truth membership function for a fuzzy set through an array space consisting of values out of the domain and their associated truth membership values. These values are spread equidistantly across the domain space and piecewise linear interpolation is performed to generate the final fuzzy set truth function. With this function you can construct fuzzy sets with any kind of morphology.

Notes:

1. A minimum of three (3) and a maximum of 128 points can be entered to define the fuzzy set truth membership.

2. Although the function does not register an error if none of the truth membership values is [1.0], it does produce a warning status code of "-1." All basic vocabulary fuzzy sets should be normal, otherwise the performance of the model may suffer (the degree of truth in the implication process will not be sufficiently strong).

Related Processes:

`FzyGetCoordinates`	This routine parses a character string containing value and membership couplets in the form "value/truth" and returns the *Values[]* and *Members[]* array with the appropriate count.

Example:

```
DomValues[]={16.0    24.0    32.0};
DomTruth[] ={ 0.0     1.0     0.0};
DomCnt=3;
FzyCoordinates(
   NearPriceFDB,DomValues,DomTruth,DomCnt,&status);
if(status!=0) return;
```

FzyCopySet

Format:

```
void FzyCopySet(FDB *inFDBptr,FDB *outFDBptr,int *statusptr);
```

Parameters:

FDB **inFDBptr*	A pointer to the incoming fuzzy set
FDB **outFDBptr*	A pointer to the outgoing fuzzy set
int **statusptr*	The status code for the copy operation

Status/Errors:

0	Copy was successfully performed.

Use:

This function performs a deep copy of the incoming fuzzy set. The outgoing fuzzy set is an exact copy of the *inFDBptr* fuzzy set except that the *FDBnext* pointer in the outgoing fuzzy set is set to `NULL`.

Notes:

1. A *deep* copy means that the copy produces two distinct structures. In C/C++ the term "shallow" copy is used to indicate a copy operation that simply reassigns the value of the fuzzy set pointers. This is equivalent to

```
outFDBptr=inFDBptr
```

and means that only one base fuzzy set structure is referenced. Shallow copies are often used to make references to a fuzzy set without actually making a distinct copy of the structure.

2. The receiving fuzzy set's FDB is not created by this function. This storage *must* be allocated and a valid pointer passed to the copy function. As an example,

```
if(!(newFDBptr=new FDB)
   {
      *statusptr=1;
         return;
   }
FzyCopySet(oldFDBptr,newFDBptr,&status);
```

A check is not made on the validity of the outgoing (or, in fact, in incoming) FDB pointer address. Illegal pointers can result in unusual program interrupts.

Related Processes:

`FzyCopyVector`	Only copies the truth membership vector of the specified fuzzy set.

Example:

```
FzyCopySet(LoPriceFDBptr,VeryLoPriceFDBptr,&status);
```

FzyCopyVector

Format:

```
void FzyCopyVector(float Fromvector[],Tovector[],int VecLen);
```

Parameters:

float *Fromvector[]*	The incoming truth membership function from any fuzzy set
float *Tovector[]*	The outgoing truth membership function from any fuzzy set
int *VecLen*	The number of elements in the truth membership function. This is normally *VECMAX*

Use:

Copies a truth membership function array into another truth membership function array for a particular number of cell positions. This function can be used to make a copy of only the truth membership array. Since the array length can be specified, you can make partial left-hand runs of the array.

Related Processes:

`FzyCopySet`	Copies a complete fuzzy set from one set to another, performing a deep copy of the FDB structures.

Example:

```
float Worktruth[VECMAX];
FzyCopyVector(FDBptr->FDBvector,Worktruth,VECMAX);
```

FzyCorrMinimum

Format:

```
void FzyCorrMinimum(FDB *FDBptr,float Plateau,int *statusptr);
```

Parameters:

FDB **FDBptr*	A pointer to the target fuzzy set's FDB.
float *Plateau*	A truth membership function value that is used to truncate the incoming fuzzy set at this value. The `Plateau` must be a value between [0,1].
int **statusptr*	The status code for the correlation minimum.

Status/Errors:

0	Correlation minimum was performed.
1	The `Plateau` is not a membership value between zero (0) and one (1).

Use:

The min-max implication mechanism for conditional fuzzy propositions involves reducing the truth of a consequent fuzzy region by the truth of the rule's premise before the solution variable's working fuzzy region is updated. The most common method of reducing the truth is by truncating the consequent fuzzy region at the truth of the premise. This is called *correlation minimum*, since the fuzzy set is minimized by truncating it at the maximum of the predicate's truth.

Notes:

1. The correlation minimum mechanism introduces the idea of an output region plateau since the top of a fuzzy region is sliced by the predicate truth value. This has the consequence of incurring a certain amount of information loss. If the truncated fuzzy set is multimodal, irregular, or bifurcated in other ways, this surface topology will be discarded when the final fuzzy region is aggregated with the output variable's undergeneration fuzzy set. The correlation minimum method, however, is often preferred over the correlation product (which does preserve the shape of the fuzzy region) since it intuitively reduces the truth of the consequent by the maximum truth of the predicate, involves less complex and faster arithmetic (an important consideration for microprocessors and microcontrollers), and generally generates an aggregated output

surface that is easier to defuzzify using the conventional techniques to composite moments (centroid) or composite maximum (center of maximum height).

Related Processes:

FzyCorrProduct	The product correction works by scaling the target fuzzy set using the predicate truth value.

Example:

```
PremiseTruth=FzyGetMembership(
   NotVryHiFDBptr,CompPrice,&Indx,statusptr);
FzyCorrMinimum(NearCPFDBptr,
      PremiseTruth,statusptr);
```

FzyCorrProduct

Format:

```
void FzyCorrProduct(FDB *FDBptr,float PredTruth,int *statusptr);
```

Parameters:

FDB *FDBptr*	A pointer to the target fuzzy set's FDB.
float *PredTruth*	A truth membership function value that is used to scale the incoming fuzzy set at this value. The *PredTruth* must be a value between [0,1].
int *statusptr*	The status code for the correlation product.

Status/Errors:

0	Correlation product was performed.
1	The *PredTruth* is not a membership value between zero (0) and one (1).

Use:

The min-max implication mechanism for conditional fuzzy propositions involves reducing the truth of a consequent fuzzy region by the truth of the rule's premise before the solution variable's working fuzzy region is updated. While correlation minimum is the most frequently used technique, correlation product offers an alternative and, in many ways, better method of achieving this goal. In the correlation product method, the intermediate fuzzy region is scaled instead of truncated. The truth membership function is adjusted by applying the following transformation:

```
for(i=0;i<VECMAX;i++)
   FDBvector[i]=FDBvector[i]*PredTruth;
```

that is, the membership is scaled using the truth of the predicate. This has the effect of shrinking the fuzzy region (when *PredTruth<1*) while still retaining the original shape of the fuzzy set.

Notes:

1. The correlation product mechanism does not introduce plateaus into the output fuzzy region, although its does increase the irregularity of the fuzzy region and could impact the results obtained from composite moments or composite maximum defuzzification. This lack of explicit truncation has the consequence of generally reducing information loss. If the intermediate fuzzy set is multimodal, irregular, or bifurcated in other ways this surface topology will be retained when the final fuzzy region is aggregated with the output variable's undergeneration fuzzy set.

Related Processes:

FzyCorrMinimum	The product correction works by truncating the target fuzzy set using the predicate truth value.

Example:

```
PremiseTruth=FzyGetMembership(
   NotVryHiFDBptr,CompPrice,&Indx,statusptr);
FzyCorrProduct(NearCPFDBptr,
      PremiseTruth,statusptr);
```

FzyCreateHedge

Format:

```
HDB * FzyCreateHedge(
   char *HedgeId,int HedgeOp,double Hedgeval,int *statusptr);
```

Parameters:

char `*HedgeId`	The hedge identifier. This can be any string from 1 to 16 characters in length. The name must conform to model identifier conventions (that is, it cannot begin with a digit or contain blanks and other system delimiters).
int `HedgeOp`	The type of operator used in the hedge. Any of the following transformer types can be specified,

`ADD`	addition
`SUBTRACT`	subtraction
`DIVISION`	division
`MULTIPLY`	multiplication
`POWER`	raise to power

You can also specify the special term *BUILTIN* if the *HedgeId* is one of the built-in hedges.

double `Hedgeval`	The value that will be used by the *HedgeOp* to modify the terrain of the fuzzy set. The permissible value of this parameter depends on the type of hedge operator specified.
int `*statusptr`	The status return code.

Returns:

HDB*	A pointer to the new hedge descriptor block (HDB) created by this function.

Status/Errors:

1	Hedge already exists.
3	*Hedgeval* is out of bounds.
5	Unrecognized *HedgeOp* value.

Use:

This procedure creates a hedge and returns a pointer to the new hedge descriptor block (HDB). A hedge changes the shape of a fuzzy set by altering the truth membership function. The fuzzy modeling system provides a complete set of built-in, context-independent hedges. The built-in hedge identifiers are located in the *fuzzy.hpp* header. The fuzzy modeling system recognizes the following built-in hedges:

`ABOUT`	general approximation
`ABOVE`	fuzzy set restriction
`BELOW`	fuzzy set restriction
`CLOSE`	narrow approximation
`EXTREMELY`	fuzzy set intensification (strong)
`GENERALLY`	contrast diffusion
`NOT`	fuzzy set complement
`POSITIVE`	contrast intensification
`SLIGHTLY`	fuzzy set intensification (weak)
`SOMEWHAT`	fuzzy set dilution
`VERY`	fuzzy set intensification
`VICINITY`	broad approximation

Context independence means that they do rely on knowledge about the underlying domain. In general, user-defined hedges based on other than the *POWER* operator are usually not context independent, that is, you must understand the domain of the fuzzy set in order to apply the hedge properly.

Notes:

1. If you specify an operator of *BUILTIN*, then the identifier of the hedge must be a character string corresponding to one of the built-in symbolic hedge constants. This will create a hedge corresponding to this built-in hedge. In this regard, while built-in hedges have their properties and actions incorporated into the *FzyApplyHedge* function, they do not actually exist until you create them.

Related Processes:

FzyApplyHedge	Modifies the membership function of a fuzzy set by applying a hedge.

Example:

```
//create the SLIGHTLY hedge
Hdbptr=FzyCreateHedge("SLIGHTLY",1.3,POWER,&status);
if(Hdb==NULL)
   {
      *statusptr=1;
      return;
   }
```

FzyCreateSet

Format:

```
FDB *FzyCreateSet(char *SetId,int SetType,
   double Domain[],double Parms[],
      int ParmCnt,int *statusptr);
```

Parameters:

char *SetId*	The name of the fuzzy set. This can be any string from 1 to 16 characters in length. The name must conform to model identifier conventions (that is, it cannot begin with a digit or contain blanks and other system delimiters).
int *SetType*	Indicates the shape of the fuzzy set. You can create a fuzzy set in any of several standard shapes (linear, S-curves, bell-shaped, etc.) using the following symbolic constants:

BETA	beta (bell) curve
DECLINE	left-facing S-curve
DECREASE	linear Decreasing
GROWTH	right-facing S-curve

`INCREASE`	linear Increasing
`MEMSERIES`	truth membership curve
`PI`	PI (bell) curve

You cannot directly create a fuzzy set with the coordinate domain/membership relationships using this function (see the *FzyCoordSeries* function for more details) The valid fuzzy types are defined in the *fuzzy.hpp* header.

double *Domain[]* — An array of two double precision numbers. The low and high edges of the fuzzy set entered in the following order:

```
Domain[0]=low edge
Domain[1]=high edge
```

This specification effectively establishes the complete "universe of discourse" for the fuzzy set. The low and high edges cannot both be the same value (including zero). The domain can include both positive and negative values.

double *Parms[]* — An array containing any additional parameters necessary to support the type of fuzzy set surface. In most—but not all—cases this array can be empty. The create process will make default assumptions about the parameters from the specified domain.

int *ParmCnt* — The extent of the *Parms[]* array.

int **statusptr* — The status return code from the create process.

Returns:

FDB* — A pointer to a new fuzzy set descriptor block (FDB). This block is allocated, initialized, and populated by this function.

Status/Errors:

0	Fuzzy set was created.
1	The low and high domains of the fuzzy set are the same (including both values being zero).
3	The low domain is greater than the high domain.
5	Insufficient memory to allocate a new FDB for this fuzzy set.
7	Unrecognized fuzzy set surface type.
10xx	The *FzyCreateSet* function calls, as required, the functions that create the various surface topologies. These functions make additional diagnostic checks and can return their own error code. In this case the error status is added to "1000" and returned. Thus, a status code of "1033" would mean that one of the curve generators returned a status of "33."

Use:

This function creates a new fuzzy set and returns a pointer to the new fuzzyset descriptor block (*FDB*). descriptor. You must indicate the generator mechanism for the fuzzy set (*SetType*), specify a domain for the base set, and, if required, indicate the parameters that will create the actual membership function.

`BETA`

A *BETA* or bell-shaped curve provides support for concepts such as *CLOSE TO* or *NEXT TO* (also see the triangular and the PI-curve fuzzy shape).The beta-curve distribution is used to generate a membership function for fuzzy propositions such as

```
Y is close to W
```

where *W* is the fuzzy region. A beta-curve topology is defined by two parameters: the center of the possibility distribution and the radius at the inflection or crossover point (that is, the half-width of the curve). A beta curve is defined by the expression

$$\mu_{closetoW}(x) = 1/(1+((x-\gamma)/\beta)^2)$$

Where γ is the center of the distribution and β is the crossover point. You should note that, unlike the PI curve and the triangular function, the beta curve goes to zero at infinity.

You can specify a maximum of two parameters indicating the center of the bell curve and the width of the inflection point:

```
Parms[0]=center of curve
Parms[1]=width of inflection point
```

The inflection width is the distance from the center of the curve (which has a value of one) to the point where the membership is at the [.50] level. Refer to the PI curve for some additional details on specifying the radius or band width of the beta curve.

DECLINE

To specify a left-facing S-curve (this is the logistic or sigmoid pattern). You can enter three parameters to specify the shape of the S-curve:

```
Parms[0]=left edge
Parms[1]=50% inflection point
Parms[2]=right edge
```

If you do not specify any parameters, then the function will automatically use the *Domain[]* values as the curve edges and will take the midpoint of the domain as the inflection point.

GROWTH

To specify a right-facing S-curve (this is the logistic or sigmoid pattern). You can also enter three surface specification parameters:

```
Parms[0]=left edge
Parms[1]=50% inflection point
Parms[2]=right edge
```

If you do not specify any parameters, then the function will automatically use the *Domain[]* values as the curve edges and will take the mid-point of the domain as the inflection point.

For S-curves the three parameters indicate not only the left and right edges of the fuzzy set domain, but also the curve's inflection point. The inflection point is the location on the domain where the value is 50% true. Note that you cannot simply specify an inflection point, you must also include the left and right domain edges, even if they are the same as the *Domain[]* arguments.

`DECREASE`

A linear decreasing fuzzy set goes from one in the left domain to zero in the left domain. No parameters are required if the fuzzy set spans the entire domain. You can specify a maximum of two parameters indicating the start and end of the fuzzy set within the domain:

```
Parms[0]=left edge
Parms[1]=right edge
```

`INCREASE`

A linear increasing fuzzy set goes from zero in the left domain to one in the left domain. No parameters are required if the fuzzy set spans the entire domain. You can specify a maximum of two parameters indicating the start and end of the fuzzy set within the domain:

```
Parms[0]=left edge
Parms[1]=right edge
```

`PI`

A `PI` or bell-shaped curve provides support for concepts such as around or near (also see the triangular fuzzy shape). You can specify a maximum of two parameters indicating the center of the bell curve and the width of the base radius:

```
Parms[0]=center of fuzzy set
Parms[1]=width of radius
Parms[2]=right edge
```

The `radius`, often called the *band width,* is the distance from the center of the curve (which has a truth membership value of one) to the point where the membership is zero. If you do not specify a center of the fuzzy set and a band width or radius for the PI curve, the function uses the center of the domain as the center of the PI curve. The radius is then taken as the distance from the center to the end of the domain.

By changing the radius of the PI curve, the curve is made steeper or broader. As an example, by moving the band width (or radius) in toward the center of the distribution the PI curve becomes sharper.

A closely related bell-shaped fuzzy set topology is the beta curve. This curve is specified with slightly different parameters. You indicate the center of the curve and then the [.50] membership inflection point. Beta distribution tails are asymptotes—they only reach zero at infinity.

`MEMSERIES`

You use the series fuzzy set generator to create a set from a equidistanced array of truth membership values.

Each truth membership value must be equal to or greater than the preceding value (that is, *Parms[n]* ≤ *Parms[n+1]*). The parameter array contains values in the form,

```
Parms[n]=truth value in range [0,1]
Parms[n+1]=truth value ≥ Parms[n]
```

You can specify a minimum of three truth values and a maximum of 256 values. The system will use piecewise linear interpolation to evenly distribute the values across the membership array. Values within the interval are automatically supplied by the process.

Notes:

1. Fuzzy sets are created at the low-level library independent of any associated variable. This means that the domain specification is not inherited from a variable (*VDB* block) but must be specified for each fuzzy set. When a collection of fuzzy sets has

been created you can associate them with a variable's *VDB* through the *VarAssociate-Sets* function. This function orders the fuzzy sets across the underlying domain and ensures that (1) the component fuzzy sets are within the variable's universe of discourse and (2) that the adjacent fuzzy set overlaps are correct.

2. In creating fuzzy sets you should understand the difference between the *universe of discourse*, the *domain*, and the *support set* of the fuzzy region. In some instances, these terms will refer to the same space, but generally, they have distinct meanings.

Universe of — This is the entire range of values over which the *discourse* fuzzy set is mapped and corresponds to the minimum and maximum values that can be assumed by the associated variable.

Domain — This is the minimum and maximum values of the specific fuzzy set itself. Often called the *domain of applicability*.

Support Set — The support set is the nonzero truth membership extent of the fuzzy set.

The *Domain[]* array specifications in the *FzyCreateSet* function refer to the underlying domain of the fuzzy set and not the universe of discourse across the variable.The support set sometimes takes on a different meaning within the solution fuzzy region and within the fuzzy sets created by the application of hedges. The support in this context refers to all the truth membership values that are above the current alpha-cut threshold.

3. Trapezoidal or shouldered fuzzy regions are not directly supported as a specific type but can be specified either as a truth membership series or through a domain and membership coordinate series.

4. You should closely manage the storage allocated through the *FzyCreateSet* function. Each fuzzy set occupies approximately *1644* bytes. For models using the conventional stack (heap) this could result in a stack overflow exception when many fuzzy sets are currently allocated.

Related Processes:

`FzyCoordSeries` — Creates a fuzzy set from a series of domain and associated truth membership values.

`FzyGetCoordinates` — Extracts the fuzzy set membership coordinates from a character string.

`FzyCreateHedge`	Creates a fuzzy set surface transformer.

Example:

```
Domain[0]=16.0;
Domain[1]=36.0;
Parms[0]=0.0;
hiFDBptr=FzyCreateSet(
   "HIGH",INCREASE,Domain,Parms,0,statusPtr);
if(*statusPtr!=0) return;
```

FzyDefuzzify

Format:

```
double FzyDefuzzify(
FDB *FDBptr,int DefuzzMethod,int *statusptr);
```

Parameters:

FDB **FDBptr*	A pointer to the fuzzy set that will be defuzzified. This is usually the fuzzy set associated with one of the fuzzy solution variable (FSV) blocks.
int *DefuzzMethod*	One of the defuzzification technique methods supported by the modeling system. This can be any of the following:
`AVGMAXIMUM`	average of maximum values
`BESTEVIDENCE`	preponderance of evidence
`CENTROID`	center of gravity
`FAREDGE`	value at right fuzzy set edge
`MAXIMUM`	maximum at set height
`MAXPLATEAU`	center of the maximums
`NEAREDGE`	value at left fuzzy set edge
	Some of these defuzzification methods are based on representational geometries that must be speci-

fied when the solution variable's VDB is initially allocated. (This is particularly true of the *BESTEVIDENCE* defuzzification, which is a special form of *CENTROID* based on a restricted region of the output fuzzy set.)

int **statusptr* — The status return code from the defuzzification process. The value also represents return codes from functions called by the defuzzification process.

Returns:

double — The expected value of the fuzzy set. This is a double precision floating-point value from the domain of the underlying fuzzy set.

Status/Errors:

0	Defuzzification succeeded.
3	Empty fuzzy set.
88	Defuzzification method not currently supported.
99	Unrecognized defuzzification method.

Use:

The result of applying a series of conditional and unconditional propositions against a fuzzy solution variable is an output fuzzy set built by the aggregation of all the component fuzzy sets. This is the fundamental fuzzy logic and approximate reasoning process.

From the final fuzzy region stored in the FSV block of the solution variable we must extract a representative value. Finding this representative or expected value is the process of defuzzification. There are several techniques for processing a final fuzzy set to derive the expected value and these are called defuzzification methods. When you create a variable VDB it has a default defuzzification technique, but depending on the model characteristics you may need to select another type of defuzzification method.

Generally speaking there are *two* classes of defuzzification techniques known as composite moments and composite maximum. Composite reflects the idea that the results are derived from a composite space. The moments reflects the first moment of inertia associated with physical bodies, while maximum reflects the idea of extracting a value based on the height or maximum value of the fuzzy set.

AVGMAXIMUM	The average of maximum values defuzzification finds the mean maximum value of the fuzzy region. If this is a single point, then this value is returned; otherwise, the average of the plateau is calculated and returned. An average value often provides a better approximation for a composite maximum approach when the plateaus tend to be clipped at the left or right edges. See the *MAXIMUM* defuzzification technique for additional details. Note that for a single-edged plateau, the *AVGMAXIMUM* does not convert to a edge-wise defuzzification as in the *MAXIMUM*.
BESTEVIDENCE	A preponderance of evidence defuzzification is a combination of a structural geometry and the application of the *CENTROID* (composite moments) defuzzification technique. It is intended to improve the generation of an expected value for some types of fuzzy models by concentrating on output fuzzy regions with the highest possibility density value. The best evidence tends to ignore single output fuzzy regions that inflate the width of the final fuzzy region, thus causing the standard *CENTROID* to generate a value that is shifted away from the region of maximum possibility density.
CENTROID	The centroid or composite moments method calculates the hypothetical center of gravity for the output fuzzy set. This center of gravity is the point where the fuzzy set would be perfectly balanced if placed on a fulcrum. The centroid technique takes the weighted sum of the domain and divides this by the sum of the truth vector to produce the expected value. Centroid defuzzification is the most widely used technique since it has several desirable properties: (1) The defuzzified values tend to move smoothly around the output fuzzy region, that is, changes in the

fuzzy set topology from one frame to the next usually result in smooth changes in the expected value. (2) It is relatively easy to calculate. (3) It can be applied to both output fuzzy set geometries: singletons and fuzzy regions.

FAREDGE

This technique selects the value at the right fuzzy set edge and is of most use when the output fuzzy region is structured as a single-edged plateau.

You will find this kind of defuzzification useful when you have only one or two highly overlapping fuzzy sets that are truncated at either end of the domains. The *FAREDGE* technique seeks for the existence of a plateau by moving toward the left from the right edge of the fuzzy region. The current truth level is not counted as a plateau.

A *FAREDGE* technique will move off the initial right-hand plateau and select the edge of the plateau at the next highest level. This keeps the defuzzification process from defaulting to the extreme edges of the fuzzy set.

MAXIMUM

This technique finds the maximum point in the solution fuzzy set. If there is no plateau then the maximum point in the fuzzy region is used. If only a single plateau is found, the process is equivalent to the standard *MAXPLATEAU* defuzzification. In general this technique provides a slightly better and more robust defuzzification method than the *MAXPLATEAU* when you are using correlation product rather than correlation minimum.

MAXPLATEAU

The maximum fuzzy set height technique returns a value from the maximum truth value of the fuzzy region. Also known as composite maximum, the maximum height is the second most popular method of defuzzification. For an output region with a double-edged plateau, the center of the plateau is used. For a single-edged plateau, the technique switches to *FAREDGE* or *NEAREDGE* depending on where the edge of the plateau is located.

MAXPLATEAU converts to a *FAREDGE* defuzzification when a single-edged plateau is discovered. Unlike the *CENTROID* technique, however, the *MAXIMUM* has some attributes that are less desirable for most models: (1) the expected value is sensitive to a single rule that dominates the fuzzy rule set, and (2) the expected value tends to jump from one frame to the next as the shape and height of the fuzzy region change.

The *MAXIMUM* technique is an important defuzzification method for applications involving risk assessment or other "high water mark" evaluations. With this technique the maximum risk extent is found among all the rules that contribute to the cumulative degree of risk.

NEAREDGE

This technique selects the value at the left fuzzy set edge and is of most use when the output fuzzy region is structured as a single-edged plateau. This technique is the complement of the *FAREDGE* defuzzification technique. See the *FAREDGE* process for additional details.

Notes:

1. All defuzzification techniques are sensitive to the current alpha-cut threshold limits either attached to the solution fuzzy set or defined globally in the system control block. All alpha cuts are *WEAK* alpha cuts, that is, the expression is evaluated as

$$\mu(x) \leq \alpha$$

Thus, an alpha cut of [.20] means that all truth values equal to or less than [.20] will be set to zero. To eliminate the affect of the alpha cut, you can set the global or output alpha thresholds to zero.

2. If you use correlation product instead of correlation maximum as an implication transfer function this can have a significant impact on some of the defuzzification techniques. In particular, since the consequent fuzzy regions are scaled rather than truncated, the output (solution) fuzzy region will lack definite plateaus. The output region can be highly modal (filled with many "peaks and valleys") so that those defuzzifications that rely on the truncation effect will not work as predicted. When you use correlation product, you should restrict the defuzzification to *CENTROID* or *MAXIMUM*.

3. Unless you have reason to believe that your model requires a more advanced or a specialized method of defuzzification, you should limit your model to either the *CENTROID* or the *MAXIMUM* defuzzification techniques. At the very least, these should be the starting point for your method of model validation and verification.

Example:

```
thisDefuzzMethod=FSVptr->FzySVdefuzzMethod;
Price=FzyDefuzzify(rskFDBptr,thisDefuzzMethod,statusptr);
fprintf(stdout,"%s%10.2f\n","Price==",Price);
```

FzyDisplayFSV

Format:

```
Void FzyDisplayFSV(FSV *FSVptr,char *DisplayTitle);
```

Parameters:

FSV **FSVptr*	A pointer to an FSV block in the *XFZYctl* control block region.
char **DisplayTitle*	A character string that is printed with the display. This allows each display to be labeled with a string identifying the state of the model.

Use:

This procedure writes the current status of an FSV block to the current model log file (*mdllog.fil*). This function is used during model development or when an error occurs in a production model to display the current fuzzy solution variable (FSV) block. As you update each generation of this control region, this procedure will show you how the current output fuzzy space is organized. The display shows the numeric constants representing the various control options rather than their symbolic names. (This allows the symbolic constants pool to increase without making constant changes to the display program). This function also calls *FzyDrawSet* to display the contents of the undergeneration fuzzy set associated with the solution variable.

Example:

```
Fuzzy Model Output Variable Generation Area
Initial Add to Work Area
Variable Name..............PRICE
   datatype................      05
```

```
   Defuzzification..........      01
   Geometry.................      01
   Implication..............      01
   Update Count.............      00
   Alfa Cut.................    0.10

FuzzySet:   PRICE
Description:
1.00
0.90
0.80
0.70
0.60
0.50
0.40
0.30
0.20
0.10
0.00
    0---|---|---|---|---|---|---|---|---|---|---|---|---|---|---|---0
  16.00   18.50   21.00   23.50   26.00   28.50   31.00   33.50   36.00
Domained:         16.00 to       36.00

-------------E  N  D      O F     D I S P L A Y-------------
```

FzyDrawSet

Format:

```
void FzyDrawSet(FDB *FDBptr,int Medium,int *statusptr);
```

Parameters:

FDB *FDBptr — A pointer to the fuzzy set whose contents will be displayed on the current model log.

int *Medium* — A symbolic constant that indicates where you want the display sent. You can use the following constants:

SYSLOGFILE	the system log file
SYSMODFILE	the system model trace log
SYSOUT	to the communications terminal
int *statusptr	the status return code from the draw function.

Status/Errors:

0 No error, the graph is produced.

1 Unrecognized medium constant.

Use:

This function draws a character-based graph showing the topology of the specified fuzzy set. With this tool you can view the results of such activities as fuzzy set creation, hedge operations, standard and compensatory operator functions, and implication functions.

Notes:

1. The graph is directed to the specified log file (*SYSMODFILE*, as an example, will send the graph to *mdllog.fil*). The file associated with the model log is established when the *XSYSctl* block is initialized.

2. You cannot change the width of the graph.

Related Processes:

`FzyPlotSets`	This related procedure graphs a collection of fuzzy sets mapped across the same universe of discourse.

Example:

```
//--Create and Draw the LOW fuzzy set
Domain[0]=16.0;
Domain[1]=36.0;
LowFDBptr=FzyCreateSet("LOW",DECREASE,Domain,Parms,0,&status);
FzyDrawSet(LowFDBptr,SMALLPLOT,&status);
```

```
    FuzzySet:   LOW
    Description:
  1.00
  0.90
  0.80..............
  0.70              ......
  0.60                    .......
  0.50                           ......
  0.40                                 .......
  0.30                                        ......
  0.20                                              .......
  0.10                                                     ......
  0.00                                                           ......
    0---|---|---|---|---|---|---|---|---|---|---|---|---|---|---|---0
    16.00   18.50   21.00   23.50   26.00   28.50   31.00   33.50   36.00
    Domained:        16.00 to       36.00
```

```
//--Create and Draw the 2*MfgCosts fuzzy set
Domain[0]=16.0;
Domain[1]=36.0;
TwiceMfgCosts=2*MfgCosts;
Parms[0]=TwiceMfgCosts;
Parms[1]=TwiceMfgCosts*.25;
ParmCnt=2;
MfgCostsFDBptr=FzyCreateSet(
"2*MfgCosts",PI,Domain,Parms,ParmCnt,&status);
FzyDrawSet(MfgCostsFDBptr,SMALLPLOT,&status);

         FuzzySet:   2*MfgCosts
         Description:
       1.00                          ..
       0.90                       ...  ....
       0.80                    ..         .
       0.70                    .           ..
       0.60                  ..               .
       0.50                 .                  .
       0.40                .                    .
       0.30               .                      .
       0.20              .                        .
       0.10            ..                          ..
       0.00...........                               ........................
           0---|---|---|---|---|---|---|---|---|---|---|---|---|---|---|---|---0
         16.00   18.50   21.00   23.50   26.00   28.50   31.00   33.50   36.00
         Domained:        16.00 to      36.00

//--Draw the current Fuzzy Solution Variable's fuzzy set
FzyDrawSet(PriceFDBptr,SMALLPLOT,&status);

         FuzzySet:   PRICE
         Description:
       1.00
       0.90
       0.80
       0.70
       0.60
       0.50                                 .
       0.40                           ...... ....
       0.30                    .......           .
       0.20              ......                   .
       0.10            ..                          ..
       0.00...........                               ........................
           0---|---|---|---|---|---|---|---|---|---|---|---|---|---|---|---|---0
         16.00   18.50   21.00   23.50   26.00   28.50   31.00   33.50   36.00
         Domained:        16.00 to      36.00
```

FzyExamineSet

Format:

```
void FzyExamineSet(FDB *FDBptr,int *statusptr);
```

Parameters:

FDB *FDBptr	A pointer to the fuzzy set whose truth membership function will be examined.

int *`*statusptr`* The function status return code.

Status/Errors:

0	Examine report was produced.
1	This is an empty fuzzy set.

Use:

This procedure builds an *MxN* matrix showing the distribution of values within the fuzzy set's truth membership function. You can use this function to display the fuzzy set's possibility distribution.

Notes:

1. The report is directed to the current model log (generally, *mdllog.fil*). The file associated with the model log is established when the *XSYSctl* block is initialized.

2. The domain values are automatically scaled to fit within the report column (see the *FzyAutoScale* function). When the value has been scaled, the scaling factor will be printed at the bottom of the report.

3. If the alpha-cut threshold associated with the fuzzy set is not zero, then the alpha-cut limit is displayed at the bottom of the report. Note that the function does not actually apply the alpha cut.

Example:

```
//--Let's see the HIGH (for price) fuzzy set
FzyDrawSet(HighFDBptr,SYSMODFILET,&status)

         FuzzySet:    HIGH
         Description: High for Price
       1.00                                                               .
       0.90                                                         ......
       0.80                                                  .......
       0.70                                            ......
       0.60                                     .......
       0.50                               ......
       0.40                         ......
       0.30                  .......
       0.20            ......
       0.10     .......
       0.00......
           0---|---|---|---|---|---|---|---|---|---|---|---|---|---|---|---0
         16.00   18.50   21.00   23.50   26.00   28.50   31.00   33.50   36.00
         Domained:       16.00 to       36.00
```

```
//--Now Let's see the HIGH truth membership function
FzyExamineSet(HighFDBptr,&status)

FuzzySet:    HIGH
 Description: High for Price
 Domain     :      16.00 to      36.00
 Plateau    :       1.00 to     256.00

  % Value                        Truth                           Value  %
  0 16.00
  1 16.20 0.01 0.01 0.02 0.02 0.02 0.03 0.03 0.04 0.04 0.04 18.00  10
 11 18.20 0.11 0.11 0.12 0.12 0.12 0.13 0.13 0.14 0.14 0.14 20.00  20
 21 20.20 0.21 0.21 0.21 0.22 0.22 0.23 0.23 0.23 0.24 0.24 22.00  30
 31 22.20 0.31 0.31 0.32 0.32 0.32 0.33 0.33 0.33 0.34 0.34 24.00  40
 41 24.20 0.41 0.41 0.42 0.42 0.42 0.43 0.43 0.44 0.44 0.44 26.00  50
 51 26.20 0.51 0.51 0.52 0.52 0.53 0.53 0.53 0.54 0.54 0.54 28.00  60
 61 28.20 0.61 0.61 0.61 0.62 0.62 0.63 0.63 0.63 0.64 0.64 30.00  70
 71 30.20 0.71 0.71 0.72 0.72 0.72 0.73 0.73 0.74 0.74 0.74 32.00  80
 81 32.20 0.81 0.81 0.82 0.82 0.82 0.83 0.83 0.84 0.84 0.84 34.00  90
 91 34.20 0.91 0.91 0.91 0.92 0.92 0.93 0.93 0.93 0.94 0.94 36.00 100

//--Let's see the intersection of the HIGH (for price)
//--and the LOW (for price) fuzzy sets using the
//--Zadeh intersection (min-max)
FzyDrawSet(HiLoFDBptr,SYSMODFILE,&status)

        FuzzySet:    HIGH&LOW
         Description: Both High and Low
       1.00
       0.90
       0.80
       0.70
       0.60
       0.50                                  .
       0.40                           ...... .......
       0.30                    .......             ......
       0.20              ......                          .......
       0.10       .......                                       ......
       0.00......                                                     ......
           0---|---|---|---|---|---|---|---|---|---|---|---|---|---|---|---0
         16.00   18.50   21.00   23.50   26.00   28.50   31.00   33.50   36.00
         Domained:        16.00 to      36.00

//--Now Let's see the HIGH&LOW truth membership function
FzyExamineSet(HiLoFDBptr,&status)

FuzzySet:    HIGH&LOW
 Description: Both High and Low
 Domain     :      16.00 to      36.00
 Plateau    :       1.00 to     256.00

  % Value                        Truth                           Value  %
  0 16.00
  1 16.20 0.01 0.01 0.02 0.02 0.02 0.03 0.03 0.04 0.04 0.04 18.00  10
 11 18.20 0.11 0.11 0.12 0.12 0.12 0.13 0.13 0.14 0.14 0.14 20.00  20
 21 20.20 0.21 0.21 0.21 0.22 0.22 0.23 0.23 0.23 0.24 0.24 22.00  30
 31 22.20 0.31 0.31 0.32 0.32 0.32 0.33 0.33 0.33 0.34 0.34 24.00  40
 41 24.20 0.41 0.41 0.42 0.42 0.42 0.43 0.43 0.44 0.44 0.44 26.00  50
 51 26.20 0.49 0.49 0.48 0.48 0.47 0.47 0.47 0.46 0.46 0.46 28.00  60
 61 28.20 0.39 0.39 0.39 0.38 0.38 0.37 0.37 0.37 0.36 0.36 30.00  70
 71 30.20 0.29 0.29 0.28 0.28 0.28 0.27 0.27 0.26 0.26 0.26 32.00  80
 81 32.20 0.19 0.19 0.18 0.18 0.18 0.17 0.17 0.16 0.16 0.16 34.00  90
 91 34.20 0.09 0.09 0.09 0.08 0.08 0.07 0.07 0.07 0.06 0.06 36.00 100
```

FzyFindFSV

Format:

```
FSV *FzyFindFSV(VDB *VDBptr);
```

Parameters:

VDB *VDBptr*	A pointer to the variable stored in the *XFZYctl* control block. The variable was inserted through the *FzyAddFZYctl* function.

Returns:

FSV*	A pointer to the fuzzy solution variable structure allocated and linked into the fuzzy control block. If this pointer is `NULL` then the FSV for the indicated variable does not exist in the control block.

Use:

Although the *FzyAddFZYctl* function returns the address of the newly allocated FSV for a variable there are programming considerations that may require your application to reacquire the address of the FSV for a variable (as an example, if you inserted the variable inside an external function that did not return the FSV address). This function finds the variable in the fuzzy control block and returns the address to its FSV.

Related Processes:

`FzyAddFZYctl`	Inserts a variable into the fuzzy control block and returns the address of the FSV and its associated output fuzzy set.

Example:

```
//--reacquire the solution structure for PRICE
PriceFSVptr=FzyFindFSV(PriceVDBptr);
```

FzyFindPlateau

Format:

```
void FzyFindPlateau(
   FDB *FDBptr,int SetEdges[],int *EdgeCnt,int *statusptr);
```

Parameters:

FDB **FDBptr*	A pointer to a fuzzy set.
int *SetEdges[]*	An array containing index offsets into the FDB vector (truth membership function) of the fuzzy set. These index values indicate the left and right edges of the plateau. This array is populated and returned by the function.
int **EdgeCnt*	The number of edges in the plateau. This is returned by the function. The value will be one of the following:

0	No plateau exists
1	A single-edged plateau
2	A double-edged plateau

No plateau exists if the maximum truth value in the fuzzy set is a single point. The use of correlation product often results in a lack of discernible plateaus in the output fuzzy set. Single point fuzzy sets also result from fuzzy models that use unconditional fuzzy propositions (that is, fuzzy assertions).

int **statusptr*	The status return code.

Status/Errors:

0	No error; plateau was found.
1	The fuzzy set is empty.

Use:

This function examines the truth membership function of a fuzzy set and finds the plateau at the maximum truth level. Plateaus result from the truncation process associated with the correlation minimum treatment of the consequent fuzzy regions.

The *FzyFindPlateau* function is used primarily during the defuzzification techniques for such methods as *MAXIMUM* and *MAXPLATEAU*. The existence of a plateau introduces a degree of ambiguity into defuzzification since the expected value can be drawn from a large number of candidate domain values, each with the same truth value. In finding a plateau, the function first calls *FzyGetHeight* to locate the maximum truth in the fuzzy set. Any plateau must be associated with this maximum truth value.

Notes:

1. **Important.** The *FzyFindPlateau* function will destroy the incoming fuzzy set. To locate a plateau, all truth membership values less than the maximum truth are set to zero. Since the routine is used mainly in defuzzification this normally has no effect on the model; however, if you are designing adaptive fuzzy systems or systems where the output fuzzy region must be preserved, then defuzzification should be performed using a copy of the FSV's fuzzy region.

Related Processes:

`FzyGetHeight`	Finds the maximum truth membership function in a fuzzy set.

Example:

```
//--locate the plateau in this fuzzy set
FzyCopySet(HiFDBptr,WrkFDBptr,&status);
FzyFindPlateau(WrkFDBptr,Edges,&EdgeCnt,&status);
```

FzyGetCoordinates

Format:

```
void FzyGetCoordinates(char *Coordstring,double Values[],
    float Members[],int *CoordCnt,int *statusptr);
```

Parameters:

char *Coordstring*	A character string containing the domain and truth membership coordinate specifications. This is a standard null terminated character string.
double *Values[]*	The array of domain values built by the function.
float *Members[]*	The array of truth membership values built by the function.

int *CoordCnt*	A count, returned by the function, of the number of entries in the *Values[]* and *Members[]* arrays.
int *statusptr*	The status return code from the function.

Status/Errors:

0	No error, the string was successfully parsed.
1	A scalar (domain) component of the coordinate is not numeric.
3	A truth membership component of the coordinate is not numeric.

Use:

This utility program is used in conjunction with the *FzyCoordSeries* function to extract and create arrays of domain and truth membership values. Coordinate specifications allow you to describe fuzzy sets topologies that cannot be approximated by the standard set shapes (also see the *MEMSERIES* option in the *FzyCreateSet* function). The coordinate string must contain specifications as pairs of values in the form

```
DomainValue/TruthValue
```

where `DomainValue` is a value from the underlying domain of the fuzzy set and `TruthValue` is its membership in the fuzzy set. The character string can contain a minimum of three (3) and a maximum of 128 coordinate specifications. As an example, the following is a coordinate specification:

```
16.0/0, 18.0/.3, 24.0/.5, 30.0/.8, 36.0/1
```

The elements of the string can be separated by blanks, tabs, or commas. Decimal places and right- or left-filled zero positions are not required (that is, *18/.3*, and *18.00/0.30* are equivalent specifications). No spaces or other delimiters can appear between the domain value, the truth membership value, and the connecting "/" character. When this function is applied against the specification character string it returns the *Values[]* array with the domain values and the *Members[]* array with the truth functions so that

```
Values[i] → Members[i]
```

The *CoordCnt* parameter will tell you how many value and membership pairs were returned by the function.

Notes:

1. You must ensure that the *Values[]* and *Members[]* arrays have sufficient extents to hold all possible specifications in the character string.

2. Negative domain specifications are allowed.

3. Note that this is a parsing utility only; it has no actual knowledge of the target fuzzy set. This means that it will perform no diagnostics on the range of the domain values that you specify. (This will be done by *FzyCoordSeries* when the arrays are actually used to create a fuzzy set truth function.) The parsing does not check for a normal fuzzy set, that is, at least one membership value is [1.0].

Related Processes:

`FzyCoordSeries`	Creates a fuzzy set membership function using an array of domain values and a corresponding array of membership values.

Example:

```
printf("Please enter a Fuzzy set Description\n");
gets(CoordData);
FzyGetCoordinates(
   CoordData,DomValues,TruthValues,&DCnt,&status);
if(status!=0) return;
FzyCoordSeries(RiskFDBptr,DomValues,TruthValues,DCnt,&status);
```

FzyGetHeight

Format:

```
float FzyGetHeight(FDB *FDBptr);
```

Parameters:

FDB **FDBptr*	A pointer to the fuzzy set.

Returns:

float	The highest truth membership value in the fuzzy set.

Use:

This function examines the truth membership function of the fuzzy set and returns the maximum truth value. This is the height of the fuzzy set.

Example:

```
memval-FzyGetHeight(PriceFDBptr);
if(memval==0) return;
```

FzyGetMembership

Format:

```
float FzyGetMembership(
FDB *FDBptr,double Scalar,int *Indexpos,int *statusptr);
```

Parameters:

FDB *FDBptr*	A pointer to the fuzzy set.
double *Scalar*	A value drawn from the domain of the incoming fuzzy set.
int **Indexpos*	The index or offset position of this scalar in the FDBvector array of the fuzzy set. This is calculated and returned by the function.
int **statusptr*	The status return code.

Returns:

float	The truth membership of this scalar in the fuzzy set.

Status/Errors:

0	No error, truth membership was returned.
1	Scalar is outside domain of the fuzzy set.

Use:

This function returns the membership value for a domain value in the fuzzy set. The actual membership value for a scalar domain value is calculated from the low (left edge) domain and the range of the fuzzy set. This calculation indicates an offset into the truth membership array containing the membership value. The function also returns this index into the truth membership function:

```
FDBptr->FDBvector[Indexpos])
```

corresponding to the actual location of the scalar in the fuzzy set.

Notes:

1. Although an error message is generated if the scalar lies outside the stated domain of the fuzzy set, a membership value is still returned. This follows the assumption that values below the zero truth value continue to have zero truth and those above the one truth level continue to have absolute membership truth. The index value will be either zero or 256 in this case, depending on whether the scalar lies below the fuzzy set's left edge or above the fuzzy set's right edge.

Related Processes:

FzyGetScalar	Returns the scalar value associated with a fuzzy truth membership.

Example:

```
double Scalar=29.0;
truth=FzyGetMembership(hiFDBptr,Scalar,&Index,&status);
```

FZYGETSCALAR

Format:

```
double FzyGetScalar(FDB *FDBptr,int IndexNo,int *statusptr);
```

Parameters:

FDB *FDBptr*	A pointer to the fuzzy set.
int *IndexNo*	The index value into the fuzzy set's truth membership function. This is a number between [0] and [256] indicating which truth value is required.

int *statusptr The status return code.

Returns:

double The scalar domain value corresponding to the center of the truth function's index position.

Status/Errors:

0 Scalar was successfully returned.

1 Index value is outside bounds of the truth membership array.

Use:

This function returns the scalar domain value for a truth value in the fuzzy set. The actual scalar value is calculated from the low (left edge) domain and the range of the fuzzy set. The modeling system stores the low and high domain for a fuzzy set as well as the proportional truth membership function for this range of values. You must use this function to find a domain value corresponding to a particular truth membership (expressed as a cell position in the *FDBvector* array).

Related Processes:

FzyGetMembership Finds the membership value associated with a scalar value from the fuzzy set domain.

Example:

```
j=HiFDBptr->FDBvector[k];
Scalar=FzyGetScalar(HiFDBptr,j,&status);
```

FzyImplMatrix

Format:

```
void FzyImplMatrix(
    FDB *FDBptr1,FDB *FDBptr2,
        double DomValues[],int DomCnt,int *statusptr);
```

Parameters:

FDB *FDBptr1 A pointer to the first fuzzy set.

FDB **FDBptr2*	A pointer to the second fuzzy set.
double *DomValues[]*	An array containing domain values from the first fuzzy set (*FDBptr1*). This array is only used when you want to produce a graph of the implication relationship.
int *DomCnt*	The count of entries in the *DomValues[]* array.
int **statusptr*	The status return code.

Use:

In a fuzzy model we are always supposing that a relationship exists between the antecedent and the consequent fuzzy set regions. This relationship is expressed as a correlation between the general topologies of the corresponding fuzzy truth membership regions. This relation implies that an implication function exists between the two fuzzy regions. As an example, in the weight estimation model, we are evaluating a logical rule,

```
if height is TALL then weight is HEAVY
```

This fuzzy association says that as an individual's height increases so does his or her weight. Further, it says that the truth of the fuzzy proposition *height is TALL* can be used to estimate a weight by restricting the proposition *weight is HEAVY* at the same truth level. The implication matrix function examines the relationship between two fuzzy sets by preparing an *MxN* report showing the minimum truth of the two fuzzy sets at each point along their respective truth membership functions. If you also include a set of domain values from the first fuzzy set, the function creates and graphs an "implication" fuzzy region. The graph is produced by truncating the second fuzzy set's topology at the truth of the first fuzzy set's domain value. This tool provides a valuable facility for exploring the relationship between multiple fuzzy regions.

Notes:

1. The report and graphs are directed to the current model log (generally, *mdllog*.fil). The file associated with the model log is established when the *XSYSctl* block is initialized.

2. The vertical and horizontal values are automatically scaled, if necessary, and the scaling factors (if not one) are displayed at the bottom of the report.

Related Processes:

FzyDrawSet	Draws the surface topology of a fuzzy set.

`FzyExamineSet`	Produces a formatted `MxN` report showing the possibility distribution of a fuzzy set's truth function.
`FzyMonotonicLogic`	Implements this form of direct point-to-point logical reasoning. Monotonic logic relies on a strong (but nevertheless approximate) relationship between the surface topologies of two fuzzy sets.

Example:

```
DomValues[0]=5.5;
DomValues[1]=6.0;
DomCnt=2;
FzyImplMatrix(
TallFDBptr,HeavyFDBptr,DomValues,DomCnt,&status);
```

```
                          Approximate Reasoning System
                      Implication Matrix for 'TALL' and 'HEAVY'

               HEAVY
TALL           12.00 13.62 15.24 16.86 18.47 20.09 21.71 23.33 24.95
 4.00 0.00 |   0.00  0.00  0.00  0.00  0.00  0.00  0.00  0.00  0.00
 4.31 0.12 |   0.00  0.03  0.12  0.12  0.12  0.12  0.12  0.12  0.12
 4.62 0.25 |   0.00  0.03  0.12  0.25  0.25  0.25  0.25  0.25  0.25
 4.93 0.37 |   0.00  0.03  0.12  0.28  0.37  0.37  0.37  0.37  0.37
 5.25 0.50 |   0.00  0.03  0.12  0.28  0.50  0.50  0.50  0.50  0.50
 5.56 0.62 |   0.00  0.03  0.12  0.28  0.50  0.62  0.62  0.62  0.62
 5.87 0.75 |   0.00  0.03  0.12  0.28  0.50  0.72  0.75  0.75  0.75
 6.18 0.87 |   0.00  0.03  0.12  0.28  0.50  0.72  0.87  0.87  0.87
 6.50 1.00 |   0.00  0.03  0.12  0.28  0.50  0.72  0.88  0.97  1.00

   'HEAVY' when 'TALL' is        5.50
    FuzzySet:   HEAVY
    Description:
  1.00
  0.90
  0.80
  0.70
  0.60                                              .............................
  0.50                                          ....
  0.40                                       ...
  0.30                                   ....
  0.20                              .....
  0.10                        ......
  0.00..............
      0---|---|---|---|---|---|---|---|---|---|---|---|---|---|---|---0
  120.00  136.25  152.50  168.75  185.00  201.25  217.50  233.75  250.00

    Domained:         120.00 to      250.00
    'HEAVY' when 'TALL' is        6.00
    FuzzySet:   HEAVY
    Description:
  1.00
  0.90
  0.80                                                      .....................
  0.70                                                 .....
  0.60                                              ...
  0.50                                          ....
  0.40                                       ...
  0.30                                   ....
  0.20                              .....
  0.10                        ......
  0.00..............
      0---|---|---|---|---|---|---|---|---|---|---|---|---|---|---|---0
  120.00  136.25  152.50  168.75  185.00  201.25  217.50  233.75  250.00
    Domained:         120.00 to      250.00
```

FzyInitCIX

Format:

```
void FzyInitCIX(CIX *CIXptr);
```

Parameters:

CIX *CIXptr	A pointer to a CIX structure.

Use:

This function initializes the CIX structure. The CIX supports the statistical compatibility index block. You will use this structure to measure the relationship between your fuzzy model and the model data.

Related Processes:

FzyStatCompIndex	Measures the statistical compatibility index in a fuzzy model.
FzyUnitCompIndex	Measures the individual or unit compatibility in a fuzzy model.

Example:

```
//--initialize the compatibility index block
FzyInitCIX(&pricingCIXblk);
```

FzyInitFDB

Format:

```
void FzyInitFDB(FDB *FDBptr);
```

Parameters:

FDB *FDBptr	A pointer to a fuzzy set descriptor block (FDB).

Use:

This function initializes a fuzzy set descriptor block (FDB). You would not normally issue this function since it is a component of the *FzyCreateSet* function.

Notes:

1. If you are using the advanced policy modeling capabilities, you must not use this function to reinitialize the properties of a fuzzy set. Doing so will destroy the linkage field in the structure that chains the fuzzy set into the current model or segment dictionary.

Related Processes:

FzyCreateSet	Allocates and creates a new fuzzy set descriptor block (FDB).

Example:

```
//--Initialize the fuzzy set block
FzyInitFDB(HiPriceFDBptr);
```

FzyInitFZYctl

Format:

```
void FzyInitFZYctl(int *statusptr);
```

Parameters:

int *statusptr*	The status return code from the initialization.

Use:

This function initializes the *XFZYctl* block structure. The fuzzy logic control block controls the approximate reasoning implication process. For each fuzzy solution variable (FSV) in your model, it maintains an entry in the fuzzy control block. This block is used to support the parallel processing capabilities of fuzzy logic on a nonparallel machine. You must issue the *FzyInitFZYctl* function at the beginning of you application model.

Related Processes:

FzyAddFZYctl	Adds a variable to the fuzzy control block, allocating a new fuzzy solution variable (FSV) block.

Example:

```
//--set up for fuzzy reasoning
FzyInitFZYctl(&status);
```

FzyInitHDB

Format:

```
void FzyInitHDB(HDB *HDBptr);
```

Parameters:

HDB *HDBptr*	A pointer to a hedge descriptor block (HDB).

Use:

This function initializes a hedge descriptor block (HDB). You would not normally issue this function since it is a component of the *FzyCreateHedge* function.

Notes:

1. If you are using the advanced policy modeling capabilities, you must not use this function to reinitialize the properties of a hedge. Doing so will destroy the linkage field in the structure that chains the hedge into the current model or segment dictionary.

Related Processes:

FzyCreateHedge	Allocates and creates a new hedge descriptor block (HDB).

Example:

```
//--Initialize the hedge block
FzyInitHedge(SlightlyHDBptr);
```

FzyInitVector

Format:

```
void FzyInitVector(float Vector[],int Veclen,float InitValue);
```

Parameters:

float *Vector[]*	A truth membership array.
int *Veclen*	The extent of the truth membership array. This is normally stored in the symbolic constant *VEC-MAX* in the *mtypes.hpp* header.

float *InitValue* — The initial value that will be stored in each array position of the truth membership array.

Use:

This function initializes a single precision floating-point array with an initial value. This is not usually a user function, but is called by many of the service routines.

Example:

```
//--initialize this vector to 1
FzyInitVector(FDBptr->FDBvector,VECMAX,ONE);
```

FzyInterpVec

Format:

```
void FzyInterpVec(float Vector[],int *statusptr);
```

Parameters:

float *Vector[]* — A truth membership vector dimensioned to *VECMAX* (see the *mtypes.hpp* header)

int **statusptr* — The status return code.

Status/Errors:

0 Interpolation was successfully done.

1 Too many intermediate values.

Use:

This function performs a piecewise linear interpolation across a truth vector with distributed truth values. To use this function, the truth membership is initially populated with "-1"—these are the unoccupied cells. Known values are then entered into the array as positive values (this method of interpolation works since truth membership values can never be negative). Linear interpolation works by finding truth vector segments bracketed by actual values. Having found an actual value, the interpolation facility loops across the truth vector until it discovers a slot that does not contain a "-1." We then calculate "*k*" as the proportional distance between the previous and the current value and use this to determine the linear distance between the two points.

This is not usually a user function. It is invoked by the utility routines that handle the creation of fuzzy set membership functions from truth series and from the coordinate memberships.

Related Processes:

FzyCreateSet	Creates a fuzzy set. The *MEMSERIES* generation type creates a fuzzy set from an equally distributed collection of membership values.
FzyCoordSeries	Creates a fuzzy set truth membership function from a collection of domain values with their truth membership grades.

Example:

```
//--initialize the working truthvector
FzyInitVector(WorkVector,VECMAX,MINUSONE);
k=VECMAX/MaxElements;
for(i=0;i<MaxElements;i++)
   {
      WorkVector[k]=TruthFunction[i];
      ++k;
   }
FzyInterpVec(WorkVector,&status);
```

FzyIsNormal

Format:

```
bool FzyIsNormal(float TruthVector[],int Veclen);
```

Parameters:

float *TruthVector[]*	A standard fuzzy set truth membership function array.
int *Veclen*	The extent of the truth membership array. This is normally dimensioned as *VECMAX* (see the *mtypes.hpp* header file).

Returns:

bool	TRUE if the fuzzy truth vector is normal, otherwise FALSE is returned.

Use:

You can use this function to determine whether or not a fuzzy set is normal. Normal fuzzy sets are those with a height of one ([1.0]). All your base fuzzy sets must be normal in order to use the implication facilities correctly.

Related Processes:

FzyGetHeight	Returns the height of a fuzzy set. If this height is [1.0] then the fuzzy set is normal.

Example:

```
if(!(FzyIsNormal(hiPriceFDBptr->FDBvector,VECMAX))
   {
      fprintf(stdout,"%s%s%s\n",
         "Warning. Fuzzy set '",hiPriceFDBptr->FDBid,
         "' is not Normal.")
```

FzyLinearCurve

Format:

```
void FzyLinearCurve(
    FDB *FDBptr,
    double LeftEdge, double RightEdge,int LineType,int *statusptr);
```

Parameters:

FDB *FDBptr	A pointer to an initialized fuzzy set descriptor block (FDB). The linear curve generator only creates the truth membership function, it does not alter any other properties of the fuzzy set. Since the linear topology generator references the domain in the FDB, this structure must be completed before invoking this function.
double *LeftEdge*	The left edge of the linear curve. This may or may not correspond to the low domain value, however,

LeftEdge ≥ FDBdomain[0]

double *RightEdge*	The right edge of the linear curve. This may or may not correspond to the high domain value, however,
	RightEdge ≤ *FDBdomain[1]*
int *LineType*	The type of linear fuzzy set. You can specify two symbolic constants as the parameter,
DECREASE	a linear curve that goes from true at the low domain to false at the high domain
INCREASE	a linear curve that goes from false at the low domain to true at the high domain.
int *statusptr	The status return code.

Status/Errors:

0	No error, the fuzzy set membership function is created.
1	An error in the domain specification. Either the high and low domains are both the same, or the low domain is greater than the high domain.
3	Unrecognized *LineType*.

Use:

This low-level surface topology generator creates a linear fuzzy set truth membership function between the limits of the *LeftEdge* and *RightEdge*. Generally, you will not use this function directly since it is incorporated into the *FzyCreateSet* function; however, the linear surface generator can be used to change the current surface of an existing fuzzy set or where the overhead of including the entire fuzzy set creation function is not necessary.

Linear fuzzy sets provide good approximations for phenomena that are modeled in a steady, proportional, and linear way. The idea of *HIGH* priced as a measure of a product (or a competitor's product) price is a good illustration. As the price increases, the degree to which it is *HIGH* also increases steadily. The same is true for a decreasing linear function (such as *LOW* for price).

You can place the linear fuzzy set within an enclosing domain by specifying different values for *LeftEdge* and *RightEdge* parameters. The actual bounding domain for the fuzzy set is taken from the domain specification in the fuzzy set descriptor block (FDB).

Related Processes:

FzyCreateSet	The general-purpose fuzzy set creation function.

Example:

```
//--create a linear fuzzy set to represent high prices
if(!HiFDBptr=new FDB))
   {
      *statusptr=1;
      return;
   }
FzyInitSet(HiFDBptr);
strcpy(HiFDBptr->FDBid,"HighPrice");
HiFdbptr->FDBdomain[0]=16.0;
HiFDBptr->FDBdomain[1]=36.0;
ledge=18.0;
redge=32.0;
FzyLinearCurve(HiFDBptr,ledge,redge,INCREASE,&status);
•
•
```

FzyMemSeries

Format:

```
void FzyMemSeries(FDB *FDBptr,float TFarray[],int TFCnt,int
                        *statusptr);
```

Parameters:

FDB *FDBptr*	A pointer to the fuzzy set descriptor block (FDB). The membership series curve generator only creates the truth membership function, it does not alter any other properties of the fuzzy set. Since the irregular topology generator references the domain in the FDB, this structure must be completed before invoking this function (see the example).
float *TFarray[]*	An array containing a collection of truth membership function values. These must be values in the range [0] to [1] and at least one of the values must be [1]. That is, the fuzzy set must have a normal surface.

int *TFCnt*	The extent of the *TFarray[]* parameter. This is the number of truth membership points in the array.
int **statusptr*	The status return code.

Status/Errors:

0	No error, the surface was created.
1	This is not a normal fuzzy set. A normal fuzzy set has a height of [1.0].
66	Too many membership function points.

Use:

As an alternative method of specifying the topology of a fuzzy set, you can directly describe its shape through a series of equidistant truth function points. This function takes the domain specification in the FDB and positions the truth function values in *TFarray[]* so that they are spread evenly across the membership array (FDB vector). Piecewise linear interpolation (using *FzyInterpVec*) is then used to complete the truth membership function.

Notes:

1. You must specify at least three (3) membership function points, and a maximum of 128 function points.

2. The fuzzy set topology must be normal. That is, you must have at least one membership function point that is [1.0]. Failure to create a normalized fuzzy set will cause an error.

Related Processes:

`FzyCoordSeries`	This is similar to the membership series function. The fuzzy set topology is created by specifying a collection of domain values and their membership truth values.
`FzyCreateSet`	The general-purpose fuzzy set creation function.

Example:

```
//--create a triangular fuzzy set
if(!AvgCostFDBptr=new FDB))
```

```
    {
        *statusptr=1;
        return;
    }
FzyInitSet(AvgCostFDBptr);
strcpy(AvgCostFDBptr->FDBid,"AvgCost");
AvgCostFdbptr->FDBdomain[0]=16.0;
AvgCostFDBptr->FDBdomain[1]=36.0;
TruthSeries[]={0 1 0};
TScnt=3;
FzyMemSeries(AvgCostFDBptr,TruthSeries,TScnt,&status);
```

FzyMonotonicLogic

Format:

```
double FzyMonotonicLogic(FDB *fromFDBptr,FDB *toFDBptr,
    double fromDomValue,int *statusptr);
```

Parameters:

FDB *fromFDBptr	A pointer to the fuzzy region that will drive the monotonic logic function.
FDB *toFDBptr	A pointer to the fuzzy region that will be used to find the corresponding domain value for the *fromDomValue* parameter.
double fromDomValue	A scalar value selected from the domain of the *fromFDBptr* fuzzy set.
int *statusptr	The status return code.

Returns:

double	A value from the domain of the *toFDBptr* fuzzy set corresponding to the degree of truth implication function in the *fromFDBptr* fuzzy set.

Use:

Monotonic logic means that an approximately linear one-to-one relationship exists between two fuzzy sets. When you know a value in one fuzzy set, you can elicit a value from the second fuzzy set based on the truth membership function of the first value. This type of reasoning is an important capability in many fuzzy models since standard forms of defuzzification will not suffice when such a direct causal dependency exists. Center of gravity and plateau-based defuzzification techniques will always return a central fuzzy set measurement rather than a point-to-point correspondence.

For each value in the base fuzzy set (the *fromFDBptr* set), we can determine a value in the dependent fuzzy set (the *toFDBptr* fuzzy set). This type of monotonic reasoning also allows the mixing of fuzzy and crisp statements in conventional decision support and expert systems.

Notes:

1. Neither the base nor the dependent fuzzy sets need be static fuzzy sets from your model's vocabulary. These fuzzy sets can be created through surface transformers (such as hedges) or by applying compensatory operators to two or more fuzzy sets.

2. You can use the *FzyImplMatrix* function to examine the implication surface between two fuzzy sets and plot the implication fuzzy region. This should be your starting point in applying monotonic fuzzy reasoning.

Related Processes:

`FzyCondProposition`	Applies a conditional (truth correlated) fuzzy region to the output fuzzy set.
`FzyDefuzzify`	Finds the expected value of an output fuzzy region.
`FzyUnCondProposition`	Applies an unconditional (truth uncorrelated) fuzzy region to the output fuzzy set.

Example:

```
//--Find weight given a particular height
Weight=FzyMonotonicLogic(
   TallFDBptr,HeavyFDBptr,Height,&status);
```

FzyNormalizeSet

Format:

```
void FzyNormalizeSet(FDB *FDBptr);
```

Parameters:

FDB *FDBptr	A pointer to a fuzzy set.

Use:

This function normalizes a fuzzy set. A set is normalized by finding its maximum truth value and making this one ([1.0]). All other values are scaled proportionally.

Related Processes:

FzyIsNormal	Indicates whether or not the fuzzy set is normal.
FzyGetHeight	Finds the maximum truth value of a fuzzy set (this is the "height" of a fuzzy set).

Example:

```
if(!(FzyIsNormal(HiFDBptr)) FzyNormalizeSet(HiFDBptr);
```

FzyOR

Format:

```
float FzyOR(float Truthval1,float Truthval2);
```

Parameters:

float *Truthval1*	A truth membership function in the range [0,1]
float *Truthval2*	A truth membership function in the range [0,1]

Returns:

float	The result of applying the Zadeh OR operation to the two truth membership values.

Status/Errors:

-1	The first truth value is not within bounds.
-2	The second truth value is not within bounds.

Use:

This function performs a standard Zadeh `OR` on the two truth values. The Zadeh `OR` returns the maximum of the two truth values.

Related Processes:

`FzyCompOR`	The compensatory `OR` processor. By selecting the *ZADEHOR* class this is equivalent to the standard Zadeh `OR` operation.

Example:

```
for(i=0;i<VECMAX;i++)
   Orvalue=FzyOR(
      HiFDBptr->FDBvector[i],LoFDBptr->FDBvector[i]);
```

FzyPiCurve

Format:

```
void FzyPiCurve(
   FDB *FDBptr,double Center,double Width,int *statusptr);
```

Parameters:

FDB **FDBptr*	A pointer to a fuzzy set. The PI curve generator only creates the truth membership function, it does not alter any other properties of the fuzzy set. Since the bell-shaped topology generator references the domain in the FDB, this structure must be completed before invoking this function (see the example).
double *Center*	The center value of the PI curve. This is the value from the domain where the truth membership function is one ([1.0]).
double *Width*	The band width or radius of the PI curve. This value is added to and subtracted from the *Center* value to give the conical diameter of the fuzzy set.
int **statusptr*	The status return code.

Status/Errors:

0	No error, bell-shaped fuzzy set was created.
1	Invalid width.
3	Center of curves lies outside the fuzzy set domain.
5	Edge of the PI curve extends beyond domain of the fuzzy set.

Use:

This low-level surface topology generator creates a bell-shaped fuzzy set truth membership function in the shape of a PI curve. Generally, you will not use this function directly since it is incorporated into the *FzyCreateSet* function; however, the PI curve surface generator can be used to change the current surface of an existing fuzzy set or where the overhead of including the fuzzy set creation function is not necessary.

The width of the PI curve is an important property of bell-shaped fuzzy sets since this determines how well the concept maps to the underlying model semantics. You can control this explicitly through the *Width* parameter. The smaller the width, the more sharply the truth function rises toward [1.0].

Notes:

1. You should also note that the width of a bell-shaped fuzzy set can be changed through the application of the approximation hedge *AROUND*, as well as the intensification and dilution hedges *VERY* and *SOMEWHAT*. The *AROUND* hedge will increase the fuzziness of the curve by broadening the curve.

Related Processes:

`FzyBetaCurve`	This low-level function generates a beta-type bell-shaped curve. A beta curve is similar to the PI curve except that it goes to zero at infinity and is specified with slightly different parameters.
`FzyCreateSet`	The general-purpose fuzzy set creation function.

Example:

```
//--create a Pi-curve fuzzy set
if(!AroundOurCostFDBptr=new FDB))
   {
```

```
        statusptr=1;
        return;
    }
FzyInitSet(AroundOurCostFDBptr);
strcpy(AroundOurCostFDBptr->FDBid,"AroundOurCost");
AroundOurCostFdbptr->FDBdomain[0]=16.0;
AroundOurCostFDBptr->FDBdomain[1]=36.0;
double Center=24;
double Width=3;
FzyPiCurve(AroundOurCostFDBptr,Center,Width,&status);
```

FzyPlotSets

Format:

```
void FzyPlotSets(
    FDB *FDBptr[],int FzySetCnt,int Medium,int *statusptr);
```

Parameters:

FDB **FDBptr[]*	An array of pointers to a collection of fuzzy sets. A maximum of 11 fuzzy sets can be specified in the array.
int *FzySetCnt*	The extent of the *FDBptr[]* array. This is the number of fuzzy sets specified in the array.
int *Medium*	A symbolic constant that indicates where the graph will be written. You can use the following constants:

SYSLOGFILE	the system log
SYSMODFILE	the system model trace log
SYSOUT	the terminal

int **statusptr*	The status return code from the draw function.

Status/Errors:

0 No error, the graph is produced.

1 Unrecognized medium constant.

3 Too many fuzzy sets.

Use:

This function draws a character-based graph showing the topology of the specified fuzzy sets. You can use this function to draw the fuzzy sets associated with a particular variable or sets that are semantically related. With this tool you can view the results of such activities as fuzzy set creation, hedge operations, standard and compensatory operator functions, and implication functions.

Notes:

1. The graph is directed to the specified log file. The file associated with the model log is established when the *XSYSctl* block is initialized.

2. You cannot change the width of the graph.

3. Each fuzzy set will be represented by a different symbol. The symbol legend is printed at the bottom of the graph. When fuzzy set surfaces overlap, the symbol for the highest level fuzzy set will be used.

4. The underlying horizontal axis for the graph is determined by taking the minimum and maximum domain edges of all the incoming fuzzy sets. If the domains are not compatible, a serious graphic output problem can occur.

Related Processes:

`FzyDrawSet`	This related procedure graphs a single fuzzy set.

Example:

```
//--Before plotting a series of fuzzy sets, we will look at
//--them individually to see their topologies.
//
FzyDrawSet(HiFDBptr,SMALLPLOT,&status);

       FuzzySet:   HIGH
        Description: High for Price
      1.00                                                          .
      0.90                                                    ......
      0.80                                             .......
      0.70                                       ......
      0.60                                 .......
```

```
      0.50                                 ......
      0.40                           ......
      0.30                     .......
      0.20               ......
      0.10         .......
      0.00......
          0---|---|---|---|---|---|---|---|---|---|---|---|---|---|---|---0
        16.00   18.50   21.00   23.50   26.00   28.50   31.00   33.50   36.00
        Domained:        16.00 to       36.00

//
//
FzyDrawSet(loFDBptr,SYSMODFILE,&status);

        FuzzySet:   LOW
        Description: Low for Price
      1.00.
      0.90 .......
      0.80        ......
      0.70              ......
      0.60                    .......
      0.50                           ......
      0.40                                 .......
      0.30                                        ......
      0.20                                              .......
      0.10                                                     ......
      0.00
                                                                     ......
          0---|---|---|---|---|---|---|---|---|---|---|---|---|---|---|---0
        16.00   18.50   21.00   23.50   26.00   28.50   31.00   33.50   36.00
        Domained:        16.00 to       36.00

//
//
FzyDrawSet(HiandLoFDBptr,SYSMODFILE,&status);

        FuzzySet:   HIGH&LOW
        Description: Both High and Low
      1.00
      0.90
      0.80
      0.70
      0.60
      0.50                                 .
      0.40                           ...... .......
      0.30                     .......             ......
      0.20               ......                         .......
      0.10         .......                                     ......
      0.00......                                                     ......
          0---|---|---|---|---|---|---|---|---|---|---|---|---|---|---|---0
        16.00   18.50   21.00   23.50   26.00   28.50   31.00   33.50   36.00
        Domained:        16.00 to       36.00

//--Now let's plot these overlapping fuzzy sets along the same
//--domain range.
//
```

```
FDBarray[0]=HiFDBptr;
FDBarray[1]=LoFDBptr;
FDBarray[2]=HiandLoFDBptr;
FDBCnt=3;
FzyPlotSets(FDBarray,FDBCnt,SYSMODFILE,&status);
      1.00*                                                          .
      0.90 *******                                             ......
      0.80        ******                                 .......
      0.70              ******                     ......
      0.60                    *******        .......
      0.50                           *****+.....
      0.40                          ++++++ +++++++
      0.30                   +++++++              ++++++
      0.20             ++++++                           +++++++
      0.10      +++++++                                        ++++++
      0.00++++++                                                    ++++++
          0---|---|---|---|---|---|---|---|---|---|---|---|---|---|---0
        16.00   18.50   21.00   23.50   26.00   28.50   31.00   33.50   36.00
      .  FuzzySet:    HIGH
         Description: High for Price
      *  FuzzySet:    LOW
         Description: Low for Price
      +  FuzzySet:    HIGH&LOW
         Description: Both High and Low
         Domained:        16.00 to      36.00
```

FzySCurve

Format:

```
void FzySCurve(
    FDB *FDBptr,
        double LeftEdge,double Inflexion,double RightEdge,
            int CurveType,int *statusptr);
```

Parameters:

FDB *FDBptr	A pointer to a fuzzy set. The S-curve generator only creates the truth membership function, it does not alter any other properties of the fuzzy set. Since the sigmoid topology generator references the domain in the FDB, this structure must be completed before invoking this function (see the example).
double *LeftEdge*	The left edge of the S-curve. This can be within the specified domain with the exception that it is restricted by the limitation

`LeftEdge ≥ FDBdomain[0]`

double *Inflexion* — The inflection point of the S-curve. This is the value from the domain where the truth membership function is ([.5]). This means that half the domain values fall above and below this value.

double *RightEdge* — The right edge of the S-curve. This can be within the specified domain with the exception that it is restricted by the limitation,

`RightEdge ≤ FDBdomain[1]`

int *CurveType* — The surface topology generator supports two kinds of S-curves and these are indicated by these symbolic constants:

`DECLINE` — a left-facing S-curve that is true at the left edge of the domain and false at the right edge of the domain.

`GROWTH` — a right-facing S-curve that is false at the left edge of the domain and true at the right edge of the domain.

Both curves are centered around an inflection point at the [.50] membership function level.

int **statusptr* — The status return code.

Status/Errors:

0 No error, S-curve fuzzy set was created.

1 The left edge is greater than the right edge, or the right edge is less than the inflection point value.

Use:

This low-level surface topology generator creates a sigmoid (or logistic) fuzzy set truth membership function in the shape of a S-curve. Generally, you will not use this function directly since it is incorporated into the *FzyCreateSet* function; however, the S-curve surface generator can be used to change the current surface of an existing fuzzy set or where the overhead of including the fuzzy set creation function is not necessary. The S-curve fuzzy surface is an important topology since it represents the growth distribution for many natural phenomena. Another important characteristic of the S-curve fuzzy set is its ability to model properties that are difficult, if not impossible, to represent in conventional first-order predicate logic.

Notes:

1. You should also note that the shape of an S-shaped fuzzy set can be changed through the application of the approximation hedge *AROUND*, as well as the intensification and dilution hedges *VERY* and *SOMEWHAT*. The *AROUND* hedge will increase the fuzziness of the curve by broadening the curve.

Related Processes:

FzyCreateSet	The general-purpose fuzzy set creation function.

Example:

```
//--create an S-curve fuzzy set
if(!RiskFDBptr=new FDB))
    {
        *statusptr=1;
        return;
    }
FzyInitSet(RiskFDBptr);
strcpy(RiskFDBptr->FDBid,"Risk");
RiskFdbptr->FDBdomain[0]=0.0;
RiskFDBptr->FDBdomain[1]=100.0;
double Redge=0;;
double Inflexion=50;
double Ledge=100
FzySCurve(RiskFDBptr,ledge,Inflexion,Redge,GROWTH,&status);
```

FZYSTATCOMPINDEX

Format:

```
bool FzyStatCompIndex(
    FDB *FDBptr,float ExpValTruth,CIX *CIXptr);
```

Parameters:

FDB *FDBptr	A pointer to an output fuzzy set, that is, a fuzzy set maintained in a fuzzy solution variable (FSV) block.

float *ExpValTruth*	The truth associated with the expected value of the output. This is the height of the output fuzzy set region at the time of defuzzification. There are two ways to calculate this height: the height of the fuzzy set and the height at the point of defuzzification. You are free to choose either metric.
CIX **CIXptr*	A pointer to the compatibility index control block (CIX). The CIX structure maintains the statistics associated with multiple invocations of the model.

Returns:

bool	`TRUE` when the statistical compatibility index is within bounds; `FALSE` when the current compatibility drops below or moves above the compatibility boundary.

Use:

Fuzzy models have an intrinsic way of measuring their compatibility with model data. This compatibility is a statistical metric, that is, it is meaningful only for a large number of model runs where we can measure the mean variations in output fuzzy set height. The idea behind the compatibility index is simple: if the mean height of an output fuzzy region is close to [0] or [1], then the model assumes the properties of a Boolean space. This very high or low height indicates that the data lie at the extremes of the fuzzy sets, causing the predicate truth to be consistently close to zero or one.

Related Processes:

`FzyInitCIX`	Initializes the CIX structure.
`FzyUnitCompIndex`	Calculates the unit or single run compatibility index and compares this to the compatibility threshold level.

Example:

```
if(!(CIXptr=new CIX))
   {
     *statusptr=1;
     MtsSendClientError(107,PgmId,"");
     return;
```

```
    }
int InCompRun=0;
FzyInitCIX(CIXptr);
•
•
if(!(FzyStatCompIndex(HiFDBptr,PredTruth,CIXptr))
        InCompRun++;
```

FzySupportSet

Format:

```
void FzySupportSet(
   FDB *FDBptr,double Scalars[],int TFIndex[],int *statusptr);
```

Parameters:

FDB **FDBptr*	A pointer to a fuzzy set.
double *Scalars[]*	An array of domain values holding the left and right edges of the support set. Two values are always returned from the function (although they may both have the same value).
int *TFIndex[]*	An array holding the index in the fuzzy set's truth membership function corresponding to the domain values in the *Scalars[]* array.
int **statusptr*	The status return code.

Status/Errors:

0	No errors, support set is returned.
1	This is an empty fuzzy set (all membership values are zero).

Use:

The support set for a fuzzy set is that region of the domain where the truth membership function is greater than zero. This function examines the membership array and finds the left and right nonzero edges (if they exist). The domain values corresponding to the membership positions are returned along with the index values into the *FDBvector[]* array.

Notes:

1. The *FzySupportSet* is insensitive to any unapplied alpha-cut threshold. It simply examines the truth membership function.

Related Processes:

`FzyGetHeight`	Finds the maximum truth value in the membership function.

Example:

```
//--find the support set for this fuzzy set
FzySupportSet(HiFDBptr,Scalars,Indexes,statusptr);
```

FzyTrueSet

Format:

```
FDB *FzyTrueSet();
```

Returns:

FDB*	A pointer to the true fuzzy set that is created and populated by this function.

Use:

This function creates a linear increasing fuzzy set of proportional truth values with a domain from one to *VECMAX-1*. This type of fuzzy set plays an important part in system calibration and in fuzzy database operations. The `TRUE` fuzzy set returns a truth proportional to the percentage of the domain.

Related Processes:

`FzyLinearCurve`	Creates a linear fuzzy set.

Example:

```
TrueFDB=FzyTrueSet();
```

FzyUnCondProposition

Format:

```
void FzyUnCondProposition(
   FDB inFDBptr*,FSV *FSVptr,int *statusptr);
```

Parameters:

FDB `*inFDBptr`	A pointer to the fuzzy set that will be used to update the undergeneration solution variable's fuzzy region (see the FSV parameter).
FSV `*FSVptr`	A pointer to the fuzzy solution variable (FSV) block stored in the *XFZYctl* structure. This pointer is returned by *FzyAddXFZYctl* when the solution variable is added to the model. You should store this pointer address for use in the *FzyCondProposition* function.
int `*statusptr`	The status return code from the conditional fuzzy proposition.

Status/Errors:

0	No error, unconditional fuzzy set is applied.
1	Incoming fuzzy set is empty.

Use:

This function updates the fuzzy solution variable block for a model output variable using the standard Zadeh min-max implication method. For unconditional fuzzy propositions, that is, a proposition that would be expressed in the Metus rule formulation language as an assertive statement, the undergeneration *FSV* is updated by taking the minimum of the current *FSV* fuzzy region and the incoming fuzzy set (*inFDBptr*). This is the same as `AND`ing the new fuzzy set with the *FSV* contents.

In an unconditional fuzzy proposition the truth of the consequent fuzzy region is unrestricted by any predicate or premise fuzzy proposition. The consequent fuzzy region is used to limit the current solution variable's output fuzzy set using the minimum functional relationship (the maximum output fuzzy set truth is limited by the minimum truth of the consequent). This function does not involve the application of a correlation method to the current consequent fuzzy set.

Unconditional propositions provide model builders with the ability to construct default support topologies for their models. If none of the conditional fuzzy propositions executes, the model space will be developed through the unconditionals.

Notes:

1. Note that an unconditional fuzzy proposition does not actually restrict the topology of the output fuzzy region. Conditional rules with predicate truths greater than the

maximum truth of any unconditional proposition will cause the output fuzzy region's height to increase above the unconditional level. This is in keeping with the general philosophy of unconditionals: they provide the underlying default surface for a model when the truth of any conditional fuzzy predicate is insufficient to execute the rule (either the predicate condition is false or the generated truth lies below the current alpha-cut threshold).

2. Since unconditional propositions are not subject to truth restrictions you should use them with care in fuzzy models. An unconditional is always applied.

Related Processes:

`FzyCondProposition`	Implements a conditional fuzzy proposition. This is equivalent to an IF-THEN statement in the Metus rule formulation language.

Example:

```
Domain[0]=16.0;
Domain[1]=36.0;
HiFDBptr=FzyCreateSet(
   "HIGH",INCREASE,Domain,Parms,0,&status);
LoFDBptr=FzyCreateSet(
   "LOW",INCREASE,Domain,Parms,0,&status);
//--Rule 1. Our price must be high
FzyUnCondProposition(HiFDBptr,PriceFSVptr,&status);
FzyDrawSet(PriceFDBptr,SMALLPLOT,&status);
//--Rule 2. Our price must be low
FzyUnCondProposition(LoFDBptr,PriceFSVptr,&status);
FzyDrawSet(PriceFDBptr,SMALLPLOT,&status);
```

FzyUnitCompIndex

Format:

```
bool FzyUnitCompIndex(FDB *FDBptr,
   float MinTFval,float MaxTFval,
      float *AvgTFval,float *DeltaTFval)
```

Parameters:

FDB **FDBptr*	A pointer to an output fuzzy set maintained in the FSV block structure.

float *MinTFval*	The minimum truth membership value threshold for true compatibility.
float *MaxTFval*	The maximum truth membership value threshold for true compatibility.
float **AvgTFval*	The average truth membership value in the output fuzzy set. This value is calculated and returned by the function.
float **DeltaTFval*	The change in truth membership within the output fuzzy set. This is the difference between the maximum and the minimum truth membership function values.

Returns:

bool	`TRUE` when the unit compatibility index is within bounds; `FALSE` when the current compatibility drops below or moves above the compatibility boundary.

Use:

Fuzzy models have an intrinsic way of measuring their compatibility with model data. This overall compatibility is a statistical metric (see the *FzyStatCompIndex*), that is, it is meaningful only for a large number of model runs where we can measure the mean variations in output fuzzy set height. Unit variations in the model performance are less critical but, for any particular model execution, can indicate whether or not the solution space is consistent with the model semantics. The concept behind the unit compatibility index is simple: IF the mean height of an output fuzzy region falls below the *MinTFval* or above the *MaxTFval*, then the model assumes some properties of a Boolean space. This very high or low height indicates that the data lie at the extremes of the fuzzy sets, causing the predicate truth to be consistently close to zero or one.

Related Processes:

`FzyStatCompIndex`	This function calculates the statistical compatibility index for a model.

Example:

```
if(!FzyUnitCompIndex(
      PriceFDBptr,MinTF,MaxTF,&AvgTF,&ChgTF))
   {
      printf(
      "Note: Model is outside Compatibility Limits\n");
      InCompRun++;
   }
```

Appendix A

The Combs Method for Rapid Inference

William E. Combs
The Boeing Company
P.O. Box 3707, MS: 6A-06
Seattle, WA 98124
(425)477-3970
william.e.combs@boeing.com

The Combinatorial Problem: Fuzzy Logic's Achilles' Heel

If you have ever tried to use fuzzy logic in complex, real-time systems, you have likely struggled with what is popularly known as the combinatorial rule explosion problem. Rules are traditionally defined by the intersection of the input subsets. As inputs increase, the number of rules can explode exponentially, quickly diminishing performance to unacceptable levels in time-critical applications.

It is one thing to produce an actuarial table with an overnight batch program or balance an inverted pendulum. It is quite another to find an optimum path in an ever-changing, on-line network or to control the dynamics of a wind tunnel.

As a quick example, suppose that each input universal set has five input subsets. If a rule matrix were generated to contain all possible rules, then a two-input, one-output system would contain 25 rules; a three input system would contain 125 rules; . . . and a six-input system would contain 15,625 rules! Since every rule is processed during each inference cycle, this exponential increase can bring an on-line application to its knees.

Rule reduction algorithms have been proposed using neural networks, genetic algorithms, and a variety of clustering techniques to select only those rules that make a significant contribution to the inference output. Though these methods can significantly reduce the number of system rules, they do not address the reason why the rules increase exponentially as inputs are added to the system.

What we would like is a rule configuration that models the entire problem space without incurring a combinatorial penalty. In order to understand how to do that, we need to go to the heart of the combinatorial problem.

We are accustomed to rules in this format: If the temperature is cool and the rate of temperature change is decreasing, then the furnace output should be moderately high. Such a rule can be generalized to the following propositional logic construct:

$$\text{If } (p \text{ and } q) \text{ then } r \tag{1}$$

Note, however, that neither p nor q has an independent relation with r. It is only the intersection of p and q that has this relation. If we define this intersection as w, then Eq. (1) can be restated as:

$$\text{If } w \text{ then } r \tag{2}$$

In fact, w can represent the intersection value of any number of antecedent elements in the rule configuration of Eq. (1). Since this rule configuration focuses on the intersection of the antecedent elements, let's call it the intersection rule configuration (IRC).

There are two significant difficulties with the IRC, the first of which affects the combinatorial problem:

- Whenever either p or q changes, w is altered as well, producing a different relation with r (a new rule).
- The relation of w with r is only as accurate as our ability to define p's intersection with q. As antecedent elements increase, this intersection definition becomes more complex.

What we would prefer in a rule configuration is a way for the antecedent elements to relate directly to their consequent counterparts. Using propositional logic, we can transform the following four rule configurations into a more desirable structure (the first line is boldfaced because we will focus on it later):

- **[(p and q) then r] is equivalent to [(p then r) or (q then r)] (3)**
- [(p or q) then r] is equivalent to [(p then r) and (q then r)] (4)
- [p then (r and s)] is equivalent to [(p then r) and (p then s)] (5)
- [p then (r or s)] is equivalent to [(p then r) or (p then s)] (6)

Of course, I don't expect you to take my word for these four constructs. So, let's prove their equivalence by using truth tables. In propositional logic, a truth table can be built on the relation between a single antecedent and a single consequent, defined as (*p* implies *r*) or (if *p* then *r*). Since it is somewhat difficult to understand what implication actually means, we can state (if *p* then *r*) in terms of its material implication equivalent, (not *p* or *r*). Then, taking the four possible values for *p* and *r*, we can build the table in Figure A.1.

p	**not *p***	***r***	**not *p* or *r***	**if *p* then *r***
T	F	T	**T**	T
T	F	F	**F**	F
F	T	T	**T**	T
F	T	F	**T**	T

Figure A.1

Looking at the implication (if *p* then *r*) another way, we can say that if *p* is the case, then *r* must also be the case. Thus, the second row of Figure A.1 is not valid. With this table as a starting point, we can test whether our alternative rule configurations are equivalent. For the tables in Figures A.2 through A.5, we see that the boldfaced columns are identical.

[(*p* and *q*) then *r*] is equivalent to [(*p* then *r*) or (*q* then *r*)]:

p	***q***	***r***	**(*p* and *q*)**	**(*p* and *q*) then *r***	**(*p* then *r*)**	**(*q* then *r*)**	**(*p* then *r*) or (*q* then *r*)**
T	T	T	T	**T**	T	T	T
T	T	F	T	**F**	F	F	F
T	F	T	F	**T**	T	T	T
T	F	F	F	**T**	F	T	T
F	T	T	F	**T**	T	T	T
F	T	F	F	**T**	T	F	T
F	F	T	F	**T**	T	T	T
F	F	F	F	**T**	T	T	T

Figure A.2

[(*p* or *q*) then *r*] is equivalent to [(*p* then *r*) and (*q* then *r*)]:

p	*q*	*r*	(*p* or *q*)	(*p* or *q*) then *r*	(*p* then *r*)	(*q* then *r*)	(*p* then *r*) and (*q* then *r*)
T	T	T	T	**T**	T	T	T
T	T	F	T	**F**	F	F	F
T	F	T	T	**T**	T	T	T
T	F	F	T	**F**	F	T	F
F	T	T	T	**T**	T	T	T
F	T	F	T	**F**	T	F	F
F	F	T	F	**T**	T	T	T
F	F	F	F	**T**	T	T	T

Figure A.3

[*p* then (*r* and *s*)] is equivalent to [(*p* then *r*) and (*p* then *s*)]:

p	*r*	*s*	(*r* and *s*)	*p* then (*r* and *s*)	(*p* then *r*)	(*p* then *s*)	(*p* then *r*) and (*p* then *s*)
T	T	T	T	**T**	T	T	T
T	T	F	F	**F**	T	F	F
T	F	T	F	**F**	F	T	F
T	F	F	F	**F**	F	F	F
F	T	T	T	**T**	T	T	T
F	T	F	F	**T**	T	T	T
F	F	T	F	**T**	T	T	T
F	F	F	F	**T**	T	T	T

Figure A.4

[p then (r or s)] is equivalent to [(p then r) or (p then s)]:

p	r	s	(r or s)	p then (r or s)	(p then r)	(p then s)	(p then r) or (p then s)
T	T	T	T	**T**	T	T	T
T	T	F	T	**T**	T	F	T
T	F	T	T	**T**	F	T	T
T	F	F	F	**F**	F	F	F
F	T	T	T	**T**	T	T	T
F	T	F	T	**T**	T	T	T
F	F	T	T	**T**	T	T	T
F	F	F	F	**T**	T	T	T

Figure A.5

If you are like me and would rather picture this equivalence visually, the Venn diagrams in Figures A.6 and A.7 demonstrate the validity of Eq. (3). First, we convert [(p and q) then r] to [not (p and q) or r)] by material implication since Venn diagrams can only display AND, OR, and NOT relations. On the left of Figure A.6 is the Venn diagram for (p and q). In the middle, NOT is applied to (p and q) so that the only area not colored is the intersection of p and q. On the right, the set r is OR'd into the Venn diagram, coloring a portion of the intersection of p and q.

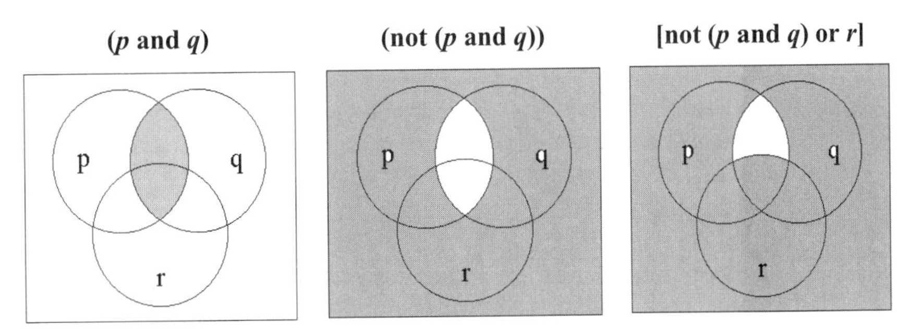

Figure A.6

Next, we convert [(*p* then *r*) or (*q* then *r*)] to [(not *p* or *r*) or (not *q* or *r*)] by material implication. On the left of Figure A.7 is the Venn diagram for (not *p* or *r*). In the middle is the diagram for (not *q* or *r*). The right diagram is the result of the first two diagrams OR'd together.

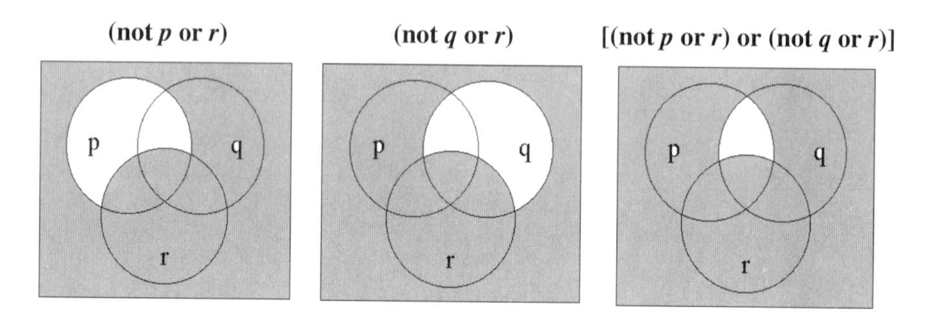

Figure A.7

Note that the completed Venn diagrams of both halves of Eq. (3) are identical. Since [(*p* then *r*) or (*q* then *r*)] focuses on the union of the rule's single input, single output relations, let's call it the union rule configuration (URC).

The formal transformation steps for Eq. (3) in propositional logic are as follows:

- (*p* and *q*) then *r*the initial IRC
- not (*p* and *q*) or *r*by material implication
- (not *p* or not *q*) or *r*by DeMorgan's Law
- not *p* or (not *q* or *r*)by association
- (not *q* or *r*) or not *p*by commutation
- (*q* then *r*) or not *p*by material implication
- ((*q* then *r*) or not *p*) or *r*by addition
- (*q* then *r*) or (not *p* or *r*)by association
- (*q* then *r*) or (*p* then *r*)by material implication
- (*p* then *r*) or (*q* then *r*)by commutation yields the URC

Now that we have rule constructs providing direct relations between antecedent and consequent elements, you might be wondering why we have couched them in propositional logic terms.

The combinatorial problem is not unique to fuzzy logic. The IRC generates an exponential rule explosion in the other forms of logic as well. Propositional logic is the foundation of all formal logic, including fuzzy logic, and defines how rule elements relate to each other. Since the combinatorial problem is a generic issue, it is appropriate to work out alternative rule constructs using propositional logic methods.

Moreover, fuzzy logic is not limited to any particular rule configuration. That is, it is able to cope with intersection and union operators in the antecedent as well as the consequent portions of any rule structure. Once our equivalent rule formats have been verified in propositional logic, fuzzy logic is free to process their fuzzy versions using normal fuzzy inference methods.

The big adjustment I have had to make with these alternative rule constructs is that they are not intuitively obvious. When I first got my driver's license, my insurance agent reminded me that since I was sixteen AND male AND single, my insurance premium would be high. Later, after college, he said that since I was in my mid-twenties AND male AND married, my insurance premium would be moderately low.

This latter statement seems to make more intuitive sense than our alternative format: since I was in my mid-twenties, my insurance premium would be moderately low, OR since I was a male, my insurance premium would be moderately high, OR since I was married, my insurance premium would be low.

No agent wanting to close a sale would utter such a seemingly conflicting statement. One of the problems with the URC is that the transformation from [(p and q) then r] to [(p then r) or (q then r)] shifts our focus from one rule to what appears to be the union of two (or more) rules. In addition, since each of these alternative rules can contain different consequent subsets, they seem to be contradicting each other: either my premium is high OR my premium is low. How can it be both?

This apparent conflict might lead us to view the OR operator as an EXCLUSIVE OR so that [(p then r) or (q then r)] becomes [either (p then r) or (q then r) but not both]. In reality, as the Venn diagram shows, the OR in the URC is not exclusive. The final value of the rule is defined by the UNION of the two (or more) relations within the rule.

Our intuition, trained as it is, forms the hurdle that must be overcome in order to trust this solution to the combinatorial problem. I remember when I first encountered Reverse Polish Notation, my initial reaction was, "How could anyone think this way?!" Yet this counterintuitive approach gives the computer a simple tool for stack operations. In the same way, the counterintuitive URC gives the computer a simple tool for solving the combinatorial problem.

How Does the URC Affect the Multiplication of Rules?

At first blush, it appears that all we have done with these alternative rule constructs is redefine what an individual rule looks like. How can such a redefinition impact rule multiplication? In order to show how the URC eliminates combinatorial rule explosion, I will use a simple, term-life insurance example. We begin with the two-input, one-output model shown in Figure A.8 since it is not easy to draw a five-dimensional IRC rule matrix.

Based on the universal sets in Figure A.8, an intersection rule matrix (IRM) may be populated with rules as shown in Figure A.9.

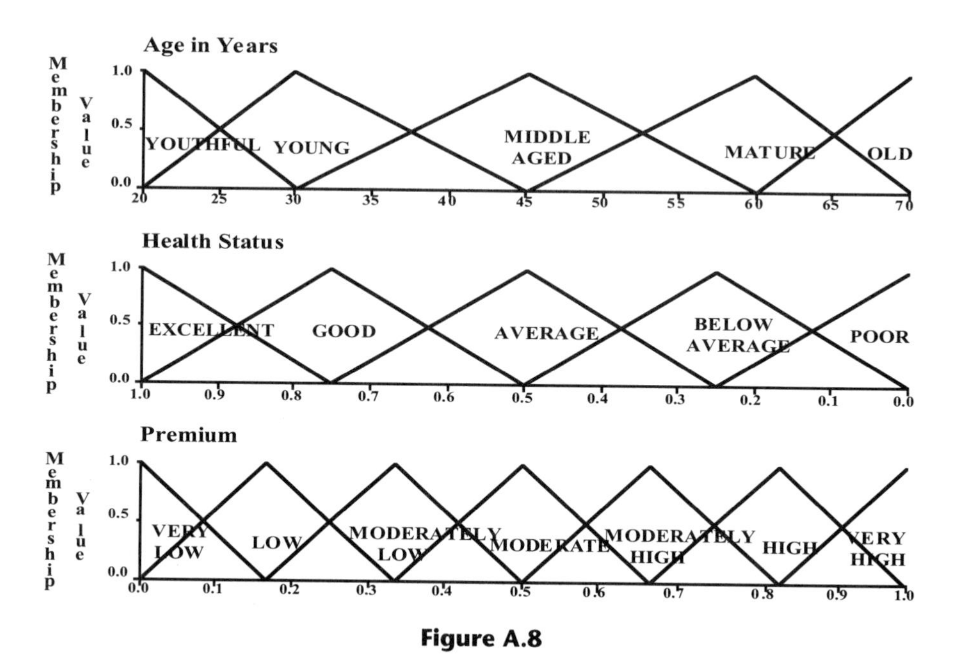

Figure A.8

Age/Health	Youthful	Young	Middle-Aged	Mature	Old
Excellent	(1) Very Low	(6) Low	(11) Mod. Low	(16) Mod. Low	(21) Moderate
Good	(2) Low	(7) Mod. Low	(12) Mod. Low	(17) Moderate	(22) Mod. High
Average	(3) Mod. Low	(8) Mod. Low	(13) Moderate	(18) Mod. High	(23) Mod. High
Below Average	(4) Mod. Low	(9) Moderate	(14) Mod. High	(19) Mod. High	(24) High
Poor	(5) Moderate	(10) Mod. High	(15) Mod. High	(20) High	(25) Very High

Note:The health status below poor and the age beyond old makes a person uninsurable.

Figure A.9

Here is a list of these IRM rules:

Rule 1: IF the person is Youthful and his/her health is Excellent, THEN the term-life insurance premium will be Very Low.

Rule 2: IF the person is Youthful and his/her health is Good, THEN the term-life insurance premium will be Low.

Rule 3: IF the person is Youthful and his/her health is Average, THEN the term-life insurance premium will be Moderately Low.

Rule 4: IF the person is Youthful and his/her health is Below Average, THEN the term-life insurance premium will be Moderately Low.

Rule 5: IF the person is Youthful and his/her health is Poor, THEN the term-life insurance premium will be Moderate.

. . . . . .

Rule 21: IF the person is Old and his/her health is Excellent, THEN the term-life insurance premium will be Moderate.

Rule 22: IF the person is Old and his/her health is Good, THEN the term-life insurance premium will be Moderately High.

Rule 23: IF the person is Old and his/her health is Average, THEN the term-life insurance premium will be Moderately High.

Rule 24: IF the person is Old and his/her health is Below Average, THEN the term-life insurance premium will be High.

Rule 25: IF the person is Old and his/her health is Poor, THEN the term-life insurance premium will be Very High.

Since each antecedent element in the URC relates directly to its consequent counterpart, we reduce the number of output subsets in Figure A.10 from seven to five, and a union rule matrix (URM) may be populated as shown in Figure A.11.

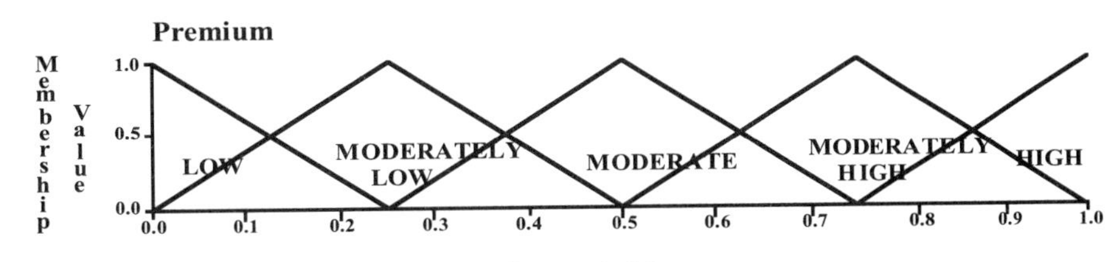

Figure A.10

Age	**Youthful Then Low**	**Young Then Mod. Low**	**Middle-Aged Then Mod.**	**Mature Then Mod. High**	**Old Then High**
Health Status	Excellent Then Low	Good Then Mod. Low	Average Then Mod.	Below Average Then Mod. High	Poor Then High

Figure A.11

For simplicity, we can state that there are ten rules in this URM. But remember these fuzzy rules grow out of the propositional logic format: [(p then r) or (q then r)] where p represents Age, q represents Health Status, and r represents Premium. Therefore, since each one of these fuzzy relations is separated logically by an OR for inference purposes, we could just as easily state that there are two rules, one for each input . . . or that there is just one rule that mirrors the one unioned propositional logic rule.

This last option emphasizes that the one fuzzy rule (composed of a series of single input, single output rule components) models the entire problem space:

Rule: [IF the person is Youthful, THEN the premium will be Low
OR
IF the person is Young, THEN the premium will be Moderately Low
OR
IF the person is Middle-Aged, THEN the premium will be Moderate
OR
IF the person is Mature, THEN the premium will be Moderately High
OR
IF the person is Old, THEN the premium will be High]
OR
[IF the person's Health is Excellent, THEN the premium will be Low
OR
IF the person's Health is Good, THEN the premium will be Moderately Low
OR
IF the person's Health is Average, THEN the premium will be Moderate
OR
IF the person's Health is Below Average, THEN the premium will be Moderately High
OR
IF the person's Health is Poor, THEN the premium will be High].

Whether you prefer to think of this alternative format as containing ten rules, two rules, or one rule (it doesn't really matter to the computer), you will note that there are just ten implications in this URM (five subsets times two inputs) instead of twenty-five implications in the IRM (five subsets raised to the power of two inputs). You will also note that there is just one antecedent element in each URM implication whereas in each IRM implication, there are two (the number of inputs).

Now let's see what happens when we add three more inputs to the URM: a person's family health history; the level of risk in one's occupation; and the level of risk if a person's occupation is in a foreign country (Figure A.12).

Age	Youthful Then Low	Young Then Mod. Low	Middle-Aged Then Mod.	Mature Then Mod. High	Old Then High
Health Status	Excellent Then Low	Good Then Mod. Low	Average Then Mod.	Below Average Then Mod. High	Poor Then High
Health History	Excellent Then Low	Good Then Mod. Low	Average Then Mod.	Below Average Then Mod. High	Poor Then High
Job Risk	Minor Then Low	Below Average Then Mod. Low	Average Then Mod.	Above Average Then Mod. High	Major Then High
Foreign Risk	Minor Then Low	Below Average Then Mod. Low	Average Then Mod.	Above Average Then Mod. High	Major Then High

Figure A.12

We are up to 25 implications for the URM as opposed to 3,125 rules for the IRM. And each implication still contains just one antecedent element, while there are five antecedent elements in each IRM rule. What is most important is that the URC/URM is modeling the entire problem space in a manner equivalent to the IRC/IRM.

Please understand that we are *not* advocating the transformation of existing rules from one format to another. Instead, we are suggesting that the problem space be modeled using a different rule configuration (URC) than the one normally used (IRC).

How Does the URC Work?

As stated earlier, fuzzy logic can work with any valid rule configuration, so all of the standard fuzzification, inference, and defuzzification techniques apply. However, since the URC reduces the basic rule structure to a series of single input, single output implications coupled by union, it lends itself to further computational streamlining. For the sake of space, I will only illustrate one simple inference method to emphasize this optimization.

First, we place the single input, single output subset relations of the URC in a matrix format such as the one in Figure A.12. Our next task is to calculate the output subset membership values from this matrix. But before we do, I want to briefly outline four processes that can help streamline URM inference: singleton output subsets, the Mamdani Method of inference calculated by SUM and PRODUCT inference operators, and Centroid defuzzification.

Singleton output subsets reduce the representation in Figure A.13 to that shown in Figure A.14.

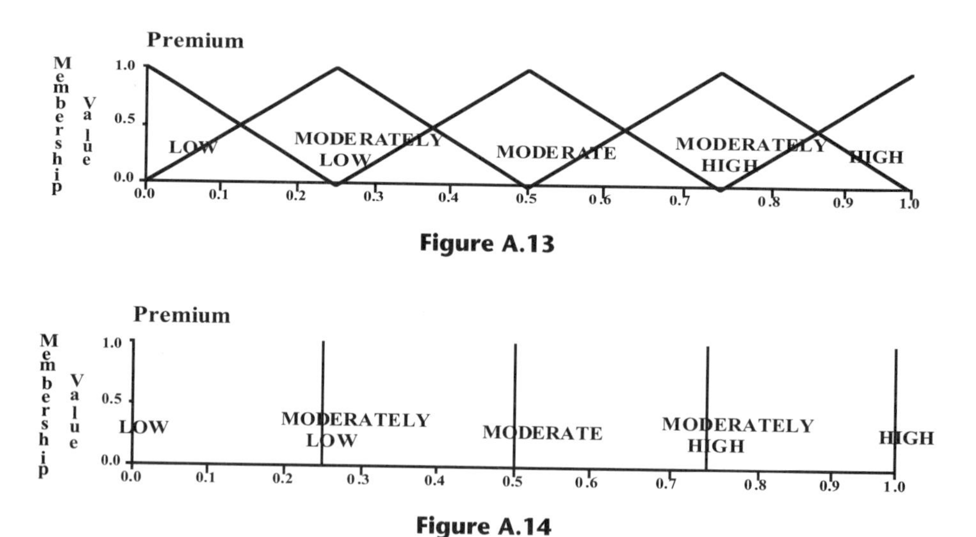

Figure A.13

Figure A.14

The only significant restrictions that singleton subsets bring to the problem space when used with Centroid defuzzification are that they cannot be shaped and the output universal set must be represented by more than one singleton subset. They are commonly used in conjunction with Centroid defuzzification to decrease calculation time.

The Mamdani Method of inference is one of the most common inference methods in fuzzy logic and interprets [*p* then *r*] as [*p* and *r*]. I personally like to use Lotfi Zadeh's SUM and PRODUCT inference operators to calculate Mamdani's implication because these operators provide a smooth, predictable output that is easily computed. (Zadeh's MAX and MIN inference operators are more often used to explain the inference process. Zadeh's operators are also among the most popular.)

The PRODUCT inference operator calculates the intersection portions of the rule. If we use Mamdani's implication [if *p* and *r*] with singleton output subsets, the output from the PRODUCT operator will always equal the antecedent membership value since the consequent membership value will always be 1.0:

Inference Membership Value =

The antecedent membership value (a value from 0.0 to 1.0)

. . . multiplied by . . .

The consequent singleton subset (whose value is always 1.0).

The SUM inference operator calculates the union portions of the rule. In using this operator, we SUM all the single input, single output relations that correspond to a given consequent subset and then normalize the membership values of all the consequent subsets to 1.0. However, if we use Centroid defuzzification, there is no need to perform the normalization step since the consequent membership values are in both the numerator and the denominator of the Centroid equation.

So, if we use these four processes and assemble the antecedent membership values in a union rule matrix format, we can sum the columns into an accumulator array representing the output subset membership values (Figure A.15).

Union Rule Matrix and Accumulator

Age	μYouthful	μYoung	μMiddle-Aged	μMature	μOld
Health Status	μExcellent	μGood	μAverage	μBelow Average	μPoor
Health History	μExcellent	μGood	μAverage	μBelow Average	μPoor
Job Risk	μMinor	μBelow Average	μAverage	μAbove Average	μMajor
Foreign Risk	μMinor	μBelow Average	μAverage	μAbove Average	μMajor

Premium	μLow	μMod. Low	μModerate	μMod. High	μHigh

Figure A.15

All that is left to do after this inference step is to plug the accumulator's membership values into the Centroid defuzzification function to obtain an output Premium value (Figure A.16).

Of course, the URC is not limited to this streamlined process. You are free to use any fuzzy logic functions because fuzzy logic is able to work with any valid rule configuration.

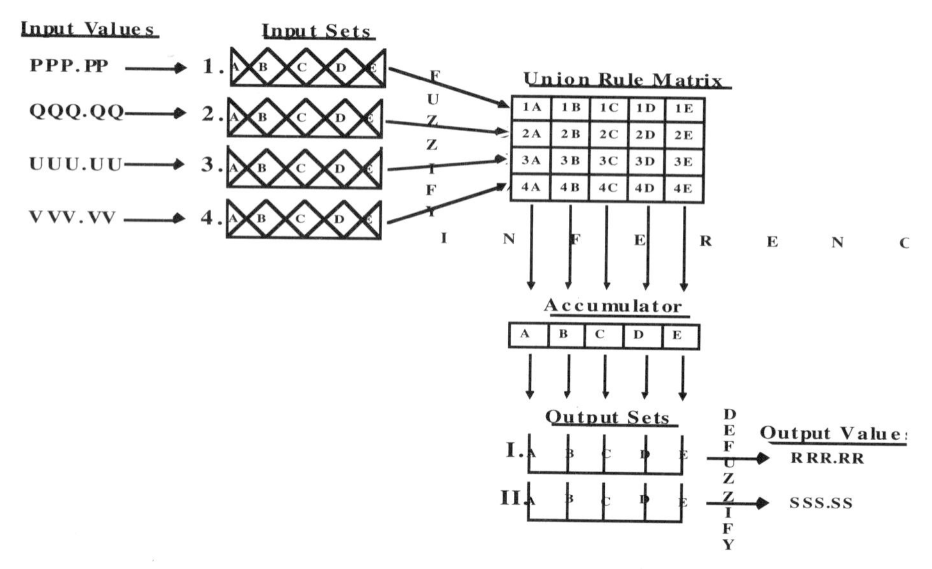

Figure A.16 Diagram of the URM inference engine

Modeling Each Input's Relative Importance

There are times when we want to modify the importance of each input universal set. One way to do this is through the use of universal and subset weights as shown in Figure A.17.

	Universal Weight					
Age	0.7	Youthful Then Low	Young Then Mod. Low	Middle Aged Then Mod.	Mature Then Mod. High	Old Then High
Health Status	1.0	Excellent Then Low	Good Then Mod. Low	Average Then Mod.	Below Average Then Mod. High	Poor Then High
Health History	0.8	Excellent Then Low	Good Then Mod. Low	Average Then Mod.	Below Average Then Mod. High	Poor Then High
Job Risk	1.0	Minor Then Low	Below Average Then Mod. Low	Average Then Mod.	Above Average Then Mod. High	Major Then High
Foreign Risk	0.9	Minor Then Low	Below Average Then Mod. Low	Average Then Mod.	Above Average Then Mod. High	Major Then High

Figure A.17

Briefly, a universal set weight is multiplied by every subset membership value in the universal set, providing a way to relate universal sets to each other. Subset weights permit us to model the relative importance of each subset within a universal set by multiplying a weight with its corresponding subset membership value.

One area where I use subset weights is in modeling rate changes such as in a temperature control problem. To solve this problem with an IRM, we could establish three input subsets for Temperature (COLD, COOL, and WARM), three for Rate of Temperature Change (DECREASING, STEADY, and INCREASING), and five subsets for furnace output (LOW, MODERATELY LOW, MODERATE, MODERATELY HIGH, and HIGH). Initially, we could arrange the rules as shown in Figure A.18.

Temp/Rate	**Cold**	**Cool**	**Warm**
Decreasing	High	Mod. High	Moderate
Steady	Mod. High	Moderate	Mod. Low
Increasing	Moderate	Mod. Low	Low

Figure A.18

However, such an arrangement means that if the temperature is Cold (say in the forties) and has been that way for some time, the furnace will not be producing at its highest level. Worse yet, when the temperature reaches 72 degrees and the rate of temperature change remains steady, the furnace will still be pumping heat into the residence even though we really want it to shut off. So, an alternative rule distribution to correct this situation might be that of Figure A.19.

Temp/Rate	**Cold**	**Cool**	**Warm**
Decreasing	(1) High	(4) Mod. High	(7) Moderate
Steady	(2) High	(5) Moderate	(8) Low
Increasing	(3) Moderate	(6) Mod. Low	(9) Low

Note: Rule numbers are in parentheses.

Figure A.19

These rules solve the immediate problem but produce jumps between noncontiguous output subsets when the system moves from rules 2 to 3, 2 to 5, 5 to 8, and 7 to 8, which in turn adversely affects the smoothness of the output.

The fact that the output subsets are identical for rules 2 and 1, 8 and 9, 5 and 3, and 5 and 7 indicates that we could eliminate the influence of the STEADY subsets altogether.

Unfortunately, if the STEADY row is cut from the IRM and the Rate input is STEADY, then the membership values of both DECREASING and INCREASING will fall to zero and no IRM rules will fire.

The URM configuration using subset weights offers a much simpler solution (Figure A.20).

	Subset Weight		**Subset Weight**		**Subset Weight**	
Temperature	1.0	Cold Then High	1.0	Cool Then Moderate	1.0	Warm Then Low
Rate	1.0	Decreasing Then High	0.0	Steady Then Moderate	1.0	Increasing Then Low

Figure A.20

Note that STEADY relates only to the MODERATE output subset, and since this rule component is associated with all other implications by union, the system is not adversely affected if we multiply STEADY's membership value by a zero subset weight.

Why should we care that the URM can provide a smoother output than the IRM in this situation? After all, will we be able to notice the difference if the furnace output jumps from LOW to MODERATE instead of to MODERATELY LOW? It is one thing to discuss temperature control for a residential thermostat that may only be sampling its environment every few minutes. It is another to consider the control requirements for an ultra-high-speed drill bit in an NC milling machine where precise control enables the drill bit to travel across the milling surface at an incredibly rapid rate.

Of course, you can use methods other than weights, such as alpha cuts, hedges, and varying subset shapes and sizes to tune your application. I have briefly discussed weights as just one tuning example.

And Speaking of Tuning . . .

The IRC is not only at the heart of the combinatorial problem, it can also adversely affect tuning. Recall Eq. (2) [If w then r] where w represents the subset intersections of the various antecedent universal sets and r represents the subsets of the consequent universal set. We said that whenever either p or q changed, w was altered as well, necessitating a different relation with r (a new rule). In this configuration, as rules become more numerous, r subsets must also increase in order to accommodate the additional implications.

But what happens when the consequent subsets do not keep pace? Consider rules 18, 19, and 23 from Figure A.9. Let's suppose that I am fully mature. That is, my age places me in the subset MATURE with a membership value of 1.0. Let's also suppose that my health is the essence of average so that it falls completely within the subset AVERAGE. With these two inputs, the IRM in Figure A.9 declares that my premium is MODERATELY HIGH.

Now, suppose that either my health deteriorates to BELOW AVERAGE [$\mu_{(19)}$ = 1.0] or my health remains stable but I live to be OLD [$\mu_{(23)}$ = 1.0]. Not to worry (unless you are my insurance company) because my premium will not change one penny according to this IRM! The reason is simple: whenever we move either vertically or horizontally to a contiguous cell in an intersection rule matrix without changing output subsets, the output values plateau.

There are several options to address this problem. First, we could increase the output subsets from seven to nine so that vertical or horizontal IRM movement would yield different outputs. Unfortunately, with this proposal, a seven-by-seven matrix would need thirteen subsets. And what about a five- or six-input IRM? Computers do not care how many output subsets we have. But as system designers, we prefer to assign some meaningful linguistic title to each one, and modifiers can only provide so much distinctive variation.

Another option is to alter the size, shape, and position of the input and output subsets, thereby modifying the shape of the plateaus to more faithfully produce a realistic output. Tuning configurations such as Figure A.21 are not uncommon.

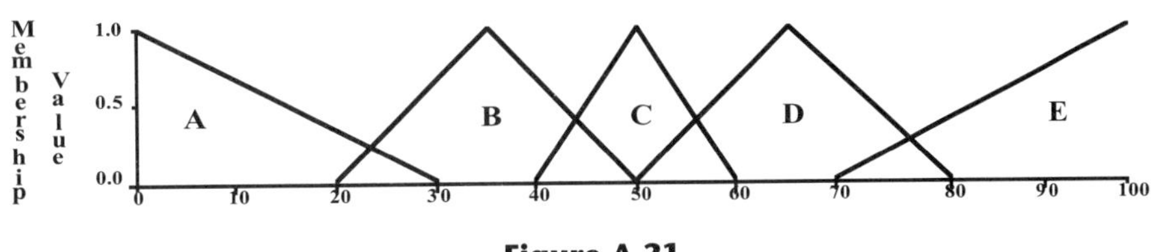

Figure A.21

To see how this option works, note that all values between 30 and 40 fall entirely within the domain of subset B. Yet only 35 has a membership value of 1.0. All other values have a lower membership in the subset and therefore have less influence on the overall system, tailoring the naturally flat output of a plateau to a more desired shape. This process imposes a kind of variable subset weighting structure on the inference engine. But the method is not straightforward, particularly for more complex systems. This makes tuning more difficult.

Why should the URC make tuning easier? Remember that the URC allows each antecedent element to have its own relation with r, independent of the other inputs, unlike the IRC where input subsets must relate to their output counterparts through their intersection value w. Since the output universal set relates directly to each input universal set, no additional output subsets are needed when inputs are added to the system in order to enable the additional implications. Thus, no plateaus are created.

Tuning, while still necessary to shape the inference engine to the desired real-world system, is greatly simplified in the URC because there are no plateaus to overcome.

Design Considerations

It is important to underscore that the combinatorial rule explosion problem is not unique to fuzzy logic. We have rooted our solution in propositional logic so that the URC can be applied to all disciplines of formal logic. Having proved its equivalency to the IRC, we can be confident that fuzzy logic will operate just as easily on URC rules as on those in the more familiar format.

As a matter of fact, the inference task for the URC can be viewed as a normal two-phase compositional rule of inference operation in a manner identical to that used for the IRC. Take Zadeh's MIN/MAX operator as an example. If we picture the URC as a series of single input, single output rules coupled by union, the first inference phase would be to apply the intersection operator (MIN) to each rule in order to obtain its weight. The second phase would be to use the union operator (MAX) to obtain the maximum membership value (weight) for each output subset of these minimized fuzzy rules.

A simpler analogy can be seen by reducing Figure A.11 to a single input, single output system shown in Figure A.22.

Age	Youthful Then Low	Young Then Mod. Low	Middle-Aged Then Mod.	Mature Then Mod. High	Old Then High

Figure A.22

The inference tasks for this system are identical regardless of whether we perceive the rules as being in an IRC or URC format.

Even though these two rule configurations are equivalent, you may not be accustomed to working with systems based on single input, single output relations. So for a moment, I would like to focus on this simple pairing. As a general rule, these input and output universal sets should be designed as monotones. That is, the values along each of the two X-axes should either only increase or only decrease.

For what appears to be an exception to this general guideline, consider the example in Figure A.23, taken from Cox's *Fuzzy Systems Handbook* (First edition), pp. 81–89.

This single fuzzy set defines the automobile insurance underwriting risk for drivers between the ages of 16 and 95, depicting young and very old drivers as high risks while placing persons between 30 and 70 in a moderately low risk category. This set is obviously nonmonotonic.

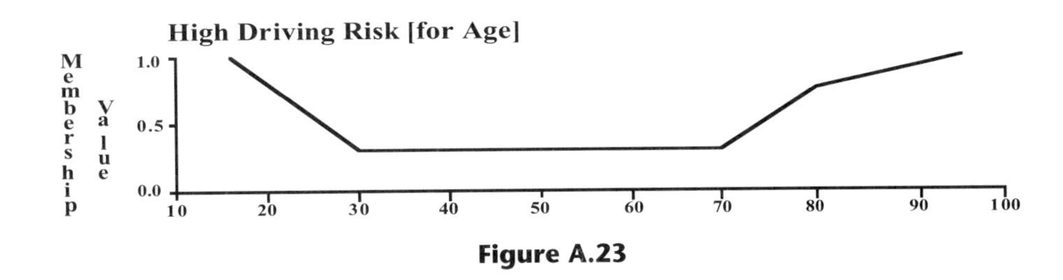

Figure A.23

Yet on closer inspection, the membership values represent the degree of risk, not the degree of membership in the universal set AGE. Cox has cleverly incorporated several risk-related factors, including driving experience, reflex reaction time, and general health, vision, and hearing into a composite set representing DRIVING RISK relative to age. Taken separately, these inputs are monotones. But as a group, they form a complex risk curve. Cox combined these elements into one set to improve performance. Now that the URC is able to manage large numbers of inputs efficiently, the need for composite curves to expedite computation should be greatly diminished.

[In the paper I wrote with James Andrews nearly two years ago for the IEEE Transactions on Fuzzy Systems, we said URC universal sets must be monotones, and their input and output universal sets must contain the same number of subsets. Since that writing, we have come to the understanding that these constraints may be too conservative. For a simple spread sheet example that extends beyond these two constraints, see the accompanying CD-ROM program entitled "URCCurve".]

An HMO Scheduling Program

An HMO (Health Maintenance Organization) differs from traditional hospitals and clinics in that its manifesto focuses on the maintenance of health rather than the resolution of illness. It functions in a manner similar to an insurance policy and hospital/clinic rolled into one. In return for a financial premium, the HMO provides for the health care needs of its members. It is in the organization's interest to keep its members healthy since chronic illness generates significant, long-term expense.

For this example, we will concentrate on a very small portion of the business. We will consider only those members entering the facility without appointments to see general practitioners. Normally, a member would have an appointment with a technician, nurse, general practitioner or specialist, depending on the circumstances.

The spread sheet on the accompanying CD-ROM entitled "HMOSched" is limited to one triage nurse and seven general practitioners. The triage nurse interviews each walk-in patient and selects an appropriate physician based on the member's condition, the doctor's competency to treat that particular condition, the doctor's patient load and his/her availability to see walk-in members. This program was adapted from my off-hour class, "Fuzzy Logic for Business Applications," at the University of Washington. Presented during the course's first computer lab, it is a simplified example that introduces the student to basic fuzzy logic functionality.

Conclusion

The information age has put mountains of data at our finger tips through corporate databases and the worldwide web. The growing challenge for computer professionals, however, is not just how to retrieve and store this information, but how to interpret it—how to use this unprecedented access to our advantage. Meaning is not a Boolean issue. It dwells in the ever-changing shades of gray of our complex world. The URC should find ready use in software agents, data mining tools, and sophisticated modeling systems in addition to the more traditional control applications in science and engineering.

This brief chapter has outlined an inference method that facilitates the creation of large-scale, time-critical fuzzy logic applications by eliminating combinatorial rule explosion from the inference process. It also cuts overall development time by simplifying tuning. We are not implying, however, that the URC should render obsolete the more conventional rule configuration. Instead, we hope you will view this alternative method as an addition to your tool kit.

Acknowledgments

I want to thank my friends: James E. Andrews, for the creation of the propositional logic proof for Eq. (3), the truth tables, and the Venn diagrams as well as his technical and stylistic support; and Earl Cox, for his encouragement and enthusiasm.

References

- J. Andrews, "Taming Complexity in Large-Scale Fuzzy Systems," *PC AI Magazine,* Phoenix: Knowledge Technology Inc., May/June 1997, pp. 39–42.
- S. Chiu, "Method and Software for Extracting Fuzzy Classification Rules by Subtractive Clustering," 1996 Biennial Conference of the North American Fuzzy Information Processing Society—NAFIPS (Cat. No.96TH8171), Berkeley, CA. pp. 461–465. NAFIPS—North American Fuzzy Inf. Process. Scc. BISC—Berkeley Initiative in Soft Computing IEEE Neural Networks Council. IEEE Systems, Man and Cybernetics Society, June 1996, pp. 19–22.

- S. Chiu, "Fuzzy Model Identification Based on Cluster Estimation," *Journal of Intelligent and Fuzzy Systems,* vol. 2, New York: John Wiley & Sons, 1994, pp. 267–278.
- W. Combs and J. Andrews, "Combinatorial Rule Explosion Eliminated by a Fuzzy Rule Configuration," *IEEE Transactions on Fuzzy Systems,* [accepted for publication February 1998].
- W. Combs, "Reconfiguring the Fuzzy Rule Matrix for Large, Time-Critical Applications," Third Annual International Conference on Fuzzy-Neural Applications, Systems and Tools, Nashua, NH: PennWell Publishing Company, November 1995, pp. 18:1–7.
- I. Copi and C. Cohen, *Introduction to Logic,* Ninth edition, New Jersey: Prentice Hall, 1994.
- E. Cox, *The Fuzzy Systems Handbook, A Practitioner's Guide To Building, Using, and Maintaining Fuzzy Systems,* Boston: Academic Press, 1994, p. 323.
- E. Cox, *Fuzzy Logic for Business and Industry,* Rockland, MA: Charles River Media, Inc., 1995.
- S. Halgamuge and M. Glesner, "Neural Networks in Designing Fuzzy Systems for Real World Applications," *Fuzzy Sets and Systems,* vol. 65, no. 1, pp. 1–12, July 1994.
- H. Ishibuchi, K. Nozaki, N. Yamamoto and H. Tanaka, "Selecting Fuzzy If-Then Rules for Classification Problems Using Genetic Algorithms," *IEEE Transactions on Fuzzy Systems,* vol. 3, no. 3, pp. 260–270, August 1995.
- G. Klir and B. Yuan, *Fuzzy Sets and Fuzzy Logic,* New Jersey: Prentice Hall, 1995, pp. 231–236, 317–322.
- G. Klir, Ute H. St. Clair, and Bo Yuan, *Fuzzy Set Theory: Foundations and Applications,* New Jersey: Prentice Hall, 1997.
- B. Kosko, *Neural Networks and Fuzzy Systems, A Dynamical Systems Approach to Machine Intelligence,* New Jersey: Prentice-Hall, 1992, pp. 322–326.
- R. Rovatti, R. Guerrieri, and G. Baccarani, "Fuzzy Rules Optimization and Logic Synthesis," Second IEEE International Conference on Fuzzy Systems (Cat. No. 93CH3136-9), San Francisco, March/April 1993, pp. 1247–1252, vol. 2.
- F. Watkins, "Building Large-Scale Fuzzy Logic Systems: Combinatorial Challenges," *PC AI Magazine,* March/April, 1995.
- F. Watkins, *Fuzzy Engineering,* Ph.D. Thesis, University of California, Irvine, 1994. UMI Order Number 9417921.
- *R.* Yager and L. Zadeh, Eds., *Fuzzy Sets, Neural Networks and Soft Computing,* New York: Van Nostrand Reinhold, 1994, pp. 2ff, 85–125.

- R. Yager and D. Filev, "Learning of Fuzzy Rules by Mountain Climbing," Proceedings of the SPIE Conference on Applications of Fuzzy Logic Technology, Boston, 1993, pp. 246–254.
- L. Zadeh and J. Kacprzyk, Eds., *Fuzzy Logic for the Management of Uncertainty,* New York: John Wiley & Sons, 1992, pp. 265–276.
- H.-J. Zimmermann, *Fuzzy Logic Theory and Its Applications,* Second edition, Dordrecht: Kluwer Academic Publishers, 1991.

Biography

William E. Combs is a computing systems architect for the Boeing Company. His department is responsible for providing configuration control of on-line CATIA CAD/CAM models worldwide on MVS, UNIX, and NT platforms for all Boeing airplanes. He also teaches an off-hour class, "Fuzzy Logic for Business Applications," at the University of Washington. His university degrees include Bachelor of Arts in Mathematics and Music from Alaska Methodist University, Anchorage, Alaska; and Master of Divinity and Doctor of Ministry from Fuller Theological Seminary in Pasadena, California. He is a member of the IEEE Computer Society.

E-Mail: william.e.combs@boeing.com

Glossary

Adaptive fuzzy associative memory (AFAM)

A fuzzy reasoning system that adjusts to the central measure of the fuzzy space as processing continues. An AFAM system can modify the topology of the underlying fuzzy sets as well as the composition of the rules themselves. One form of the AFAM, which uses the Wang–Mendel algorithm, produces the rule set from approximated fuzzy sets and actual data streams. See also *fuzzy associative memory.*

Alpha-level

A measure that establishes the minimum degree of membership in a fuzzy set or fuzzy region. The alpha-threshold level (or "alpha cut") is a truth membership value. There are two types of alpha thresholds, strong and weak. A strong threshold is indicated by:

$$\mu_A[x] > A$$

while the weak threshold is defined as

$$\mu_A[x] \geq A$$

This slight change in the definition plays a significant role in the definition of other fuzzy control parameters, most notably the idea of the underlying support set (where the alpha level is set to zero, and the strong alpha cut is applied to the fuzzy space).

Ambiguity

Ambiguity is generally distinct from fuzziness. Our everyday experiences tell us that an ambiguous reference involves a control variable, statement, or other symbol whose meaning has several distinct yet possible interpretations. An ambiguous statement might be "that car is hot," where the term hot might refer to the temperature of the car, to the power and speed of the car, or to the overall desirability of owning the car. Yet, the conceptual definition of a fuzzy concept, such as tall, involves little direct ambiguity, only imprecision or uncertainty. In a broader sense, however, ambiguity is at the heart of fuzzy ontology since it is in composite fuzzy regions that we experience true ambiguity. This ambiguity is directly related to the idea of undecidability in these regions. See also *fuzzy space, vagueness, imprecision,* and *undecidability.*

Antecedent

The conditions that precede the current model state. While rule-based expert systems often use this term to define the premise of a rule, in a fuzzy system, the antecedent defines the combined truth state of the current fuzzy region for a particular control or solution variable.

Approximate reasoning

A reasoning process incorporating fuzzy sets, hedges, fuzzy operators, and rules of decomposition in an implication process, such as:

$$\begin{aligned} \mu_c[x] &\Leftarrow \overset{n}{\underset{i=1}{M}} \, comp(\mu_{P_L}[y] \cdot \mu_{P_R}[z])_i \\ \Re &\Leftarrow decomp(C) \end{aligned}$$

That is (with some apologies to the logical symbolism), the final membership function associated with a fuzzy consequent space (*M*) is developed by processing each premise fuzzy assertion with its combining operators (•) and applying the appropriate rule of composition. The "combining operators" are AND and OR, obeying either Zadeh's min–max rules or one of the compensatory family rules. Using the method of decomposition, this fuzzy space is used to derive the solution variable ($\Re$), also known as the expected value. This process is also called *defuzzification*.

Approximation

The process of forming a fuzzy set from a scalar or non-fuzzy continuous variable. An approximation of a scalar generally results in a bell-shaped fuzzy space (such as a PI or Weibul curve). Approximations can be applied to both fuzzy sets and fuzzy regions as well as scalars. The hedges *around*, *near* {*to*}, *close* {*to*}, and *about* are used to approximate the fuzzy set or scalar.

Assertions

A fuzzy system model consists almost completely of assertions (or fuzzy propositions). These can be conditional or unconditional. Unconditional assertions establish a fuzzy region that acts as the support space for the solution space. A statement such as *our price must be high* is an unconditional assertion. On the other hand, conditional fuzzy assertions update the solution space with a strength or degree that depends on the truth of their predicate (the "conditional" part). A statement such as *if competition_price is not very high then our price should be near the competition_price* is a conditional fuzzy assertion.

Bayes theorem

A theorem of probability that says if you know the *a priori* probability of a hypothesis before examining the results of some test, the future test results can be used to modify the cumulative probability to predict the *a posteriori* probability. In practice, however, Bayes theorem has become a measure of subjective probability based on a scale of "rational belief" in the probability that an event will occur. The scale ranges from [0], absolute disbelief, to [1], absolute belief. Bayesian probabilities often diverge from frequentist or parametric probabilities in the sense that *a priori* observational histories are not required to establish the total probabilities.

Belief system

A belief system operation establishes a plausibility state for an object or a particular model state, thus projecting its domain into a set of candidate values. A belief system corresponds to the subjective intentions of the observer in confirming or denying the validity of a model assertion or proposition.

Cartesian product

The product of *n*-state spaces and *m*-state spaces, yielding a state space with [mxn] values. We usually represent the Cartesian product as {*N*,*M*,...*S*} where *N*, *M*, etc., are fuzzy sets or fuzzy regions. Simply put, the Cartesian product is a set of all the combinations of two or more value sets. In New York City, as an example, if we want to find places where games of 3-card monty are in progress, the Cartesian product of 12 midtown avenues (running north and south) and 10 midtown streets (running east and west) will produce a set space of 120 intersections.

Centroid method

See *composite moments*.

Compatibility

A central idea of fuzzy set theory representing the degree to which a control variable's value is compatible with the associated fuzzy set. In the view of some theorists, compatibility is the essence of fuzzy logic and possibility theory. When we encounter a fuzzy statement such as *if the temperature is hot then increase cool air about 2 units*, the conditional fuzzy proposition *the temperature is hot* is a measure of the compatibility between the value for temperature and the fuzzy set *Hot*. The degree of membership value returned by evaluating the assertion is a measure of this compatibility.

Compatibility index

A real number between [0] and [1] in a fuzzy model that indicates the degree to which the control variables are compatible with the underlying fuzzy sets. This index is the truth membership function associated with the consequent fuzzy space. The compatibility index has the property of central moments, that is, it provides a measure of fuzzy compatibility when the index ranges generally between [.2] and [.8]. Numbers above or below this threshold indicate that the control variables lie at the extremes of the combined fuzzy sets.

Compensatory operators

Connective operators that compensate for Zadeh's strict rules of combination for the fuzzy intersection and union. Under Zadeh's fuzzy algebra, the OR (union) is performed by talking the maximum of the two fuzzy sets; the AND (intersection) is performed by taking the minimum of the two fuzzy sets. These rules for connecting fuzzy assertions are often too strict—a single absolutely true fuzzy proposition will propagate through a chain of propositions. Compensatory operators provide alternative rules for combining fuzzy propositions or assertions with AND and OR operators.

Complement

The negation of a fuzzy set is the complement of the fuzzy set. The complement of a fuzzy space is usually produced by the operation, $1-\mu_A(x)$. The fuzzy complement indicates the degree to which an element (x) is not a member of the fuzzy set (A). Unlike conventional crisp sets, an element is not either in or out of a fuzzy set, thus the complement also contains members that have partial exclusions. See also *law of the excluded middle*.

Composite mass

A defuzzification method that produces the expected value for a consequent variable by examining the area of the fuzzy consequent that has the highest intersection density of premise fuzzy sets. This will be the area where the preponderance of the rules are executed, thus establishing the "most votes" for a value from this region. The composite mass decomposition technique applies the rules of evidence in determining a value for the solution variable.

Composite maximum

A defuzzification method that produces the expected value for a consequent variable by examining the edges of the fuzzy space across the fuzzy region's domain. Composite maximum takes the point with the maximum truth value along this edge and uses the domain value at that point as the solution value. If the region is a double-edged plateau, the center of the plateau is selected. If the region is a single-edged plateau, the point at the left-most edge is selected. This defuzzification method responds to the maximum truth value of any rule that fires in the model.

Composite moments

A defuzzification method that produces the expected value for a consequent variable by calculating the center of gravity (or first moment of inertia) for the consequent fuzzy region. This is also called the *centroid method*. The composite moments method is widely used in process control and robotics since it tends to smooth out the solution variable's fuzzy region, eliminating the abrupt value jumps that are often associated with boundary movements in the composite mass and maximum methods.

Conjunction

The process of applying the intersection operator (AND) to two fuzzy sets. For each point on the fuzzy set, we take the minimum of the membership functions for the two intersected sets.

Consequent

The action or right-hand side part of a rule. In a fuzzy modeling system, the consequent defines each control fuzzy region that updates the specified solution variable. As an example, in the rules

```
our price must be high

if rotor.temperature is hot
   then rotor_speed must be reduced
```

the control fuzzy sets *high* and *reduced* are consequent sets. The variables *price* and *rotor.temperature* are solution variables (and are represented during the model processing as solution fuzzy sets).

Consistency principle

A principle from possibility theory that says the possibility of event *X* is always at least as great as its probability, or stated formally:

$$\text{poss}(X) \geq \text{prob}(X)$$

Continuous variable

A control variable whose possible values are not discrete. In statistical and probability models, these are usually called *continuous random variables*, and they play an important role in expectation based on their probability density function (PDF). Generally speaking, fuzzy spaces are continuous variables.

Contrast

A hedge operation that moves the truth function of a fuzzy set around the [.5] degree of membership. The contrast intensification process increases membership values above [.5], while the dilution process defuses or decreases membership below this value. The hedge *positively* implements the intensification capability and the hedge *generally* implements the defusing capability.

Correlation minimum rule

The method of implication that correlates the consequent fuzzy region with the antecedent (premise) fuzzy truth value by truncating the consequent fuzzy set at the limit of the premise's truth value. The formal implication space is defined as:

$$\mu_R[xi] = \min(\mu_R[xi], \mu P_{tv})$$

That is, the membership value of the current consequent fuzzy region (*R*) is the minimum of the fuzzy region and the truth of the premise (P_{tv}). The correlation maximum maintains the general dimensions of the fuzzy region but forms a plateau across the membership plane corresponding to the truth of the premise.

Correlation product rule

The method of implication that correlates the consequent fuzzy region with the antecedent (premise) fuzzy truth value by taking the outer product of the membership values. The formal implication space is defined as:

$$\mu_R[xi] = \mu_R[xi] \bullet \mu P_{tv}$$

That is, the membership value of the current consequent fuzzy region (R) is the product of the fuzzy region and the truth of the premise (P_{tv}). This effectively scales the fuzzy space. See also *correlation minimum rule.*

Correlation rules

Also called *correlation encoding,* these rules couple the truth of the consequent assignment fuzzy region with the truth of the rule's premise. Predicated on the principle that the truth of consequent cannot be greater than the truth of the premise. In the fuzzy rule space.

$$\text{if } FR_p \text{ then } FR_c \text{ is } FR_a$$

it is the fuzzy region FR_a that is correlated with the premise fuzzy expression FR_p—the result of this correlated space is used to update the consequent fuzzy region FR_c. Both correlation rules reduce the height of the fuzzy region.

Crisp set

The term, in fuzzy logic and approximate reasoning, usually applied to classical (Boolean) sets where membership is either [1] (totally contained in the set) or [0] (totally excluded from the set). As an example, the crisp representation for the concept *tall* might have a discrimination function,

$$T = \{x \in U \mid \mu_t(x) \geq 6\}$$

meaning that anyone with a height greater than or equal to six feet is a member of the *tall* set. Crisp sets, unlike fuzzy sets, have distinct and sharply defined membership edges. Note that the vertical line connecting nonmembership with membership, is dimensionless, unlike a fuzzy set. An intrinsic property of a crisp set is the well-defined behavior of its members. In particular, crisp sets obey the geometry of Boolean and Aristotelian sets. This means that the universe of discourse for a set and its complement is always disjoint and complete. Thus the relation

$$A \cap \sim A \equiv \varnothing$$

(also known as the *law of the excluded middle*) is always obeyed.

Decomposition methods

The methods used to derive the expected value of a model solution variable from a consequent fuzzy region. There are several common types of decomposition available in fuzzy system modeling, including composite moments, composite maximum, composite mass, reduce entropy, and plateau positioning. These methods are discussed elsewhere in this glossary. The process of decomposition is an important component of fuzzy modeling since it provides one part of the required symmetry between scalars and fuzzy sets. See also *defuzzification.*

Defuzzification

The process of deriving a scalar, representing a control variable's expected value, from a fuzzy set. Following one of several rules for selecting a point on the edge of the fuzzy region, the defuzzification process isolates a value on the fuzzy set's domain. Defuzzification is primarily a matter of selecting a point on the fuzzy region's boundary and then dropping a "plumb line" to the domain axis. The point of contact on this axis is the scalar's value. See also *decomposition methods, composite mass, composite maximum,* and *composite moments.*

Degree of membership

In fuzzy set theory, this is the degree to which a variable's value is compatible with the fuzzy set. The degree of membership is a value between [0] (no membership) and [1] (complete membership) and is drawn from the truth function of the fuzzy set. While the values in the domain of a fuzzy set always increase as you go from left to right, the degree of membership follows the shape of the fuzzy set's surface—so that, as an example, membership values associated with a bell curve rise to a maximum value (usually [1]) and then fall back toward the zero [0] membership point. The term *truth function* is often used interchangeably with degree of membership.

Delphi method

A consensus gathering method used predictively or prescriptively (as a diagnostic and analysis vehicle). The Delphi method uses blind voting, convergence, and threshold analysis to evaluate a premise (hypothesis) and its possible configurations (outcomes). When knowledge is distributed among many experts or the knowledge is represented among different interlocking fields of expertise, the Delphi method is one important technique in knowledge acquisition and validation. A fuzzy model that uses the Delphi multi-expert resolution method is composed of many independent policies. Each policy is, in fact, a fuzzy model. Each model predicts the answer for a specific solution variable (or a set of solution variables). The fuzzy regions for each solution variable are combined using linked compensatory weighted operators for the strength of each policy. That is, some policies represent expert judgments with a higher degree of credibility, so they contribute more to the possible density space of the combined policy spaces.

A Delphi fuzzy region is defuzzified by finding the center of mass associated with the highest rule density for each policy. This is equivalent to finding the region of maximum possibility as supported by the evidence. From this region, a value is defuzzified ($\Re_k$) and an accompanying compatibility index ($\mu[c_i]$) is found. A more profound, but more difficult

Delphi region to calculate is sometimes used by considering the composite fuzzy region as a second-order fuzzy space (i.e., an ultra-fuzzy set). From this region, another fuzzy region is extracted, representing the approximate possibility that the recommendation is contained within this space. Such fuzzy spaces, instead of scalars, provide a more judgment-tolerant environment for simulation, Monte Carlo, or other stochastic-based models.

Dempster–Shafer theory

A mathematical model of plausible belief systems based on frames of discernment, probability ranges, and evidential reasoning. The Dempster–Shafer theory assumes that, in response to a hypothesis, a mutually exclusive set of possible solution states exist. In practice, the Dempster–Shafer (DS) approach accumulates evidence much in the same manner as Bayesian systems, except that here the belief in the evidence itself is under analysis. The total accumulated evidence or evidence measure (*m*) is treated as a mass function. The DS theory differs significantly from probability in its handling of uncertainty and especially of ignorance. Unlike probability which assigns the value ($1/N$) to an event with N outcomes—even if we have no information about the statistical behavior of the system—the DS model treats the unknown state of a system as simple nonbelief. This means that if we only know S_1 with a probability of .80, then $p(\sim S_1)$ is held in a state of reserved judgment. Thus, under Dempster–Shafer:

$$p(S_1) + p(\sim S_1) \geq 1$$

Because this expression of probabilities is very close to the nature of fuzzy possibilities, and since both fuzzy systems and DS processes incorporate rules of evidence, much research has recently gone into reconciling Dempster–Shafer's rules of evidence processing and possibility theory.

Dilution

A hedge operation, generally implemented by *rather*, *quite*, or *somewhat*, that moves the truth function of the fuzzy set so that its degrees of membership values are increased for each point along the line. This means that, for a given variable's membership rank, its value in a diluted fuzzy set will increase. The most common dilution mechanism changes the fuzzy surface by taking the square root of the membership function at each point along the original fuzzy set's limits. This has the effect of moving the surface to the left.

Disjoint fuzzy space

In fuzzy modeling, a consequent fuzzy space with a discontinuous region. Disjoint solution sets produce "prohibited" zones that require special means of defuzzification.

Disjunction

The process of applying the union operator to two fuzzy sets. For each point on the fuzzy set, we take the maximum of the membership functions for the two combined sets.

Domain

The range of real numbers over which a fuzzy set is mapped. A fuzzy set domain can be any set of positive or negative monotonic numbers. See also *fuzzy sets*.

Expected value

The singleton value derived from "defuzzifying" a fuzzy region that represents the model's value for a solution or control variable. An expected value is derived from a fuzzy set by applying one of the methods of decomposition. This value represents the central measure of the fuzzy space according to the type of defuzzification technique. See also *composite moments, composite maximum,* or *composite mass.*

Fuzziness

The degree or quality of imprecision (or, perhaps, vagueness) intrinsic in a property, process, or concept. The measure of the fuzziness and its characteristic behavior within the domain of the process is the semantic attribute captured by a fuzzy set. Fuzziness is not ambiguity nor is it the condition of partial or total ignorance; rather, fuzziness deals with the natural imprecision associated with everyday events. When we measure temperature against the idea of hot, or height against the idea of tall, or speed against the idea of fast, we are dealing with imprecise concepts. There is no sharp boundary at which a metal is precisely cold, then precisely cool, then precisely warm, and finally, precisely hot. Each state transition occurs continuously and gradually, so that, at some given measurement, a metal rod may have some properties of warm as well as of hot.

Fuzzy associative memory (FAM)

An [MxN] array of fuzzy conditions and actions. FAM arrays are used principally in process control environments where a correspondence exists between a [2x2] antecedent array and a solution variable. As an example, consider the following fuzzy control rules that adjust the consequent fuzzy region *Y* based on the fuzzy propositions x_i and z_i, where the controlling fuzzy rules have a form like

```
if c1i is hot and c2i is fast
then si is positively decreased
```

where *c* is a control variable, and *s* is a solution or consequent variable in a fuzzy space. A FAM bank (or array) is a set of fuzzy rules that are selected and run in parallel. The following is a set of FAM rules controlling the state of fuzzy region *Y*:

```
if x1 and z1 then Y1if x3 and z1 then Y2
if x1 and z2 then Y2if x3 and z2 then Y2
if x1 and z3 then Y3if x3 and z3 then Y3
if x2 and z1 then Y1if x4 and z1 then Y3
if x2 and z2 then Y3if x4 and z2 then Y2
if x2 and z3 then Y3if x4 and z3 then Y1
```

The actual representation of the FAM matrix would appear as:

		Fuzzy Z Region		
		z_1	z_2	z_3
	x_1	Y_1	Y_2	Y_3
Fuzzy X Region	x_2	Y_1	Y_3	Y_3
	x_3	Y_2	Y_2	Y_3
	x_4	Y_3	Y_2	Y_1

When the fuzzy propositions x_i and z_j are true, then the fuzzy consequent region is updated by Y_{ij}.

Fuzzy logic

A class of multivalent, generally continuous-valued logics based on the theory of fuzzy sets initially proposed by Lotfi Zadeh in 1965, but which has its roots in the multivalued logic of Lukasiewicz and Gödel. Fuzzy logic is concerned with the set theoretic operations allowed on fuzzy sets, how these operations are performed and interpreted, and the nature of fundamental fuzziness. Most fuzzy logics are based on the min–max or the bounded arithmetic sum rules for set implication.

Fuzzy membership

See *truth function*.

Fuzzy numbers

Numbers that have fuzzy properties. Models deal with scalars by treating them as fuzzy regions through the use of hedges. A fuzzy number generally assumes the space of a bell curve with the most probable value for the space at the center of the bell. Fuzzy numbers obey the rules for conventional arithmetic but also have some special properties (such as the ability to subsume each other or to obey the laws of fuzzy set geometry). See also *approximation*.

Fuzzy operators

The class of connecting operators, notably AND and OR, that combines antecedent fuzzy propositions to produce a composite truth value. The traditional Zadeh fuzzy operators use the min–max rules, but several other alternative operator classes exist, such as the classes described by Yager, Schweizer and Sklar, Dubois and Prade, and Dombi. Fuzzy operators determine the nature of the implication and inference process and thus also establish the nature of fuzzy logic for that implementation.

Fuzzy region

See *fuzzy space.*

Fuzzy sets

A fuzzy set differs from conventional or crisp sets by allowing partial or gradual memberships. A fuzzy set has three principal properties: the range of values over which the set is mapped, which is called the *domain* and must have monotonic real numbers in the range $[-\infty,+\infty]$; the degree of membership axis that measures the value's membership in the set; and the actual surface of the fuzzy set—the points that connect the degree of membership with the underlying domain.

The fuzzy set's degree of membership value is a consequence of its intrinsic truth function. This function returns a value between [0] (not a member of the set) and [1] (a complete member of the set) depending on the evaluation of the fuzzy proposition "X is a member of fuzzy set A." In many interpretations, fuzzy logic is concerned with the compatibility between a domain's value and the fuzzy concept. This can be expressed as "How compatible is X with fuzzy set A?"

Fuzzy space

A region in the model that has intrinsic fuzzy properties. A fuzzy surface simply differs from a fuzzy set in its dynamic nature and composite characteristics. In modeling terms, we view the fuzzy sets created by set theoretic operations, as well as the consequent sets produced by the approximate reasoning mechanism, as fuzzy spaces. This distinction remains one of semantics, enabling the model builder to draw a distinction between the sets that are created as permanent parts of the system and those that have a lifetime associated with the active model.

Gödelian logic

A continuous-valued logic proposed by Kurt Gödel that recognizes the inadequacies of, but still retains the flavor of, Boolean logic. The Gödelian logic uses the standard fuzzy min–max operators for AND and OR; however, set complement and implication are handled quite differently. The complement or negation of a Gödelian set is described by:

$$\mu_a[x] = 1 \text{ when } \mu_a[x]=0 \text{ otherwise } 0$$

Similarly, the implication function in Gödelian logic is defined as:

$$X \supset Y \; 1 \text{ when } \mu_y[x] > \mu x[x] \text{ otherwise } \mu_y[x]$$

This adaptation of continuous logic supports a system of logical implication that retains the conventions of the excluded middle and the law of noncontradiction.

Hedge

A term, basically linguistic in nature, that modifies the surface characteristics of a fuzzy set. A hedge has an adjectival or adverbial relationship with a fuzzy set, such that the natural specification of a hedge with a fuzzy set acts in the same fashion as its English language counterpart. Hedges can approximate a scalar or another fuzzy set (*near*, *close to*, *around*, *about*, *approaching*), intensify a fuzzy set (*very*, *extremely*), dilute a fuzzy set (*quite*, *rather*, *somewhat*), create the complement of a fuzzy set (*not*), and intensify or diffuse through contrasting (*positively*, *generally*). Examples of hedged fuzzy statements include:

```
our price must be very high
the price should be close to 2*mfg_costs
if the temperature is positively very high
   then reduce speed by about 10 units
if costs are high
   but not very high then inflation has increased
when salinity is quite high,
   buoyancy is very very high
```

The definition and calibration of hedges play an important part in the construction and validation of fuzzy systems. In functional fuzzy systems, users define their own hedge vocabularies and the way each hedge will change the surface of a fuzzy set.

Implication rules

Implication is a formal method of assigning truth to fuzzy propositions and assertions involved in a logical implication process. This sounds a little like a tautology, but, in fact, it means that when we use the fuzzy implication operators—AND and OR—There are rules about the nature of truth they imply. In most fuzzy systems there are two rules of implication—the min–max and the bounded arithmetic sum.

Imprecision

The degree of intrinsic imprecision associated with an event, process, or concept is a measure of the overall system fuzziness. In this respect, imprecision is a characteristic of fuzzy systems. We note that fuzziness and imprecision are intransitive phenomena; that is, a fuzzy system is always imprecise, but an imprecise system is not always fuzzy. The latter definition of imprecision is based on the evident role of granularity in our measurements. Within some systems, the resolution of the control variables will move toward greater and greater precision as we increase our level of detail. As an example, the imprecision associated with measuring the edge of a cube will diminish as we change measuring tools from our finger, to a ruler marked in 1/16 inches, to a ruler marked in millimeters, to an optical sensor [laser] that can read in angstrom units. At some arbitrary level of detail, the measurement falls below the smallest unit of measurement in our model, thus becoming, for all purposes, absolutely precise. See also *fuzziness.*

Intensification A hedge operation, usually implemented by *very* or *extremely*, that moves the truth function of a fuzzy set so that its degrees of membership values are decreased for each point along the line. This means that, for a given variable's membership rank, its value in the intensified fuzzy set will decrease. The most common intensification mechanism changes the fuzzy surface by squaring the membership function at each point along the original fuzzy set's limits. This has the effect of moving the surface to the right.

Intersection The conjunction of two fuzzy sets.

Intrinsic fuzziness The property of fuzziness that is an inseparable characteristic of a concept, event, or process. The universe at large consists of mostly fuzzy processes, that is, measurable events that cannot be calibrated or separated into distinct groups. The simplest and most commonly used example is the concept of tall associated with height. At what precise point does a height measurement move from short, to average, to tall? We can demonstrate the ineluctable notion of imprecision by attempting to separate tallness from non-tallness. Let set T be the set of all tall people. At any chance division point $[S]$, we note that the adjoining value $[S]-\in$ (where $\in$ is any arbitrarily small number approaching $[S]$) is excluded from the set, yet this point, if sufficiently close to $[S]$, shares nearly all the properties of set T. We can solve this problem by moving the division point left by $\in$. Applying the same separability analysis to this new division point $[S_1]$, we find the same indistinctness.

Now, we can resolve this by moving the separation point continuously to the left. At some point, however, the membership of set T no longer represents its semantic property of tall. Thus, in attempting to resolve imprecision, we are forced to abandon any precise separability. This means that tallness is an intrinsic, imprecise, or fuzzy characteristic of height. See also *imprecision.*

Law of the Excluded Middle In classical set theory, an element has two states respective to its membership in a set: true [1] (complete membership), or false [0] (complete nonmembership). This means that the intersection of set *A* with its complement ~*A* is the null or empty set, expressed symbolically as:

$$A \cap \sim A \equiv \varnothing$$

This equation indicates that an element cannot be, at the same time, a member of a set and its complement (which, by definition, says that ~*A* contains all the elements not in *A*). This is also the law of noncontradiction.[1] Since fuzzy sets have partial membership characteristics we might expect that they behave differently in the presence of their complement. This is correct. Fuzzy logic does not obey the law of the excluded middle. To see this clearly, consider the case of the fuzzy set *High* (①) and its complement, the fuzzy set *Low* (②). The set *Low* is produced by the normal fuzzy complement,

$$\text{low} = \sim\text{high} = 1-\mu_H[x]$$

The membership transition graph for these sets appears as in Figure G.1.

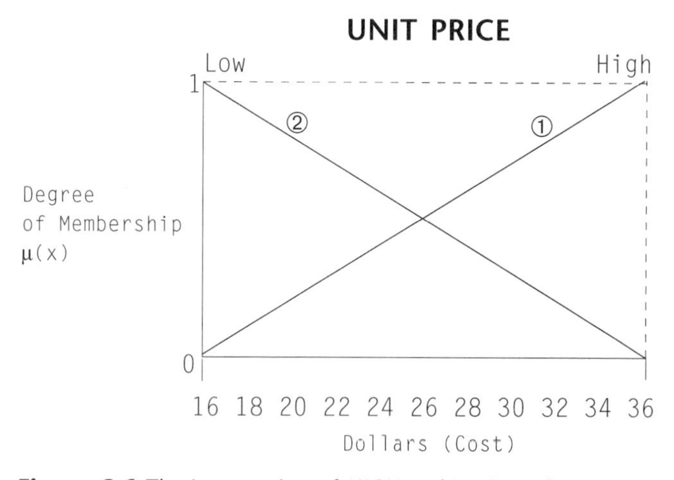

Figure G.1 The Intersection of *HIGH* and Its Complement *LOW*

The intersection of the two fuzzy sets is found by applying the rules for the intersection or conjunction of two fuzzy sets:

$$\text{high} \cap \text{low} = \min(\mu_H[x], \mu_L[x])$$

The intersection of *High* and *Low*, the triangular area subtended by their membership functions, represents a region that has some properties of both *High* and *Low*.

Linguistic Variable

In the rule formulation language of fuzzy systems, the term applied to a fuzzy set or the combination of a fuzzy set and its associated hedges. Thus the rule:

```
our price must be positively very high
```

The term *positively very high* is a linguistic variable. Since the effect of a hedged fuzzy set is a single fuzzy space, it is, conceptually, a single fuzzy set.

Membership function

See *truth function*.

Methods of decomposition

The techniques employed to derive the expected value from a consequent fuzzy region. The methods of decomposition vary according to the type of expectation associated with the composite fuzzy region. See *composite moments, composite maximum,* and *composite mass.*

Min–max rule

The basic rule of implication and inference for fuzzy logic that follows the traditional Zadeh algebra of fuzzy sets. There are two statements of this rule pertaining to different elements in the fuzzy logic process.

The min–max rule of implication. This rule specifies how fuzzy unions and intersections are performed. When two fuzzy sets are combined with the intersection operator (AND), the resulting fuzzy space is found by taking the minimum of the truth functions across the compatible domains. When two fuzzy sets are combined with the union operator (OR), the resulting fuzzy space is found by taking the maximum of the truth functions across the compatible domains.

The min–max rule of inference. This rule specifies, in a fuzzy system, how conditional and unconditional fuzzy assertions (propositions) are combined. An unconditional assertion is applied to the consequent fuzzy region under generation by taking the minimum of the unconditional's fuzzy space and the consequent's fuzzy space at each point in the region. A conditional fuzzy proposition first has its truth reduced to the maximum truth of the rule's premise, then it is applied to the consequent fuzzy region under generation by taking the maximum of the conditional's fuzzy space and the consequent's fuzzy space at each point in the region.

Modus ponens

A form of implication in both classical and fuzzy logic used to infer the existence of a consequent state from an antecedent or premise state. The rules of modus ponens follow the reasoning paradigm:

$P \supset Q$	P (the premise) implies Q (a consequent)
P	We know P
$\therefore Q$	Therefore, we can infer Q

This is the conventional logical syllogism formally and initially defined by Aristotle. This kind of syllogism is traditionally expressed through such logical reasoning chains as:

All men are mortal
Socrates is a man
∴ Socrates is mortal

In fuzzy logic, modus ponens is concerned with the degree of truth between the premise and the consequent:

$\mu_p \supset \mu_q$	Membership in P implies membership in Q
$\mu_p[x]$	We know P's degree of membership (truth)
$\therefore \mu_q[y]$	Therefore, we can infer Q's membership (truth)

An example of a fuzzy implication using modus ponens is:

Tall men are Heavy
Peter is Tall
∴ Peter is Heavy

This implication says that, to the degree that *Peter is Tall*, he is also *Heavy*. This is a consequence of the basic rules for implication in fuzzy logic. Fuzzy modus ponens is calibrated to the continuous linguistic space of the fuzzy regions. Thus, the reasoning chain can also admit hedges and dispositional logics, such as:

Tall men are Heavy
Peter is very Tall
∴ Peter is very Heavy

Middle aged men are usually Heavy
Professors are usually middle aged
∴ Professors are usually^2 Heavy

Note that these kinds of logical inferences involving fuzzy logic and linguistic variables cannot be expressed by classical logic, notably, the predicate calculus. This is especially true for statements involving fuzzy quantifiers such as *many*, *mostly*, *usually*, *few*, *some*, and *nearly all*.

Modus tollens

An alternative form of the logical implication process used to infer the lack of a premise state given the negation of a consequent state. Modus tollens, like modus ponens, is a syllogism with the following paradigm:

$P \supset Q$	P (the premise) implies Q (a consequent)
~Q	We know that Q does not exist
∴ ~P	Therefore, we can infer P does not exist

This kind of syllogism is traditionally expressed through such logical reasoning chains as:

All Rainy days are Cloudy
It is not Cloudy
∴ It is not Raining

In fuzzy logic, modus tollens, like modus ponens, is concerned with the degree of truth between the premise and the consequent:

$\mu_p \supset \mu_q$	Membership in P implies membership in Q
$\sim\mu_p[x]$	We know Q's degree of membership (truth)
$\therefore \sim\mu_q[y]$	Therefore, we can infer P's membership (truth)

Since the complement of a fuzzy set contains a truth value graded on the degree to which value [x] is not a member of set Q, we can use the same transitive implication function we used for modus ponens (see also *fuzzy complement*). An example of a fuzzy implication using modus tollens is:

Tall men are Heavy
Peter is Not Heavy
∴ Peter is Not Tall

This implication says that, to the degree that *Peter is Not Heavy*, he is also *Not Tall*. Like modus ponens, modus tollens is calibrated to the continuous linguistic space of the fuzzy regions. Thus the reasoning chain can also admit hedges and dispositional logics, such as:

Tall men are Heavy
Peter is not very Heavy
∴ Peter is not very Tall

See also *modus ponens*.

M-order Fuzzy Set A fuzzy set whose characteristics are themselves, to some degree, fuzzy. First-order fuzzy sets are mappings of truth functions (degrees of membership) to a monotonically increasing range of real numbers. Both of these values are scalars. Second-order fuzzy sets, or ultrafuzzy sets, are sets in which the membership function on the domain is itself a fuzzy space. In these sets, the result of fuzzy implication is itself a fuzzy region (that is, the degree of membership of value [x] in fuzzy set [*T*] is "about .48" or "near .88").

Necessity The degree to which we can say that an event must occur, that is, an inevitable event. Necessity is related to the Markov dependency in both probability and fuzzy systems in describing the extent that one event is dependent on another event. Bayesian systems often employ the concept of logical necessity (as well as logical sufficiency). Necessity in this context is the extent to which an observed phenomenon is said to be essential in order for another phenomenon or condition to exist.

Normalization The process of ensuring that the maximum truth value of a fuzzy set is one [1]. Fuzzy sets are normalized by readjusting all the truth membership values proportionally around the maximum truth value, which is set to [1]. If the fuzzy sets defined as part of the model vocabulary are not in normal form, they cannot contribute properly to the implication and inference process. This not only will affect the compatibility index associated with the final control variable regions, but also will dampen the fuzzy implication strength, thus producing spurious results.

Possibility theory Possibility theory evolved from the concept of fuzzy sets and provides a calculus of possibility analysis similar to the crisp calculus of probability (where the degrees of possibility can take only true or false values [1,0]). Developed and solidified by such researchers as Dider Dubois and Henri Prade, possibility theory embraces the entire spectrum of fuzzy logic and approximate reasoning by codifying the relationship between fuzzy interval analysis, event phenomena, and fuzzy spaces into a comprehensive, mathematically sound discipline. What possibility theory attempts to do is provide a mathematical framework for the intrusion of fuzzy logic into a wide spectrum of disciplines such as game theory, linear programming, operations research, and general decision theory.

Predictive precision

A concept used in the fuzzy verification and stability metrics; indicates the degree to which we can falsify or disprove a prediction (or proposition) of the system. Precise predictions are, naturally, highly falsifiable since they make specific predictions. On the other hand, imprecise or approximate claims are less verifiable.

Premise

Also called *the predicate,* the conditional or left-hand side part of a rule. In fuzzy modeling, the premise is a series of fuzzy assertions (or propositions) connected by AND and OR operators. As an example, in the rules

```
if height is tall and waist is large
   then weight is very heavy
if the competition_price is not very high
   and the mfg_costs are low
      then our price should be low but not very low;
if temperature is hot or rotor_speed is high
   then increase lubrication by about 3 units
```

the statements preceded by the *if* and terminated by the *then* are premise assertions. A fuzzy rule-based system will evaluate each premise proposition to find its truth value. These will be combined according to the nature of the AND or OR operators (using Zadeh's algebra, as an example, this will be the min–max rule) to produce a final and total truth for the rule premise. See also *rules*.

Principle of incompatibility

Zadeh's intuitive appeal to linguistic system modeling based on his belief that, as systems are defined in ever more rigorous mathematical terms, there is a point beyond which our understanding of the system as a whole and the mathematical representation are incompatible. Zadeh says:

> *As the complexity of a system increases, our ability to make precise yet significant statements about its behavior diminishes until a threshold is reached beyond which precision and significance (or relevance) become almost mutually exclusive characteristics.*

Quote from L.A. Zadeh, "Outline of a New Approach to the Analysis of Complex Systems and Decision Processes," *IEEE Transactions on Systems, Man, and Cybernetics,* Vol. SMC-3, No. 1, 1973.

Probability

A quantitative method of dealing with uncertainty based on the measurement of behavior patterns for some selected set of properties in a large population. Probability provides a statistical inference process, since if we know the *a priori* chance of an event occurring, we can predict (within a prescribed confidence limit) the chance of that event occurring in the future. The basic classical probability formula is defined as:

```
p(x) = [e/N]
```

The probability of x is the number of observed x events (e) divided by the total number of equally possible events N. As an example, when you roll a single die, the probability of getting a one is:

```
P(1) = [1/6] = .16666666667
```

Probability, concerned with deterministic and nondeterministic (random) systems, attempts to quantify the behavior of these systems as we gain more and more knowledge about the underlying populations. See also *Bayes theorem*.

Randomness

A characteristic of a property or event such that, when we sample that property, any of its possible values is equally likely to occur. The principal characteristic of such a random event is that there is no way to predict its outcome. One view of fuzziness is coupled closely with randomness in the sense that fuzzy sets are considered as measurements "corrupted" by white or random noise. If we consider the bell-shaped fuzzy set *MiddleAged*, one interpretation of this is the number 40 surrounded by random noise. One way to view a fuzzy set is to consider the membership function as a random variable with noise around or near the true truth function values.

Rules

Statements of knowledge that relate the compatibility of fuzzy premise propositions to the compatibility of one or more consequent fuzzy spaces. The rule

```
if temperature is high then friction is decreased;
```

is interpreted as a correlation between two fuzzy states such that the rule should be read as

```
[to the degree that] temperature is high
   friction is [proportionally] decreased
```

or, perhaps, in a slightly more formal statement considering the idea of fuzzy compatibility

```
[to the degree that] temperature
   is [compatible with the concept] high
   make friction [compatible with the concept]
     decreased
```

Of course, the proportionality need not be linear. In fact, the correspondence or compatibility function between an antecedent fuzzy region and a consequent fuzzy state is determined by (1) the shape of the fuzzy sets, (2) the connector (implication) operators, (3) the preponderance of truth in the antecedent, and (4) the technique for transforming the consequent fuzzy region from the current fuzzy state. This means that fuzzy rules operate in a different manner than rules in conventional knowledge-based systems (which are concerned with pattern matching and the logical evaluation of discrete expressions).

Support set

That part of a fuzzy set or fuzzy region defined by the strong alpha-cut threshold:

$$\mu_A[x] > 0$$

The support set, since it contains the actual membership region for the set, is the area that actually participates in the fuzzy implication and inference process.

Truth function

A view that the degree of membership axis of a fuzzy set or region acts as a function:

$$\mu_A = T(x)$$

For each unique value selected from the domain, the function returns a unique degree of membership in the fuzzy regions. We call this a "truth" function since it reflects the truth of the fuzzy proposition *x is a member of fuzzy set A*. For second-order sets, the truth function returns a vector containing the fuzzy region approximating the truth value. For third-order sets, the truth function returns an [NxM] matrix of vectors representing the possibility density at the truth region.

Ultra-fuzzy set

See *M-order fuzzy set*.

Undecidability

The degree to which the state of a control variable cannot be determined from the fuzzy space context. Undecidability is most generally associated with the fuzzy region bounded by two intersecting, complementary fuzzy sets (or fuzzy regions).

Union

The disjunction of two fuzzy sets.

Vagueness

A property closely allied with imprecision and thus a characteristic of fuzzy systems. Vague concepts approach fuzziness as we begin to calibrate them in terms of their compatibility with an overlying idea. To say

```
Bill is Big
```

may be vague rather than fuzzy, if the idea underlying *Big* is not particularly correlated to any metric from which we could sample individual property points. On the other hand, if the attributes of Bill were defined in terms of height (*Tall*) and weight (*Heavy*), then *Big* becomes imprecise rather than vague. We should note that vagueness is distinct from ambiguity.

Notes

1. The laws of the excluded middle and noncontradiction are really separate but very closely related statements of logic originally postulated by Aristotle as his three rules of thought in the *Organon*. The law of noncontradiction is actually a statement of the excluded middle $(A \cap \sim A \equiv \varnothing)$, while the law of the excluded middle $(A \cup \sim A \equiv U)$ says that the union of a set with its complement yields the universal set. Both of these statements are generally equivalent for operations on crisp sets.

Bibliography

This is not a complete listing of all the papers and books currently available on fuzzy logic—since most of the literature deals almost exclusively with the control engineering aspects of fuzzy logic. If you want the references to the hundreds of research papers appearing monthly on fuzzy neural approaches to balancing a pendulum or backing up a truck—apparently the only fuzzy logic applications that control scientists and academics ever actually build, or at least write about—then I apologize for this glaring lack of completeness. However, with the exception of my own published papers on fuzzy information systems and a few pivotal papers that address the issues of fuzzy decision support, I have limited the bibliography to a selection of the important books that cover the entire spectrum of fuzzy logic, fuzzy systems, and applications.

Cox, Earl D. (1993). "Adaptive Fuzzy Systems." *IEEE Spectrum* (February), pp. 67–70.

——— (1992). "Applications of Fuzzy System Models." *AI Expert* (October), pp. 34–39.

——— (1991). "Approximate Reasoning: The Use of Fuzzy Logic in Expert Systems and Decision Support." Proceedings of the Conf. on Expert Systems in the Insurance Industry (April 24–25), Institute for International Research, New York.

——— (1992). "A Close Shave with Occam's Razor: Fuzzy-Neural Hetero-Genetic Object-Oriented Knowledge-Based Nano-Synthetic Reasoning Models: Throwing the Kitchen Sink at Problem Solving." *A Workshop in the Industrial Applications of Philosophy and Epistemology to AI.* Proceedings of the Third Annual Symposium of the International Association of Knowledge Engineers (November 16–19), Software Engineering Press, Kensington, MD.

——— (1991). "Company Acquisition Analysis: Formulating Queries with Imprecise Domains." Proceedings of the First Intl. Conf. on Artificial Intelligence Applications on Wall Street (October 9–11), IEEE Computer Society Press, Los Alamitos, CA. pp. 194–199.

——— (1992). "Effectively Using Fuzzy Logic and Fuzzy Expert System Modeling—in Theory and Practice." Proceedings of the Conf. on Advanced Technologies to Re-Engineer the Insurance Process (September 17–18), Institute for International Research, New York.

——— (1992). "Fuzzy Fundamentals." *IEEE Spectrum* (October), pp. 58–61.

——— (1993). "Fuzzy Information Systems with Multiple Conflicting Experts." Proceedings of *Computer Design* Magazine's Fuzzy Logic '93 Conference (March), pp. M223 1–13.

——— (1992). "Fuzzy Logic and Fuzzy System Modeling." Proceedings of the Fourth Annual IBC Conf. on Expert Systems in Insurance (October 28–29), IBC USA Conferences, Southborough, MA.

——— (1993). "A Fuzzy Systems Approach to Detecting Anomalous Risk Behaviors in Portfolio Management Strategies." Proceedings of the Second Intl. Conf. on Artificial Intelligence Applications on Wall Street (April 19–22), Software Engineering Press, Gaithersburg, MD. pp. 144–148.

——— (1992). "The Great Myths of Fuzzy Logic." *AI Expert* (January), pp. 40–45.

——— (1992). "Integrating Fuzzy Logic into Neural Nets." *AI Expert* (June), pp. 43–47.

——— (1993). "A Model-Free Trainable Fuzzy System for the Analysis of Financial Time-Series Data." Proceedings of *Computer Design* Magazine's Fuzzy Logic '93 Conference (March), pp. A124 1–7.

——— (1992). "Solving Problems with Fuzzy Logic." *AI Expert* (March), pp. 28–37.

Cox, Earl D., and Martin Goetz (1991). "Fuzzy Logic Clarified." *Computerworld* (March 11), pp. 69–71.

Dubois, Didier, and Henri Prade (1980). "Fuzzy Sets and Systems: Theory and Applications." *Mathematics in Science and Engineering.* Vol 144. Academic Press, San Diego, CA.

——— (1988). *Possibility Theory, An Approach to Computerized Processing of Uncertainty.* Plenum Press, New York.

Jamshidi, M., N. Vadiee, and Timothy J. Ross, eds. (1993). *Fuzzy Logic and Control.* Prentice Hall, Englewood Cliffs, NJ.

Jones, Peter L., and Ian Graham (1988). *Expert Systems: Knowledge, Uncertainty, and Decision.* Chapman and Hall, London.

Klir, George J., and Tina A. Folger (1988). *Fuzzy Sets, Uncertainty, and Information.* Prentice Hall, Englewood Cliffs, NJ.

Kosko, Bart (1992). *Neural Networks and Fuzzy Systems, A Dynamical Systems Approach to Machine Intelligence.* Prentice Hall, Englewood Cliffs, NJ.

Kosko, Bart, and Satoru Isaka (1993). "Fuzzy Logic." *Scientific American* (July), pp. 76–81.

Masters, Timothy (1993). *Practical Neural Network Recipes in C++.* Academic Press, San Diego, CA.

Pedrycz, Witold (1993). *Fuzzy Control and Fuzzy Systems.* 2nd ed. John Wiley and Sons, New York.

Schmucker, Kurt J. (1984). *Fuzzy Sets, Natural Language Computations, and Risk Analysis.* Computer Science Press, Rockville, MD.

Smets, P., E.H. Mamdani, D. Dubois, and Henri Prade (1988). *Non-Standard Logics for Automated Reasoning.* Academic Press Limited, London.

Smithson, Michael (1987). *Fuzzy Set Analysis for Behavioral and Social Sciences.* Springer-Verlag, New York.

Terano, T., K. Asai, and Michio Sugeno (1991). *Fuzzy Systems Theory and Its Applications.* Academic Press, San Diego, CA.

Wang, Li-Xin, and Jerry M. Mendel (1991). "Generating Fuzzy Rules from Numerical Data, with Applications." USC-SIPI Report No. 169. Signal and Image Processing Institute, University of Southern California, Los Angeles, CA.

Wang, Zhenyuan, and George J. Klir (1992). *Fuzzy Measure Theory.* Plenum Press, New York.

White, David A., and Donald A. Sofge (1992). *Handbook of Intelligent Control: Neural, Fuzzy, and Adaptive Approaches.* Van Nostrand, Reinhold, NY.

Yager, Ronald E., and Lotfi A. Zadeh, eds. (1992). *An Introduction to Fuzzy Logic Applications in Intelligent Systems.* Kluwer Academic Publishers, Norwell, MA.

Yager, R.R., Ovchinnikov, S., Tong, R.M., and Nguyen, H.T. (1987). *Fuzzy Sets and Applications: Selected Papers by L.A. Zadeh.* John Wiley and Sons, New York.

Zadeh, Lotfi A., and Kacprzyk, Janusz, eds. (1992). *Fuzzy Logic for the Management of Uncertainty.* John Wiley and Sons, New York.

Zimmerman, Hans J. (1985). *Fuzzy Set Theory—and Its Applications.* Kluwer Academic Publishers, Norwell, MA.

——— (1987). *Fuzzy Sets, Decision Making, and Expert Systems.* Kluwer Academic Publishers, Norwell, MA.

Zurek, Wojciech H., ed. (1990). *Complexity, Entropy, and the Physics of Information.* Vol. VIII. Santa Fe Institute, Studies in the Sciences of Complexity. Addison-Wesley Publishing Company, Redwood City, CA.

Index